Robotic Fabrication in Architecture, Art and Design 2018

Jan Willmann · Philippe Block
Marco Hutter · Kendra Byrne
Tim Schork
Editors

Robotic Fabrication in Architecture, Art and Design 2018

Foreword by Sigrid Brell-Çokcan and Johannes Braumann,
Association for Robots in Architecture

Springer

Editors
Jan Willmann
Faculty of Art and Design
Bauhaus-Universität Weimar
Weimar, Germany

Philippe Block
Department of Architecture
Swiss Federal Institute of Technology
Zurich, Switzerland

Marco Hutter
Department of Mechanical and Process Engineering
Swiss Federal Institute of Technology
Zurich, Switzerland

Kendra Byrne
San Francisco, CA, USA

Tim Schork
Faculty of Design, Architecture and Building
University of Technology, Sydney
Sydney, NSW, Australia

Funded by KUKA Robotics Germany and the Association for Robots in Architecture

ISBN 978-3-030-06400-6 ISBN 978-3-319-92294-2 (eBook)
https://doi.org/10.1007/978-3-319-92294-2

This Springer imprint is published by the registered company Springer Nature Switzerland AG
The registered company address is: Gewerbestrasse 11, 6330 Cham, Switzerland

Foreword by the Association for Robots in Architecture

Since the beginning of the Association for Robots in Architecture, it has been our goal to promote collaboration within the community, but also with partners from industry and cross-disciplines. Robots in Architecture is proud to be associated with leading research organizations within architecture, such as ACADIA, or eCAADe but the scope of work that is being done by the community is starting to exceed the field of architecture into many other new domains, and we feel that it is important to accompany such steps. One measure to ensure the exchange across disciplines and other domains is to open the community to disciplines such as engineering and robotics by establishing common platforms where people can meet and exchange their ideas and research. Together with Springer, the Association for Robots in Architecture has therefore established a new Journal for Construction Robotics with the first published issue at the end of 2017 to foster collaborative papers and high-quality research in architecture.

Another goal for the Association was to join forces with associations in Robotics. In 2016, Robots in Architecture joined euRobotics AISBL, the largest public–private partnership involving robotics in Europe. Sigrid Brell-Çokcan co-established a new Topic Group on Construction Robotics, acting alongside the other established 30 topic groups within euRobotics, ranging from wearables, bio-inspired robotics, health care, mining to infrastructure. In 2018, Sigrid has joined the board of directors.

Through networks such as euRobotics, it is not only possible to promote a field of research, but also to actively shape policy, so that the importance of Construction Robotics is recognized, and appropriate research funding is allocated to relevant topics. Through the Multi-Annual Roadmap in research for the EU and its Horizon 2020 programme, these initiatives not only reach a chosen few researchers in academia, but also a wide range of commercial and non-commercial research institutions, robotics developers and users alike in Europe and beyond.

This year, we recognize the importance of such public–private partnerships and euRobotics by presenting euRobotics chairman Bernd Liepert with the Rob|Arch Community Contribution award. It is our goal that more researchers within the Robots in Architecture community will reach out to large-scale research and believe strongly that euRobotics AISBL is a prime example on how to combine economic with academic and political interests, fostering the common goal of creating robotic innovation.

Innovation is also one of the core qualities we are looking for when selecting awardees. This year's Pioneering Achievement and Pioneering Industry Award goes to two very different architectural companies, who have greatly facilitated the potential of robotic processes in their work.

The internationally highly reputed architectural office Snøhetta was one of the first architectural companies to invest in robotic arms, joining us for our first KUKA|prc workshop in 2010 at the Advances in Architectural Geometry Conference in Vienna. For them, the robot has been an important tool for prototyping new approaches and design, placed closely to the open office in Oslo.

The second awardee is Branch Technology from Tennessee in the USA, who have gone even further by not just using robots as CNC machines, but by developing their own robotic processes for large-scale robotic 3D printing. What we feel is special about Branch is that it is a company by architects who develop products for architects. As such, they did not stem from academia but from practice and therefore had to find investors to fund their ideas. Today they have a team with a wide variety of backgrounds and have realized a number of large-scale projects, collaborated with companies such as Foster + Partners and even won an award for their joint vision of future construction on Mars by the NASA. We believe that this drive embodies the true spirit of the community. We see in Branch a perfect role model where robotics lies at the core to enable technology-driven creativity in new business models.

By Yu Lei's research at Tsinghua University in China and his own professional workshop, we do not just honour a single person, who has made significant contributions to architectural robotics in China, but also the entire Chinese community of companies and researchers, where the past years have seen great developments and a surge of new ideas and initiatives, as was demonstrated at this year's CAADRIA conference organized by Prof. Xu Weiguo and the DADA community. While the potential for Construction Robotics in China is huge, there is also a great need for education and research and thus educators like Yu Lei are important trailblazers by sharing and starting new business models in architecture.

We also believe that it is important to recognize the researchers, without whose work into robotics we would not have today's sophisticated methods and machines at hand, and who create tomorrow's tools and processes today. Jonas Buchli, director of the ETH Zurich Agile & Dexterous Robotics Lab, is one such pioneer whose work spans across many disciplines—from neurobiology and human locomotion to service robotics and also architecture, where has been a Principal Investigator of the NCCR Digital Fabrication, developing the In situ Fabricator (IF), with the goal of enabling the machine to autonomously perform precise mobile manipulation tasks in unstructured and unpredictable environments. We recognize his work with a Pioneering Research Award.

In 2018, the Rob|Arch conference is back in Europe for the first time since Vienna in 2012 and is being hosted at ETH Zurich. ETH Zurich, in particular through the work of Gramazio Kohler Research, has been a central part of the research community and further solidified that status with the creation of the NCCR Digital Fabrication, a multi-disciplinary research cluster with more than 14 professors. We see ETH Zurich and the NCCR Digital Fabrication as a prime example how research can happen on a very large scale, with highly interdisciplinary and diverse research teams. At the same time, we believe that innovative research can also be done at a smaller scale, as it is demonstrated by this year's 15 Rob|Arch conference workshops, involving tutors from more than 25 institutions, that are hosted centrally in ETH Zurich's Robotic Fabrication Laboratory.

An effort like Rob|Arch 2018 is only possible when many people work together towards a common goal. We would like to thank our local hosts at ETH Zurich and the NCCR Digital Fabrication, the Scientific Board led by Jan Willmann and all minds and hands involved in setting up such a big event.

We are grateful to Matt Jezyk (Autodesk) and Alois Buchstab (KUKA) for their continuing and enthusiastic support of the community through their respective companies, and we would like to thank all sponsors of this conference – Arup, BCG, Sika, Erne, Moog, and Bachmann Engineering – for this year's collaboration in making Rob|Arch 2018 a success.

Sigrid Brell-Çokcan
Association for Robots in Architecture

Johannes Braumann
Association for Robots in Architecture

Preface

New Scientific Frontiers

The emergence of robotics with the creative sectors has led to an entirely new epistemology of collective making that is inextricably open and future-oriented. Challenged by increasingly complex technological and environmental problems, architects, designers, civil and process engineers, and roboticists are seeking novel practices of collaboration and exchange that deliberately overcome and dissolve traditional disciplinary boundaries. This collective approach to working with robots is not only revolutionizing how things are designed and made, but is fundamentally transforming the culture, politics and economics of the creative industries as a whole.

What distinguishes contemporary industrial robots from their industrial predecessors—and indeed from other contemporary computer-controlled devices—is their versatility. Like computers, today's robotic arms are suitable for a wide variety of tasks: they are "generic", open-ended, adaptable and not restricted to any particular application or disciplinary focus. This versatility allows them to be readily customized and programmed to suit a wide range of specific intentions, both at the material and conceptual levels. It has also allowed us to shift our perception of robots as mechanistic, utilitarian devices suited to standardized serial production, towards understanding them as creative tools for exploring, designing and realizing physical objects and the built environment. If the first robotic age—the age of industrial automation—vastly improved our physical productivity, the second robotic age will surely come to distinguish itself as a driver of creative capacity.

The present moment is ripe for connecting robot technology with imagination and materialization, inspiring new fundamental discoveries and opening new scientific frontiers. In fact, we have within reach access to volumes of information and centuries of knowledge about how to design and realize the material world. Aided by global digital connectedness, open-source ideals and collective encounters, robotics rejuvenates traditional disciplinary wisdom with entirely new practices of scientific collaboration and knowledge transfer. Now, more than ever, we are coming to understand that robotics research should not be bound by constricting disciplinary standards, constraints or ideologies lest we limit its potential. Yet to explore this unprecedented potential requires not only a technical grasp of robots' capabilities and limitations, but also an in-depth understanding of the disciplinary consequences of robotics research. With its theme of "Radical Cross-disciplinarity", Rob|Arch 2018 facilitates this understanding by encouraging novel scientific approaches, applications and collaborations, not just in robotics, but beyond.

Closing the Loop

The Rob|Arch conference series was first launched in Vienna, Austria, in 2012 by Sigrid Brell-Çokcan and Johannes Braumann, the founders of the Association for Robots in Architecture. Their purpose was to make industrial robots more accessible to the creative industries—including art, design and architecture—by sharing ideas, research results and technological developments. The series has since become a biannual tradition in the international community (travelling to Michigan, US, in 2014 and to Sydney, Australia, in 2016) and has decisively boosted the exchange and dynamics within.

In 2018, Rob|Arch lands at the Swiss Federal Institute of Technology in Zurich (ETH Zurich), marking an important milestone for the digital fabrication community: ETH Zurich is not only one of the leading international universities for technology and science, it is also the institution where the first industrial robotic fabrication laboratory for non-standard architectural fabrication processes was installed in 2005. Closing this loop gives us the opportunity to foster novel explorations and state-of-the-art knowledge, techniques and methods, while consolidating and advancing our collective understanding of the evolution and impact of robotics in art, design and architecture.

It is no coincidence that Rob|Arch 2018 is also co-hosted by the Swiss National Centre of Competence in Research (NCCR) in Digital Fabrication. Launched in 2014, the NCCR Digital Fabrication is itself a truly cross-disciplinary research platform meant to foster the seamless combination of digital technologies and physical building processes through cooperation and exchange beyond disciplinary boundaries.

Content and Contributions

The Rob|Arch 2018 publication features the most important contributions to the conference. Rather than featuring merely formalist or technicist robotic adventures, this publication goes beyond pure built outcome to forward fresh approaches to scientific innovation, knowledge exchange and cross-disciplinary collaboration.

This includes designers, artists and architects, and also—and increasingly—computation and robotics experts and builders, materials scientists and engineers, process and systems specialists and manufacturers, to name just a few. As a consequence, this book gathers exceptional, scientifically rigorous projects that not only transform the way we design and make, but which also build collaborative capacity in the field of robotic fabrication.

The structure of this publication addresses this "new territory" of collaborative research. Stepping beyond theoretical observation, it outlines five distinct epicentres of practical research, which range from design and simulation research to automated assembly and real-world applications. Robotics and material and structural engineering play an integral role in each of these five areas.

Chapter 1 ("Design and Simulation") discusses new computational approaches to image classification using neural networks, stochastic assembly and deep learning for robotic construction; it also presents procedural fabrication workflows and haptic programming techniques, automatic path planning methods, visual feedback techniques, and function representation models.

Novel materials and material processes for robotic fabrication are introduced in Chapter 2 ("Material and Processes"), including thermally tuned concrete panel printing, time-based material deposition, and digitally controlled concrete injection processes. This is complemented by research into the robotic manipulation of filament material and the automated control of material behaviour for spatial extrusion processes.

In Chapter 3 ("Construction and Structure"), the emphasis is on new robotic construction processes and structural applications, for example bespoke concrete reinforcement, highly versatile wood processing, automated band-saw cutting for complex timber structures, fabrication-aware methods for the realization of non-standard timber shells, and an advanced hybrid subtractive-additive approach to robotically construct double-curved concrete shells. Finally, the chapter presents a novel approach to the construction of jammed architectural structures.

Robotic control, machinery, tooling and fabrication are discussed in Chapter 4 ("Control and Fabrication"), involving tubular composite fabrication with the aid of robotic swarms, automated manufacturing of natural composites, 3D printing with clay on freeform moulds, choreographic robotic wood manipulation, aerial construction using a cyber-physical macro-material system, as well as adaptive robotic carving. Also outlined in this chapter are approaches for multi-mode hybrid fabrication, robotic extrusion of functionally graded building components, as well as of elements with non-standard topology, on-site robotic construction and additive manufacturing techniques for non-woven textiles.

The transfer to larger scales of real-world applications and practices is addressed in Chapter 5 ("Application and Practice"). Here we present automated slipforming for façade elements, robotic brick printing and stacking, robotic sewing of wooden shells, additive manufacturing of truss-shaped concrete pillars, and the realization of topology-optimized concrete structures using abrasive techniques. Large-scale bespoke timber frame construction and cooperative robotic brick assembly are also discussed.

Workshop Activities

Rob|Arch 2018 features a variety of formats and sessions to encourage creative dabbling and encounters with different research topics, practices and field-wide issues. Led by experts from academia, practice and industry, Rob|Arch 2018 workshops empower participants to learn and practise hands-on skills, and discuss cutting-edge fabrication techniques and trends with their peers in a collaborative environment.

This year's workshops offer a broad range of topics, including multiple robotic fabrication, industry-grade robotic programming using HAL, robotic real-time control using Grasshopper, robotic fabrication through the COMPAS framework, chainsawed wood joinery, cooperative robotic assembly of spatial timber structures, large-scale robotic construction, hybrid robotic 3D printing of concrete shell structures, autonomous robotic swarm systems, adaptive spatial 3D printing of space frame structures, automated assembly in constrained sites, mixed reality environments for complex steel structures, mixed reality simulation for collaborative design exploration, as well as an introduction to KUKA|prc for Dynamo.

Beyond Boundaries

Rob|Arch 2018 aims to bring the community ground-breaking approaches to robotic fabrication from the most innovative research laboratories in the world, all while illuminating alternative pathways to boosting cross-disciplinary research and exchange. This publication therefore highlights contributions that not only substantially advance the state-of-the-art in robotic fabrication, but also challenge the reputedly clear division between research, practice and industry.

It is our belief that effective knowledge transfer and exchange between different disciplines is crucial for the development of truly innovative and high-impact research in robotics, a priori, rather than a posteriori. Specifically, Rob|Arch 2018 looks at new paradigms of scientific collaboration, along with the challenges, risks and dynamics within this process. Given that our collective expertise includes autonomous control systems, advanced construction, collaborative design tools, computerized materials and structures, adaptive sensing and actuation, on-site and cooperative robotics, machine-learning, human–machine interaction, large-scale robotic fabrication and networked workflows (the list goes on), we can no longer discuss cross-disciplinarity, cooperation and collaboration in abstract terms. Doing so would be utterly inadequate to address the manifold cultures and practices of robotics that have emerged to master the increasingly complex technological and environmental challenges we face today.

While we have observed a growing capacity for knowledge transfer and exchange in Rob|Arch submissions with each subsequent edition of the conference, this year the blurring of disciplinary boundaries between creative-, scientific- and practice-based domains is particularly significant. We view this as a sign that complex problems cannot be dealt with from a single disciplinary perspective alone.

Yet, while this blurring has yielded many new robotic explorations and real-world applications, these have not taken place uniformly. For example, the fields of intelligent computational design and simulation systems are particularly benefiting from an expanded set of collaborations and exchange between researchers and industry practitioners. Other areas that have especially benefitted from collaborative exchange include: advanced robotic control systems, and feedback processes that enable robots to adapt to different material conditions and changing environments. In all these cases, constant interaction and knowledge transfer between architects, designers, engineers and roboticists are pivotal, both as a result and as a catalysing instrument.

The fast pace of creative and scientific research documented by Rob|Arch is no doubt a result of the bringing-together of diverse disciplines, competences and cultures. Perhaps the emerging cross-disciplinary culture of robotic fabrication research will, through the collaboratively built future environment, one day yield a generational change in how we view the collaborative creative process more broadly. As Richard Sennett once described it: it stimulates a gathering of creative explorations similar to collective encounters that in the pre-machinic age used to be related with, and venerated for, all things man-made.

Acknowledgments

The Scientific Chairs would like to express their gratitude to the Conference Chairs, Fabio Gramazio and Matthias Kohler, for entrusting us with the development of Rob|Arch 2018. We would like to extend our gratitude to the Association for Robots in Architecture, namely Sigrid Brell-Çokcan and Johannes Braumann, for their invaluable support and commitment, and, above all, for the forming of a global (and cross-disciplinary) creative robotics community through the development and promotion of Rob|Arch. In addition, we would also like to thank Autodesk, KUKA, ARUP, Boston Consulting Group, Sika, ERNE, Moog and Bachmann Engineering who financially supported Rob|Arch 2018. Our sincere appreciation goes out as well to the Paper Committee; this conference and publication would not have been possible without their timeless effort and support.

The Scientific Chairs also wish to thank the National Centre of Competence in Research (NCCR) Digital Fabrication for co-hosting and supporting Rob|Arch 2018. The engagement of the NCCR Digital Fabrication, including its management staff, technicians and researchers, has been decisive in making this conference and publication possible. As such, a special thanks goes to Russell Loveridge, Orkun Kasap and Kaitlin McNally for their extraordinary commitment and work in coordinating and pushing Rob|Arch 2018 forward. We would also like to thank our Workshop Chair, Romana Rust, and all our workshop partners for their exceptional engagement. And, we would also like to extend our gratitude to ETH Zurich and the

Department of Architecture for the generous opportunity to pursue Rob|Arch 2018 in Zurich at the Hoenggerberg Campus.

Last but not least, we would like to thank all our research partners and peer institutions, our local supporters and colleagues at ETH Zurich, University of Technology, Sydney, and Bauhaus-Universität Weimar. Finally, we would also like to thank Springer Engineering for their kind support in editing and publishing this scientific publication.

June 2018

Jan Willmann
Philippe Block
Kendra Byrne
Marco Hutter
Tim Schork

Paper Committee

Sigrid Adriaenssens	The Department of Civil and Environmental Engineering, Princeton University, USA
Mania Aghaei Meibodi	Chair for Digital Building Technologies, ETH Zurich, Switzerland
Francis Aish	Foster + Partners, UK
Shajay Bhooshan	Block Research Group, ETH Zurich, Switzerland/CODE, Zaha Hadid Architects, UK
Tobias Bonwetsch	ROB Technologies AG, Switzerland
Johannes Braumann	Association for Robots in Architecture, Austria
Jonas Buchli	Agile & Dexterous Robotics Lab, ETH Zurich, Switzerland
Michael Budig	Architecture and Sustainable Design, Singapore University of Technology and Design, Singapore
Jane Burry	School of Design in the Faculty of Health Arts and Design, Swinburne University, Australia
Xavier De Kestellier	HASSELL, UK
Karola Dierichs	Institute for Computational Design and Construction, University of Stuttgart, Germany
Benjamin Dillenburger	Chair for Digital Building Technologies, ETH Zurich, Switzerland
Leda Dimitriadi	Department of Digital Knowledge, ENSA Paris-Malaquais, France
Thomas Feix	Adidas FUTURE, Germany
Jelle Feringa	Aectual, The Netherlands
Mary Franck	ESI Design, USA
Fadri Furrer	Autononous Systems Lab, ETH Zurich, Switzerland

Norman Hack	Institute of Structural Design, Technical University of Braunschweig, Germany
Volker Helm	Department of Architecture, University of Applied Sciences and Arts Dortmund, Germany
Mats Isaksson	Department of Mechanical Engineering and Product Design Engineering, Swinburne University, Australia
Roderick Jackson	National Renewable Energy Laboratory, USA
Jason Kelly Johnson	Architecture Division, California College of Arts, USA
Hanif Kara	AKT II, UK
Steve Keating	Mediated Matter Group, Massachusetts Institute of Technology, USA
Axel Kilian	School of Architecture, Princeton University, USA
Toni Kotnik	Department of Architecture, Aalto University, Finland
Torsten Kroeger	Intelligent Process Automation and Robotics Lab, Karlsruhe Institute of Technology, Germany
Nitish Kumar	Computational Robotics Laboratory, ETH Zurich, Switzerland
George Legendre	Graduate School of Design, Harvard University, USA
Katharina Lehmann	Blumer-Lehmann AG, Switzerland
Christiane Luible-Bär	Department of Fashion & Technology, University of Art and Design Linz, Austria
Areti Markopoulou	Institute for Advanced Architecture of Catalonia, Spain
Iain Maxwell	School of Architecture, University of Technology, Sydney, Australia
Wes McGee	Taubmann College of Architecture and Urban Planning, University of Michigan, USA
Marek Michalowski	X: The Moonshot Factory, USA
Ammar Mirjan	Chair of Architecture and Digital Fabrication, ETH Zurich, Switzerland
Stefanie Mueller	Computer Science and Artificial Intelligence Laboratory, Massachusetts Institute of Technology, USA
Paul Nicholas	Schools of Architecture, Design and Conservation, The Royal Danish Academy of Fine Arts, Denmark
Andy Payne	Autodesk, USA
Marshall Prado	College of Architecture and Design, The University of Tennessee, Knoxville, USA

Mette Ramsgaard Thomsen	Schools of Architecture, Design and Conservation, The Royal Danish Academy of Fine Arts, Denmark
Dagmar Reinhardt	Sydney School of Architecture, Design & Planning, The University of Sydney, Australia
Matthias Rippmann	Block Research Group, ETH Zurich, Switzerland
Christopher Robeller	Digital Timber Construction, Technical University of Kaiserslautern, Germany
Romana Rust	Chair of Architecture and Digital Fabrication, ETH Zurich, Switzerland
Fabian Scheurer	Design-to-Production, Switzerland
Jonatan Schumacher	Konstru, USA
Tobias Schwinn	Institute for Computational Design and Construction, University of Stuttgart, Germany
Claire Sheridan	Brick, USA
Asbjørn Søndergaard	Aarhus School of Architecture, Denmark
Hanno Stehling	Design-to-Production, Switzerland
Bratislav Svetozarevic	Chair of Architecture and Building Systems, ETH Zurich, Switzerland
Paul Tierman	Morris Adjmi Architects, USA
Brian Trump	Gehry Technologies, USA
Jaime Valls Miro	Centre for Autonomous Systems, University of Technology, Sydney, Australia
Teresa Vidal-Calleja	School of Mechanical and Mechatronic Engineering, University of Technology, Sydney, Australia
Timothy Wangler	Institute for Building Materials, ETH Zurich, Switzerland
Aaron Willette	WeWork, USA

Contents

Material and Processes

Construction and Structure

Design and Simulation

Image Classification for Robotic Plastering with Convolutional Neural Network

Joshua Bard(✉), Ardavan Bidgoli, and Wei Wei Chi

Carnegie Mellon University, Pittsburgh, PA 15213, USA
jdbard@cmu.edu, {abidgoli,weiweic}@andrew.cmu.edu

Abstract. Inspecting robotically fabricated objects to detect and classify discrepancies between virtual target models and as-built realities is one of the challenges that faces robotic fabrication. Industrial-grade computer vision methods have been widely used to detect manufacturing flaws in mass production lines. However, in mass-customization, a versatile and robust method should be flexible enough to ignore construction tolerances while detecting specified flaws in varied parts. This study aims to leverage recent developments in machine learning and convolutional neural networks to improve the resiliency and accuracy of surface inspections in architectural robotics. Under a supervised learning scenario, the authors compared two approaches: (1) transfer learning on a general purpose Convolutional Neural Network (CNN) image classifier, and (2) design and train a CNN from scratch to detect and categorize flaws in a robotic plastering workflow. Both CNNs were combined with conventional search methods to improve the accuracy and efficiency of the system. A web-based graphical user interface and a real-time video projection method were also developed to facilitate user interactions and control over the workflow.

Keywords: Architectural robotics · Machine learning
Convolutional neural networks · Image classification

1 Motivation

Surface finishing is an essential domain in the architectural construction practice, which requires high-skilled workers and demand accurate quality control procedures. By way of example, the authors have developed a robotic workflow to use industrial robots for decorative plastering techniques (Bard et al. 2016a, b). One of the remaining challenges in this workflow is to implement an automated, precise, and reliable quality control pipeline to guarantee satisfying results through a touch-up scenario. The touch-up procedure would let the user automatically inspect the surface and detect any unwanted fabrication artifact and command the robot to correct it.

Researchers have developed a wide range of scanning systems using different combination of sensory data, including, but not limited to, multiple RGB cameras/view (Vasey et al. 2014), RGB cameras combined with pattern projection (Rocchini et al. 2001; Zhang et al. 2002), RGB-D sensory data (Amtsberg et al. 2015), and depth data (Bard et al. 2016a, b) to reconstruct a digital representation of the physical models that can be used in the feedback loop.

J. Willmann et al. (Eds.): ROBARCH 2018, *Robotic Fabrication in Architecture, Art and Design 2018*, pp. 3–15, 2019.
https://doi.org/10.1007/978-3-319-92294-2_1

Construction tolerances standards may vary from rough to finish application resulting in different level of accuracy in each of these phases (Bard et al. 2016a, b). The achievable level of accuracy by the above-mentioned techniques with respect to the construction tolerances might not be desirable for such a delicate task as surface finishing and touch-up tasks.

Our approach requires a vision-based solution to detect texture flaws (i.e., scratches, bubbles, …) and small-scale 3D finishing issues (i.e., holes, unfinished patches). It proposes a single-camera solution without 3D reconstruction as the main input for the quality check workflow. This will result in simpler hardware setup, faster workflow, and lower costs. This approach can also be useful for other fabrication workflows, for example subtractive and deforming manufacturing.

The proposed system takes advantage of a state-of-the-art computer vision method based on *Convolutional Neural Network* (CNNs or *ConvNets*) for image classification and object detection.

Recently CNNs have dominated the image processing field, outperforming other image processing and computer vision methods by a large margin. Since 1990's CNNs have been used for different applications, including, but not limited to optical character recognition, medical image processing, feature extraction, object recognition, image understanding, and optimization (Egmont-Petersen et al. 2002; LeCun et al. 1989), thanks to their robust and real-time performance even in noisy spaces (Pal and Pal 1993).

The breakthrough advancements in computational hardware (efficient GPU architectures, possibility of distributed/multi-core/cloud-based/parallel processing, and dramatic cut in the hardware and service prices), alongside the open-source and widely accessible software platforms for machine learning, simplified and enhanced the implementation of CNNs in different contexts. CNNs can now be deployed with a reasonable budget and fewer technical challenges. These factors render CNNs as the method of choice for image classification and object detection in the past few years.

2 Methodology

The authors tested two approaches to implement CNN for this task, (1) transfer learning on *Google*'s *Inception v3* model and (2) design and training a CNN from scratch. The resulting CNN were tested for accuracy and efficiency in a robotic plastering workflow as a vision-based feedback loop for surface touch-ups.

Since its public release, Inception v3 has been used as an almost off-the-shelf image classifier for different use-case scenario. Several research teams leveraged this architecture for scientific studies, possibly most notably, for example detecting skin cancer classification (Esteva et al. 2017). In its purest form, it only requires users to organize the training dataset in a folder structure and run the provided script for a desired number of epochs. This doesn't necessitate any substantial machine learning knowledge or advanced programming skills. On the other hand, design and training a model from scratch demands for both but may provide simple and optimized results.

2.1 Apparatus Setup

Hardware. The hardware setup consists of two different end-of-arm-tool assembly (EOAT) and the connected computer unit. The first EOAT was designed to apply plaster and the second one was dedicated to the quality control and feedback loop. It includes a camera and a video projector to capture images and project results on the probing surface (see Fig. 1).
Despite the resiliency and scalability, CNN algorithms are computationally complex and expensive. While embedding computing solutions from *nVIDIA*, i.e. *Jetson* series, are capable of running such CNNs, authors decided to centralize all the computational process on a connected computer, leveraging GPU acceleration.

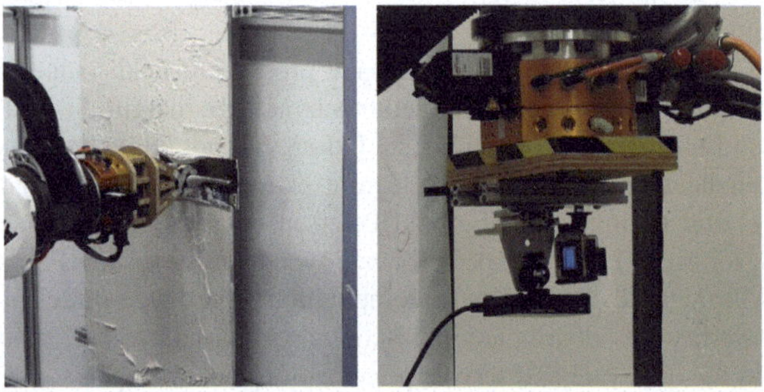

Fig. 1. End-of-arm-tools, left: plastering tool, right: camera and projector

Software.[1] A user interface was developed as a web-based application using *Django* framework. This architecture makes it possible to use *HTML/JavaScript* interactions at the front-end while leveraging *Python* scripts on the back-end. User can interact with the robot for motion commands, triggering the image processing workflow, and monitor the results.

To control the cameras, an in-house library was developed leveraging GoPro's built-in Wi-Fi protocol that provides full control over the camera functionalities.

The robotic control module was developed based on project *Open-ABB* (Dawson-Haggerty, n.d.). It communicates with the *IRC5* controller to transfer motion commands and inquire robot's status (see Fig. 2).

2.2 Data Set

The data set consists of images taken from a series of plaster finishes applied by a robot on drywall test panels. To collect the training samples, the GoPro camera was used to take 5 mega-pixel images of available plastered panels. Due to the GoPro camera significant lens distortion an image calibration method was applied using *OpenCV*, and

[1] Code for this project are available on GitHub (https://github.com/Ardibid/RoboticPlasteringCNN).

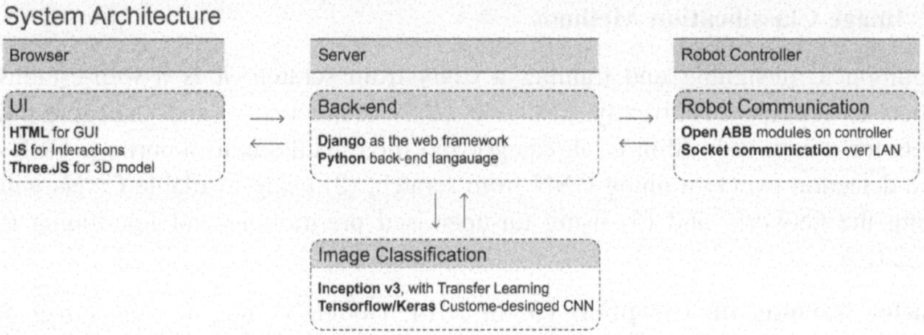

Fig. 2. System architecture

only the central 1024 px × 1024 px region of each image was used. Images were manually cropped and labeled into smaller sections as one of the three main classes: (1) pass, (2) fail, (3) markup (see Table 1). Then the same data set was categorized in five classes; perfect or near-perfect plaster regions were labeled as (1) pass, while images containing fabrication flaws including: (3) holes, (4) scratches, and (5) unfinished surfaces were categorized as fail. The markup class was left intact (see Table 2).

The markup class was dedicated to hand drawn characters that users could sketch on the work surface to communicate with the robot. Markup training samples were taken from hand drawn marks on a white surface in the same lighting condition as the plastered panels.

Table 1. Training data set for three classes

Class	Training samples	Validation samples	Test samples
1 Pass	77	23	26
2 Fail	135	42	52
3 User markup	91	28	36
Total = 510	303 (∼%60)	93(∼%18)	114(∼%22)

Table 2. Training data set for five classes

Class	Training samples	Validation samples	Test samples
1 Bad finish	26	9	10
2 Holes	29	8	11
3 Rough finish	78	22	29
4 User markup	87	29	36
5 Pass	68	23	27
Total: 492	288 (∼%59)	91(∼%18)	113(∼%23)

2.3 Image Classification Methods

In addition to designing and training a CNN from scratch, it is a well-established practice to repurpose currently available CNN architectures and their pre-trained models for a new task. Shin et al. categorized three methods to repurpose CNNs for image detection as: (1) training CNN from scratch, (2) using available CNNs without training the network, and (3) using unsupervised pre-training and fine-tuning (Shin et al. 2016).

Transfer learning on Inception v3. In 2014, *Google*'s entry for the *Large-Scale Visual Recognition Challenge* (ILSVRC2014), titled *GoogleNet* demonstrated astonishing performance (Russakovsky et al. 2015; Szegedy et al. 2015; Szegedy et al. 2016). The core model behind *GoogleNet* was called *Inception* (Szegedy et al. 2016) (Fig. 3) which adopted a relatively simpler architecture compared with other competitors and was computationally less expensive. Since then, the inception model has been used in cutting-edge research for object classification in different contexts, notably medical image processing (Esteva et al. 2017).

What makes the *Inception* model an ideal platform for object categorization in different contexts is its flexibility to be retrained for relatively similar tasks. This approach, called *transfer learning*, is a well-practiced method to fine-tune and repurpose CNN models for new tasks with very small training data set to train a deep CNN (Donahue et al. 2014).[2]

By only modifying the second to last layer of the model, transfer learning on inception v3 eliminates the need for training a whole new model from scratch for every new set of classes. Transfer learning could be a proper choice for this use case scenario since (1) it has already been trained to detect a wide range of features and there is no need to train it from scratch, theoretically this will save substantial amount of time; (2) it performs well when the size of training dataset is relatively small; (3) it requires a

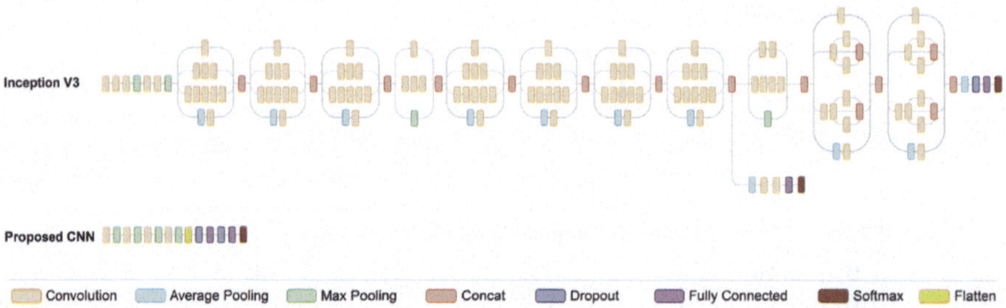

Fig. 3. Inception v3 architecture compared with the proposed architecture (Inception architecture diagram is reproduced from ("Inception v3," 2018))

[2] In the same year, at CVPR2014, Oquab et al. and Sharif et al. also addressed transfer learning and representation. For further information please look at (Oquab et al. 2014; Sharif Razavian et al. 2014).

simple workflow to repurpose the model. In light of these three advantages, we decided to exploit transfer learning technique as a possible back-end methods for our feedback loop system.

Proposed CNN. A significant trade-off of using transfer learning is the heavy model that it entails. Trained to classify one thousand classes of objects, the CNN trained model occupies hundreds of megabytes on the system storage and requires expensive computation to process a single image.

However, in our case, most of the captured images are of low contrast with primarily white backgrounds and subtle changes in color. This color space requires different feature layers for an efficient classification. Accordingly, the authors designed and trained a sequential multi-layer CNN. This architecture has already been proved its performance in several state-of-the-art models, including *AlexNet* (Krizhevsky et al. 2012) and later *VGGNet* (Simonyan and Zisserman 2014).[3] The proposed architecture is significantly simpler than of the *Inception*, resulting in a speed boost.

The authors designed and tested a series of CNNs using *Keras* with *Tensorflow* back-end to find an optimum architecture. Several combinations of convolutional, dropouts, and fully connected layers have been tested. In each architecture, all models have been trained for a fixed number of epochs and the model with the highest f1 score were selected. The results from each architecture were then compared with each other to select the optimum architecture. The selected architecture demonstrated the highest f1 score on both 5 and 3-class classification, while the others failed to demonstrate same f1 score or took longer epochs to converge to the same score.

The proposed architecture consists of four convolutional layers (3 $\times$ 3 kernel) paired with *Relu* activation function, and *maxPooling* (2 $\times$ 2), followed by three fully connected layers and *softmax* at the end. To reduce the effects of overfitting, it also leverages *dropout* to prevent inter-dependencies between hidden layer nodes (see Fig. 4 and Table 3).

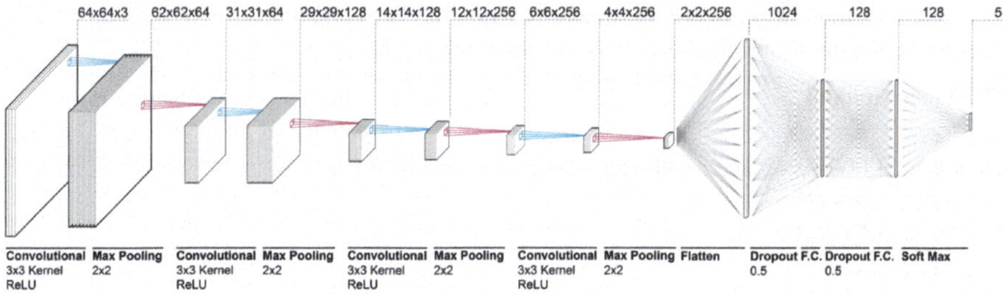

Fig. 4. Our CNN architecture for 5-class model

[3] AlexNet architecture might be confusing at the first sight since it has two parallel pipelines. However, the reason behind this dual pipeline is to train the model on two separate GPU simultaneously.

Table 3. Proposed CNN layers

Layer (type)	Output shape	Param #
conv2d_1 (Conv2D)	(None, 62, 62, 64)	1792
max_pooling2d_1 (MaxPooling2)	(None, 31, 31, 64)	0
conv2d_2 (Conv2D)	(None, 29, 29, 128)	73856
max_pooling2d_2 (MaxPooling2)	(None, 14, 14, 128)	0
conv2d_3 (Conv2D)	(None, 12, 12, 256)	295168
max_pooling2d_3 (MaxPooling2)	(None, 6, 6, 256)	0
conv2d_4 (Conv2D)	(None, 4, 4, 256)	590080
max_pooling2d_4 (MaxPooling2)	(None, 2, 2, 256)	0
flatten_1 (Flatten)	(None, 1024)	0
dropout_1 (Dropout)	(None, 1024)	0
dense_1 (Dense)	(None, 128)	131200
dropout_2 (Dropout)	(None, 128)	0
dense_2 (Dense)	(None, 128)	16512
dropout_3 (Dropout)	(None, 128)	0
dense_3 (Dense)	(None, 5)	645

3 Training Results

3.1 Transfer Learning

We used the model from the Tensorflow GitHub repository and followed the steps described in Tensorflow documentation ("Image retraining Tutorial," n.d.). The result after 4000 epochs are reported in Table 4.

Table 4. Training results, transfer learning (The training results in this chart directly result from the script provided on TensorFlow's GitHub repository after introducing new training samples. No fine-tuning, modification, or custom loss function has been applied to the re-training process. F1 score is measured on a test data set after the training phase.)

Three-class model	Ave. time to process batch of 114 patches of 64 × 64
Train acc.: 1.000 | Val. acc.: 1.00 (N = 100) Test. acc.: 0.964 (N = 112) Test F1 score: 0.9469	10–11 s
Five-class model	
Train acc.: 0.990 | Val. acc.: 0.79 (N = 100) Test. acc.: 0.904 (N = 104) Test F1 score: 0.8938	12–13 s

3.2 Our Model

In this model, the authors leveraged data augmentation to increase the data set size and improve the model's resiliency against small variations of the input data. Training and test samples where reshaped to the same size (28 × 28 × 3) beforehand. The model

Table 5. Training results, our model

Three-class model | F1 Score	Ave. time to process batch of 114 patches of 64 × 64
Train: 0.9826 | Validation: 0.9892 | Test: 0.9736	19 ms
Five-class model | F1 Score	
Train: 0.8559 | Validation: 0.9333 | Test: 0.9292	19 ms

was trained in two scenarios, one with (1) pass, (2) fail, and (3) markup labels[4] and the second one trained to define different types of fail including (1) bad finish, (2) hole, (3) rough finish (see Table 5 and Fig. 5).

3.3 Method Comparison

Comparing the speed and accuracy of the two approaches, signifies that the lighter and less complex architecture of the proposed CNN is on par and even better than what we could obtain using transfer learning. Transfer learning is significantly less complicated method, from the user point of view, that doesn't require significant understanding of machine learning. However, it is not an efficient method for fast image classification which is essential in this use-case-scenario.

On the other hand, designing, fine-tuning, and testing a CNN from scratch requires additional skills and substantial amounts of time in advance. But it pays off with the accuracy and efficiency it brings to the inspection process. Accordingly, we decided to continue with this proposed CNN.

4 Testing Implementations Scenarios

The proposed CNN was used as the image-classification back-end for a robotic plastering feedback loop, which consist of classification tool, user interface, and user interactions.

4.1 Image Surveying Methods

To survey each image, the algorithm divides it into a gird of 64 × 64 px patches that could be fed into the classifier in batches.[5] With the hardware setup described above, it took 50 ms on average to process a 256 × 64 × 64 batch of data, equivalent of a 1024 × 1024 image.

[4] Markups are simple user-defined drawings, i.e. circles and crosses, that can be used to communicate with the system.

[5] Although the authors first implemented Quad Tree search algorithm to compensate for the possible slow classification pipeline, the final model performance was good enough to provide near real-time experience. Accordingly, we opted for a grid search algorithm and avoided potential challenges that a Quad Tree search would introduce. The biggest drawback being its tendency to ignore small features in the initial steps of the search process when surveying large areas of the given image.

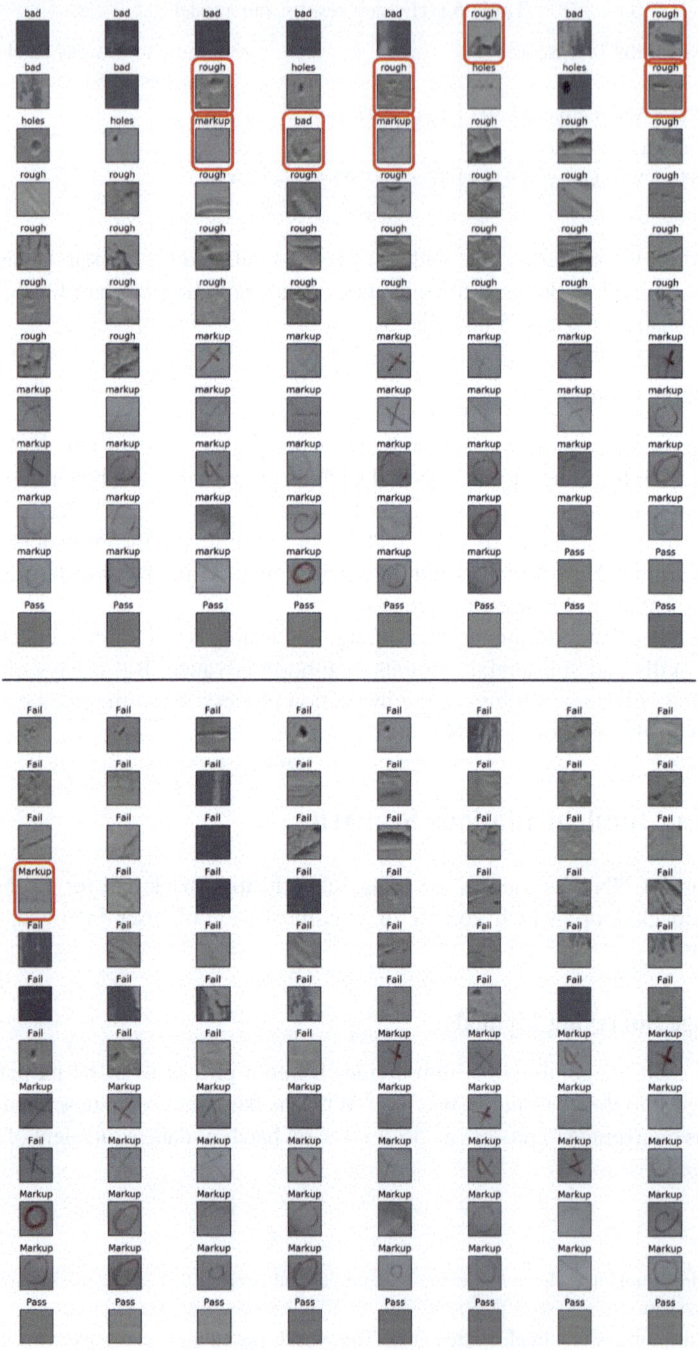

Fig. 5. Test cases for five-class (Top), three-class (Bottom) models. Errors are highlighted in red

4.2 User Interface/Interaction

Two models of interaction were designed and implemented for user interaction, first a web-based graphical user interface (see Fig. 6). In this interface, the user can send motion commands, make queries on the robot status and trigger the feedback loop process. Moreover, the user can command the robot to trowel a new layer of plaster and/or compensate for the detected flaws in a localized area of the surface.

Second interaction model consists of a real-time image projection over the working area. Users can jog the robot to a specific region of the surface and trigger the feedback process. In this case, a mounted projector highlights the detected flaws directly on the surface in real-time (see Fig. 7). In this model of interaction, users do not need to use a separate computer unit, making it more useable for on-site applications. Users can also interact with the robot by drawing a series of pre-defined mark ups on the surface. Using such markups, user can override the feedback loop results and force the robot to add/remove specific regions to error/pass cases (se Fig. 8).

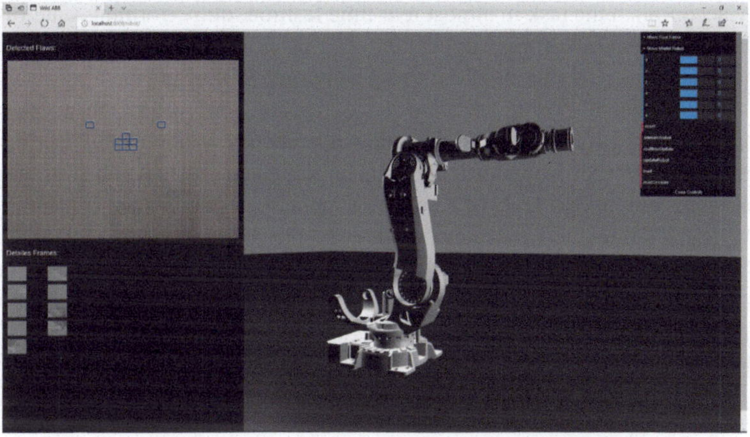

Fig. 6. User interface as running on a web browser

Fig. 7. Feedback-loop highlights on the finished surface (Left: scanning area, Center: camera/projector calibration, right: projected results, highlighting fail regions)

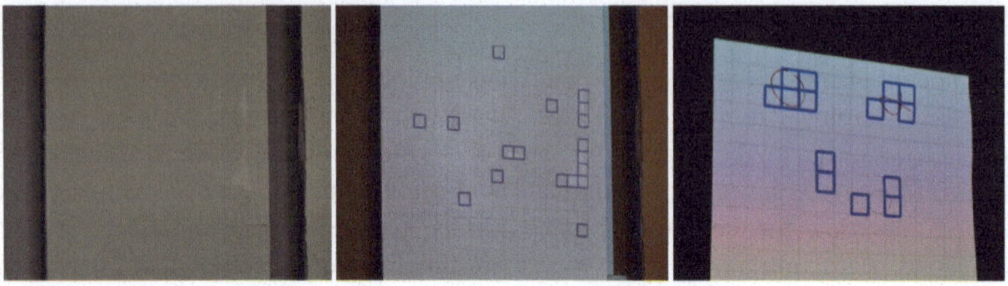

Fig. 8. Projection results on a surface with fine finish (Left: Original surface, Center: Highlighted flaws, Right: Markup detection)

5 Conclusions and Next Steps

In this work, we proposed two vision-based feedback loops based on CNNs and measured their performance in classification accuracy and speed. Due to the specific visual properties of the scanned surfaces, a custom-made shallow model could successfully accomplish the task in almost real-time fashion, outperforming state-of-the-art deep CNN architectures in this application. The proposed models performed well on both rough surfaces and fine finishes, using only one RGB camera.

As additive manufacturing techniques for construction mature, there will be an increased demand for more robust finishing approaches when manufacturing viable architectural components. In automated robotic finishing and touchup scenarios, inspection feedback loops with a high level of hardware simplicity, accuracy, and capability of categorizing detected flaws are of great importance. As the authors look to develop this work, we anticipate extending the application of the proposed model to more complex component geometries (i.e. (Bidgoli and Cardoso Llach 2015)) and different finishing/touch up scenarios using a variety of building materials (e.g., concrete, metal, or wood).

References

Amtsberg, F., Raspall, F., Trummer, A.: Digital-material feedback in architectural design. In: Ikeda, Y., Kaijima, S., Herr, C., Schnabel, M.A. (eds.) Proceedings of the 20th International Conference of the Association for Computer-Aided Architectural Design Research in Asia: Emerging Experience in Past, Present and Future of Digital Architecture, CAADRIA 2015, pp. 631–640. Daegu, South Korea (2015)

Bard, J., Blackwood, D., Sekhar, S., Brian, N.: Reality is interface: two motion capture case studies of human–machine collaboration in high-skill domain. Int. J. Architectural Comput. **14**(4), 398–408 (2016a)

Bard, J., Tursky, R., Jeffers, M.: RECONstruction. In: Reinhardt, D., Saunders, R., Burry, J. (eds.) Robotic Fabrication Architecture, Art and Design 2016, pp. 262–273. Springer, Switzerland (2016b)

Bidgoli, A., Cardoso Llach, D.: Towards a motion grammar for robotic stereotomy. In: Ikeda, Y., Kaijima, S., Herr, C., Schnabel, M.A. (eds.) Proceedings of the 20th International Conference of the Association for Computer-Aided Architectural Design Research in Asia: Emerging Experience in Past, Present and Future of Digital Architecture, CAADRIA 2015, pp. 723–732. Daegu, South Korea (2015)

Dawson-Haggerty, M. (n.d.): Open ABB Driver. https://github.com/robotics/open_abb. Accessed 5 Oct 2017

Donahue, J., Jia, Y., Vinyals, O., Hoffman, J., Zhang, N., Tzeng, E., Darrell, T.: DeCAF: a deep convolutional activation feature for generic visual recognition. In: Proceedings of the 31st International Conference on Machine Learning, PMLR, vol. 32(1), pp. 647–655. ACM, New York (2014)

Egmont-Petersen, M., de Ridder, D., Handels, H.: Image processing with neural networks—a review. Pattern Recogn. **35**(10), 2279–2301 (2002)

Esteva, A., Kuprel, B., Novoa, R.A., Ko, J., Swetter, S.M., Blau, H.M., Thrun, S.: Dermatologist-level classification of skin cancer with deep neural networks. Nature **542**(7639), 115–118 (2017)

Image retraining Tutorial (n.d.). https://www.tensorflow.org/tutorials/image_retraining#training. Accessed 20 Nov 2017

Inception v3 (2018). https://www.kaggle.com/pytorch/inceptionv3. Accessed 20 Feb 2018

Krizhevsky, A., Sutskever, I., Hinton, G.E.: ImageNet classification with deep convolutional neural networks. In: Proceedings of the 25th International Conference on Neural Information Processing Systems, NIPS 2012, vol. 1, pp. 1097–1105. Lake Tahoe, CA (2012)

LeCun, Y., Boser, B., Denker, J.S., Henderson, D., Howard, R.E., Hubbard, W., Jackel, L.D.: Backpropagation applied to handwritten zip code recognition. Neural Comput. **1**(4), 541–551 (1989)

Oquab, M., Bottou, L., Laptev, I., Sivic, J.: Learning and transferring mid-level image representations using convolutional neural networks. In: Proceedings of the 2014 IEEE Conference on Computer Vision and Pattern Recognition (CVPR 2014), pp. 1717–1724. Columbus, OH (2014). https://doi.org/10.1109/CVPR.2014.222

Pal, N.R., Pal, S.K.: A review on image segmentation techniques. Pattern Recogn. **26**(9), 1277–1294 (1993)

Rocchini, C., Cignoni, P., Montani, C., Pingi, P., Scopigno, R.: A low cost 3D scanner based on structured light. Comput. Graph. Forum **20**(3), 299–308 (2001)

Russakovsky, O., Deng, J., Su, H., Krause, J., Satheesh, S., Ma, S., Bernstein, M.: ImageNet large scale visual recognition challenge. Int. J. Comput. Vis. **115**(3), 211–252 (2015)

Sharif Razavian, A., Azizpour, H., Sullivan, J., Carlsson, S.: CNN features off-the-shelf: an astounding baseline for recognition, In: Proceedings of the 2014 IEEE Conference on Computer Vision and Pattern Recognition Workshops (CVPR-W 2014), pp. 806–813. Columbus, OH (2014). https://doi.org/10.1109/CVPRW.2014.131

Shin, H.-C., Roth, H.R., Gao, M., Lu, L., Xu, Z., Nogues, I., Summers, R.M.: Deep convolutional neural networks for computer-aided detection: CNN architectures, dataset characteristics and transfer learning. IEEE Trans. Med. Imaging **35**(5), 1285–1298 (2016)

Simonyan, K., Zisserman, A.: Very deep convolutional networks for large-scale image recognition. arXiv Preprint arXiv:1409.1556 (2014)

Szegedy, C., Liu, W., Jia, Y., Sermanet, P., Reed, S., Anguelov, D., Erhan, D., Vanhoucke, V, Rabinovich, A.: Going deeper with convolutions. In: Proceedings of the 2015 IEEE Conference on Computer Vision and Pattern Recognition (CVPR 2015), pp. 1–9. Boston, MA (2015)

Szegedy, C., Vanhoucke, V., Ioffe, S., Shlens, J., Wojna, Z.: Rethinking the inception architecture for computer vision. In: Proceedings of the 2016 IEEE Conference on Computer Vision and Pattern Recognition (CVPR 2016), pp. 2818–2826. Las Vegas, NV (2016)

Vasey, L., Maxwell, I., Pigram, D.: Adaptive part variation. In: McGee, W., Ponce de Leon, M. (eds.) Robotic Fabrication in Architecture, Art and Design 2014, pp. 291–304. Springer International Publishing Switzerland (2014)

Zhang, L., Curless, B., Seitz, S.M.: Rapid shape acquisition using color structured light and multi-pass dynamic programming. In: Proceedings of First International Symposium on 3D Data Processing Visualization and Transmission (3DPVT 2002), pp. 24–36. Padova, Italy (2002)

Designing Natural Wood Log Structures with Stochastic Assembly and Deep Learning

Kaicong Wu(✉) and Axel Kilian

Princeton University, Princeton, USA
{kaicongw, akilian}@princeton.edu

Abstract. Advances in contemporary 3D scanning and bespoke robotic technologies have enabled architectural structures to be directly constructed from naturally grown wood. However, design precedents using natural wood logs are still dominated by design approaches using predefined geometric models. The limit of this approach lies in the necessity to model the form of every structural member based on the captured geometries of all the materials before design begins. Moreover, human designed rules for joining irregular components are limited and solutions are prone to be limited by empirical knowledge. In this paper, we introduce a method for assembling natural wood log structures with higher goals autonomously using robotic stochastic assembly and deep learning. The novelty of this method is that the design of structures does not rely on prior-knowledge of the to-be-assembled materials but is generated by assembling materials iteratively. A vision system with a position suggestion network based on convolutional neural networks (CNNs) was implemented and trained to drive an industrial robotic arm for negotiating between the topological changes from potential connections and the local assembly constraints of the log. A robotic hand-eye coordination database recording the assembly of birch logs has been established and small-scale wood structures were built by the trained robot. Results show that our robot can find desired structural configurations autonomously and can assemble unfamiliar batches of wood logs. The cost and gain of using stochastic assembly and deep learning as a design strategy are discussed and future research on using different learning strategies and large-scale implementations are laid out.

Keywords: Natural wood structures · Stochastic assembly
Convolutional neural networks · Unprocessed material sensing
Generative design

1 Introduction

Despite the exploration of digital construction of timber structures in many directions such as additive assembly, subtractive fabrication, topological optimization and automated large-scale spatial assembly, most of these studies use standard industrial wood products such as plywood. Recently, researchers have shown that industrial material processing is not necessary for developing spatial structures directly from unprocessed natural wood using 3D scanning and bespoke robotic CNC technologies [21]. However, the design intent for these natural wood structures was still predetermined by geometric

J. Willmann et al. (Eds.): ROBARCH 2018, *Robotic Fabrication in Architecture, Art and Design 2018*, pp. 16–30, 2019.
https://doi.org/10.1007/978-3-319-92294-2_2

models using prior knowledge of all the collected materials and human-designed fabrication toolpath. This approach will fail when the to-be-assembled materials are unknown at the beginning of construction.

Deep learning (DL) studies in artificial intelligence (AI) have opened up promising approaches for controlling robots to pick up and manipulate irregular-shaped objects or interact with unknown environments only through visual data. Studies have proven that machines can develop better control policies than human-designed rules [11, 14, 18]. However, designing structures with natural materials goes several steps further than picking and handling unknown objects and is an open-ended design problem. The geometric constraints of naturally grown materials influence possible assemblies and how to position parts are not easily understood. Although path-planning for positioning and grasping objects has been extensively researched [25], research on how to assemble objects into meaningful configurations is still missing.

In this paper, we introduce a method for simultaneously designing and constructing natural wood log structures using stochastic assembly. Our ambition is to enable an architectural robot to autonomously search for potential structural configurations by learning from the previous experiences of assemblies without following a prior geometric design. The goal of the learning agent is not achieving the path-planning control for positioning human-designed assemblies but generating the overall design intention of the structures. To realize this, we built our own convolutional neural networks (CNNs) to control an ABB IRB 120 Industrial Robotic Arm to negotiate between the various elements of structural compositions and the local constraints of log assemblies. Different from precedents, 3D scanning and fitting a picked-up log have been developed into one process so that decisions can be made immediately after a material is scanned. A database containing the hand-eye coordination between the 2D images, the 3D volumes of scanned birch logs and robotic positions was established, different training approaches were tested, and small-scale wood structures were constructed by the trained robot. Results show that learning from the combination of multi-dimensional data enables our robot to make decisions on finding desired assembly positions even with material never seen before. The two approaches human-designed geometric modeling and learning-enabled stochastic assembly are compared for their potential as design strategies and the future implementations of large-scale structures using other learning strategies such as reinforcement learning are discussed.

2 Related Works

Designing and Fabricating Natural Wood Structures

One method of using unprocessed natural wood in fabrication is 3D scanning and reconstructing the geometries of collected pieces. In the "Woodchip Barn" and "Hooke Park Biomass Boiler House" projects, tree trunks were laser-scanned to collect available 3D geometric resources for design [21, 26]. From the laser scanning data, axis representations of the tree trunks and branches were extracted and the angles of branch crotches were calculated for measuring the moment-bearing capacities [20]. Different fabrication and design methods have been explored after collecting well-reconstructed natural wood geometries. After flattening the tree trunk sides, the curved trunks were arranged into a curved wall in the Biomass Boiler House [26]. Robotic CNC milling of notches was applied in creating customized joints for connecting to adjacent wall section built from other materials [21]. Features of a wood plank have been studied by cutting it into strips with minimal waste, and reassembling them edge to edge into a wood surface approximating a designed target surface [12]. One key idea in the precedent research is reducing the complexity of natural wood to a level that can be understood with our current design knowledge. The workflow can be abstractly summarized as planning irregular materials to best fit into a geometric design. However, approaches based on human-designed geometric modelling of untreated wood are limited, therefore we explored alternative approaches through deep learning.

Deep Learning (DL) for Robotic Manipulation
Deep learning has been used for defining robotic toolpaths without a geometric model with promising results for manipulating irregular objects. The Robotics Institute of Carnegie Mellon University has developed methods for driving robotic arms to grasp objects with CNNs [18]. Further promising results have been released from Google's Robotic Arm Farm for successfully grasping objects with hand-eye coordination data [15]. Recently, goal-directed path-planning control has been explored for training robots to efficiently learn new tasks with visual data [22]. Inspired by this research, our research focus became the translation of design problems into learning-enabled robotic behaviors. But a key difference is that the above robotic studies concentrate on manipulation or motion planning alone rather than understanding the configurations of the manipulated objects.

3 Method - Part I: Stochastic Assembly

3.1 Design Modeling

Although the possible combinations for designing natural wood log assemblies are infinite, we define the design task by computing only the next step joint positions of a robotic arm according to the captured robotic hand-eye coordination data and build our machine setup (Fig. 1). A widely adopted way of constructing wood structures is to use a geometric model to mockup desired structural configurations and then include a path-planning process to fix every piece [9]. In this direction, however, the approach is limited to the way human designers understand the design task. We believe a prior defined geometry as an indicator of the design intent is not always necessary, but alternatively we propose an iterative process of generating the construction guided by goal criteria of the design. Thus, our design task is defined as a robot trying all the potential structural configurations stochastically to reach two bases and improving its behaviors to create more desired configurations (Fig. 2).

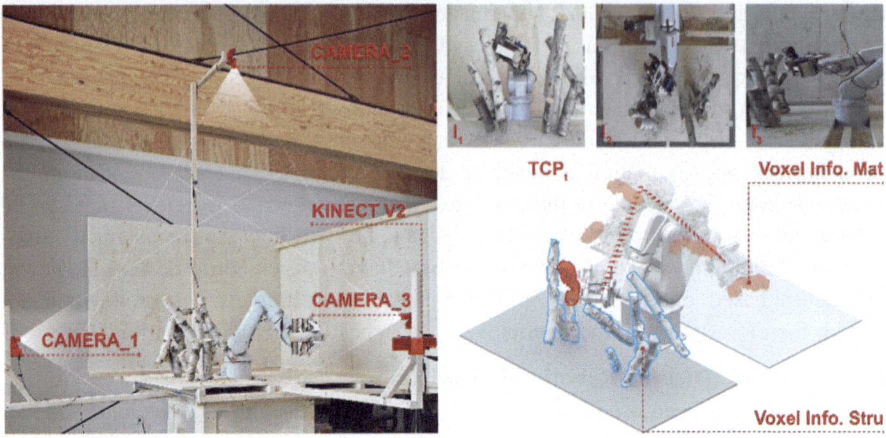

Fig. 1. Robotic hand-eye coordination between assembly positions and the images of materials.

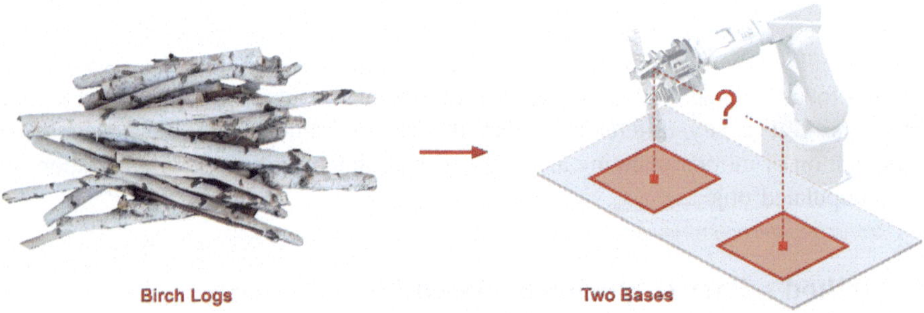

Fig. 2. Generalization of designing natural wood structures following simple design goals.

3.2 End-Effectors, Scanning and Reconstruction

Unlike industrialized materials, naturally grown wood components such as the chosen birch logs are unpredictable and there are endless geometric variations. To conform to this variation in shape, we designed a three-finger gripper (Fig. 3) for tightly gripping the irregular components. The fingers were designed, and 3D printed similar to Festo's "BionicCobot" [1] with Silicon-cast [2] soft grip pads and driven by linear actuations to bend to adjust to the logs. The three-finger gripper has the freedom of rotating its fingers so that it can adapt to local material variations (Fig. 3).

A Microsoft Kinect [3] sensor has been calibrated for partially scanning an attached log and a customized scanning programme has been developed in Processing [4] for generating the point cloud of the pieces (Fig. 6). The point clouds are automatically reconstructed into a mesh geometry using the Grasshopper add-on "MeshEdit" [5] by tracking the positions that were used for holding the log in front of the scanner (Figs. 4 and 6). Further cleaning operations were manually applied to reduce the noise of the reconstructed mesh. The robotic arm then continuously holds the scanned log during assembly to not lose the relative position between the material and the gripper.

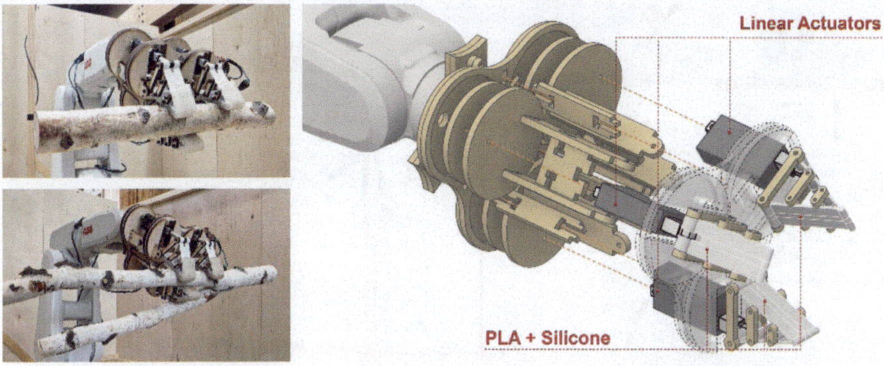

Fig. 3. The chosen birch logs and a three-finger gripper with Silicone-cast grip pads.

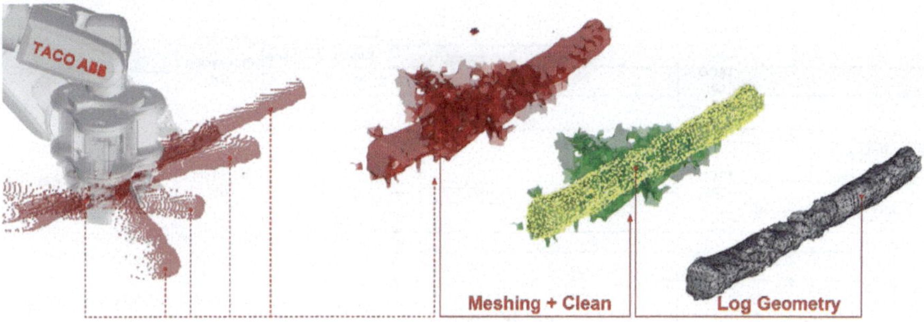

Fig. 4. Scanning birch logs and reconstructing mesh geometries.

3.3 Stochastic Assembly

A human operator feeds a log into the gripper hand perpendicular to the fingers. After that, the robotic arm drives the piece into a scanning position and next stochastically positions it within the existing structures. Previous studies in computer graphics have developed methods for finding the most relevant fitting features among broken fragments [10, 24]. However, those fitting problems are solved by finding the most relevant geometric features and image similarities to fit in. For designing structures with unprocessed natural materials, there is no given solution to achieve a construction. Another concern is avoiding collisions in both the finished state and in any in-between states. Researchers have solved path-planning problems based on the orientations and positions of to-be-assembled components [9, 13, 23]. But using naturally grown wood increases the complexity of orientations and the environment changes when the structures are updated with additional components. An alternative way of computing both the topological connections and the in-between assembly states is necessary. Our solution is to use evolutionary algorithms (EA) to reduce the cost $\boldsymbol{C_t}$ created by the conflicts between a random target position and the assembly constraints of approaching the target from a defined start position (Fig. 5).

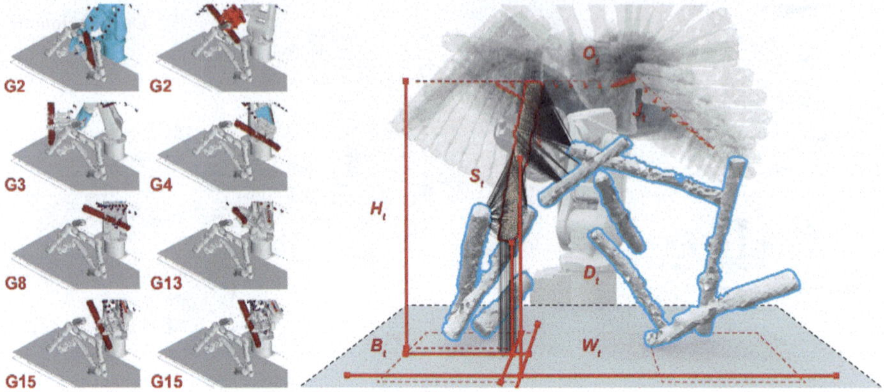

Fig. 5. Evolutionary computing with Galapagos to maximize the total of $\boldsymbol{T_t = R_t - C_t}$.

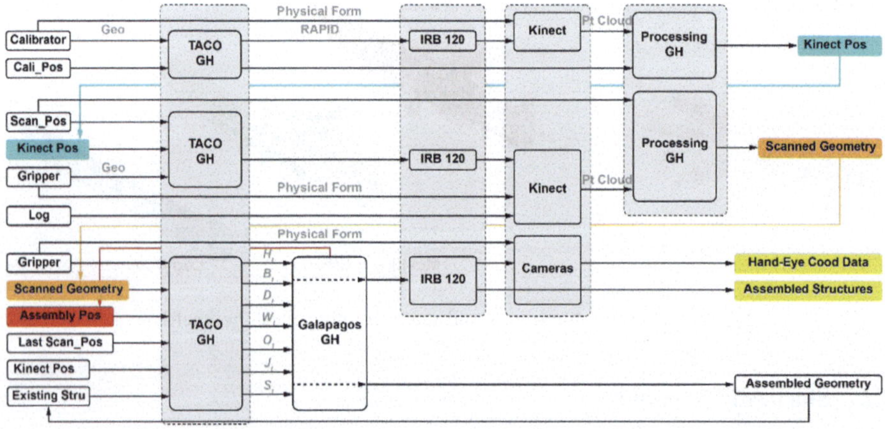

Fig. 6. Workflow of the calibration, scanning and assembling process.

The robotic arm is set to assemble every log in 40 steps, during which the boundary check of the construction base $\boldsymbol{B_t}$ (final step), the distance to the base for a log $\boldsymbol{D_t}$ (final step), the range check of working area $\boldsymbol{W_t}$ (final step), the collision check $\boldsymbol{O_t}$ (per step), the joint out- of-range check $\boldsymbol{J_t}$ (per step) are added to the cost function $\boldsymbol{C_{(B,D,W,O,J)}}$ with different parameters. For interconnecting log constructions, they cannot be fixated in midair, so the distance between a to-be-assembled piece and the existing structures $\boldsymbol{S_t}$ (final step) is also added to the cost function $\boldsymbol{C_{(B,D,W,O,J,S)}}$. The overall goal of the system is set to increase the height of the constructed structures $\boldsymbol{H_t}$, and the height is calculated into a reward function $\boldsymbol{R_{(H)}}$ (Fig. 5). The evolutionary algorithm is computed in the Grasshopper component "Galapagos" [19] for finding the maximum of the total $\boldsymbol{T_t = R_t - C_t}$. The resulting positions are thus collision free (per step) and best fitting into the structures (final step). Once the path is computed, the generated Rapid code using "TACO" [6] is updated to the IRC5 controller (Fig. 6). The robotic arm then drives the log through all the calculated positions leaving a final 1/8 to 1-1/2 in. gap, and a human operator fixes the log with an electric drill and 3-in. screws (Fig. 7).

Fig. 7. Fixating a log into a generated position.

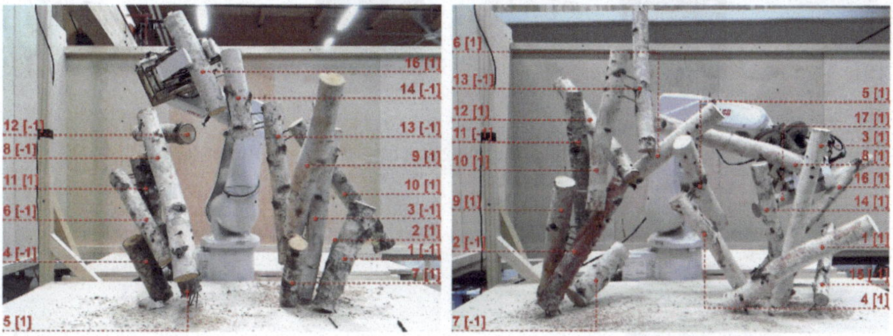

Fig. 8. (a) Construction one (b) Construction two – a series of labeled log constructions.

4 Method – Part II: Position Suggestion with Convolutional Neural Networks (CNNs)

Each of the three cameras captures 40 images from three angles for the visual appearance of all the steps and the 40 positioning coordinates are stored. To learn innovative assembly principles from the previous sampling process, 33 birch logs were assembled into two structures using the stochastic approach outlined above without any guiding geometries (Fig. 8). The next stage concentrated on formatting the collected training database and developing a position suggestion network for finding desired assembly positions based on the stochastic trials.

4.1 Data Collections

We understand the challenge of designing a wood log topology as increasing the probability of assembling logs into more structurally relevant positions and decreasing the chance of making irrelevant trials. Although positioning a log unconnected in midair has been eliminated from the stochastic sampling process, many component positions in assemblies resulting from the stochastic process are still not desirable.

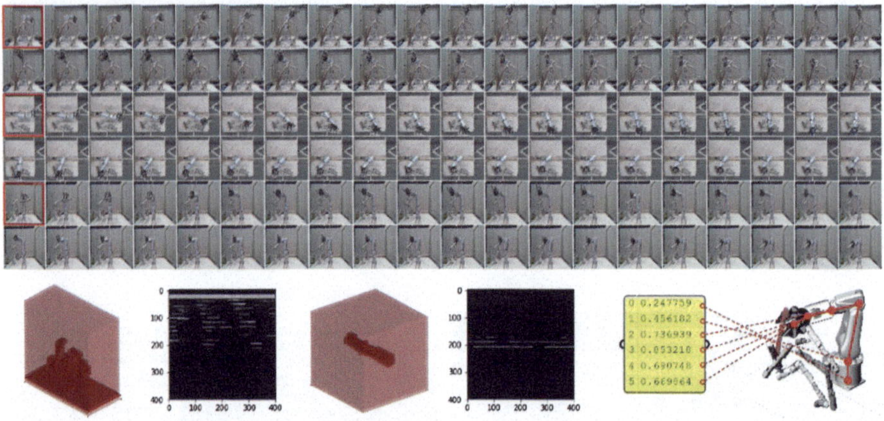

Fig. 9. Illustration of the dataset of assembling a birch log.

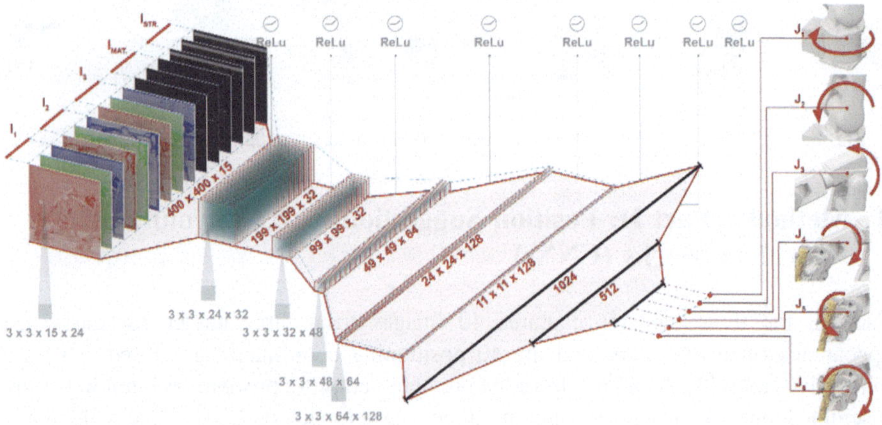

Fig. 10. Illustration of our convolutional neural network for generating future positions.

To capture those qualities, each log in the constructed assemblies was manually labeled with a score of **−1** or **+1** based on if it is meeting the goal of bridging the bases (Fig. 8). To increase the chances of generating desired configurations, the data with positive label was duplicated 3 times more than the negative ones in training. To enlarge the assembly solution space of the robotic arm in the second structure, a key step was generated randomly by evolutionary computing every five step of all the in-between positions in a robotic arm-path and it resulted in a more spatially distributed log assembly (Fig. 8b). The original database of the log construction process consists of 40 × 3 × 33 = 3960 images registering an image for 40 samples of the assembly process from 3 cameras for each of the 33 logs. Each recorded image is also linked with the joint positions of the robotic arm holding the piece. The geometries of a to-be-assembled component and the space occupied by the existing structure are recorded in voxel form (25 × 25 × 24 and 25 × 25 × 48) for each step. For a better learning efficiency, the joint angles have been recorded as radians in each different working range (Fig. 9).

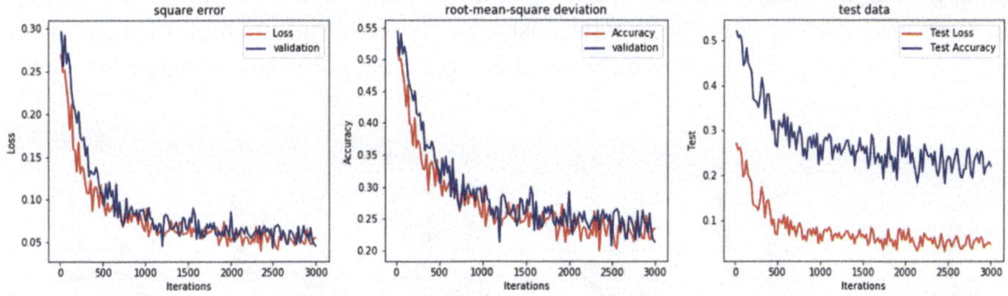

Fig. 11. Training the CNNs with the collected database.

4.2 Position Suggestion with Regression Model

The position suggestion network has been implemented for computing a target position for where the robotic arm is moving. The first approach is to directly map the pixel and voxel data in the current step to the joint values of the robotic arm in the next step. The meaning of this mapping is to consider the design generation as a statistic model and the robot positions a log where similar material features and assembly conditions have been recorded. This network takes the three input images $\boldsymbol{I}_t$, the target positions taken at that step $\boldsymbol{P}_t$, the voxel of the scanned log $\boldsymbol{I}_t$ and the existing structures $\boldsymbol{S}_t$ as input. To improve the learning efficiency, the input images with 3 RGB channels (1080 × 1080 × 3) were reduced to the dimension 400 × 400 × 3. The 3D voxel information was flattened into a one-dimensional list, repeated and reshaped into the same dimension as shown in Fig. 10. The different reshaped input information was further passed through five convolutional layers and three fully connected layers. To avoid overfitting, validation (15% database) and testing (5% database) datasets were constructed and the training progress is illustrated above (Fig. 11). Having high dimensional data, the network was developed in Python 3.6 [7] with the TensorFlow GPU computing library [8].

4.3 Position Suggestion with Classification Model

Inspired by the recent studies on applying convolutional networks in predicting the positions of a controller for playing Atari games [16, 17], we tried to divide each robotic joint domain into 50 positions and the collected positions in the database were attached accordingly. Thus, the output of our network has $50 \times 6 = 300$ classes linking the joint angles. The recorded positions were reformatted after being subtracted from the next step. Thus, we understand how much relative joint movement is in every step and the amount of change for the last step in a loop is 0.

5 Experiment: Generative Design Using Robotic Assembly

Through training, our CNN has learned a control policy that takes the real-time captured images, the current robotic joint positions, the voxel information of the scanned material and the existing structures as the input and generates a suggested future position.

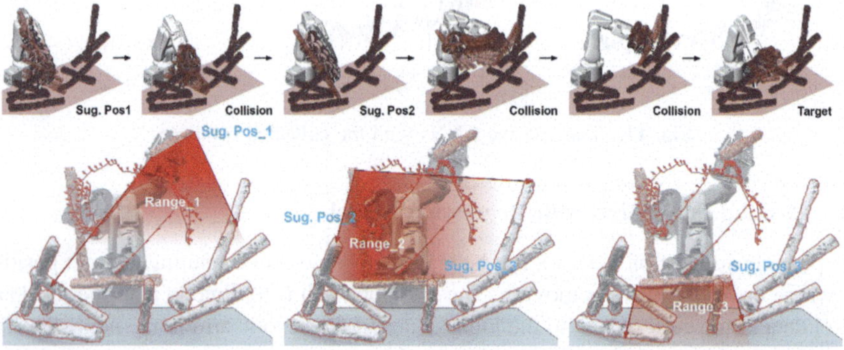

Fig. 12. Different solution accessibilities of suggested positions with classification networks.

Ideally, this procedure can drive the robotic arm to iteratively position a log in the structures along a collision free tool-path. We experimented with the idea by using only the learned control policy and by combining it with local evolutionary computing.

5.1 Combining the Classification Model and Evolutionary Computing

We started by investigating the combination of evolutionary sampling process with the learned classification model. With the size of our current database, the learned model cannot generate a human-level controlled tool-path towards a fixating position. Generating the assembly tool-path with the learned control policy was not satisfactory. Reasons such as overfitting to the training dataset due to lack of data, error aggregations as well as real-time environmental noises can all cause those undesired behaviors. Thus, evolutionary computing was integrated. At every step, the evolutionary computing process finds an optimized local configuration, but if there is a collision ahead, the robotic arm follows the suggested positions until the direct working range is clear (Fig. 12). This is a way of finding a stable position by balancing random trials in a solution space and an existing collected knowledge base. New structures have been constructed which also creates new database with the data collection procedure (Fig. 13).

5.2 Controlling Robotic Assembly with the Regression Model

Results show that the target positions generated by the regression model are more satisfactory. Although the final distance between an assembled log and the existing structures are still large, the learned control policy directs assembly positions for each log to iteratively stay at a specific spot without collisions. In every step, the suggested position changes every joint within a limited magnitude until a stable position is noticed by a human operator. The evolutionary computing has still been included but only for optimizing the touching positions of the final steps locally, so the logs can be connected by screws. To generalize the experiment, branched logs which were never used during training were feed to the gripper, scanned and assembled. In contrast to the classification model, the assembled logs are better aligned with their adjacent members (Fig. 14).

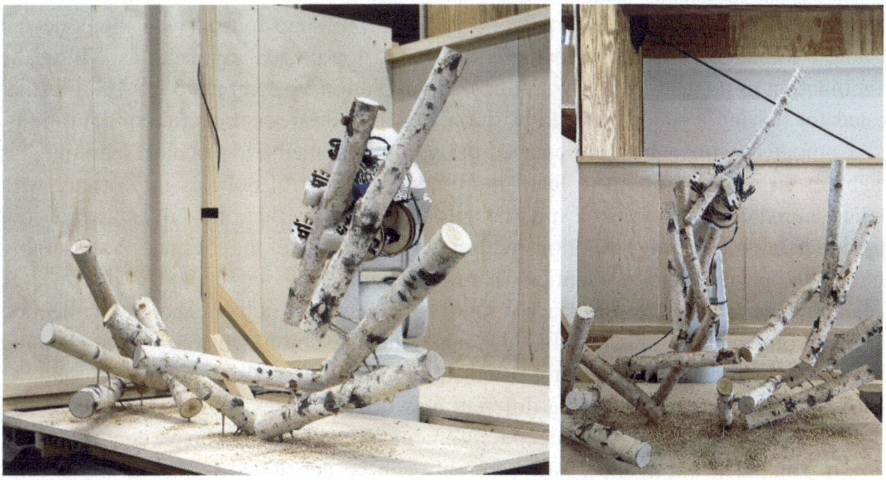

Fig. 13. Construction by combining learned control policy with evolutionary computing.

Fig. 14. Construction using regression model and branched logs.

6 Conclusion

In this research, we presented an innovative design strategy for generating spatial structures from natural wood logs with the applications of stochastic assembly and deep learning. First, a sampling procedure has been developed through evolutionary algorithms for stochastically generating tool-paths for robotically assembling birch logs. Second, a convolutional neural network has been built and trained with the collected database to suggest assembly positions of desired topological connections. Third, the combination of the learned control policy and evolutionary computing has been further tested for driving a robotic arm to position logs into desired structural configurations.

Compared with a human-designed programming method, machine learning has enabled our robotic system to learn more complex control policies in working with unknown natural materials. Experimental results show that the advantage of our method is to find inspiring solutions within the constraints of the logs and the assembly setup without any human intention. The disadvantage of our method is that a large database must be collected to learn a more robust control policy. The data collection takes not only considerable time but also raises the question of what information should be recognized for designing natural wood structures. Future works should explore other learning techniques such as using different network structures, various batch sizes and other learning strategies such as deep reinforcement learning for a better control policy.

To conclude, robotic stochastic assembly can be developed further as a powerful design strategy for solving structural configurations without planning out the geometric design beforehand. This human design intuition has not been fully explored on autonomous machines. Architectural robotics has the potential for the development of novel creative principles by improving the assembly decision making with deep learning. Compared to using target geometries as design drivers, now the structural compositions can be less defined. To influence the appearance of an assembled structure, a designer now concentrates on providing evaluation criteria rather than directly modifying the construction. Such a design control shift can further stimulate designers to discover previously unknown design criteria and include them in the conceptualization stage for finding innovative solutions. We believe the advances in artificial intelligence and robotics can not only reactivate the applications of natural materials, but also the possibility of becoming design generators to enhance our design knowledge.

Acknowledgements. We would like to specially thank Andy Zeng and Shuran Song from the Computer Science department at Princeton University for their inspiring discussions at the early stage of the research. We would also like to thank staff William Tansley and Grey Wartinger for their constant lab support and the research funding provided by the School of Architecture of Princeton University during summer.

References

1. BionicCobot. https://www.festo.com/group/en/cms/12746.htm. Accessed 1 Jun 2018
2. Smooth-on. https://www.smooth-on.com/products/mold-star-20t/. Accessed 1 Jun 2018
3. Windows Kinect. https://developer.microsoft.com/en-us/windows/kinect. Accessed 1 Jun 2018
4. Processing. https://processing.org/. Accessed 1 Jun 2018
5. Grasshopper Meshedit. http://www.food4rhino.com/app/meshedit/. Accessed 1 Jun 2018
6. Taco v0.70. http://blickfeld7.com/architecture/rhino/grasshopper/Taco/. Accessed 1 Jun 2018
7. Python Software Foundation. https://www.python.org/. Accessed 1 Jun 2018
8. TensorFlow. https://www.tensorflow.org/. Accessed 1 Jun 2018
9. Eversmann, P., Gramazio, F., Kohler, M.: Robotic prefabrication of timber structures: towards automated large-scale spatial assembly. Constr. Robot. (2017). https://doi.org/10.1007/s41693-017-0006-2
10. Funkhouser, T., Shin, H., Toler-Franklin, C., Castañeda, A.G., Brown, B., Dobkin, D., Rusinkiewicz, S., Weyrich, T.: Learning how to match fresco fragments. J. Comput. Cult. Herit. (JOCCH) **4**(2), 7 (2011)
11. Goodfellow, I., Bengio, Y., Courville, A., Bengio, Y.: Deep Learning, vol. 1. MIT Press, Cambridge (2016)
12. Johns, R.L., Foley, N.: Bandsawn bands. In: McGee, W., Ponce de Leon, M. (eds.) Robotic Fabrication in Architecture, Art and Design 2014, pp. 17–32. Springer International Publishing Switzerland (2014)
13. Landa, Y., Galkowski, D., Huang, Y.R., Joshi, A., Lee, C., Leung, K.K., Malla, G., Treanor, J., Voroninski, V., Bertozzi, A.L.: Robotic path planning and visibility with limited sensor data. In: Proceedings of the American Control Conference, ACC, pp. 5425-5430. Piscataway, NJ (2007)
14. Levine, S., Finn, C., Darrell, T., Abbeel, P.: End-to-end training of deep visuomotor policies. J. Mach. Learn. Res. **17**(39), 1–40 (2016)
15. Levine, S., Pastor, P., Krizhevsky, A., Quillen, D.: Learning hand-eye coordination for robotic grasping with large-scale data collection. Int. J. Rob. Res. **37**(45), 421–436 (2018)
16. Mnih, V., Kavukcuoglu, K., Silver, D., Graves, A., Antonoglou, I., Wierstra, D., Riedmiller, M.: Playing Atari with deep reinforcement learning. arXiv preprint arXiv:1312.5602 (2013)

17. Mnih, V., Kavukcuoglu, K., Silver, D., Rusu, A.A., Veness, J., Bellemare, M.G., Graves, A., Riedmiller, M., Fidjeland, A.K., Ostrovski, G.: Human-level control through deep reinforcement learning. Nature **518**(7540), 529–533 (2015)
18. Pinto, L., Gupta, A.: Supersizing self-supervision: Learning to grasp from 50 k tries and 700 robot hours. In: Proceedings of the IEEE International Conference on Robotics and Automation, ICRA 2016, pp. 3406-3413. Stockholm (2016)
19. Rutten, D.: Evolutionary principles applied to problem solving. In: AAG10 Conference, Lecture at the International Conference on Advances in Architectural Geometry (AAG) 2010, Vienna (2010).
20. Schindler, C., Tamke, M., Tabatabai, A., Bereuter, M., Yoshida, H.: Processing Branches: reactivating the performativity of natural wooden form with contemporary information technology. Int. J. Archit. Comput. **12**(2), 101–115 (2014)
21. Self, M., Vercruysse, M.: Infinite variations, radical strategies. In: Sheil, B., Menges, A., Glynn, R., Skavara, M. (eds.) Fabricate: Rethinking Design and Construction, pp. 30–35. UCL Press, London (2017)
22. Srinivas, A., Jabri, A., Abbeel, P., Levine, S., Finn, C.: Universal planning networks. arXiv preprint arXiv:1804.00645 (2018)
23. Steinhagen, G., Braumann, J., Brüninghaus, J., Neuhaus, M., Brell-Çokcan, S., Kuhlenkötter, B.: Path planning for robotic artistic stone surface production. In: Reinhardt, D., Saunders, R., Burry, J. (eds.) Robotic Fabrication in Architecture, Art and Design 2016, pp. 122–135. Springer International Publishing Switzerland (2016)
24. Toler-Franklin, C., Brown, B., Weyrich, T., Funkhouser, T., Rusinkiewicz, S.: Multi-feature matching of fresco fragments. ACM Trans. Graph. (TOG) **29**(6), 1–12 (2010)
25. Van Den Berg, J., Abbeel, P., Goldberg, K.: LQG-MP: optimized path planning for robots with motion uncertainty and imperfect state information. Int. J. Robot. Res. **30**(7), 895–913 (2011)
26. Wang, Y.: Hooke park biomass boiler house. In: Menges, A., Schwinn, T., Krieg, O.D. (eds.) Advancing wood architecture: A computational approach, part 4, chapter 12. Routledge, London (2016)

Mockup Method: Heuristic Architectural Fragments as Central Models in Architectural Design

Kevin Pazik(✉)

University of California, Los Angeles, CA 90016, USA
kevin.pazik@aud.ucla.edu

Abstract. Standardization and increased specialization have slowly begun to separate the means and methods of making from the process of architecture. The introduction of digital tools towards the latter half of the century have functioned to further this divide, removing any remaining traces of materiality and scale. Accordingly, architectural design exploration primarily resides in the creation and modification of digital objects, which must then be translated into the physical world. This positions built architecture in a curious position of constant catch up, chasing the impossible ideal of its digital counterpart. However, the tools predominant in architectural design and fabrication today (CAD, CAM) may be appropriated, along with sensory feedback, towards the development of a new material workflow. This paper presents a prototypical workflow which combines computational methods and robotic fabrication techniques with the spontaneity of the human and the messiness and contingency of material. The workflow is tested through the design of 1:1 heuristic architectural fragments.

Keywords: Human-robot collaboration · 1:1 · Procedural fabrication

1 Introduction

The contemporary architectural workflow includes a stark division between design, performance parameters, and fabrication techniques. Designs are iterated through the use of CAD technologies, featuring the creation of geometry based surfaces and meshes without reference to scale, performance, or material. These designs then acquire performative, structural, thermal, or infrastructural, capabilities through the application of predefined assembly systems, and are finally problem-solved through a set of fabrication protocols, mediating the translation from digital to physical. This positions built architecture within a state of constant comparison to an idealized digital counterpart. Within the contemporary workflow, performance and fabrication may be seen as unfortunate aspects which corrupt the intention of the original design. However, an earlier integration of performance parameters and fabrication methods into a threaded workflow, operating simultaneously or in quick succession, may produce an integrated design process resulting in designs which embody their performative constraints and fabrication procedures.

J. Willmann et al. (Eds.): ROBARCH 2018, *Robotic Fabrication in Architecture, Art and Design 2018*, pp. 31–43, 2019.
https://doi.org/10.1007/978-3-319-92294-2_3

2 Related Work

Previous work in the weaving of design data into the material building process has focused primarily on the augmentation of material with information [2, 3]. This transforms modelling from geometry into digital organization of complex material processes and phenomena. The integration of material, structure, and form into the design processes is also examined in Oxman's thesis Material Based Design Computation [8]. The move away from modeling with geometry, to modelling with constraints in recursive explorations is discussed in Kilian's thesis Design Exploration through Bidirectional Modeling of Constraints [6]. This transforms the process of design from geometry towards the design of dynamic rules and systems [7]. While this methodology focuses on means of integration and coordination of all internal constraints and external influences [4], the interface and medium of design remains mediated through a digital screen.

The use of a physical material as an interface for controlling a robotic manipulator has been explored in a number of projects [5, 9]. These projects, however, generally employ a hierarchical human-robot relationship, with the human at the head supervisor and manager while the robot performs the labor. This work has been motivated by a desire to leverage both the computational precision of automated labor and the gestural hand of manual labor through an equal human-robot collaboration.

3 Digital Collaboration

Contemporary trends of architectural collaboration place emphasis on the digital model, trading it back and forth between team members and consultants, with much of the design's history, development, detail, and specification embedded within a central master model. While this may streamline collaboration across team members and consultants, allowing multiple people to exchange and work on the model simultaneously, it emphasizes the conclusion of a design exploration/development exercise is a digital model, increasingly separating design from fabrication.

Further compounding this separation is the increasingly linearized flow of information throughout the design process. Design decisions cycle seamlessly around the digital model, however once fabrication begins the design is frozen and become designated read only. Design, therefore ends at the 3D model (Fig. 1).

4 Architectural Fragments as Central Models

The proposed method involves the fabrication of 1:1 architectural fragments as the primary medium of design. This positions built architecture at the center of the design process, the physical artifact is in constant refinement as design development, clarification, and production mediates through it. The architectural fragment takes on the role similar to BIM's central model, an aggregator and coordinator of all design inputs and collaborators.

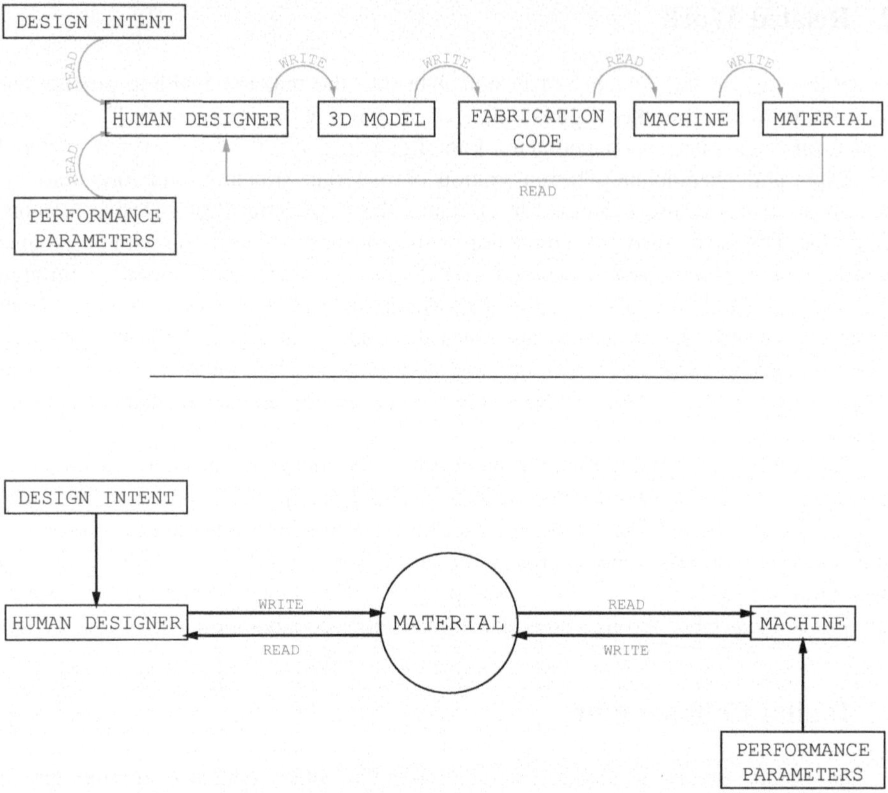

Fig. 1. Stereotypical architectural workflow (top) vs. Proposed recursive real time modeling workflow (bottom).

In the production of the design artifact are two collaborators: the human designer and the robotic manipulator. The human designer works according to design intent, mitigating design ideals and material reactions. While the robotic manipulator compares sensed data against defined performance parameters, working to adjust areas of non-compliance.[1] Comparison to performance parameters occurs through either physical testing of the fragment or through direct correlation with a simulated digital model.

Communication between the collaborators happens through the physical material of the fragment, as each actor's work modifies the work of its predecessor. Design exploration concludes upon the simultaneous cooperation of its actors: the human designer, the material reaction, the robotic manipulator, and the performance parameter. This methodology allows the design agents to collaboratively work in succession,

[1] The robotic manipulator is an ABB IRB 7600 with a S4CPlus controller.

taking turns throughout the design - fabrication process, working towards a design artifact which is the embodiment of design intent, fabrication limits, material properties, and performance requirements.

5 Case Study: The Design-Fabrication of Insulated Concrete Tilt-Up Walls

A full-scale test case was developed to show the coordination of digital simulation, human design decisions, robotic manipulation, and material properties. Insulated concrete tilt-up walls provided the necessary constraints to test the prototypical workflow in various configurations. Supplying a format which exemplifies both robotic responses to human proposals, and human responses to robotic execution, while allowing for the use of both digital simulations and sensed values to be compared to performance parameters. Additionally, the tilt-up wall typology includes its own set of fabrication processes and limitations, including self-weight and righting counter-weighting, which have been included in this case study.

Here, the steps within the design/fabrication narrative are presented sequentially, however in reality much of the process happens simultaneously and disarranged, with tasks operating in both foreground and background. This process repeats itself until the human designer, digital simulation, fabrication process, and material are fully coordinated.

5.1 Structural Loading

The structural component is primarily an investigation into the coordination of digitally simulated, robotic and material responses to human proposals for the mold of a monolithic concrete tilt up wall. This researched employed granular materials, in this case sand, as the mold material for the concrete cast, providing a formative, rather than additive or subtractive manufacturing process. The sand presents a material interface which is easily and intuitively sculpted by both the human designer and robotic manipulator without the need for complex machinery or tools.

The primary constraints in this segment is the human's design of the surface and section of the mockup and its structural capacity, this positions the human designer in constant conversation with digital structural simulations through the manipulation of the mold by either party (Fig. 2).

This method begins with the definition of a bounding box for the overall wall, setting maximums for width, height, and thickness and a tilt-up direction. The robotic manipulator, equipped with a purpose built end-effector, consisting of a material depositor, material extractor, and a 3D scanner positions itself at a home position with full view of the build volume. The 3D scanner provides a 3D point cloud of the build

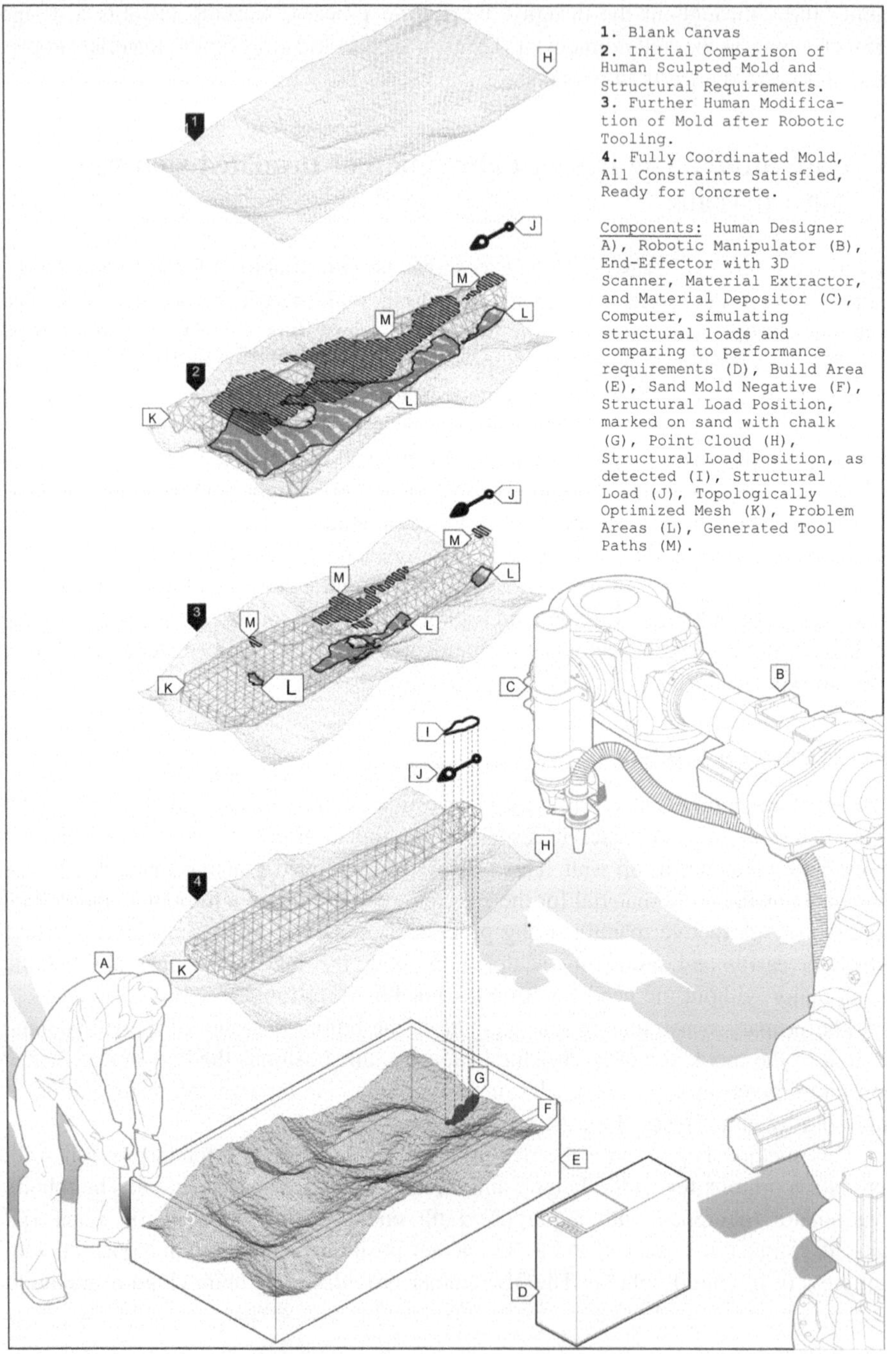

Fig. 2. Diagram of the coordination process between digitally simulated structural loading and human sculpted exterior surface.

area, which is converted from a mold negative into a digital positive and then structural analysis begins.[2]

The structural analysis includes two steps: equilibrium of the structure when tilted up, and the optional capacity to carry load. Identification of downward load placement occurs through human drawn colored markings on the sand, which is then added into the finite element analysis simulation.[3,4] Considering the tilt-up direction, the software reorients the 3D scan, places a virtual load, if applicable, and calculates the regions of material which are most, or least, important for structural integrity using topological optimization.[5]

While the computer is calculating structural necessity, the human designer may sculpt the sand, adding, removing, or shaping material according to design intent. When the human concludes design, an updated 3D scan is compared to the topologically optimized output mesh, locating problem areas with insufficient materials, or excess materials. Insufficiency implies structural weakness, while excess refers to manufacturing limits, such has material quantities or weight restrictions. These problem areas are then translated into robotic toolpaths.[6] While the robotic manipulator is executing its tool paths, the computer works to further optimize the structural load, while the human is invited to contemplate design decisions,

Once the execution of the tool path is finished, the robotic manipulator heads back to its home position, scans its work, and once again compares the 3D scan to the most recent topologically optimized mesh. The tool path may be stopped at any point in time, as new tool paths are constantly written as 3D scans are being updated. This allows for the human designer to change the load location, add or remove structural loads, or re-sculpt the mold at any time in the design-fabrication process (Fig. 3).

5.2 Casting Concrete - Center of Mass/Tilt up Operations

Once the mold is satisfactory, concrete is mixed and poured into the mold. While the human is mixing and pouring concrete, the computer works to calculate the location of

[2] A Microsoft Kinect V2 was used as a 3D scanner in tandem with Hojoong Chung's Grasshopper Plugin: Hojoong Chung, *Project Owl.* Accessed: 20 February, 2017. https://github.com/hodgoong/grasshopper-kinect2.

[3] The colored markings are detected by their Hue range using HSV color space in Processing, a programming language, with the OpenCV for Processing Library and the Microsoft Kinect V2's Color Stream option as the webcam. Greg Borenstein, *OpenCV.* Accessed: 5 March 2017. https://github.com/atduskgreg/opencv-processing.

[4] Communication between Grasshopper and Processing occurred through OSC, the contour's vertices were sent as an array from Processing and reconstructed as a curve in Grasshopper using *oscP5* library for Grasshopper and the Grasshopper Plugin *gHowl*. The center of this closed curve was used as the load location for the topological optimizations. Andreas Schlegel, *oscP5*. Accessed: 10 March, 2017. http://www.sojamo.de/libraries/oscP5/. Luis Fraguada, *gHowl*. Accessed 10 March, 2017. http://www.grasshopper3d.com/group/ghowl.

[5] Topological optimization calculations are processed in Swapan's Grasshopper Plugin: Panagiotis Michalatos and Sawako Kaijima, *Millipede*. Accessed: 15 February, 2017. http://www.sawapan.eu/.

[6] Toolpaths are generated using Greyshed's Grasshopper Plugin: *Mussel*. Accessed: 14 January, 2017. http://www.grasshopper3d.com/group/mussel.

Fig. 3. A fully coordinated upright concrete tilt-up wall.

the lifting hooks, used to demold, maneuver, and tilt-up the concrete wall. The calculation, assuming uniformity in the concrete, divides the most recent 3D scan into X amount of equal volumes for equal weight distribution using an evolutionary solver.[7] When ready, the robotic manipulator drives to each point, dropping a small amount of sand, marking the placement of the lifting hook for the human to insert into the wet concrete. Before lifting, the robotic manipulator places itself in a lift position, locating itself between all four lift points for equal weight distribution upon initial lift off.

5.3 Thermal Performance

As a direct response to the design-fabrication act that precedes it, thermal insulation is applied as a result of the thermal performance of the structural layer. The architectonic elements in design at this stage are the wall's overall interior section, interior surface articulation, and the wall's thermal resistance (Fig. 4).

[7] The evolutionary solver used is Galapagos for Grasshopper. While Galapagos slowly works to equalize the volume into X equal parts, the human designer has plenty of time to begin mixing and pouring concrete.

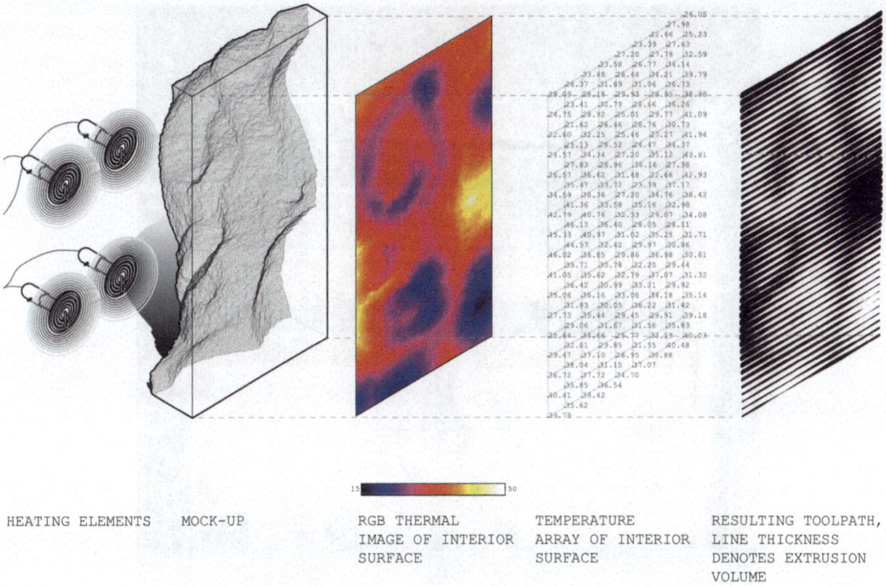

Fig. 4. Diagram of thermal response to the preceding the structural act.

Once the concrete has cured, the wall is demolded and lifted to a heating bed where its exterior face is heated and its interior face is monitored by a thermal camera.[8] Thermal readings are saved as both temperature arrays and color maps. The temperature arrays are correlated to their relative spatial position on the fragment. As the rate of extrusion for the insulation foam remains relatively constant, volume is built up by varying the speed of the robotic manipulator, slowing down in areas with high thermal transmissions to increase their thermal resistance.

Whereas the temperature array is used by the computer, the color map provides a legible interface for the human designer to read, making legible the parameters that the computer and robotic manipulator will be responding to. The human designer may also intervene in the deposition process by indicating areas of more or less insulation by drawing directly on the material with two different colored chalks. Once these colors are recognized through computer vision, the robotic manipulator's speed is adjusted accordingly within these boundaries. Additionally, the human designer may also add insulation manually, adjusting the interior surface articulation or overall section. This process also repeats itself, checking the insulated concrete for thermal weak points throughout the process (Fig. 5).

[8] A FLIR Lepton Micro Thermal Camera Module was used and controlled with an Arduino Due using Josep Bordes Jové's sketch, *FLiR-lepton.* Accessed: 13 March 2017 https://github.com/josepbordesjove/FLiR-lepton.

Fig. 5. Sectional relationship between the structural concrete exterior and the insulated interior.

6 Standards

Rather than the application of dimensional standards onto idealized geometric forms, we propose contractual standardization in the form of constraints that exist throughout a recursive design process. Minimizing the process of "rationalization" and "optimization" of drawn or modeled architectural forms, and producing an end result which embodies the parameters from the beginning. The embodiment of constraints throughout a design process allows architectural design to move away from problems of translation from drawing, or 3D models, to building.

The move from dimensional standards to contractual standards suggests a shift from representation to verification. Whereas dimensional standards rely on thicknesses and products, contractual standards confide in on-site verification. Favoring the sensing and testing of materials and assemblies over drawings and installation instructions.

7 Holistic Design

While the experiment was mostly tested within the realms of the architectural fragment, a similar process may be applied across an entire building. The parameters from the preceding sections would be taken into account when constructing the next, incorporating a set of trade-offs and accommodations. If none of the preceding fragments can accommodate the necessary structural load the later walls must get thicker, if insulation isn't sufficient in a certain section, another may make up for it. As structural load, thermal requirements, and infrastructure move around the plan of a building, walls acquire or lose thickness. Throughout the design process the building is simultaneously all of its possibilities and none. Possible results are removed as the design is implemented, and the final form is narrowed down (Fig. 6).

Fig. 6. Plan diagrams demonstrating the transformation of poche based on varying structural load configurations. Load bearing walls shown in black, insulation shown in grey.

8 Discussion

The nature of the mock-up in this project is not to confirm specified materials or test contractor craft, rather the mock-up is treated as an exercise to explore the cooperation and coordination of design intent, material properties, fabrication protocols, and performance requirements. It proposes the architectural mock-up as the site of design research and investigation. It grants physical material artifacts the role of aggregator and coordinator of all design inputs, parameters, and collaborators, a maquette which is in constant refinement throughout a recursive design process.

Insulated concrete tilt-up walls were the primary case study in this project, exploring the coordination of human design intent, structural requirements, thermal performance, and sand cast concrete. Expanding the potential of this workflow would require greater experimentation in both assemblies and their requirements. Additionally, this project focused design development from exterior inward, however this relationship is also ripe for development.

By working directly with a material artifact, fabrication and construction is not considered as a separate phase, but as a critical piece intertwined with the design process. An integration of fabrication protocols directly into the design process provides additional room for the discovery of architectural form through the contingency of material.

Fig. 7. Exterior and interior faces of an insulated concrete tilt-up wall.

This then transforms the nature of standards, moving towards the use of on-site verification, shifting from dimensional standards to contractual standards, allows for the settlement of performance parameters to be developed and negotiated rather than applied or solved. The sensing and testing of materials and assemblies would also allow architectural components to appropriate other roles, for example conduit could take on the role of structure.

The proposed workflow breaks out of the paradigm of monotasking within the architectural design process towards a multi-task, recursive, approach where each part is fabricated and compared to a series of parameters (Fig. 7).

Acknowledgements. This research owes much to the faculty of Princeton University School of Architecture, specifically, Axel Kilian, Forrest Meggers, Ryan Luke Johns, Paul Lewis, Liz Diller, and Jaffer Kolb, and to the cohort of classmates in my thesis class for their equal support and skepticism.

References

1. Brooks, H.: The Tilt-UP Construction and Engineering Manual. Tilt-Up Concrete Association, Mount Vernon (1988)
2. Gramazio, F., Kohler, M.: Digital Materiality in Architecture. Lars Müller, Baden (2008)
3. Gramazio, F., Kohler, M., Oesterle, S.: Encoding material. In: Oxman, R., Oxman, R. (eds.) New Structuralism: Design, Engineering and Architectural Technologies, Architectural Design, vol. 80(4), pp. 108–115. Wiley, London (2010)
4. Hensel, M., Menges, A., Weinstock, M.: Emergent technologies and design, pp. 43–68. Routledge, Oxford (2010)
5. Johns, R.L: Augmented reality: modelling with material indeterminacy. In: Gramazio, F., Kohler, M., Langenberg, S. (eds.) Fabricate: Negotiating Design & Making, pp. 216–223. UCL Press, London (2014)
6. Kilian, A.: Design exploration through bidirectional modelling of constraints. Ph.D. thesis, Massachusetts Institute of Technology (2006)
7. Mark, E., Gross, M., Goldschmidt, G.: A perspective on computer aided design after four decades. In: Muylle, M. (ed.) Proceedings of the 26th International Conference on Education and Research in Computer Aided Architectural Design in Europe (eCAADe): Architecture 'in computro': Integrating methods and techniques, pp.169–176. Antwerp (2008)
8. Oxman, N.: Material-based design computation. Ph.D. thesis, Massachusetts Institute of Technology (2010)
9. Willis, K., Xu, C., Wu, K., Levin, G., Gross, M.: Interactive fabrication: new interfaces for digital fabrication. In: Proceedings of the fifth international conference on tangible, embedded, and embodied interaction (TEI 2011), pp. 69–72. ACM, New York (2011)

Haptic Programming

Sven Stumm(✉) and Sigrid Brell-Çokcan

Chair for Individualized Production in Architecture,
RWTH Aachen University, Aachen, Germany
office@ip.rwth-aachen.de

Abstract. Current industrial robotics focuses on the utilization within clearly defined and structured production environments. However due to increasing product variety, a paradigm shift away from repetition of static task towards dynamic human-robot collaboration is noticeable. Due to the fact that static automation can only be achieved at a prefabrication level within the construction industry, this shift towards adaptable robotics can be utilized for new concepts for on-site robotic assistance. We extensively illustrate our approach towards robotics that adapts to changing environmental conditions and material features, while retaining a degree of predictability necessary for effective collaboration. Furthermore, by integrating human-robot collaboration with parametric modelling a feedback to design is established. The term haptic programming is coined in order to illustrate the direct interconnection between parametric model and human-robot collaboration. First application examples are shown to illustrate the use of a priori knowledge from the design phase in combination with haptic interaction primitives to enable intuitive human-robot collaboration. Haptic programming allows the exchange of knowledge between the user and a robot on a physical level.

Keywords: Robot programming · Visual programming · Haptic programming Construction robotics · Human-robot collaboration

1 Introduction

The current degree of automation within the construction industry is still low especially compared to industrial production [1]. Figure 1 illustrates some of the problems employing industrial solution on-site. Nevertheless, a number of previous attempts at automation were made using specialized machines for specific tasks [2, 3].

These approaches however failed for a number of reasons. Specialized machines are mainly low in flexibility, while construction tasks can be highly complex, especially if only single tasks are automated, while the rest are manually executed. This leads to necessity for construction work being adapted towards automated processes, which causes additional cost and can be time consuming (see Fig. 1). Additionally, a number of high cost result out of the development, implementation and maintenance of a solution, which can only be used for a small specific and individual application. While these might be acceptable within the field of industrial mass production, the required time for setup, removal and the resulting cost is often too high for the field of construction.

J. Willmann et al. (Eds.): ROBARCH 2018, *Robotic Fabrication in Architecture, Art and Design 2018*, pp. 44–58, 2019.
https://doi.org/10.1007/978-3-319-92294-2_4

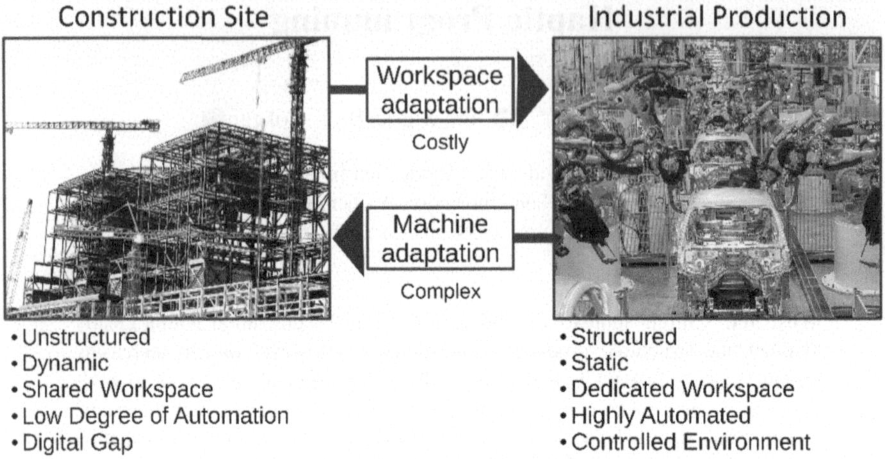

Fig. 1. Problems in converting a solution for industrial production to the construction site.

The applicability of a solution highly depends on its market range. However, there are a number of factors which reduce the market range of specialized solutions with demanding requirements such as separation of work area, workspace constrictions (e.g. straight walls, leveled floor, low dirt tolerance, etc.), exact storage of work material and more. In addition to the small lot sizes within construction, high tolerances and variation are very common especially when working with natural materials.

Building components are therefore fitted manually on-site. Even though industrial robotics is not specifically suited for such environments, they allow a free workspace layout and can be seen as a general purpose machine. As robotics is able to process digital data towards physical manipulation it can act as a mediator between design and construction. Having said this, a number of adaptations towards uncertainties within the unstructured nature of the construction environments are still necessary.

Efforts are made to mediate this through tighter process control and detailed digital representation within Building Information Modeling (BIM) [4, 5]. BIM allows the parameterization and placement of intelligent objects, containing a number of properties, in a common environment for different parties involved in the planning process. However, to inform construction workers accordingly these models still need to be broken down to physical plans. This leads to a very loose correlation of the current model and the actual unstructured environment of the construction site. These deviations have to be taken into account for on-site automation approaches.

Rather than creating a specialized machine for a static execution of specific construction tasks and letting the workers compensate material tolerances and planning deviations, we want to achieve an environment using flexible and intelligent machines to compensate the conditions on-site. Yet for a universal solution this requires a level of intelligence within the programming, that takes into account all possible and often unpredictable events and is therefore beyond the scope of current technologies.

These difficulties can however potentially be mediated through new developments within human-robot collaboration. Tellex et al. [6] demonstrated the idea of "asking for help" if the robot fails at an assembly task. This however also requires a common knowledge base between robot and operator. By shifting information from the planning phase directly towards the robot controller we try to use the design model as a knowledge base for the robot. Throughout the multitude of planning phases and through the experts integrated into BIM a very intensive a priori knowledge of the environment and the task at hand is created in the form of a digital model.

This knowledge can therefore be employed to create an assembly plan for the robot. Through the concept of Haptic Programming we create a physical interaction with this data, as well as create contingency strategies to compensated deviations through human-robot collaboration.

For our test cases we use an industrial robot with six Degrees of Freedom (DoF) on a mobile platform with three DoF as a flexible machine. Specifically, we are using the KUKA iiwa robot mounted on the KUKA mobile platform kmr. The kmr uses an advanced robot controller, where the KUKA SunriseOS allows us to use object oriented programming via JAVA. Additionally, the KUKA iiwa hardware has force torque sensors in every axis, which makes it suitable for human-robot collaboration, due to the fact that this allows the detection of contact on every part of the robot.

As this maximal load of the robot is 14 kg and the mobile platform is not suited for dusty environments the relevance of this concept for actual construction tasks still needs to be verified. Within our future work we will transfer our approach of haptic programming towards other kinematics and controllers, which are better suited for construction.

2 Design to Construction

While the use of BIM is still not as widespread as desired, the use of Computer Aided Design (CAD) is by now a standard practice within Architecture Engineering and Construction (AEC) industries. CAD models encompass different degrees of detail ranging from simple geometric models, over meta data pertaining to materials and part spacing, towards intelligent objects containing parametric digital models that allow immediate dimensioning of objects according to the required specifications. In order to achieve an efficient workflow from the design phase to part production it becomes necessary to link geometry and fabrication process [7].

Most Computer Aided Manufacturing (CAM) software allows for the path generation of gantry kinematics based on CAD. However, in this case fabrication parameters are often connected to specific motion commands and not to a parametric design model. Fabrication with articulated robot arms with six DoF increases the complexity, while simultaneously allowing the fabrication of new geometries as well as surface aligned fabrication. Even though gantry robots are less likely influenced by external forces and therefore achieve higher accuracies, the flexibility of articulated six DoF robot arms results in different fields of application.

Parametric robot control systems such as KUKA|prc [8, 9] work within a parametric design environment and directly link CAD geometries and fabrication through direct KRL code generation out of the model. This allows the dynamic fitting of the design model towards specific conditions. KUKA|prc builds upon CAD software

Rhinoceros from Robert McNeel & Associates and in particular the visual programming environment Grasshopper 3D. KUKA|prc allows the definition of fabrication strategies through visual programming linking geometry and robotic toolpath through a parametric model. Additional components allow simulation and analysis of the toolpath. Changes to the source geometry propagate through the parametric model and are directly reflected by the simulation result. Although CAD environments work with idealized data this still allows the fabrication of a high number of product variations enabling mass customization. Building on top of this approach we were able to identify a number of advantages within parametric robot control.

Employing parametric models as a basis for robot path planning allows the reusability of existing definitions for similar tasks within the same topology. The parametric definitions allow the introduction of deviations and modification of the source geometry.

However, CAD is unable to fully represent physical reality and its complexity. Although high resolution scans might be able to create an extensive geometric representation they are unable to accurately predict and recreate material properties especially for inhomogeneous materials. Furthermore, a detailed analysis of every part is often too time consuming to be applicable.

While this idealized planning environment might be useful for single part fabrication at assembly level material tolerances become even more influential and create a higher complexity. These simplified models based on human perception and the ability to create abstract representations are however very useful for the decision making process. Therefore, building upon these idealized models gives us the possibility to create an intuitive interface for task planning.

The field of constraint-based robot programming [10–12] introduced a first approach towards handling complex sensor-based robotic tasks, which consider geometric uncertainties. The approach employs the description of geometric features to create a dependent constraint space. Within this approach the actual path planning becomes a multi-criterial optimization problem. However, constraint based programming describes a model control theory based approach for robotic adaptivity rather than a programming methodology. A major disadvantage of this approach concerns its usability. Planning of task sequences is very time consuming. Definition of the constraints also requires detailed knowledge of control theories.

Alternatively, methods of learning by demonstration can be used, however automated generalization of tasks and the efficient usage of the recorded data is still as major challenge within this field [13]. The field of Kinesthetic Teaching showed the intuitive nature of teaching robotic task through direct physical interaction [1314-16]. As part of learning by demonstration it faces similar difficulties of reusability. Additionally, the required knowledge lies entirely by the operator, as the robot acts completely passive. In most cases even kinematic constraints are not indicated before they are reached.

For these reasons our approach towards haptic programming started out as a way to combine the advantages of these different methods [17].

We use an idealized CAD environment to plan our assembly sequence using parametric part representations. Simultaneously we can employ KUKA|prc to create a priori simulations of the toolpath. Parametrizing the CAD model in a number of variations, we are able to analyze reachability and possible collision scenarios to optimize our sequence towards flexibility. The resulting CAD based description of the assembly sequence in terms of a robot toolpath also creates a certain predictability, which is often required within industrial robotics for reproducibility of results.

In contrast to parametric robot control however we link the geometry and the resulting toolpath not to a fabrication logic but towards parameterizable assembly skills. Meaning that instead of simple trajectory planning we plan assembly operations (see Fig. 2).

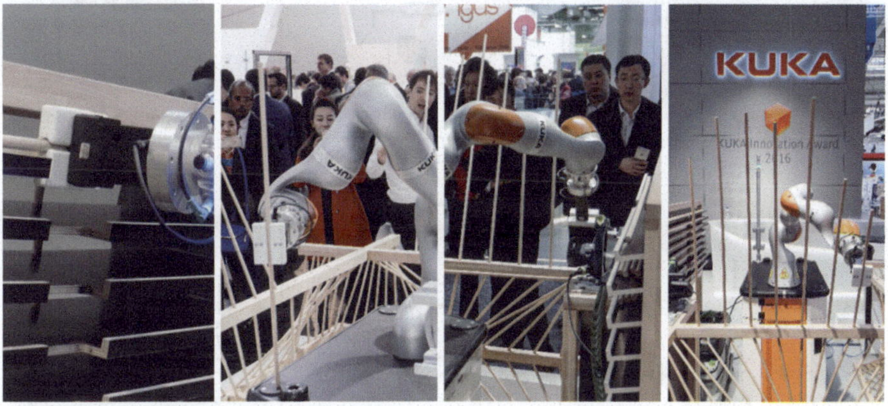

Fig. 2. Demonstrator implemented as part of the Hannover trade fair 2016 illustrating basic assembly operation pertaining to the main categories of assembly. From left to right: handling; checking; adjusting; joining.

As the assembly process are defined parametrically, we can use this information to define constraints as part of the parameter space. We can then very easily introduce deviations and re-evaluate the process based on adaptations. While simultaneously transferring the core parametric of the geometric design model towards the controller.

In turn this leads to a shift in the knowledge basis towards the robot allowing for a basic understanding of the design intent and its constraints. We therefore differentiate between the robot coordinate space and the parameter space generated through the parametric design model (see Fig. 3). However, in worst case scenarios on site deviations might lead to an oversized search space. We therefore added constraints for timely execution, if the search and adaption to the environment takes longer than acceptable or deviations exceed the parameter space human help is requested for faster adaptation.

Furthermore, on-site adaptation of construction requires an approach that allows people without prior experience in robotics to work with the robot. Due to changing lighting and dust the use of vision systems on constructions sites is very difficult.

For the use of robotics on-site it is also very important that the configuration of a robotic setup is fast and easy. Therefore, we used the internal force torque sensors of the robot to extend the assembly operations through haptic interaction primitives.

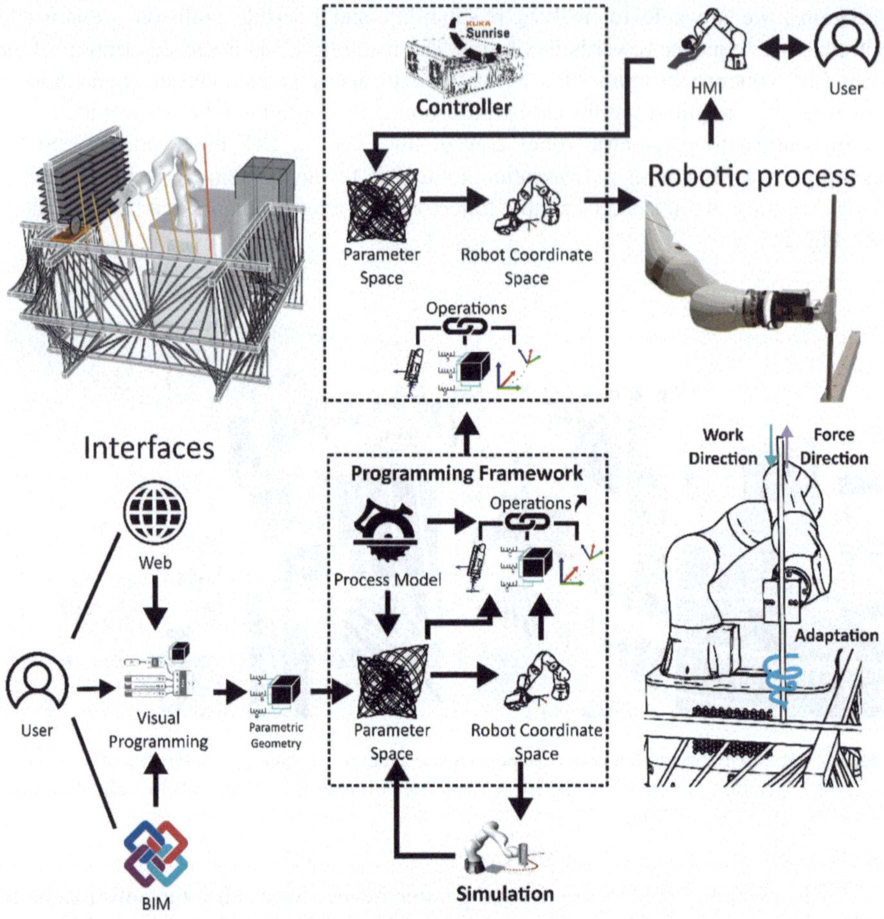

Fig. 3. The main programming concept employs an abstract parametric design model that is interconnected with the robot coordinate space through robotic operations. While this link is created within the visual programming framework the actual operation execution is implemented on the controller side. This allows the execution of operations based on the abstract parametric model contained within the robot controller. Operations are further enriched through haptic interaction primitives for configuration, adaptation and contingency planning. The feedback of the haptic interaction primitives can in turn influence the original abstract parametric model.

3 Haptic Interaction Primitives

Using the described concept, assembly sequences are planned based on the parametric design data – similar to offline robot programming. However, the generated positions do not define the entire movement of the robot, instead they only provide the underlying programme model that is the basis for the adaptive assembly process.

If this is combined with the kinematic constraint space, by overlaying both models a description of both adaptivity and design constraints is achieved for each operation,

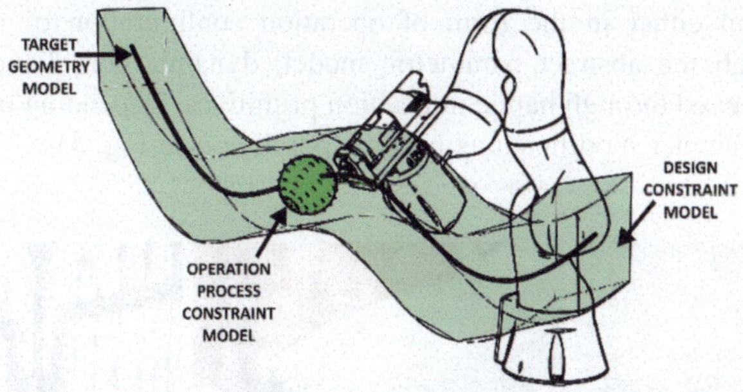

Fig. 4. Example constraint space visualization. Constraints which describe the outer hull of the parameter space are superimposed from different model inputs. These are constraints from the abstract parametric design model, process constraints and kinematic constraints.

as illustrated in Fig. 4. In order to achieve an easily verifiable kinematic constraint model a spherical description of possible orientation and position based deviation is precalculated for each operational position.

We use haptic programming of robots in order to adapt model parameters as well as positions. If time constraints are met or further information is required, the haptic programming mode is triggered. Visually the robot signals their coworker for help and is then manually adjusted through direct contact with the robot.

Similar means of programming such as Kinesthetic Teaching differ from this approach. For one the taught motion is not used as a demonstration for imitation, but the required information is extracted in the form of parameters. Also, the robot cannot be moved around arbitrarily, but instead leads the user passively, within its parameter space through dynamic compliance. Dynamic compliance is achieved by increasing joint stiffness of the robot close to the parametric and kinematic constraints.

Following a quick confirmation, the information regarding the current position of the robot is used to adapt the current parameterization of the model. This in turn can affect the following robot operations. This is specifically the case if the haptic programming employs force-torque feedback not only for the human-robot collaboration but to adapt model parameter through direct interaction with the environment (e.g. dynamic measurements of a work-object/base affecting later motions).

Another way in which we use haptic programming is the initial parametric configuration of the task on-site. In these cases, we use haptic programming instead of CAD software to define parameter. For this the user manually guides the robot in dynamic compliance. This is in turn used to inform the assembly process, allowing each produced structure to be unique.

As the robot controller does not contain a geometric modelling library only simple geo-metric models can be used as a basis. Therefore, only planes, curves, surfaces (described through two perpendicular curves) as well as boxes, cylinders and spheres are used as an abstract model basis.

This is interconnected through operations, which describe the geometry based path as well as the adaptation strategy. This creates a direct link between robot coordinate space and the abstract parametric model. Operations are in turn interconnected with haptic interaction either in the form of operation configuration or as contingency strategy. Through the abstract parametric model, dynamic compliance within each operation is achieved through haptic interaction primitives. Depending on the operation a number of parameter modifications become possible (see Fig. 5).

Fig. 5. Examples from different implementations illustrating haptic interaction primitives. From left to right: initial geometry parameterization, positional adaptation and material handover.

For the initial parameterization this means for example that the control points within a curve can be moved in accordance with the defined parametric. Each assembly operation is defined using a target frame, a type of operation, an origin frame as well as a number of parameter.

However positional information within each operation is seldom adapted directly. As each operation is linked to the parametric design model we are able to modify the parametric model directly. Within the controller structure a meta controller for operation handling was created. Each operation consists of the following four parts:

- Direct execution of the motion resulting from the parametric model in the robotic coordinate space
- Overlay of a sensor based control strategy for online adaption of the motion
- Adaption strategy of the dynamic search space through modification of the parameter space resulting in a motion change

- Contingency strategy for constraint violation handling (including timing constraints) through haptic programming

Accepting the restrictions on the side of the robot and communicating these visually and haptically can improve the human experience of robotics, when working together. Moreover, the robot is able to lead the user along its task, which helps humans to understand the tasks as well as the capabilities and restrictions of the robot directly and can therefore be naturally used to improve knowledge. If the digital a priori knowledge does not correspond to the constructional reality a reciprocal transfer of information between operator and robot becomes possible.

4 Adaptation for Future Task Execution

Current adaptation is limited to a few process parameters; However, every operation execution creates a significant amount of data. The integration of further data into the model could potentially allow for a learning process creating new operations.

These operations can also be combined with optimization approaches as described within the field of constraint based programming. The advantage here is for one is an easier modeling of constraints through the abstract parametric model. Splitting tasks into operations also reduces optimization to single operations, which reduces the search space and therefore leads to faster optimization results.

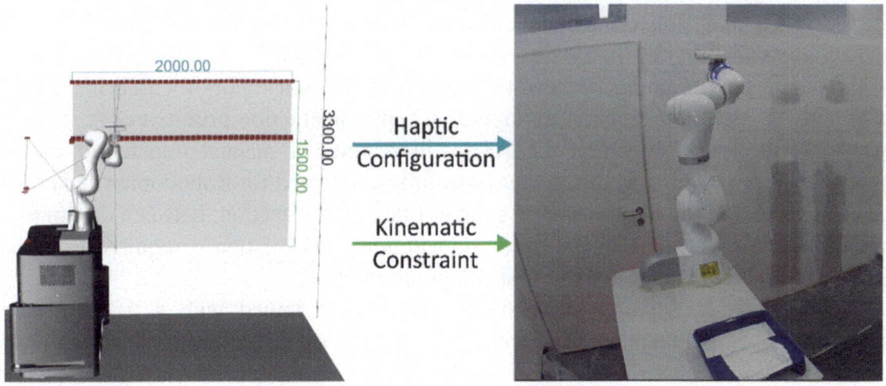

Fig. 6. Example digital model and real world application in a wall painting demonstrator. This illustrates the abstract parametric model allowing for wall length configuration.

Example applications of haptic programming (see Fig. 6) were shown open to the public in a first version at the Hannover Fair 2016 (HMI) as part of the KUKA Innovation Award (see Fig. 2), the Robotic Art Festival Robodonien 2016 (Robodonien) in Cologne (see Fig. 7), as well as at the Aachener Werkzeugmaschinen Kolloquium 2017 (AWK) (see Fig. 8), which employed the HMI setup and extended it towards the use of mobile robotics. These first results showed that the adaptation

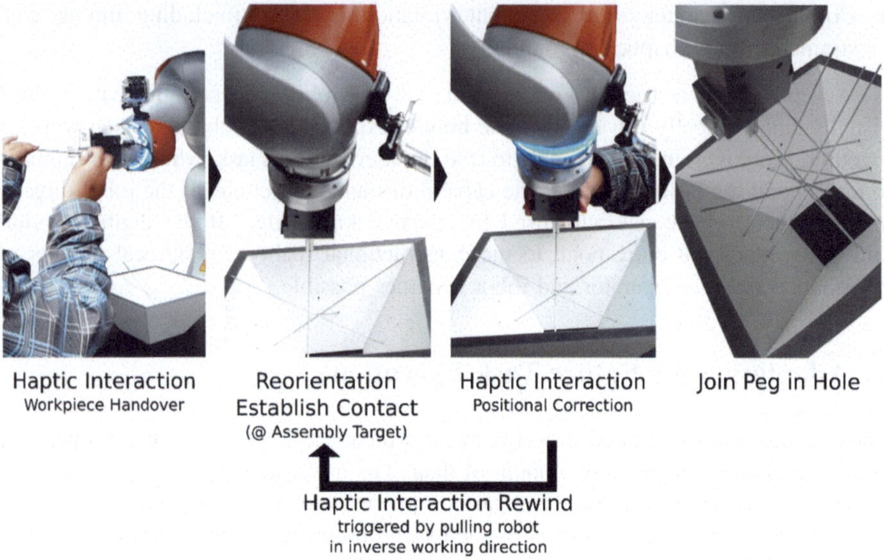

Fig. 7. Process sequence and haptic interaction within the Robodonien 2016 Demonstrator.

process is easily understood and executed. Layman where able to grasp the concept quickly and help the robot in the construction of interwoven foldable elements (Robodonien) and wooden rod structures (HMI).

The HMI and Robodonien demonstrator worked in different environments and showed the scalability of the task execution due to the parametric model. Modification to the haptic programming framework between the demonstrators focused mainly on data transfer, and the extension through new haptic interaction primitives.

While the work shown at the HMI still required a manual transfer of model parameter to the controller, this process was fully automated for Robodonien and could be achieved either via file transfer or a direct Ethernet connection, between design tool and robot. However, for task execution no additional information is required from the CAD environment except for the first design abstraction.

With the HMI demonstrator wooden rods where joined with a wooden base structure. The initial on-site parmeterization of the tasks consisted of manually guiding the robot along a curve atop the assembly structure with dynamic compliance. The required constraint space is similar to the one illustrated in Fig. 4.

By capturing the movement, a curve is parametrized that informs the fabrication process, making every generated structure unique. Based on this data, a list of assembly operations is generated, consisting of the following steps further described in [14]:

- Separation and gripping of a rod from a supply station, where rods of different length are stored within a range of about 50 mm.
- Measurement of rod length through establishing contact between the rod and the base platform, and using the distance between the toolframe and the static measurement frame on the base platform.
- Placement of the rod in a processing station.

- Cutting the rod to length using a common circular saw, with the length corresponding to the previously parametrized target curve.
- Re-gripping the rod from the processing station.
- Joining the rod to the base station, this is achieved by moving the rod to the approximate position of each mounting spot as known from the design. Due to the high material tolerances a force torque based search for the exact position is triggered. Once a significant change is detected in the force values, it can be concluded that the target position was found and the joining operation is concluded.
- Repetition of the steps for a number of assembly positions, continued by retrieving the next rod.

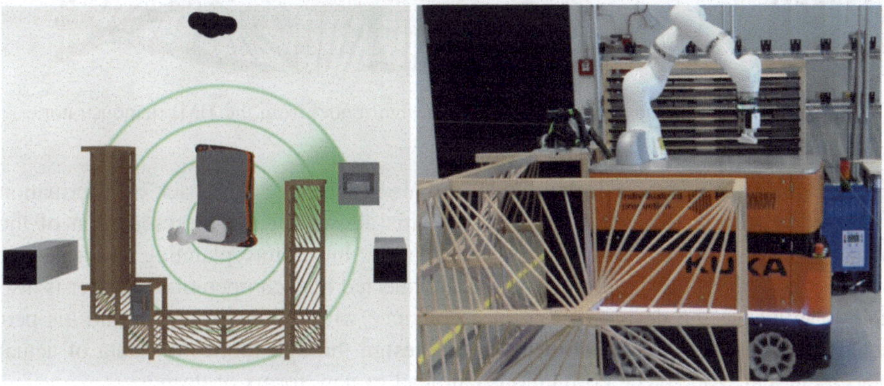

Fig. 8. Demonstrator implementation for the AWK 2017 illustrating the positioning of the mobile robot. The workstation is identified through laser scanner data. The motion within a larger workspace is implemented through human guidance. Positional deviations are compensated through sensor based adaption and haptic programming.

Haptic interaction primitives allow for an adaptation if problems occur, meaning collisions, position or sensor data values outside the expected range, etc., or if timing constraints are met for any operation, human interaction is requested by the robot, by turning on an easily visible light on the flange of the robot.

Within the Robodonien exhibition instead of using wooden rods with a length of 400 mm up to 1000 mm and a diameter of 12 mm we used metal and wooden rods of approximately between 180 mm and 240 mm length and diameters ranging from 2 to 3 mm. These where then interwoven in a specific pattern to create foldable parts.

In addition to the joining and handling operations from the previous exhibit, we created the possibility for material handover through haptic programming. By pressing onto the robot with the material the robot grips the material and continues the assembly process. If the robot is pressed without any material the missing material is detected in the next joining operation and the handover operation is restarted. Due to the interweaving pattern the necessity to rewind the programme in order to fix the pattern was added.

Fig. 9. Possible haptic parameterization of design model from the HMI demonstrator.

The employed abstract design model can also enable the flow back of information from the execution to design (see Fig. 9). This allows for the representation of the actual process executed and gives us further inside into future planning scenarios.

Due to the haptic interaction primitives connected to each operation the user is also able to influence the design on-site in accordance with the environment and his personal expectations. Directly influencing the design through fabrication data of actual placement and work-piece correlations can lead to new forms of design.

From the perspective of construction automation, the repetition of patterns and taught motions on-site enables an intuitive interface for robotic assistance. Most applications of automation in construction still require an adaptation of the workers and the environment. Our approach tries to adapt robotics towards humanly taught input.

5 Synopsis and Future Work

Within this work we described some of the inhibiting factors towards the application of automation in construction. A direct transfer of approaches used for structured production environments leads to high costs and requires the compensation of inflexibility by the worker on-site. However, by trying to take advantage of more flexible methods for robot programming, which allow for degrees of uncertainty, through the combination and extension of these methods we introduced the field of haptic programming.

We took first steps in compensating for material tolerances and uncertainties through adaptive operations and human-robot collaboration. Through this approach we try to bridge the gap between the digital model and the complexity of practical realization, without the necessity for robot reprogramming on-site.

However, the creation of the employed operations is still quite complex. Additionally, a simplified abstraction of parametric design model need to be created and interconnected with these operations.

In our Future work we will focus on extending the adaptability as well as creating a more extensive set of haptic interaction primitives. The definition of operation interfaces and design parameter still requires simplification. Furthermore, design implications of the information feedback from the construction process needs to be analyzed.

References

1. Arntz, M., Gregory, T., Zierahn, U.: Digitalisierung und die Zukunft der Arbeit: Makroökonomische Auswirkungen auf Beschäftigung Arbeitslosigkeit und Löhne von morgen, Zentrum für Europäische Wirtschaftsforschung, Mannheim (2018)
2. Taylor, M., Wamuziri, S., Smith, I.: Automated construction in Japan. In: Proceedings of the Institution of Civil Engineers - Civil Engineering, vol. 156, no. 1, pp. 34–41 (2003)
3. Saidi, K.S., Bock, T., Georgoulas, C.: Robotics in construction. In: Siciliano, B., Khatib, O. (eds.) Springer Handbook of Robotics, pp. 1493–1520. Springer, Heidelberg (2016)
4. Eastman, C., et al.: BIM Handbook: A Guide to Building Information Modeling for Owners, Managers, Designers, Engineers and Contractors, 2nd edn. Wiley, Hoboken (2011)
5. Becerik-Gerber, B., Kensek, K.: Building information modeling in architecture, engineering, and construction: emerging research directions and trends. J. Prof. Issues Eng. Educ. Pract. **136**, 139–147 (2010)
6. Tellex, S., Knepper, R.A., Li, Rus, A.D., Roy, N.: Asking for help using inverse semantics. In: Robotics: Science and Systems X, Berkeley, CA (2014)
7. Jaillon, L., Poon, C.S.: Life cycle design and prefabrication in buildings: a review and case studies in Hong Kong. Autom. Constr. **39**, 195–202 (2014)
8. Brell-Çokcan, S., Braumann, J.: Robotic production immanent design: creative toolpath design in micro and macro scale. In: Gerber, D., Huang, A., Sanchez, J. (eds.) Acadia 2014: Design Agency, Proceedings of the 34th Annual Conference of the Association for Computer Aided Design in Architecture, Los Angeles, pp. 579–588. Los Angeles, CA (2014)
9. Braumann, J., Brell-Çokcan, S.: Parametric robot control: integrated CAD/CAM for architectural design. In: Taron, J.M. (ed.): Acadia 2011: Integration through Computation, Proceedings of the 31st Annual Conference of the Association for Computer Aided Design in Architecture, pp. 242–251. Calgary/Banff (2011)
10. De Schutter, J., et al.: Constraint-based task specication and estimation for sensor-based robot systems in the presence of geometric uncertainty. Int. J. Robot. Res. **26**, 433–455 (2007)
11. Hayes, B., Scassellati, B.: Discovering task constraints through observation and active learning. In: Proceedings of the IEEE/RSJ International Conference on Intelligent Robots and Systems, pp. 4442–4449. Chicago, IL (2014)
12. Borghesan, G., et al.: A constraint-based programming approach to physical human-robot interaction. In: Proceedings of the 2012 IEEE International Conference on Robotics and Automation, pp. 3890–3896. Saint Paul, MN (2012)
13. Welschehold, T., Dornhege, C., Burgard, W.: Learning mobile manipulation actions from human demonstrations. In: Proceedings of the IEEE/RSJ International Conference on Intelligent Robots and Systems, pp. 3196–3201. Vancouver, BC (2017)
14. Lutscher, E., et al.: Hierarchical force and positioning task specification for indirect force controlled robots. IEEE Trans. Robot. **34**(1), 280–286 (2018)

15. Ruffaldi, E., Di Fava, A., Loconsole, C.: Vibrotactile feedback for aiding robot kinesthetic teaching of manipulation tasks. In: Proceedings of the 26th IEEE International Symposium on Robot and Human Interactive Communication (RO-MAN), pp. 818–823. Lisbon (2017)
16. Huang, Y., Zhang, X., Chen, X., Ota, J.: Vision-guided peg-in-hole assembly by Baxter robot. Adv. Mech. Eng. **9**(12) (2017)
17. Stumm, S., Braumann, J., Brell-Çokcan, S.: Human-machine interaction for intuitive programming of assembly tasks in construction. In: 6th CIRP Conference on Assembly Technologies and Systems (CATS) Procedia CIRP, vol. 44, pp. 269–274. Gothenburg, Sweden (2016)
18. Stumm, S., von Hilchen, M., Braumann, J., Brell-Çokcan, S.: On-site robotic construction for assembly using a-priori knowledge and human robot collaboration. In: IFToMM/IEEE/euRobotics 25th International Conference on Robotics in Alpe-Adria-Danube Region, RAAD 2016, Belgrade (2016)

Z+

Towards Automatic Path Planning for Robotically Assembled Spatial Structures

Augusto Gandia(✉), Stefana Parascho(✉), Romana Rust(✉), Gonzalo Casas(✉), Fabio Gramazio(✉), and Matthias Kohler(✉)

ETH, Zurich, Switzerland
{gandia, parascho, rust, casas, gramazio, kohler}@arch.ethz.ch

Abstract. This paper discusses the integration of automatic robot path planning into the computational design environment. A path planning software interface is presented that allows to support fabrication-aware design of robotically assembled structures with discrete elements. Using the large-scale Robotic Fabrication Laboratory (RFL) as test-bed, the software interface is validated through three experiments, in which building members need to be guided around obstacles and which are fabricated using two cooperative robotic arms. Specific focus of this paper is the investigation of strategies to narrow the path search by adjusting design and path planning parameters in order to achieve a calculation time that is suitable for design applications. A close integration of automatic path planning and design is presented, which does not only enable the negotiation between design intention and fabrication feasibility, but allows for an understanding of the constraints present in robotically fabricated spatial structures. Thus, this research contributes to expand these structures' design and fabrication space.

Keywords: Automatic path planning · Computational design
Spatial structures · Cooperative robotic processes

1 Introduction

Today, several robotic simulation and visualization tools for computational design environments (e.g. HAL [1], PowerMill Robot [2], MORSE [3]) are assisting architects in designing robotically fabricated structures. These tools however, require the designer to manually draw and adjust auxiliary curves (toolpaths) until the robot can follow the curves without any collision. This is a tedious and time-consuming process, especially if the structure consists of a high number of building elements that need to be maneuvered to intricate positions or into gaps. Apart from such CAD-based robot programming tools, there exist several offline robot-programming tools [4] (e.g. Moveit! [5] with Gazebo [6]) that are based on the Open Motion Planning Library (OMPL) [7]. These libraries could enable automatic path search while taking robot disruptions, such as collisions and robot axes limits into account, but are not easily accessible from within computational design environments used in architecture. Thus, this hinders architects to fully explore the design space offered through the integration

J. Willmann et al. (Eds.): ROBARCH 2018, *Robotic Fabrication in Architecture, Art and Design 2018*, pp. 59–73, 2019.
https://doi.org/10.1007/978-3-319-92294-2_5

of automated path planning, which limits their possibilities to work with over-simplified models of fabrication.

To solve these restrictions, this paper introduces a Python software interface, which connects the design environment Rhino/Grasshopper [8] with specific capabilities of the OMPL. It is developed and evolves through several experiments, from a tool for post-rationalization of a computationally designed structure, towards a tool that can be tightly integrated in the structure's generation. This enables fabrication-aware design, which in particular considers cooperative robotic assembly with obstacle avoidance. The research is conducted in collaboration with the projects "Multi-Robotic Prefabrication of Spatial Lightweight Structures" [9] and "Spatial Timber Units" [10], which physically validate the approach through three experiments and their respective 1:1 prototypes. Experiment 1 and 3 are conducted within the first project and Experiment 2 within the second project. Using the large-scale multi-robotic laboratory RFL [11], the projects build upon the same assembly logic, in which two cooperating robotic arms alternate their function along the fabrication process. While one robot supports the structure temporarily by holding a building member, the other robot positions the next member (see Fig. 1), and conversely in the following step [9]. After placement, the member is manually joined to the existing structure through either welding (Experiment 1 and 3) or screwing the connection (Experiment 2).

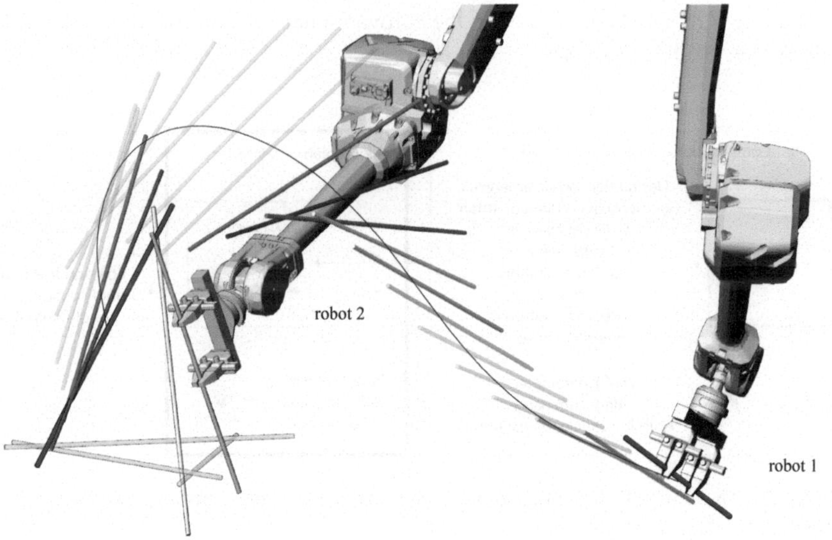

Fig. 1. Diagram of a cooperative robotic assembly process showing robot 1 avoiding collisions with robot 2 and with the built structure in order to assemble a building member (blue), while robot 2 stabilizes the structure

2 Path Planning Software Interface and Integration of Collision-Free Robot Path Generation into Computational Design

The main challenge in the development of the software interface between the computational design environment and the path planning setup is the achievement of a calculation time suitable for design applications. It is also important to define the interface's in- and outputs, which consist of path planning parameters controlled from within the design environment, and of information from the automated path planner that is useful to feed back.

The software interface builds upon a library (OMPL) containing algorithms for robot path planning (e.g. RRT [12] and RRT* [13]), which, however, is independent of a specific collision checker or visualization tool [7]. Therefore, the software interface accesses the library through the simulation platform V-REP [14], which extends the functionality with fast collision detection through the definition of collidable entities [15]. A GhPython component [16] in Rhino/Grasshopper allows to interact with the software interface[1]. Additionally, Grasshopper can be used to visualize the robot paths calculated by the planner. At a later moment in the development (see Experiment 2), the software interface was enhanced with simultaneous search capabilities through containerized instances of V-REP using the Docker platform [17] and a TCP/IP coordinator (see Fig. 2). This allowed to improve the workflow by increasing the path search success rate and contributed to the software interface' practicality, since V-REP could run as headless software.

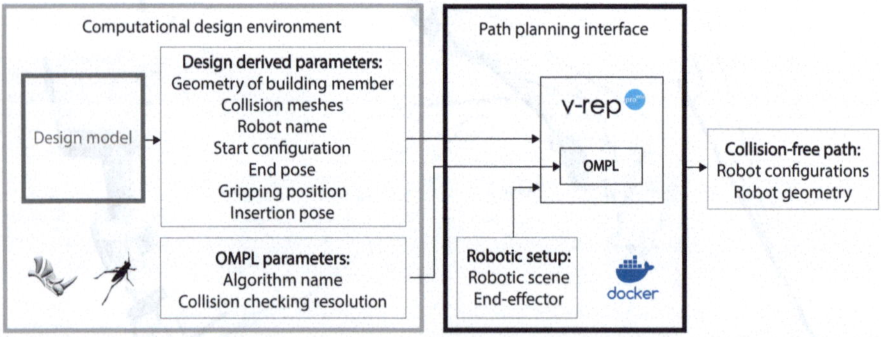

Fig. 2. Visualisation of the integration of the path planning software interface into the computational design environment

[1] The software interface could also be integrated in any CAD software that includes a Python interpreter (e.g. Blender [24] and Maya [25]) or through frameworks supporting algorithmic design (e.g. COMPAS) [26].

In order to simulate the pick and place assembly process of a computationally designed structure within a specific robotic setup, several parameters have to be set in advance. This includes (1) the computational representation of the robotic setup, (2) OMPL path planner parameters and (3) design-derived parameters.

In V-REP, the *robotic setup* (see Fig. 2.) is defined by collision meshes including scripts that access the geometrical objects programmatically by specifying their joints and respective types in a hierarchical structure [18]. In the presented experiments, the robot model consists of four six-axis robots (ABB IRB 4600-40/2.55) mounted on a gantry system (Güdel) that adds three linear axes to each robot, resulting into nine DOF. Additionally, the geometry of the respective *end-effector* including its tool centre point (TCP) has to be defined[2].

Through the software interface it is possible to control three OMPL parameters (see Fig. 2): the name of the path planning *algorithm*, the *collision checking resolution* and the *robot axes metrics*. In OMPL, the 25 available path planning algorithms are categorized into geometric planners and control-based planners [19]. For the presented experiments, only geometric planners have been used since it was only necessary to consider the geometric and kinematic constraints of the system and not its dynamics (e.g. acceleration of the robot). The distance between the steps on a calculated robot path derive from the parameter *collision checking resolution* [20]. Furthermore, the *robot axis metric* can be defined for each joint as a floating-point value between 0 and 1 (from not constrained to blocked), which allows to constrain its movement.

Finally, design derived parameters (see Fig. 2) include the *building member's geometry*, which is picked at a defined *gripping position* with a *start configuration* (robot's joint and axes values) by the *robot name* (as defined in the robot model) and placed at the *end pose* (frame of reference). Additionally, other *collision meshes* can be added for collision checks, such as for example the representation of the already assembled structure. If the search with the described parameters is successful, the collision-free path, which consist of angular and translational values (joints and linear axes) can be visualized in Rhino/Grasshopper. On the contrary, if the path search is not successful, a message indicates if the problem occurred during the initialization of the planner (e.g. the robot in the start configuration is colliding with an obstacle), or if no solution was found due to collisions along the path between the start and the end configuration. Once all feasible paths are calculated, fabrication starts by sending each path via a custom robot control interface to the IRC5 ABB robot controller, which executes the received commands.

Finding collision-free paths is a time-intensive task since it is a highly complex problem [21] that requires to recurrently check for collisions [22]. Therefore, independent of the path planning algorithm, the resolution of the meshes, or the collision checking resolution, it is important to evaluate if design parameters can already constrain the search space of the path. Therefore, in this paper a strategy is presented, in which the path planner does not calculate the full path from the building-members' picking station, to its final position in the structure, but only to a certain offset from it that will be later referred to as *insertion pose(s)*.

---

[2] Flexible parts such as cables and hosepipes are not included in the model.

While for certain computationally designed structures path planning can be applied as post-processing step for fabrication (see Experiments 1 and 2), for other structures, deadlocks might occur. This means that for a given geometrical configuration no possible path can be found, which would require redesigning/regenerating the complete structure and starting the path search from the very beginning. Experiment 3 presents a case, in which the path planner is already called in the design generation process, after defining a part of a structure. This approach allows reacting locally by adjusting the geometry of one building element once a path search was not successful, therefore omitting deadlocks through fabrication-aware design.

3 Experiment 1: Post-rationalization of Spatial Metal Structures Using Sequential Robot Path Search

The first experiment presents a workflow for post-processing the computational design of a spatial metal structure to enable its robotic construction using two cooperating robotic arms. The structure is generated in a computational design setup [9] and comprises 72 tubes (tube size: 16 mm ø, 1000–1200 mm length) positioned in space that are connected through a reciprocal interlocking node, creating 23 non-regular tetrahedra that concatenate into a spiral configuration (see Fig. 3).

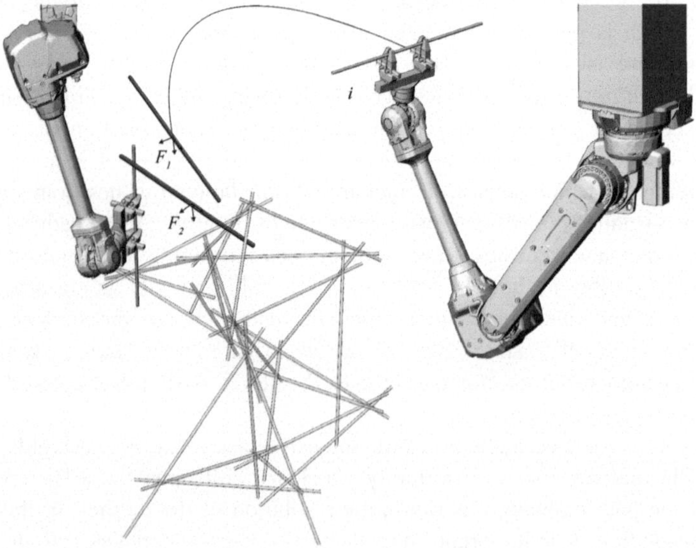

Fig. 3. Visualization of placing one tube into the spatial metal structure indicating robot start configuration i, insertion pose F_1, final pose F_2

For each tetrahedron, the computational design tool assigns a preliminary fabrication sequence, which describes the order of how to stably assemble the tetrahedron. An insertion pose (see Fig. 3, F_1) is calculated for each metal tube, representing the end pose of the path planner's robot path. The insertion pose is a robot end-effector position in an offset (20 cm) to the structure that defines the last section of the robot path before reaching the building member's final position. It is calculated by taking possible collisions of the inserted building member with the existing structure into account. In the fabrication process, the movement from the insertion pose to place the tube in its final pose (see Fig. 3, F_2) is performed through a linear movement in the robot's tool space. Despite this movement is not considered in the collision check of the planned path, collisions were not registered during the experiment.

The procedure to validate the structure's fabrication feasibility consists of iterating over the tetrahedra along the fabrication sequence and searching for collision-free assembly paths for the tube within each tetrahedron (see Fig. 4), considering the abovementioned robot alternation of stabilizing the structure and placing a building member.

Each path starts with the robot start configuration (see Fig. 3, *i*) for picking the precut tubes and ends at the robot insertion pose (see Fig. 3, F_1). Within the path search, the path planner does not only check for collisions of the manipulated building member with the existing structure, but also with the other robot that is temporarily stabilizing the structure. The path search is automatic, however, assigning the respective robot to each building member, selecting the next tube within the fabrication sequence and changing the algorithm name, etc. is performed manually.

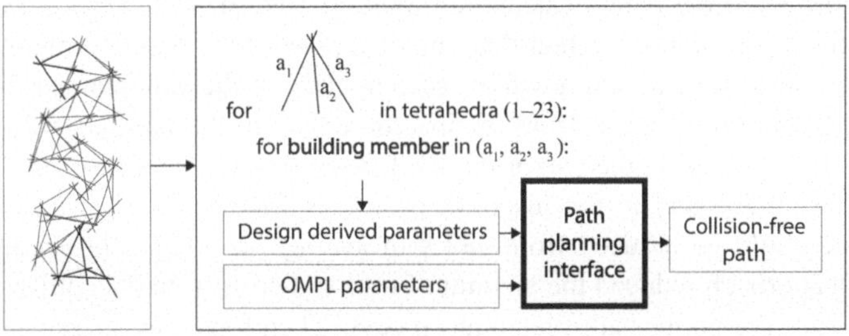

Fig. 4. Diagram of the parameters and processes involved during the sequential path planning for the first prototype

Finding paths is occasionally not possible, for example, if the selected robot workspace is highly constrained by the other robot's configuration or the insertion pose is almost inaccessible. Therefore, it is necessary to change the path planner's parameter settings and restart the search. Preliminary empirical tests hint towards the use of fast algorithms, such as RRT, for paths planned in a less constrained search space and algorithms, such as SBL [19], for intricate paths for which the robot can hardly manoeuvre building members without colliding. However, changing only the algorithm does not always lead to success. In order to counteract such deadlocks it is also necessary to adjust the fabrication sequence, the gripper position/orientation along the tube and the insertion pose as well. In this experiment the total of 72 paths were found by a combined manual adjustment of the fabrication sequence, the gripping position and the insertion poses.

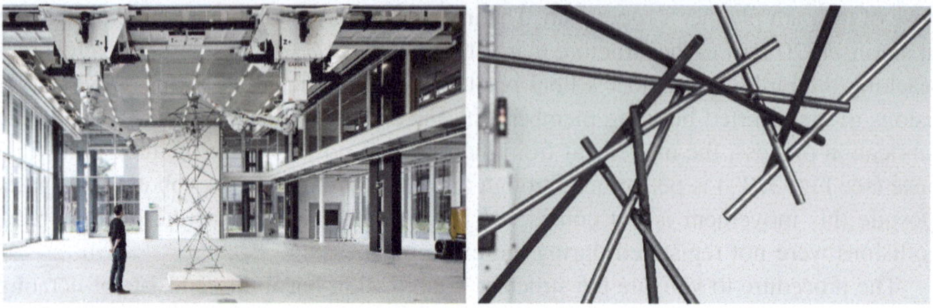

Fig. 5. Assembled structure of Experiment 1 during the AAG Conference in the RFL (left), detail of the structure showing a reciprocal welded connection (right)

The successful realization of the structure proved the usability of the software interface to find collision-free assembly paths for a multi-robotic spatial assembly process (see Fig. 5). However, during the experiment several problems could be identified that require further improvements. The workflow was very slow due to the considerable amount of manual steps that required multiple adjustments until finding valid paths.

In average, it took 30 min of testing and adjusting different parameters settings to find one path, while the path's calculation took approx. 30 s. Nevertheless, using the strategy of the insertion pose, rather than the final pose of the building member, proved advantageous since it narrowed down the search space of the path planner, by avoiding to calculate the insertion move from the insertion pose to the building member's final position. A test comparing the use of the RRT algorithm to calculate paths to the final pose (see Fig. 3, F_2) and to the insertion pose (see Fig. 3, F_1) showed that the calculation time could be reduced from 487 s in average to 34.86 s in average. Consequently, this approach reduced the amount of trial and error to find a collision-free path and speeded up the entire path planning process. Therefore, this strategy was further used in the following experiment in a different design and material system.

4 Experiment 2: Post-rationalization of Spatial Timber Frame Structures Using Parallel Path Search

As discussed in Experiment 1, finding collision-free paths depends on multiple parameters and often requires multiple trial and error steps that considerably slow down the workflow. Thus, the path planner was enhanced with simultaneous search capabilities, which allowed checking in parallel for several paths (two paths per processor core) with different parameter settings. The validity of this approach was tested in the Experiment 2, which investigates the post-processing of a spatial timber structure to enable its construction. The structure represents 81 freely oriented beams of different lengths (2000–4600 mm) and cross-sections (200 × 80 mm, 120 × 100 mm, 120 × 60 mm).

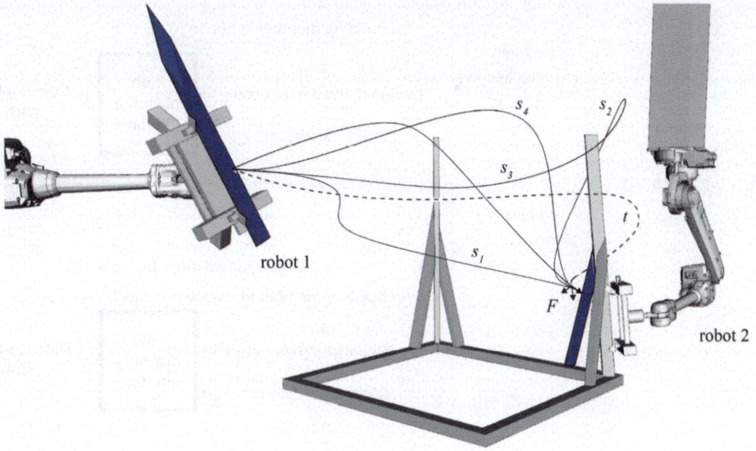

Fig. 6. Visualization of four collision-free paths (s_1–s_4) and one path t colliding with robot 2 and the insertion pose F as part of the planning needed to collaboratively assemble a beam (blue) of a corner triangulation

Besides the geometry of the structure (meshes oriented in space), and the additional robot poses for cutting, drilling and milling the beams, the computational design model calculates a preliminary fabrication sequence and an insertion pose for each beam. The fabrication sequence describes to assemble the floor (beams c_1–c_3) first, then the walls (beams a1–a3, b1–b3, c4 and c5) and finally the ceiling (beam c_6) (see Fig. 7). In this experiment, all beams can be placed with only one robotic arm, except for the two corner triangulations (beams a_1–a_3 and b_1–b_3), which require a cooperative assembly process with two alternating robots. Similar to Experiment 1, the insertion pose is calculated with an offset to the structure (20 cm) and describes the last section of the robot path before reaching the building member's final position (Fig. 6).

In the path search process the parallelization can be used to calculate the assembly paths for beams c_1–c_6 all at once, since they are placed with only one robotic arm and the other robot does not need to be considered as a potential obstacle. The robotic paths to assemble the beams of the corner triangulations (a_1–a_3 and b_1–b_3) have to be sequentially searched for, considering the fabrication approach with the two alternating robots (see Fig. 7).

The construction of the timber prototype (Fig. 8) validated the approach for post-rationalizing the spatial structure, but in contrast to Experiment 1, for a different construction system with elements of different size and geometry. The average calculation time for a sequentially planned path was 22.5 s. This however cannot be compared with Experiment 1 due to different parameters settings and differently sized building elements. In a separate benchmark test 103 beams were tested by searching 4 paths per beam for which at least one valid path was found. Therefore, the parallel search speeded up the general planning workflow by saving several manual trial and error steps. This strategy could be further developed in an automated approach for

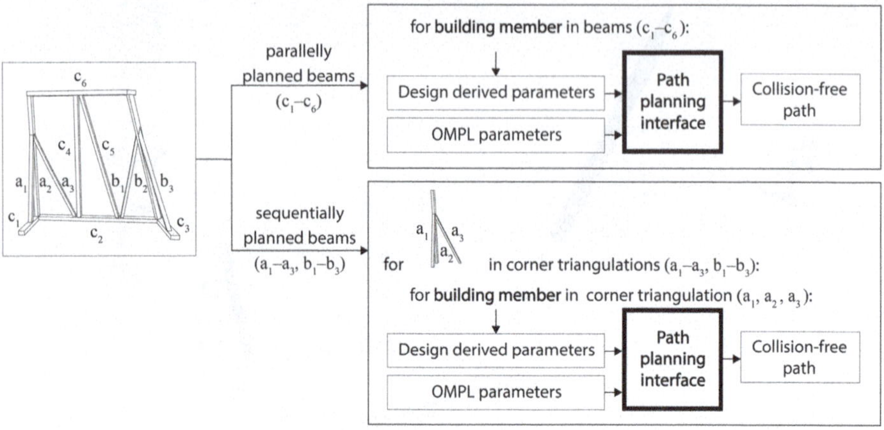

Fig. 7. Diagram of the parameters and processes involved to sequentially and parallel planned robot assembly paths for the beams of Experiment 2

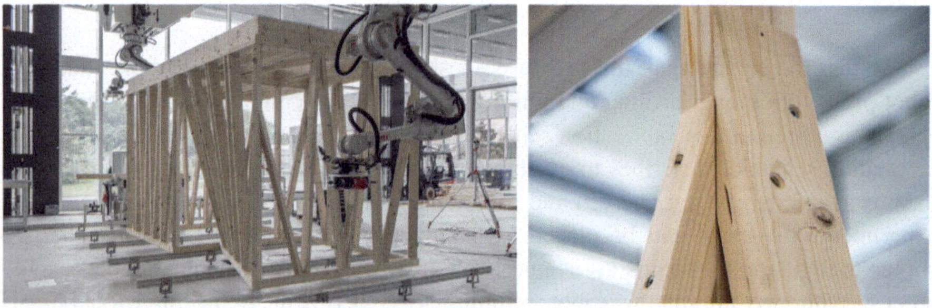

Fig. 8. Physical prototype of Experiment 2 (left) and a detail of the cooperatively assembled corner triangulation (right) with screwed t-butt joint.

testing and evaluating parameter combinations to achieve a better performance, which is partly implemented in the next experiment.

5 Experiment 3: Fabrication-Aware Design of Spatial Structures

In Experiment 3, the computationally designed structure builds upon the same reciprocal node configurations as in Experiment 1. However in this case, the 10 tetrahedra are arranged around one point in space (see Fig. 9). This results into a highly complex structure of 33 tubes (tube size: 25 mm ø, 1070–1900 mm length), which, in order to avoid deadlocks, requires fabrication feasibility checking already in the design generation process. Here, in comparison to the previous experiments, the path planner is tightly integrated in the computational design tool of the structure.

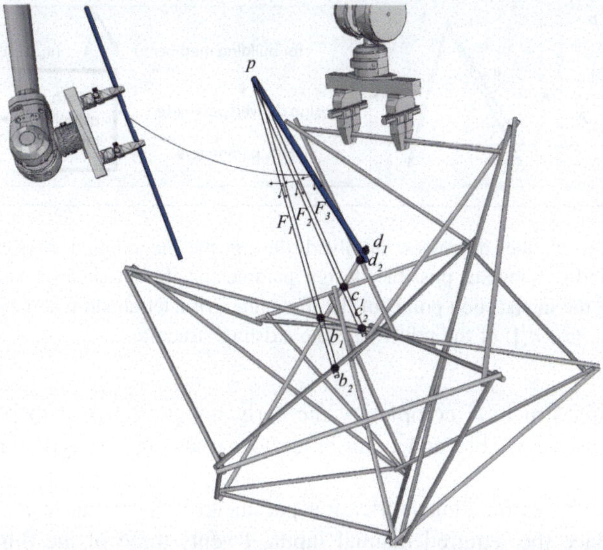

Fig. 9. Possible design adjustments for one element (blue) showing potential poses F_1, F_2, F_3 for a tube, respective connecting points ([b_1, b_2], [c_1, c_2], [d_1, d_2]) and vertex position p that can be tested in order to find a valid path

The computational design tool iteratively generates only three tubes with the respective robot insertion pose and a preliminary fabrication sequence before running the path planner. The insertion pose defines the position and orientation of the gripper at a distance of 20 cm before reaching the final position of the building member. This is calculated considering the movement to interlock a tube in the reciprocal connection without colliding with the structure and with the other robot.

Similar to Experiment 1, solution paths are searched for each tube in the order of the fabrication sequence inside the tetrahedron considering the robot alternation (see Fig. 10). However, this time the adjustment of parameters is automated, except for adjustments of the insertion pose. Also, the path planning algorithm was not changed since SBL provided a high success-rate for intricate path planning with high resolution. If even the automated search was not successful, the designed geometry was adjusted. In the context of this experiment this was performed manually through changing design parameters of the tetrahedron, such as the vertex (the approximate position of the tubes' meeting point, see Fig. 10, p) or the connection points (see Fig. 10, [a_1, a_2], [b_1, b_1], [c_1, c_2]), which are geometrically defined within the design script by considering requirements of the connection system[3] [9]. At a later state of implementation, it is foreseen to automate these changes.

The experiment shows the potential of using the software interface to explore fabrication-aware design of cooperatively assembled spatial structures. It exemplifies

[3] Adjustments of the design do not take structural considers into account.

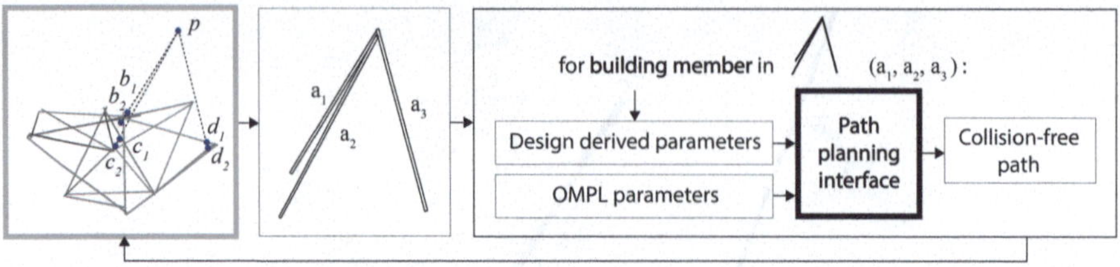

Fig. 10. Diagram of the processes involved during the negotiation between design and fabrication feasibility showing possible design parameters: the tetrahedron vertex position p, which represents the intersection points of the three tubes of a tetrahedron and connecting points ($[b_1, b_2]$, $[c_1, c_2]$, $[d_1, d_2]$) of the tubes with the existing structure

that for a certain structural complexity the early integration of path planning in the design process is inevitable and contributes to expand the design space of feasible structures.

The strategy of automating fabrication parameter adjustments was successful and allowed to reduce the required manual input. Twenty-three of the thirty-three paths were found automatically through adjustments in the fabrication parameters, such as fabrication sequence and gripping position on the tube. In four cases, the path search had to be paused since no path was found, but could continue after adjusting the insertion pose. Finally, six tubes required changes in the overall design, since no other measures led to a successful path generation. The changes involved choosing one of the different connection pairs (see Fig. 10, $[b_1, b_2]$, $[c_1, c_2]$, $[d_1, d_2]$) which allowed to find a valid path. Further advancement could comprise the integration of evaluation criteria for the effects of parameter changes to speed up the search with more effective adjustments. Additionally, a modular set-up of the search procedure could facilitate a faster identification of high-impact parameters and optimize the number of iterations needed to find a path (Fig. 11).

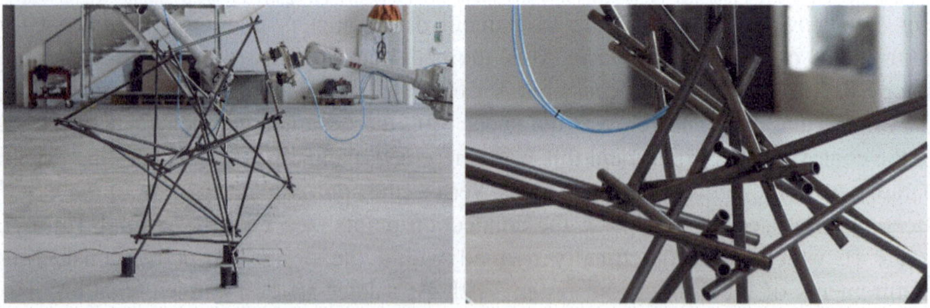

Fig. 11. Physical prototype of Experiment 3 (left) and detail of the reciprocal welded node (right)

6 Conclusion

The presented work demonstrates the use of automatic path planning for enabling the robotic assembly of spatial structures showcased through three experiments. The fabricated prototypes built with different assembly logics (individually and cooperatively assembled members) and consisting of building members in varied scales, show its successful application for constructive systems with linear building members, thus expanding the design and fabrication space of robotically assembled spatial structures.

However, several aspects of the path planning interface still require further development. While over the course of development the software interface was already enhanced by simultaneous search capabilities (see Experiment 2) and the possibility to automate parameter combinations (see Experiment 3), the performance could be improved further. Instant design adjustments according to the interface's feedback requires a more radical reduction of calculation time. This could be pursued through the mentioned strategy for the identification of high-impact parameters for specific design scenarios. Additionally, the definition of multiple optimization objectives for the optimal planners available in OMPL (e.g. RRT* and PRM*) would allow to automate the adjustments of the OMPL parameters. More radical approaches towards improving performance would include the optimization of OMPL path planner algorithms through fine-grained parallelism [23]. Additionally, to improve the interface's usability, the ability to specify the robotic setup from the computational design environment would allow using the software without any knowledge of V-REP.

The conceptualization and materialization of robotically assembled spatial structures that are truly aware of fabrication require the consideration of tolerances that sum up not only vertically as for layer-based assemblies, but also in multiple directions. In Experiment 2 sensors measured deviations of material deformations and of the structure during fabrication. These tolerances were fed back into the design environment and the design was adapted accordingly. Topic of further investigation is to simulate tolerances and to visualize deadlocks in the design in order to investigate strategies to counteract thereon.

Acknowledgements. This research presented here is part of and supported by the research programme NCCR, funded by the Swiss National Science Fundation. Special thanks goes to Ammar Mirjan for his advice during the early stages of the PhD and for the people that contributed to the realization of the first experiment: Gergana Rusenova, Petrus Aejmelaeus-Lindström, Marco Palma and Martin Rusenova. Also thanks for the people that were part of the second experiment: Andreas Thoma, Arash Adel and Matthias Helmreich. Additional thanks goes to Marc Freese for collaborating as consultant of robotic simulation and for the integration of OMPL in V-REP and to Philippe Fleischmann and Michael Lyrenmann for their support in the RFL.

References

1. HAL Robotics: HAL Robotics Framework (2017). http://hal-robotics.com/. Accessed 18 May 2018
2. Autodesk: Autodesk Robotics Features. https://manufacturing.autodesk.com/solutions/robotics/features/index.asp. Accessed 18 May 2018
3. Laboratoire d'Analyse et d'Architecture des Systèmes: What is MORSE? The MORSE Simulator Documentation. https://www.openrobots.org/morse/doc/stable/what_is_morse.html. Accessed 18 May 2018
4. Neto, P., Mendes, N.: Direct off-line robot programming via a common CAD package. In: Berns, J.O.K., Gini, M. (eds.) Robotics and Autonomous Systems, pp. 896–910. Elsevier, Amsterdam (2013)
5. Şucan, I.A., Chitta. S.: MoveIt! Motion Planning Framework (2011). http://moveit.ros.org/about/. Accessed 18 May 2018
6. Howard, A., Koenig, N.: Open Source Robotics Foundation, "Gazebo" (2002). http://gazebosim.org/. Accessed 18 May 2018
7. Şucan, I.A., Moll, M., Kavraki, L.E.: The Open Motion Planning Library (2012). http://ompl.kavrakilab.org/citations.html/. Accessed 18 May 2018
8. D. Rutten at Robert McNeel & Associates: Grasshopper 3D | algorithmic modeling for Rhino (2007). http://www.grasshopper3d.com/. Accessed 18 May 2018
9. Parascho, S., Gandia, A., Mirjan, A., Gramazio, F., Kohler, M.: Cooperative fabrication of spatial metal structures. In: Sheil, B., Menges, A., Glynn, R., Skavara, M. (eds.) Fabricate: Rethinking Design and Construction, pp. 24–29. UCL Press, London (2017)
10. Thoma, A., Adel, A., Mirjan, A., Gramazio, F., Kohler, M.: Robotic fabrication of Bespoke timber frame modules. In: Robotic Fabrication in Architecture, Art and Design 2018. Springer International Publishing, Switzerland (2018, accepted for publication)
11. Bonwetsch, T., Lyrenmann, M.: Gramazio Kohler Research, Robotic Fabrication Laboratory (2016). http://dfab.arch.ethz.ch/web/e/forschung/186.html. Accessed 18 May 2018
12. Kuffner, J.J., LaValle, S.M.: RRT-connect: an efficient approach to single-query path planning. In: Proceedings of the IEEE International Conference on Robotics and Automation, ICRA 2000 (Millennium Conference), Symposia Proceedings. pp. 995–1001. San Francisco, CA (2000)
13. Şucan, I.A., Moll, M., Kavraki, L.E.: OMPL | RRT* Class Reference (2012). https://ompl.kavrakilab.org/classompl_1_1geometric_1_1RRTstar.html#gRRTstar. Accessed 18 May 2018
14. Coppelia Robotics: V-REP | Create, Compose, Simulate, any Robot (2010). http://www.coppeliarobotics.com/. Accessed 18 May 2018
15. Coppelia Robotics: Collision detection (2010). http://www.coppeliarobotics.com/helpFiles/en/collisionDetection.htm. Accessed 18 May 2018
16. Piacentino, G.: GhPython | Food4Rhino (2012). http://www.food4rhino.com/app/ghpython. Accessed 18 May 2018
17. Docker: Docker | Build, Ship, and Run Any App, Anywhere (2013). https://www.docker.com/. Accessed 18 May 2018
18. Coppelia Robotics: V-REP models (2010). http://www.coppeliarobotics.com/helpFiles/en/models.htm. Accessed 18 May 2018
19. Şucan, I.A., Moll, M., Kavraki, L.E.: OMPL Available Planners (2012). http://ompl.kavrakilab.org/planners.html. Accessed 18 May 2018
20. Şucan, I.A., Moll, M., Kavraki, L.E.: OMPL Graphical User Interface (2012). http://ompl.kavrakilab.org/gui.html. Accessed 18 May 2018

21. Moll, M., Şucan, I.A., Kavraki, L.E.: Benchmarking motion planning algorithms: an extensible infrastructure for analysis and visualization. In: Vanderborght, B. (eds.) IEEE Robotics and Automation Magazine, pp. 96–102 (2015)
22. Li, T.-Y., Chen, J.-S.: Incremental 3D collision detection with hierarchical data structures. In: Proceedings of the ACM symposium on Virtual reality software and technology (VRST 1998), pp. 139–144. Taipei, Taiwan (1998)
23. Murray, S., Floyd-Jones, W., Qi, Y., Sorin, D., Konidaris, G.: Robot motion planning on a chip. In: Berman, S., Jacobs, S., Hsu, D., Amato, N. (eds.) Robotics: Science and Systems XII, Ann Arbor, MI (2016)
24. Blender Foundation: Home of the Blender Project | Free and Open 3D Creation Software (1998). https://www.blender.org/. Accessed 18 May 2018
25. Autodesk: Maya | Computer Animation; Modeling Software (1998). https://www.autodesk.eu/products/maya/overview. Accessed 18 May 2018
26. Block Research Group: The COMPAS framework (2018). https://compas-dev.github.io/. Accessed 18 May 2018

ABB
HENN
DDU
Design
Distributed Design

Communication Landscapes

Giovanni Betti[1(✉)], Saqib Aziz[1], Andrea Rossi[2], and Oliver Tessmann[2]

[1] HENN, Alexanderstraße 7, 10178 Berlin, Germany
{giovanni.betti, saqib.aziz}@henn.com
[2] Technical University Darmstadt, El-Lissitzky-Straße 1, 64287 Darmstadt, Germany
{rossi, tessmann}@dg.tu-darmstadt.de

Abstract. In this paper we present an installation which explores a robotic fabrication process through real time human-machine interaction using natural interfaces. Aiming to contribute to the debate regarding automatization processes described by the industry 4.0, we suggest a future collaborative approach to distributed and participatory design. The installation invites participants to shape a physical object through a robot that interprets the acoustic signal of their voices. The object is then fabricated in near real-time through robotic hotwire cutting. The visitor is gifted with the result of its exploration to take home as a souvenir of a possible future. The negative traces of voices are aggregated through an algorithm in a sculptural wall.

Keywords: Natural interfaces · Collaborative fabrication · Hotwire cutting Parametric design

1 Introduction

1.1 Aim

The installation Communication Landscape was commissioned for the exhibition "Imminent Commons" at the first Seoul Biennale of Architecture and Urbanism. Among the stated aims of the biennale is to open the debate about the dramatic challenges global cities are facing - may they be social, ecological or technological. On this occasion HENN presented together with the Technical University Darmstadt an interactive installation which explores the idea of industry 4.0, digital craft and human-machine interfaces.

This installation aims to place itself in the current debate about industry 4.0 and digital craft, offering a vision for future production and an optimistic view of the automatization of production.

We try to argue that innovative, lean and clean production methods would enable a departure from the centralized industrial model of the XX Century and create smaller, decentralized production units that can be better integrated in the urban and social fabric. In this sense this would be a return to the craftmanship based model prevalent prior to the industrial revolution (Gershenfeld 2005).

J. Willmann et al. (Eds.): ROBARCH 2018, *Robotic Fabrication in Architecture, Art and Design 2018*, pp. 74–84, 2019.
https://doi.org/10.1007/978-3-319-92294-2_6

In such context, production would no longer reside within the artisan, but with a machine. It is here postulated that the definition of interaction strategies between machines and users will enable the creativity of individual production to be distributed among the wider population rather than being concentrated with a single individual. This shift would create a new category, the prosumer -in equal measure producer and consumer of its own goods (Ritzer et al. 2012).

Emerging Fabrication Labs (FabLabs) around the world and Tech Start-up companies such as the membership based TechShop in San Francisco are substantiating these assumptions that access to industrial manufacturing processes acts as a beacon in democratizing invention. It allows young start-ups ventures and the so-called "Do it yourself" community to explore new ideas accentuated by state-of-the-art technology and weaves the industrial fabrication site back into the social fabric of the inner cities.

1.2 Natural Interfaces for Fabrication

Naturally, intuitive man-machine interfaces are crucial to this process as they would provide people with easy and direct access to production.

In recent years, with ever more ubiquitous computing, a lot of research has gone in the topic of natural interfaces, where the computer itself seems to disappear and more intuitive, human-like interactions take place (Francese et al. 2012).

The apparent disappearance of the computer (at least in it classical mouse-keyboard-terminal configuration) actually promotes a closer relationship with the machine (Willis et al. 2011), enabling different interaction modes, which can be integrated with logics of gamification of such interactions (Savov et al. 2016).

Creating custom software and interfaces especially for computer-aided manufacturing software (CAM) further liberates the end user from the necessity to understand programming and native robot code and allows him/her to focus on the design tasks at hand and further amplifies additional design of custom robotic tools or methodology development referred by Brell-Çokcan as production-immanent design (Brell-Çokcan et al. 2013).

Such an advancement naturally comprises educational benefits and has the potential to release some pressure addressed by the demographical changes in the upcoming future. Less skilled labour will be present in the future and therefor the need for intuitive and collaborative interfaces and interaction for/with robotic fabrication is relevant (Petereit et al. 2012).

The Commonly known linear fabrication approach, where an object will be defined inside a CAD environment and then exported for robotic fabrication only scratches the surface of the capability this technology contains. More responsive and dynamic investigations using physical scanning of the environment with devices such as the leap motion, Kinect and depth sensors and real-time feedback loops, possible by machine learning processes allow for natural or conscious fabrication behavior (Rossi et al. 2017; Mueller et al. 2012). Further encouraging progress is made in the field of virtual and mixed reality (VR/MR). Using the visual and haptic sensation and immersion enabled by AR/MR, dynamic and in-situ modelling in the user natural environment highlights more creative and hands-on manufacturing methodologies. (Peng et al. 2018).

Additional Voice user interfaces (VUI) explorations of audio-visual design strategies, where the form generation is concluded by an audio input emphasis an additional humane sensorial parameter (Lim 2014).

The vocal tract is the source of human sound and while the vocal chords are the primary source of sound generation, other mechanisms further feature sound making abilities such as whistling. In conjunction with contextual and emotional meta level meaning of the spoken sound, the human voice offers a plethora of filterable entry parameters. It is not surprising that VUI in combination with artificial intelligence (AI) are one of the leading frontiers when discussing commercial user interfaces showcased by VUI products such as Apple's Siri or Amazon's Alexa.

1.3 Robotic Hotwire Cutting

Robotic hot-wire cutting processes offer the possibility of quickly and efficiently generate 3-dimensional shapes, due to the fact that the process of material removal is volumic, allowing on to process an entire ruled surface with a single motion, avoiding the need for layer-by-layer processing, typical of other processes, such as milling (McGee 2012). Additionally, hot-wire cutting produces two parts where both sides are usable, avoid the production of waste from the removal of the upper part when a block is milled (Ruttico et al. 2016). A variety of projects (McGee 2012; Schwartz 2012a, b; Brell-Çokcan and Braumann 2014, Bidgoli and Cardoso-Llach 2015; Søndergaard et al. 2016) have demonstrated the efficiency and reliability of hot-wire cutting as method for fast processing of large volumes, offering a high level of accuracy and repeatability, as well as the ease of integration with other production methods (Sousa 2017).

2 Methodology

2.1 Installation Set-Up

With this background, the installation is an attempt to explore a future collaborative approach to distributed and participatory design. It consists primarily of a microphone, a video screen and a robotic arm.

The visitors are invited to speak in the microphone. This simple act triggers a custom algorithm to generate in real time an evolving 3-dimensional representation of the speaker's voice. The live feedback on the screen creates an immediate learning loop, where the visitors, almost instinctively, learn how to shape the virtual object by modulating their voice. Once the visitor is satisfied with the object displayed on the screen, they can save their configuration. This automatically activates the robot arm to carve out the desired shape out of a block of foam through a custom attachment.

The process of extracting the "shape of a voice" out of a foam block creates simultaneously the desired object and its negative form. The visitor is gifted with the result of its exploration to take home as a souvenir of a possible future. The negative traces of voices are aggregated through an algorithm in a sculptural wall.

The result is an endless range of individually formed pieces, composed of a patterned surface that represents the physical/digital translation of the visitor's sound input and individual voice, whether it's a word, a scream, a song or simply a breath.

Also the role of the designer is redefined: as the end user gains access to more powerful customization tools and easily accessible interfaces, the designer no longer needs to determine the final form or configuration of his creation but only its initial state and the limits of what's possible are left to the individual user to explore.

This overall fabrication and design process evokes a more top-level and tangible meditation about the cognitive design making choices from both the producer and consumer (Sharif 2013; Sharif et al. 2015). Recent scientific assumptions about the dynamic and distributed nature of cognitive processes as in interplay of the brain, body tool, material, product and society illustrate the complexity of the act of design making (Malafouris 2004). This premise is interesting because it promotes the notion that the learning process between human and machine can be multidirectional and that by learning more about human cognition and design choices the digital processes can factor more actively in the production of artifacts (Fig. 1).

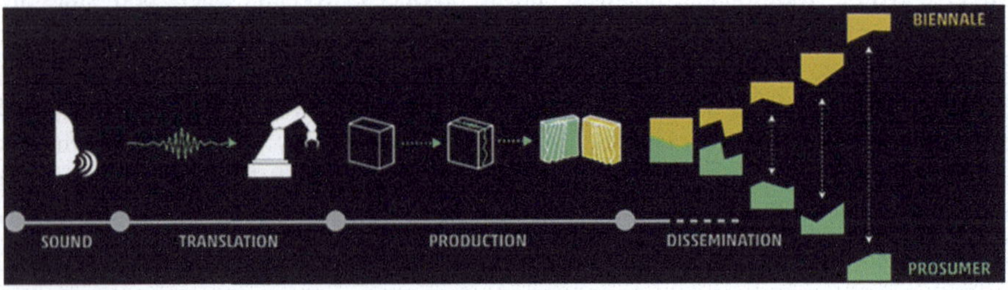

Fig. 1. Installation workflow

2.2 Voice-Controlled Modelling

It was crucial to establish a resilient workflow allowing to extract the voice pattern and translate it into numeric data that can be modified within the computational logic. To accomplish this, we used the plug-in Firefly. The plug-in primarily enables real-time data flow between Grasshopper and micro-controllers such as Arduino.

One of the digital tools in the firefly audio library is the *Frequency Spectrum*. It allows the user to feed real-time sound patterns via a microphone.

We extract the real-time values representing the volume and the corresponding frequency levels.

We limit both the minimum and maximum boundary conditions of the ranges to cull outlier values and ensure that geometric properties do not exceed fabrication constraints. One of the main challenges in this process was to control the rescaling of the voice recordings caused by the firefly component. If one would imagine the filtered and unique voice trace as a scatterplot illustration, then depending on the input the two numeric dimensions would always adjust according to the new input range. By a more hands-on evaluation process we had to determine those ranges manually. The two numeric lists are used to generate two curves illustrating the volume and frequency wave pattern over the recorded time, subsequently building the guiding ruled surface for the robotic cutting motion. The ruled surface is allowed to turn in a 90° angle. The level of rotation hereby is controlled by the participants peak frequency value (Fig. 2).

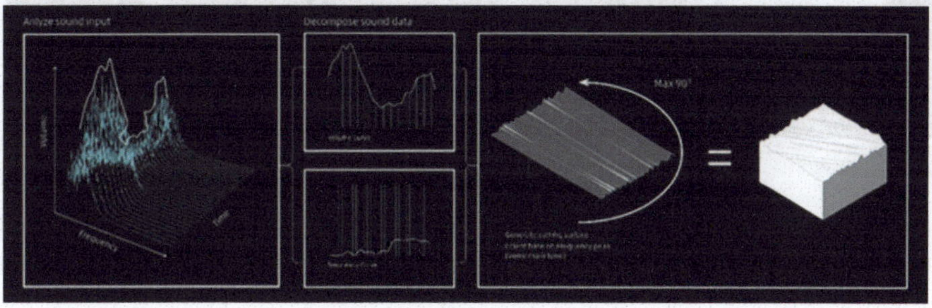

Fig. 2. Base module generation

2.3 Custom User Interface (CUI)

Once the core logic of the set-up was operating, we investigated a variety of options how to visualize the user interfaces. The main focus hereby was to ensure that the participant understands fast and intuitively how to modulate the displayed digital object with his/her voice. A combination of the Grasshopper plug-ins Human and Human UI enabled us to design a CUI. The plug-in Human UI enables the design of interactive user interfaces within Grasshopper.

The main element displayed in the interface is a revolving 3-dimensional representation of the digital foam block. As soon as the user presses and holds the record button, a trigger is activated in the software and the block is shaped according to the volume, frequency and duration of the recording. If no sound is emitted into the microphone the block will be presented as a flat white box. The more diverse the audio input the more the deformation and rotation of the block becomes readable.

On the right side of the interface we located three infographics and the interactive control station. The placement map on the top of this section highlights in red where the current block will be placed. We will explain the placement logic more in depth in Sect. 4. This approach enables the user to choose not only the geometry of the object, but also gives him the freedom to choose the placement of his negative foam piece. The user can also assess the graphs of the guiding curves and change the recording if he is not satisfied with the outcome. Once he is satisfied the user then writes his name and presses the save configuration button, hereby activating the robotic fabrication (Fig. 3).

2.4 Real-Time Data Broadcasting

Caused by the programmatic nature of the installation, two scripts had to run simultaneously on two different PCs, one controlling the interface and the other one for robotic fabrication. The installation depends on the real-time input of the user and a

Fig. 3. Interface interaction

method had to be defined that enabled the interface button to trigger events in both scripts synchronously. Speckle Streams is a plug-in for Grasshopper that enables the broadcast of real-time data via the internet through a sender and receiver methodology. The data can be sent to any receiver within any Grasshopper script, that is connected to the internet. Utilizing the capabilities of Speckle the interface script could now send the data via live stream to the fabrication script. This liberating ability to send real-time data inside the Rhinoceros and Grasshopper framework and even across platforms increased the speed of idea and fabrication flow drastically. One of the central aspects of our installation was precisely the ability to send the command for fabrication just by the act of triggering a digital button. The profound process behind the translation of the desired geometry into robotic path motion, evaluation and debugging of the same, executive instruction generation for the physical robot arm and the fact that this digital framework acted independent from the actual operating system of the form generation interface and could have been located anywhere globally are all hidden and condensed for the visitor into one embodied act.

3 Robotic Hotwire Cutting

In order to create a direct link between the geometry, generated by the participants' voices, and robotic cutting motion, the Hal Robotics plug-in for Grasshopper (Schwartz 2012) was used to translate the incoming curves first into ruled surfaces, and then into planes to be followed by the robot motion. The combination of constraints on the generation of shapes, and a custom-designed hot-wire end-effector, specifically dimensioned in order to allow maximum flexibility in the cutting possibilities, allowed to ensure the feasibility of the cutting within robot and fabrication setup constraints.

The direct integration of constraints within geometry generation and its direct translation into robotic toolpaths, allows to remove the complexity of robotic programming from the interaction space of the user, and hence enables the design of natural interfaces, where voice is directly translated into robotic motion, creating a more direct connection between user and fabrication process. This allows to robotically fabricate infinite design variants, each adapted to the user's voice parameters, avoiding the need to pass through multiple export/import steps from CAD to CAM to the robot (Fig. 4).

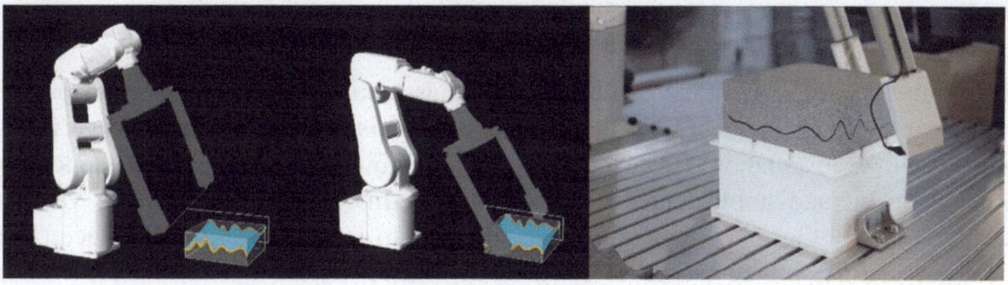

Fig. 4. Digital to physical translation of the hotwire cutting process

4 Panel Clustering and Installation Assembly

The placement of the randomly generated foam pieces had to be derived by a superordinate logic in order to establish a correlation to each other.

The placement logic is assigning each individually generated piece of foam with a vector representing its properties. The vector itself is calculated by summing up the properties of the two guiding curves and the rotation value that generates each foam piece. In its initial state the placing map generates a default map of values as shown in Fig. 5. The rectangles labelled with a -1 are to be understood as empty containers resembling the entire capacity of locatable objects. In order to initialize the placing process we feed a hidden row of values to serve as an orientation for the bottom-up construction.

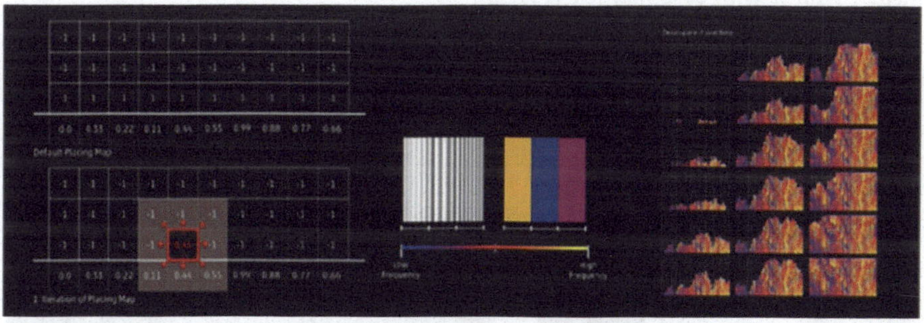

Fig. 5. Simulated placement logic

As soon as a new vector is generated the algorithm iterates through all the values in the default map and locates the new value by calculating the immediate neighbours sharing similar traits. Once this process is finalized the default map gets replaced by the newly generated placing map, containing the previous located vector. Subsequently the location of every new value is defined in such a way, that it can only be placed in immediate proximity to other existing vectors.

5 Results

Before the opening day of the exhibition we pre-assembled 50 pieces to encourage the participation. Over the course of the installation 50 visitors per day on average interacted with the installation. It was quite remarkable how well overall the methodology was perceived and implemented by the visitors. Almost intuitively curious visitors would come by the microphone and instantly start to interact with the Interface. The need for clarification of the process was low. Visitors across all ages would within minutes learn to modulate their desired artifact and activate the fabrication. In order to ensure the safety on site we closely monitored the fabrication process. This monitoring quickly transformed into an in-situ observation of the nuances each individual voice trace played on the form generation and to what degree presumably the personality factored substantially into the equation. Due to the sensitivity and even limitations of our set-up in respect to the ambient noise of the environment and the preset boundary constraints of the volume and frequency range the visitor often was compelled to swiftly overcome reservations and really engage with the microphone. If the visitor was too hesitant the result of his/her exploration was more modest in contrast to the more extravert form generations by other visitors. Unsurprisingly the more the visitors would depart from linguistic patterns and start to playfully generate abstract sound the more exceptional the resulting form would be.

Within minutes the robotic fabrication was finished and together with the designer we then manually installed the piece onto the evolving sculpture. Perhaps the most notable interactions took place between children and the robotic process. Proving that even a small child was able to intuitively comprehend the design task and successfully interact with it with ease and joy (Fig. 6).

Fig. 6. Installation interaction and final result

6 Conclusion and Outlook

In retrospect the calibration of the volume and frequency boundaries that where preset for this installation would have benefitted from a more generous field testing. This also transfers to the geometric variability of the formable ruled surface that our interface could offer and the geometric equity of the foam pieces that when aggregated resulted into a simple vertical 3-dimensional extrusion. It would be of interest to explore more spatial and contextual sculptural forms. In addition one could try to aggregate the sculpture using a set of differently shaped foam pieces. The most important improvement however would be to orchestrate the assembly process more efficient and maybe even utilizing other means of interface technology such as mixed reality.

This installation offers an optimistic view of the automatization of production. New lean and clean production methods enable a departure from the centralized industrial model of the XX Century and create smaller, decentralized production units that can be better integrated in the urban and social fabric. In this sense this would be a return to the craftsmanship based model prevalent prior to the industrial revolution. The production would no longer reside with the artisan, but with a machine. It is here postulated that this will enable the creativity of individual production to be distributed among the wider population rather than being concentrated with a single individual.

References

Bidgoli, A., Cardoso-Llach, D.: Towards a motion grammar for robotic stereotomy. In: Ikeda, Y., Kaijima, S., Herr, C., Schnabel, M.A. (eds.) CAADRIA 2015: Emerging Experience in Past, Present and Future of Digital Architecture, Proceedings of the 20th International Conference of the Association for Computer-Aided Architectural Design Research in Asia, pp. 723–732. Daegu, South Korea (2015)

Brell-Çokcan, S., Braumann, J.: Industrial Robots for Design Education: Robots as Open Interfaces beyond Fabrication. In: Zhang, J., Sun, C. (eds.) Global Design and Local Materialization. CAAD Futures 2013. Communications in Computer and Information Science, vol. 369, pp. 109–117. Springer, Berlin, Heidelberg (2013)

Brell-Çokcan S., Braumann, J.: Stack it Workshop. Rob|Arch 2014 Conference, University of Michigan, USA (2014)

Francese, R., Passero, I., Tortora, G.: Wiimote and Kinect: gestural user interfaces add a natural third dimension to HCI. In: Proceedings of the International Working Conference on Advanced Visual Interfaces (AVI 2012), pp. 116–123. Capri Island, Italy (2012)

Gershenfeld, N.: Fab. The Coming Revolution on Your Desktop. From Personal Computers to Personal Fabrication. Basic Books, New York (2005)

Lim, J.: Live robot programming. J. Prof. Commun. **3**(2), 165–175 (2014)

Malfouris, L.: The cognitive basis of material engagement: where brain, body and culture conflate. In: DeMarrais, E., Gosden, C., Renfrew, C. (eds.) Rethinking Materiality: the Engagement of Mind with the Material World, pp. 53–62. McDonald Institute for Archaeological Research, Cambridge (2004)

McGee, W., Feringa, J., Søndergaard, A.: Processes for an Architecture of Volume, Robotic wire cutting. In: Brell-Çokcan, S., Braumann, J. (eds.) ROB|ARCH 2012: Robotic Fabrication in Architecture, Art and Design, pp. 62–71. Springer, Vienna (2013)

Mueller, S., Lopez, P., Baudisch, P.: Interactive construction: interactive fabrication of functional mechanical devices. In: Proceedings of the 25th ACM Symposium on User Interface Software and Technology, UIST, pp. 599–606. Cambridge, MA (2012)

Peng, H., Briggs, J., Wang, C., Guo, K., Kider, J., Mueller, S., Baudisch, P., Guimbretiere, F.: RoMA: interactive fabrication with augmented reality and a robotic 3D printer. In: Proceedings of the 2018 CHI Conference on Human Factors in Computing Systems, CHI 2018, paper No. 579, Montreal, QC (2018)

Petereit, S., Albert, A., Jeridi, N., Schoenleber, R., Rebmann, C., Vallery, H.: Work life in the light of the demographic change: case study force assistive device for craftsmen. In: Wichert, R., Eberhardt, B. (eds.) Ambient Assisted Living. Advanced Technologies and Societal Change, vol. 2, pp. 45–60. Springer, Heidelberg (2012)

Ritzer, G., Dean, P., Jurgenson, N.: The coming of age of the prosumer. Am. Behav. Sci. **56**(4), 379–398 (2012)

Rossi, A., Tessmann, O.: Collaborative assembly of digital materials. In: ACADIA 2017 (2017)

Ruttico, P., Rossi, A., Panahikazemi, L., Andaloro, M.: Innovative methods for mold design and fabrication. In: Bock, T., Georgoulas, C., Linner, T., Guettler, J., Lee, S. (eds.) Proceedings of the CIB IAARC W119 CIC 2015 Workshop, Advanced Construction and Building Technology for Society, p. 36. Munich (2016)

Savov, A., Tessmann, O., Nielsen, S.A.: Sensitive assembly: gamifying the design and assembly of façade wall prototypes. Int. J. Architectural Comput. **14**(1), 33–48 (2016)

Schwartz, T.: Automated Foamdome #2, Exhibition Synthetic 2012: Foam. ESBA TALM, Le Mans, France (2012)

Schwartz, T.: HAL – extension of a visual programming language to support teaching and research on robotics applied to construction. In: Brell-Çokcan, S., Braumann, J. (eds.) ROB|ARCH 2012: Robotic Fabrication in Architecture, Art and Design, pp. 92–101. Springer, Vienna (2013)

Sharif, S.: Material cognition: designer's perception of material. In: Proceedings of the XVII Conference of the Iberoamerican Society of Digital Graphics (SIGraDi): Knowledge-based Design, vol 1, pp. 23–26. Valparaiso, Chile (2013)

Sharif, S., Gentry, T.R.: Design cognition shift from craftsman to digital maker. In: Ikeda, Y., Kaijima, S., Herr, C., Schnabel, M.A. (eds.) CAADRIA 2015: Emerging Experience in Past, Present and Future of Digital Architecture, Proceedings of the 20th International Conference of the Association for Computer-Aided Architectural Design Research in Asia, pp. 683–692. Daegu, South Korea (2015)

Søndergaard, A., Feringa, J., Nørbjerg, T., Steenstrup, K., Brander, D., Graversen, J., Markvorsen, S., Bærentzen, A., Petkov, K., Hattel, J., Clausen, K.: Robotic hot-blade cutting. In: Reinhardt, D., Saunders, R., Burry, J. (eds.) Robotic Fabrication Architecture, Art and Design 2016, pp. 150–164. Springer International Publishing Switzerland (2016)

Sousa, J.P.: Robotic technologies for non-standard design and construction in architecture. Nexus Netw. J. **19**(1), 73–83 (2017)

Willis, K.D., Xu, C., Wu, K.J., Levin, G., Gross, M.D.: Interactive fabrication: new interfaces for digital fabrication. In: Proceedings of the fifth international conference on tangible, embedded, and embodied interaction (TEI 2011), pp. 69–72. ACM, New York (2011)

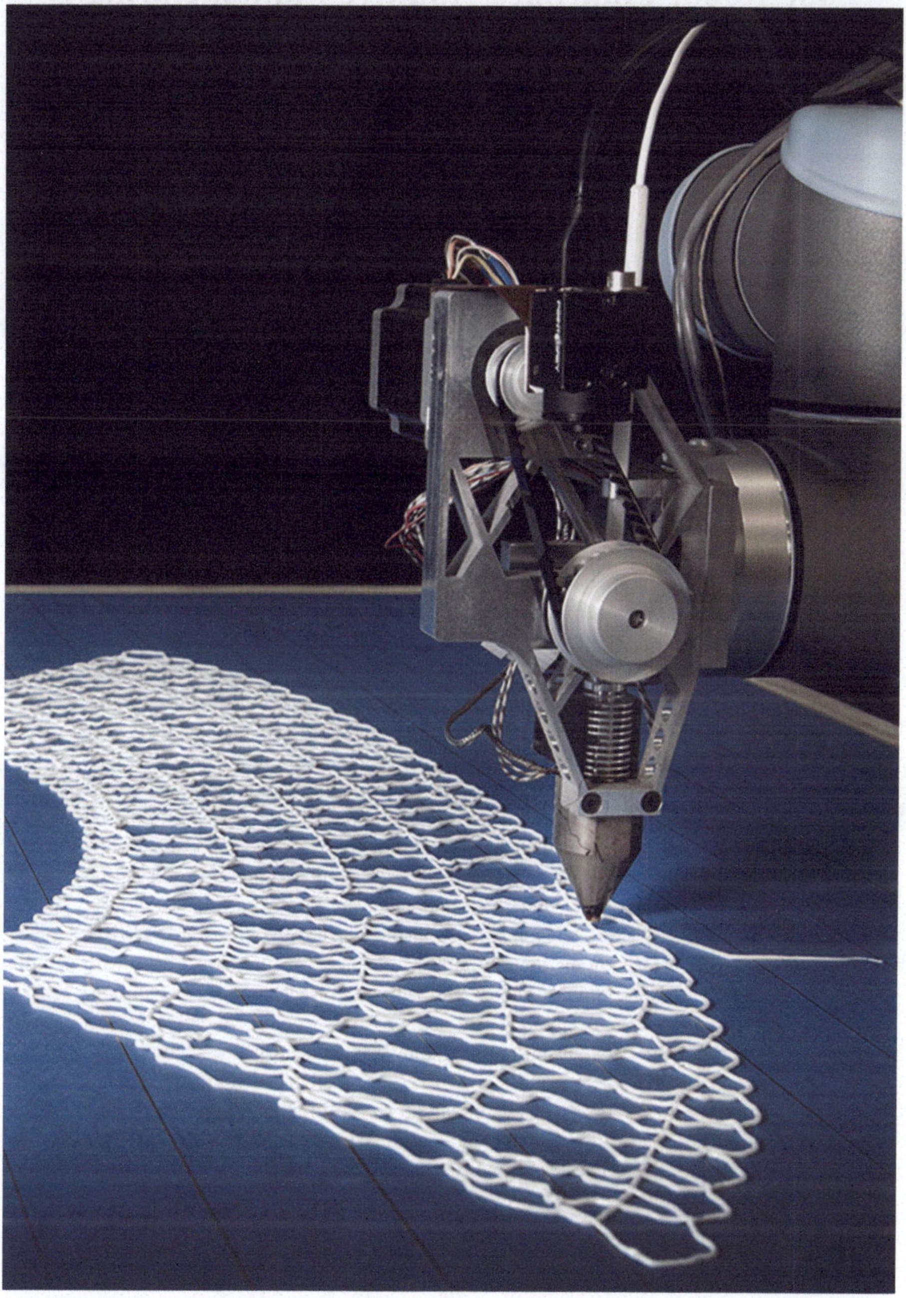

Towards Visual Feedback Loops for Robot-Controlled Additive Manufacturing

Sheila Sutjipto[1], Daniel Tish[2], Gavin Paul[1], Teresa Vidal-Calleja[1], and Tim Schork[2(✉)]

[1] Centre for Autonomous Systems, Faculty of Engineering and Information Technology (FEIT), University of Technology, Sydney, PO Box 123, Broadway, NSW 2007, Australia
Sheila.Sutjipto@student.uts.edu.au, {Gavin.Paul-1, Teresa.VidalCalleja}@uts.edu.au

[2] School of Architecture, Faculty of Design, Architecture and Building (DAB), University of Technology, Sydney, PO Box 123, Broadway, NSW 2007, Australia
Daniel.N.Tish@student.uts.edu.au, Tim.Schork@uts.edu.au

Abstract. Robotic additive manufacturing methods have enabled the design and fabrication of novel forms and material systems that represent an important step forward for architectural fabrication. However, a common problem in additive manufacturing is to predict and incorporate the dynamic behavior of the material that is the result of the complex confluence of forces and material properties that occur during fabrication. While there have been some approaches towards verification systems, to date most robotic additive manufacturing processes lack verification to ensure deposition accuracy. Inaccuracies, or in some instances critical errors, can occur due to robot dynamics, material self-deflection, material coiling, or timing shifts in the case of multi-material prints. This paper addresses that gap by presenting an approach that uses vision-based sensing systems to assist robotic additive manufacturing processes. Using online image analysis techniques, occupancy maps can be created and updated during the fabrication process to document the actual position of the previously deposited material. This development is an intermediary step towards closed-loop robotic control systems that combine workspace sensing capabilities with decision-making algorithms to adjust toolpaths to correct for errors or inaccuracies if necessary. The occupancy grid map provides a complete representation of the print that can be analyzed to determine various key aspects, such as, print quality, extrusion diameter, adhesion between printed parts, and intersections within the meshes. This valuable quantitative information regarding system robustness can be used to influence the system's future actions. This approach will help ensure consistent print quality and sound tectonics in robotic additive manufacturing processes, improving on current techniques and extending the possibilities of robotic fabrication in architecture.

Keywords: Robot control · 3D printing · Vision-based sensing

J. Willmann et al. (Eds.): ROBARCH 2018, *Robotic Fabrication in Architecture, Art and Design 2018*, pp. 85–97, 2019.
https://doi.org/10.1007/978-3-319-92294-2_7

1 Introduction

With the advent of new, robotically enabled fabrication methods comes the ability to design new materials and geometries with new functionalities. Additive manufacturing methods have been rapidly developed for several new materials from the starting point of thermoplastics to silicones (Rodrigue et al. 2015), concrete (Lloret et al. 2015), and hydrogels (Barry et al. 2009), and many more. Approaches to robotic 3D printing have also started to go beyond in-plane, 2D layer-based methods towards freeform 3D material depositions (Hack and Lauer 2014; Laarman et al. 2014; Soler et al. 2017). These advancements bring about the ability to design not only novel forms, but also new performative material systems that produce variations according to local stresses or to respond to environmental conditions. These new material systems are possible in mono-material additive manufacturing and their potentials are greatly expanded with a move towards multi-material 3D printing. However, to fully leverage these functional capacities, a greater control of the material deposition is required.

Additive manufacturing processes comprise a complex ecology of interactions between a diverse set of parameters, such as the rheological characteristics of the material, the rate of material flow from the extrusion nozzle, the rate of cooling or curing of the material, and the structural capacities of the pre-cooled or -cured material as additional material is deposited. The complexity and non-linear interactions of these parameters make it challenging to predict and ensure the accurate deposition of material during the 3D printing process without extensive and computationally intensive simulations of the entire process prior to commencing the print.

In order to accommodate some of these parameters and overcome the challenges several researchers have taken an open-loop approach of explicit tool-pathing and control of the robotic processes. Hack et al. (2013) implemented a series of explicit robot motions including amplifying the z-direction movements, short stops for cooling, and air pressure changes to increase the level of control and predictability over the material behavior. McGee et al. (2017) implemented an explicit "pressing" motion path with additional tolerances embedded within it to ensure fully-fused joints between all 3D printed connections in a tensile mesh.

Approaches to robotic systems can be divided into two categories; open-loop systems and closed-loop systems (Vidal-Calleja et al. 2010). To a large extent, open-loop systems are the predominantly used approach in the field of architecture and design. As an alternative to the previously described open-loop approaches, there exists an opportunity to equip the robotic system itself with sensing capabilities and decision-making agency (Paul et al. 2009) to ascertain the accuracy of previously deposited material and adjust future toolpaths (Paul et al. 2015). Research suggests that closed-loop systems, are critical to the success of robotic fabrication processes in which the material does not behave predictably or in situations where conditions can change or lead to inaccuracies (Giftthaler et al. 2017). In robotic additive manufacturing processes, surveying the position of the previously deposited material, would give the system agency to take corrective action to these inaccuracies by recalibrating robot trajectories and accurately executing tectonic motion paths like the pressing motion developed previously.

Vision-based feedback systems are generally classified depending on the number of cameras, the position of the camera with respect to the robot or the design of the error minimization function used to control the robot (Hutchinson et al. 1996). When considering the position of the camera, two main configurations exist; end-effector mounted (eye-in-hand) or fixed in the workspace (bird's eye). Feedback approaches are further classified into position-based, image-based, and hybrid-based and motion-based feedback systems. In a position-based feedback system, the online analysis of the scene is performed in what is referred to as the task space, whereby image information is used to reconstruct or map the scene via the *a priori* calibrated camera model (Zhang 1999). This provides geometric information of the scene in the industrial robot's base coordinates.

To leverage geometric information of the scene, metric maps are utilized to build accurate representations of the environment. An algorithm proposed by Elfes (1989) and Moravec (1988) known as Occupancy Grid (OG) mapping, utilizes grids to model the environment's free and occupied space. In 3D grid map representations, techniques such as voxel hashing and octrees discretize space into voxels. Each voxel possesses a position in space and a probability of occupancy; occupied, free-space or unknown (Paul et al. 2015). It is common to combine probabilistic approaches in conjunction with mapping methods as sensors are subject to measurement noise. Algorithms such as extended Kalman filters, Bayesian filters, and particle filters have been widely explored to provide improved state estimates (Thrun et al. 2005).

This paper proposes a framework for online visual-feedback in robotic additive manufacturing processes (shown in Fig. 2) via image analysis and probabilistic OG maps (Elfes 1989). The framework sequentially takes an image as its input, segments the current area, utilizes the *a priori* calibrated camera model to build a map in task space and probabilistically fuses the information into an overall OG map of the printed areas.

2 Methodology

The research described here seeks to verify the accuracy of previously deposited material during additive manufacturing processes. The design process generates a set of task space waypoints for the Tool Center Point (TCP) to track. The robot is controlled so that it tracks the trajectory while images are acquired via a calibrated camera. As shown in Fig. 4, gathered images are processed such that the deposited material is segmented from the background. The classified images are transformed into task space, and sequentially fused into a probabilistic OG map.

2.1 Robot Eye-to-Extruder Calibration

The eye-in-hand configuration proposed in this work possesses a camera located on the filament extruder system as shown in Fig. 1. As the TCP moves through a given trajectory, the images gathered from the camera are unable to be related unless the transformation between the camera and robot base is known.

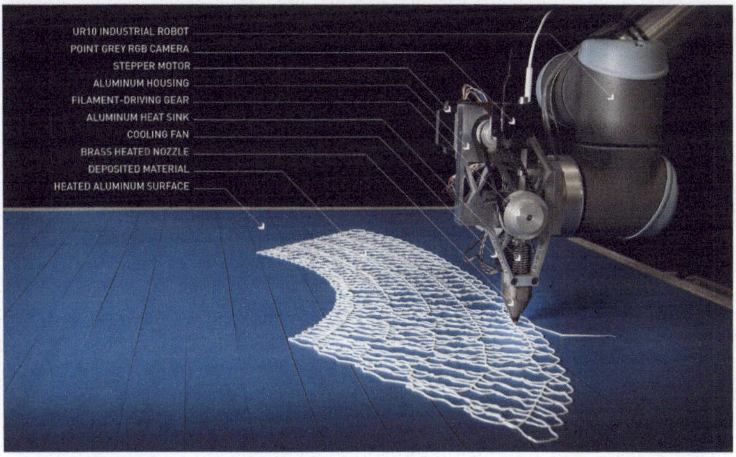

Fig. 1. UR10 robot with wrist-mounted filament extruder system, Point Grey Blackfly RGB camera and heat bed (in blue) while printing part B30.

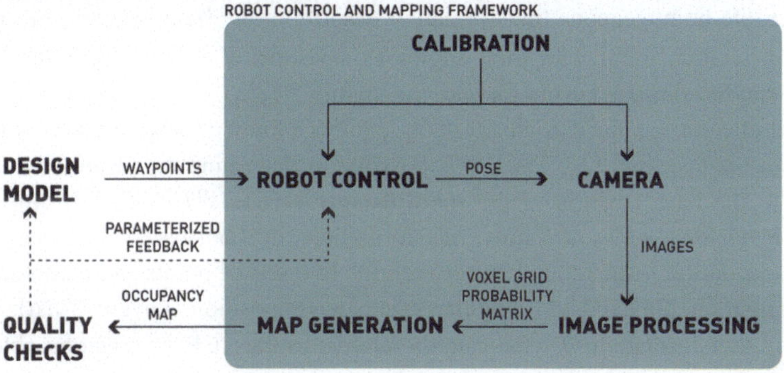

Fig. 2. Overview of the proposed robot control and mapping framework for closed-loop additive manufacturing processes.

Accurate geometric information of the environment obtained from the camera requires two types of calibration, intrinsic and extrinsic. Intrinsic calibration determines the intrinsic matrix that represents the projective transformation from the 3D camera coordinate system into a 2D image coordinate system. This calibration utilizes the perspective projection camera model and procedure proposed by Zhang (1999) and implemented by Bouguet (2000). Extrinsic calibration is used to determine the camera's pose (i.e. rotation and translation) in a defined coordinate system, or robot base, ${}^{0}T_{c}$.

Hand-eye calibration is the process that ascertains the camera coordinate system relative to the reference frame of the robot's TCP. Well-known approaches can be based on determining the rotation and translation consecutively (Daniilidis 1999), simultaneously (Strobl and Hirzinger 2006), or considering time-offsets (Furrer et al. 2018).

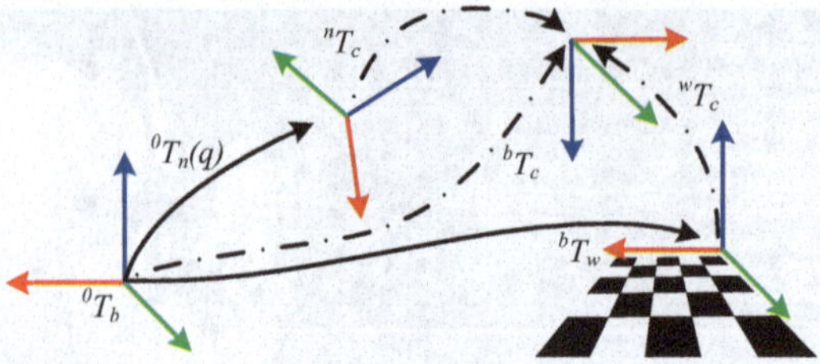

Fig. 3. Eye-in-hand calibration to determine the extrinsic calibration of a camera so the pixels in the camera coordinate frame can be transformed to world or robot base coordinate frame.

The hand-eye calibration implemented in this work requires a set of known points, $P \in R^3$, defined in robot base coordinates and corresponding points defined in camera coordinates. Determining the rotation and translation can be formulated as an optimization problem. As shown in Fig. 3, a calibration target with known dimensions is needed to obtain the 3D points in camera coordinates from the points in the 2D image. To recover the points defined in camera coordinates requires P to be visible from a known camera location.

Given the homogeneous transformation describing the base location, ^0T_b of the 6 degree-of-freedom robot with joint angles as a vector, $\mathbf{q} = [q_1, \ldots q_6]^T$, the last link location can be computed using forward kinematics, $^bT_f(\mathbf{q})$. Consequently, the addition of an end-effector, i.e. the extruder system, requires a known transformation of the TCP relative to the last link, ^fT_n. Thus, it is possible to determine the location of the TCP relative to the base location of the robot using, $^0T_n(\mathbf{q}) = {}^0T_b{}^bT_f(q){}^fT_n$. Moving the TCP to P, yields a series of known transformations in robot base coordinates.

The translation is the difference between the two sets of points and incorporates the optimal rotation obtained via singular-value decomposition. Given $^0T_n(\mathbf{q})$ and the calculated transformation ^0T_c, the camera relative to the TCP, ^nT_c, can be obtained.

2.2 Deposited Material Segmentation via Image Processing

To segment the image into deposited material and background, the image undergoes a series of image processing steps. Initially, the image undergoes a color space conversion. Since the RGB color space suffers from variations in light intensity, the HSV color space is used as data represented in HSV is less susceptible to these changes. A threshold is applied to the image, then normalized to effectively occupy its range of pixel intensity values. Median filtering is applied to the image to remove noise from normalization. The image is then converted to a binary representation and undergoes median filtering again to remove noise that may have been a product of conversion.

The output of the process is a segmented image that categorizes the original values from the data into deposited material or background, ultimately represented as a binary matrix. This output is then fused into one probability map for further analysis as discussed in the next section.

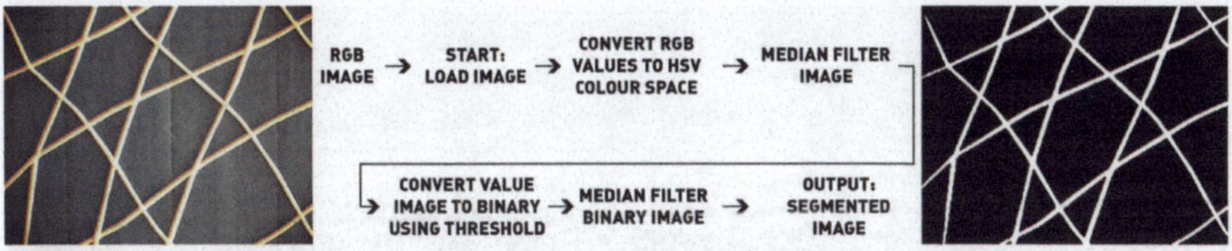

Fig. 4. Image processing procedure, takes in color images, processes them and then outputs the classified deposited material and background.

2.3 Deposited Material Occupancy Mapping

Mapping in robotics is the process of building a representation of the environment, commonly based on sensors. Given the intrinsic calibration and transformation, ${}^{0}T_{c}$ (Sect. 2.1), images can be sequentially fused over time to infer whether a location in task space is occupied by deposited material. A Deposited Material Occupancy Map (DMOM) is used to represent the printed space.

The images obtained for fusion are a mapping of 3D points to a 2D surface, forming a 2D representation of the environment. Since this process involves the loss of depth information, depth to the printing surface can be obtained as the camera pose is known. Extending this approach to 3D map representations like octrees can be achieved using the method described by Paul et al. (2015).

The Occupancy Grid (OG) mapping method addresses the problem of generating a consistent map from noisy or incomplete sensor data and possesses potential use for data fusion (Stepan et al. 2005). As the eye-in-hand configuration limits the field of view of the camera, the probabilistic nature of OG mapping allows the images captured to be fused. Each grid cell in the OG map, DMOM in this case, is individually treated as having a mutually-exclusive probability of existence of deposited material. Since there is overlap in the data observed by the camera, the certainty of deposited material occupying a grid cell is dependent on whether deposited material is observed at a location, and the frequency it appears at that location. The probability is updated using a Bayes update (Thrun et al. 2005), where higher trust is given to more recent measurements. The probability that a grid cell is occupied is independent of the location of the image pixel and distance from the camera. Due to this independence, it is assumed that at each point that is classified, regardless if it contains deposited material or not, is equally trustworthy.

3 Experiment Setup

The approach is tested in a pilot study towards the production of a site specific, partially dynamic tensile surface installation covering over $25\,\mathrm{m}^2$ as shown in Fig. 5. The installation was designed and simulated using the Kangaroo Physics plug-in for Grasshopper, developed by Piker (2013). The overall form of the installation is divided into 59 panels to facilitate the printing of each panel within the reach of the robot.

Fig. 5. Detail of robotic 3D printed functionally graded net installed on site.

The topology of the tensile mesh is functionally graded to respond to the simulated differences in tension forces across the surface. By employing a bespoke computational method, the resulting graded mesh is flattened to embed the 3D geometry within the 2D pattern for printing McGee et al. (2017).

3.1 Physical Hardware Setup

To facilitate the manufacturing of the tensile meshes, a custom work cell was constructed. The setup consists of a Universal Robot (UR10) robot mounted to a frame housing a custom heat bed measuring $1.2\,m^2$. The extruder end-effector consists of a stepper motor that feeds filament to the nozzle, where it is melted by two heat cartridges embedded in the nozzle, with a thermocouple for temperature regulation.

A downward-facing global shutter Point Grey Blackfly RGB camera is mounted to the extruder. The resolution of the camera is 648×488 pixels, which results in an area of 138×104 mm when the print nozzle is in contact with the heat bed. The offset position of the camera relative to the TCP, ${}^{f}T_c$ was estimated as [−43, −72, 187] mm.

3.2 Software

The toolpaths are developed from the form-found digital mesh output from the Kangaroo physics system. These meshes are translated into the final, continuous line toolpaths through a series of scripts which eliminate odd-valence vertices by replacing each edge in the mesh with a vertex, making connections between the mesh faces that share the original edge, a process which always results in an even-valence at the new vertices. The toolpaths are also constructed to ensure that the order of lines produces full cross over joints at each intersection for improved structural performance, rather than joints where the two polyline segments attempt to meet at a point which are failure prone.

The UR10 toolpath for each panel is initially solved through Robots Grasshopper plug-in. A list of target joint angles and their corresponding TCP locations are produced from the plug-in and passed to ROS to communicate with the robot. Additionally, the commands are distributed through ROS to overcome the limited number of points that can be sent through the Grasshopper plug-in. ROS enables the state of the hardware to be queried at 125 Hz providing a means to have positional feedback, whilst the images are acquired at 7 Hz and processed in MATLAB.

3.3 Measurement Test Using Generated Map

To test the validity of the map, quality experiments have been conducted including: intersection detection, filament thickness, and alignment assurance.

The detection of intersections between lines is necessary so that the additional "pressing" process can occur to create structurally sound joints at each crossing. To perform intersection detection, a point in the image is found and denoted as the start of the line. From this point, a series of smaller image segments are generated in order to locate the lines in 3D space. The result is a vector between a start and end point that can be iteratively checked to determine if there are intersections with any of the previous lines found before.

The thickness of the extruded filament affects the strength of the overall print, and should remain consistent throughout the print. However, unintended variances can occur due to issues with extruder speed or filament feed, heating element inconsistency, or imperfections in the original filament. The map can hence be analyzed to check that the extruded filament line width is within expected tolerances.

During the printing process, it is possible that the extruded filament fails to fully adhere to the heat bed surface, resulting in misaligned or otherwise incorrect geometries. To detect if there are any misalignment issues and to incorporate quality control, the map can be compared against the expected location of the print in the map. If there are large variances, particularly global orientation differences, then it is likely there is an alignment issue and the print should be aborted and the user notified.

4 Results

Several experiments have been conducted to collect data during the printing process. The intersection points are taken from the geometry generation process and form the baseline ground truth from which the observed intersection points can be compared.

Varying illumination sources on the printed filament and background presented challenging conditions for developing a robust method of detection. Two image segmentation results are shown in Fig. 6. The lighting in the room was a mixture of natural and artificial light that was not specifically controlled.

The proposed approach demonstrates the segmentation of RGB images into binary representations of deposited material and printing surface. However, it has been shown that the process is unable to distinguish between the current print and residual filament from previous prints due to similarities in color properties. Noise from varying illumination sources challenges the robustness of the image processing procedure. Thus,

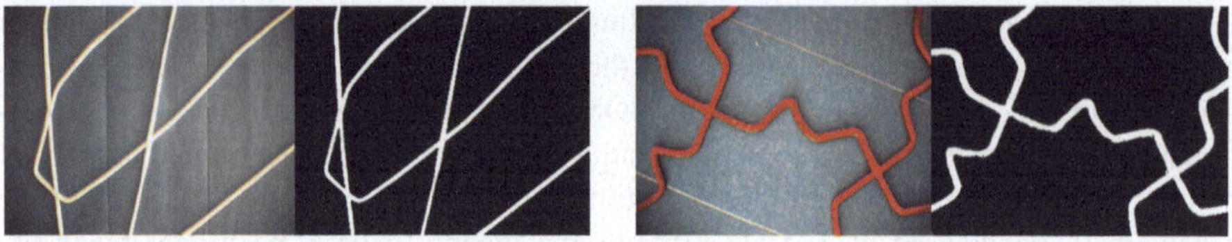

Fig. 6. Given two input images of deposited materials: cream-colored and red with varying lighting conditions on the background, the image processing results are shown.

probabilistic fusion during map generation allows each pixel to possess a better estimate of the state (as areas of the print are observed multiple times).

A series of images shown in Fig. 7 demonstrate the progression of the print based upon the observed data fused by using the Bayes update. Each grid cell in the DMOM represents 0.2132 mm^2, with each possessing an intensity ranging from a maximum value of 1 (white) and a minimum value of 0 (black), and the initial state and final state of unseen grid cells possess a value of 0.5 (gray). The DMOM after 35 min of printing and 1575 processed images is shown in Fig. 7. As more images are fused, the certainty about the existence of printed material at that location is increased, converging upon known to be empty (i.e. a value of 0), or known to be occupied (i.e. a value of 1).

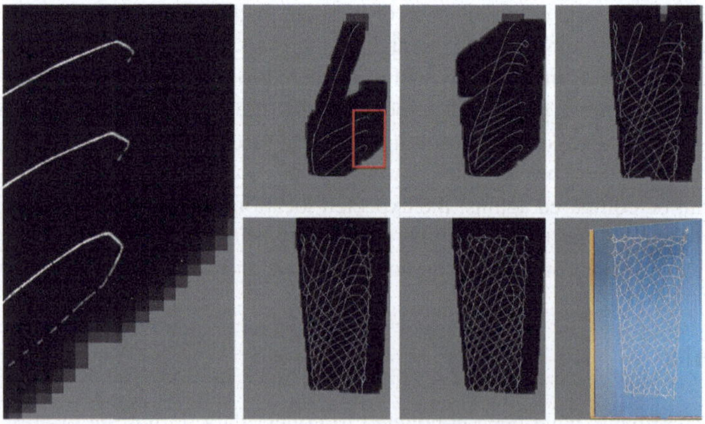

Fig. 7. Part A28. Results of a series of extrusion classifications as they are fused together into the task space DMOM. (Left) Detail of DMOM resolution. (Right, Top row from left) After 225 images and 5 min of printing; After 450 images and 10 min of printing; After 900 images and 20 min printing; (Right, Bottom row from left) After 1350 images and 30 min printing; Final DMOM; the actual complete print.

The printed piece shown in Fig. 8 contains a total of 466 intersections, with 15.88% of the intersections considered as unseen due to the trajectory, 5.15% of the intersections considered as unseen due to misalignment and 7.73% of the intersections considered as seen, although incorrectly mapped. Since the eye-in-hand configuration

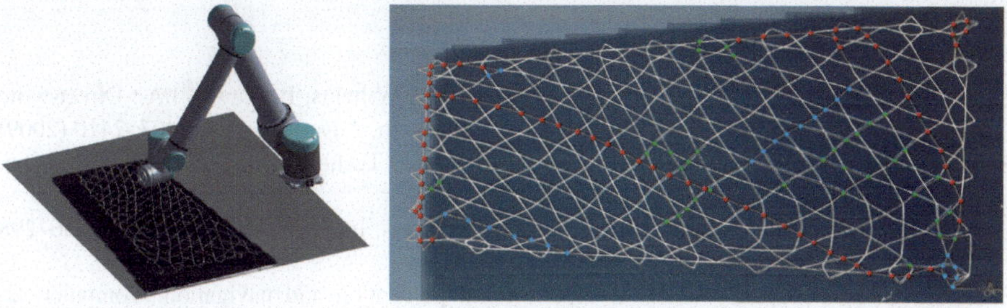

Fig. 8. (Left) Simulation of the robot with the final DMOM; (Right) Actual photo of final print overlaid with top-down view of the DMOM. Red markers indicate unseen intersections due to camera field of view and trajectory, blue markers indicate unseen intersections due to misalignment, green markers indicate detected intersections at incorrect locations due to misalignment.

relies on the given trajectory to observe the print, Fig. 8 illustrates an incomplete DMOM shown with missing, incorrect and misaligned intersection points, due to incorrect mapping and unobserved locations.

The average thickness of the print is determined using the probabilistic OG map. A set of print thicknesses were obtained by randomly sampling 30 locations of the print, then measuring the thickness. The average thickness is 9.03 grid cells with a standard deviation of 1.20 grid cells, translating to 1.93 mm and 0.26 mm respectively.

Applying a binary classification test, the statistical measures of performance: accuracy, sensitivity, and specificity are obtained by comparing the classified pixels to the ground truth. A true (or false) positive occurs when a grid cell is correctly (or incorrectly) identified as containing; a true (or false) negative occurs when a grid cell is correctly (or incorrectly) identified as not containing deposited material. The accuracy of the print is 91.8%, indicating a high percentage of correctly classified grid cells. The sensitivity (40.2%) indicates a relatively weak ability to correctly identify grid cells that contain deposited material, and the specificity is 95.8% which indicates a strong ability to identify grid cells that are free of deposited material.

5 Discussions and Conclusions

This paper has presented an approach to online feedback using a map-building technique that utilizes data from the robot and camera. The system probabilistically fuses the obtained data to construct an overall map of the print during the additive manufacturing process, enabling quality assurance processes to be applied. Throughout the research, the team has gained several insights towards this process. First, vibration of the extruder and camera system was noted during the extrusion process due to robot dynamics. This produces slight inaccuracies and misalignments with the camera's captured images but has no noticeable effect on the print quality. Future work is to close the control feedback loop by triggering pre-programmed localized actions, such as the pressing motion necessary for the tensile mesh extrusion when the system predicts an upcoming intersection.

References

Barry, R.A., Shepherd, R.F., Hanson, J.N., Nuzzo, R.G., Wiltzius, P., Lewis, J.A.: Direct-write assembly of 3D hydrogel scaffolds for guided cell growth. Adv. Mater. **21**, 2407–2410 (2009)

Bouguet, J.Y.: Matlab camera calibration toolbox. Caltech Technical report, http://www.vision.caltech.edu/bouguetj/calib_doc/ Accessed 1 Jun 2018

Daniilidis, K.: Hand-eye calibration using dual quaternions. Int. J. Robot. Res. **18**(3), 286–298 (1999)

Elfes, A.: Using occupancy grids for mobile robot perception and navigation. Computer **22**, 46–57 (1989)

Furrer, F., Fehr, M., Novkovic, T., Sommer, H., Gilitschenski, I., Siegwart, R.: Evaluation of combined time-offset estimation and hand-eye calibration on robotic datasets. In: Hutter, M., Siegwart, R. (eds.) Proceedings of the 11th International Conference on Field and Service Robotics, Zurich, pp. 145–159. Springer, Cham (2018)

Giftthaler, M., Sandy, T., Dörfler, K., Brooks, I., Buckingham, M., Rey, G., Kohler, M., Gramazio, F., Buchli, J.: Mobile robotic fabrication at 1:1 scale: the In situ Fabricator. Constr. Robot. **1**(1–4), 3–14 (2017)

Hack, N., Lauer, W.V.: Mesh mould: robotically fabricated spatial meshes as reinforced concrete formwork. Archit. Des. **84**(3), 44–53 (2014)

Hack, N., Lauer, W.V., Langenberg, S., Gramazio, F., Kohler, M.: Overcoming repetition: robotic fabrication processes at a large scale. Int. J. Architectural Comput. (IJAC) **3**(11), 285–299 (2013)

Hutchinson, S., Hager, G.D., Corke, P.I.: A tutorial on visual servo control. IEEE Trans. Robot. Autom. **12**(1), 649–774 (1996)

Laarman, J., Jokic, S., Novikov, P., Fraguada, L.E., Markopoulou, A.: Anti-gravity additive manufacturing. In: Gramazio, F., Kohler, M., Langenberg, S. (eds.) Fabricate: Negotiating Design & Making, pp. 192–197.UCL Press, London (2014)

Lloret, E., Shahab, A.R., Linus, M., Flatt, R., Gramazio, F., Kohler, M., Langenberg, S.: Complex concrete structures: merging existing casting techniques with digital fabrication. Comput. Aided Des. **60**, 40–49 (2015)

McGee, W., Thun, G., Velikov, K., Tish, D.: Infundibuliforms: kinetic systems, additive manufacturing, and tensile surface control. In: Sheil, B., Menges, A., Glynn, R., Skavara, M. (eds.) Fabricate: Rethinking Design and Construction, pp. 84–91. UCL Press, London (2017)

Moravec, H.: Sensor fusion in certainty grids for mobile robots. AI Mag. **9**(2), 61–74 (1988)

Paul, G., Kirchner, N., Liu, D.K., Dissanayake, G.: An effective approach to simultaneous mapping and surface-type identification of complex 3D environments. J. Field Robot. **26**(11–12 SI), 915–933 (2009)

Paul, G., Quin, P., To, A.W.K., Liu, D.K.: A sliding window approach to exploration for 3D map building using a biologically inspired bridge inspection robot. In: Proceedings of the IEEE International Conference on Cyber Technology in Automation, Control, and Intelligent Systems, pp. 1097–1102. Shenyang (2015)

Piker, D.: Kangaroo: form finding with computational physics. Archit. Des. **83**(2), 136–137 (2013)

Rodrigue, H., Bhandari, B., Wang, W., Ahn, S.H.: 3D soft lithography: a fabrication process for thermocurable polymers. J. Mater. Process. Technol. **217**, 302–309 (2015)

Soler, V., Retsin, G., Jimenez Garcia, M.: A generalized approach to non-layered fused filament fabrication. In: Nagakura, T., Tibbits, S., Mueller, C., Ibañez, M. (eds.) Acadia 2017: Disciplines & Disruption, Proceedings of the 37th Annual Conference of the Association for Computer Aided Design in Architecture, pp. 562–571. Cambridge, MA. (2017)

Stepan, P., Kulich, M., Preucil, L.: Robust data fusion with occupancy grid. IEEE Trans. Syst. Man Cybern. Part C **35**(1), 106–115 (2005)

Strobl, K.H., Hirzinger, G.: Optimal hand-eye calibration. In: Proceedings of the IEEE/RSJ International Conference on Intelligent Robots and Systems, pp. 4647–4653. Beijing (2006)

Thrun, S., Burgard, W., Fox, D.: Probabilistic Robotics. MIT Press, Cambridge (2005)

Vidal-Calleja, T., Andrade-Cetto, J., Sanfeliu, A.: Action selection for single camera SLAM. IEEE Trans. Syst. Man Cybern. Part B **40**(6), 1567–1581 (2010)

Zhang, Z.: Flexible camera calibration by viewing a plane from unknown orientations. In: Proceedings of the 1999 IEEE International Conference on Computer Vision, vol. 1, pp. 666–673. Kerkyra, Greece (1999)

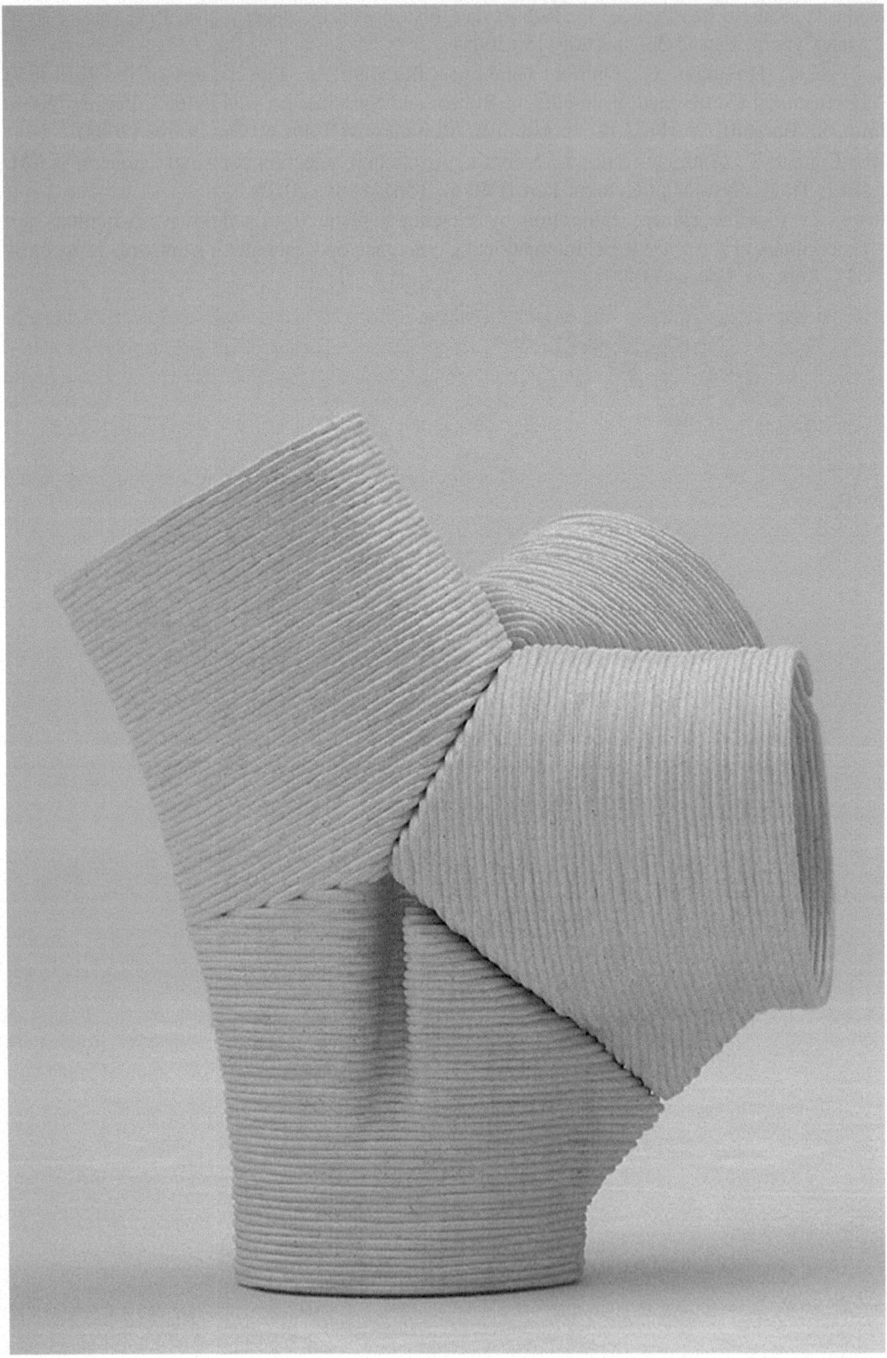

Function Representation for Robotic 3D Printed Concrete

Shajay Bhooshan[1,2(✉)], Johannes Ladinig[3,4], Tom Van Mele[1], and Philippe Block[1]

[1] Institute of Technology in Architecture, Block Research Group, ETH Zurich, Stefano-Franscini-Platz, HIB E 45, 8093 Zurich, Switzerland
bhooshan@arch.ethz.ch
[2] Zaha Hadid Computation and Design Group, London, UK
[3] Incremental 3D, Innsbruk, Austria
[4] Structure and Design, Institute of Design, University of Innsbruck, Innsbruck, Austria

Abstract. The use of Function Representation (FRep) to synthesise and specify geometries for 3D printing is finding renewed interest. The usefulness and extension of this representation in the synthesis and analysis of geometries for the process of large-scale, layered concrete 3D printing has been previously articulated by the authors. This paper fully extends the implicit representation used previously in shape-design to fabrication-related processing of compressive skeletal structures for realisation by robotic 3D printing of concrete. In particular, we use an initial value formulation of a propagating front to process the nodes and bars of a given funicular spatial structure.

Keywords: 3D printing · Implicit modelling · Function representation
Shape design · Funicular skeletons · Robotic concrete printing

1 Introduction

The use of Function Representation (FRep) to synthesise and specify geometries for 3D printing is finding renewed interest (Keeter 2013; Lu et al. 2014). The extension of this representation for the synthesis and analysis of geometries for layered process of large-scale concrete 3D printing has been previously articulated by the authors (Bhooshan et al. 2018). The shape-design framework described there is motivated by the rapid growth in large-scale, robotic 3D printing and lack of appropriate shape-design tools. Further, the framework is based on a novel insight regarding the applicability of design methods used in unreinforced, compressive masonry to layered 3D printing with materials of relatively low tensile capacity such as concrete. This paper further extends the FRep, used previously in shape-design, to fabrication-related processing of geometries. In particular, the framework is applied to process compressive skeletal structures for realisation by concrete printing. Lastly, physical results demonstrating the application are shown.

J. Willmann et al. (Eds.): ROBARCH 2018, *Robotic Fabrication in Architecture, Art and Design 2018*, pp. 98–109, 2019.
https://doi.org/10.1007/978-3-319-92294-2_8

1.1 Compressive Skeletal Structures and 3D Printing

Skeletal structures are commonly realised in steel and timber (Rian and Sassone 2014). Historically, they have also been realised in stone, prominently at the Sagrada Familia cathedral in Barcelona. Their realisation in cast concrete is perhaps most famously executed by Architect-Engineer Luigi Nervi (Perugini and Andreani 2013) and Engineer Robert Malliart (Zastavni 2008). However, the use of concrete for their realisation is nowadays often hindered by the high cost associated with the use of custom-tailored moulds and formwork (West 2001). There have been some efforts to alleviate this short-coming with the use of knit and fabric formworks (Popescu et al. 2016; West 2016). Of relevance to the current discussion however, are recent developments in the use of large-scale 3D printing with concrete to produce such skeletons. These efforts are currently restricted to the production of outer casings into which regulation-approved concrete is subsequently poured (XtreeE 2017).

1.2 Contributions and Outline

In considering a shape representation that is appropriate for both design and fabrication processing of skeletal shapes for concrete printing, we may highlight the following: (1) An ideal shape representation includes both the centre-lines of structural action and the layers of printing; (2) the shape representation makes no or little *a priori* assumptions regarding the topological complexity of the skeletal networks; and (3) the shape representation is amenable for fabrication-related processing addressing the particular constraints of concrete 3D printing such as heuristics related to maximum overhang, preferably planar assembly interfaces etc. The main contributions of this paper relate to outlining the extension of our FRep-based framework to address the first two aspects above (Sect. 3.1), whilst detailing its application to address the third (Sects. 3.2 and 3.3).

2 Related Work

Given the dual intention of both previous and current work, to support interactive and exploratory shape design whilst ensuring adequate representation of critical fabrication-related aspects, relevant domains of precedent work include the areas of computer graphics and computer-aided-geometric-design (CAGD). Medical imaging with its history of representing topologically complex and solid 3D shape via a stack of 2D images resonates with the requirements of representing the process of layered 3D printing of concrete. It thus provides another rich, if unusual, domain of precedent work.

2.1 3D Printing and Implicit Shape Representation

Implicit representation has widespread use in various domains including design and animation of articulated characters (Ji et al. 2010), modelling of biological shapes (Sherstyuk 1999a), computer-aided-design (Rossignac 1985), physically based animation (Desbrun and Gascuel 1995), rendering of fonts (Green 2007) etc. However, to the best of our knowledge, their use has not yet been adapted to synthesise shapes for concrete printing, apart from our previous work and its application by our collaborators (Bhooshan et al. 2018).

All forms of implicit representation share the following advantages (Opalach and Maddock 1995): (1) efficient to check for whether a given point is inside or outside the object, (2) efficient boolean operations, (3) easy-to-make topological changes, (4) efficient blending between objects, etc. We thus believe it is a natural representation scheme to be adapted for shape design for robotic and layered 3D printing of concrete, as well as its fabrication-related processing.

2.2 Implicit Modelling

In implicit representations, the shape of an object is represented by defining a scalar field (also called potential field or field function) that assigns positive values to points in the field that are outside the represented object and negative values for those inside. Thus, the surface of the object is implied as the boundary between the inside and outside of the object so represented. Technically, the surface of the object is the zero iso-surface of the scalar field. In contrast, Boundary Representations (BRep) parametrically define the surface of the object. For more on implicit surfaces, we refer the reader to the seminal paper by Bloomenthal and Bajaj (1997).

2.3 Skeleton Based Implicit Modelling and Function Representation

Given the requirements of the shape representation in relation to and its manipulation via a skeletal, centre-line graph of vertices and edges, skeleton-based implicit modelling becomes relevant precedent work (Bloomenthal and Shoemake 1991; Sherstyuk 1999a). Here, the scalar fields are defined in relation to the distance from a given skeletal graph with distances on the inside of the object being negative and those outside positive. Thus, sometimes a related terminology of signed distance fields (SDF) is used in this context (Hubert and Cani 2012). Lastly, various models of implicit representations are unified into the so-called Function Representation (FRep) (Pasko and Adzhiev 2002).

2.4 Medical Imaging and Layered Printing

The trajectories that the print head of the 3D printer has to traverse - the so-called task graphs (Khoshnevis et al. 2006) – are usually generated by processing BRep surfaces (Fig. 1). Alternatively, the layered process of 3D printing, like brick masonry, can be considered as advancing or propagating ground-upwards, one layer at a time. Furthermore, this view is particularly instructive in understanding the stability of the layers whilst printing, apart from foregrounding the layers themselves as opposed to the geometry from which they are derived. In this context, precedent work from the field of medical imaging becomes relevant. Front propagation or advancing front can be modelled as a parametrically represented boundary curve that evolves in time (boundary value formulation) or as the zero iso-contour of a scalar field that itself moves or evolves through time (initial value formulation) (Kass et al. 1988; Sethian 1998). Both methods

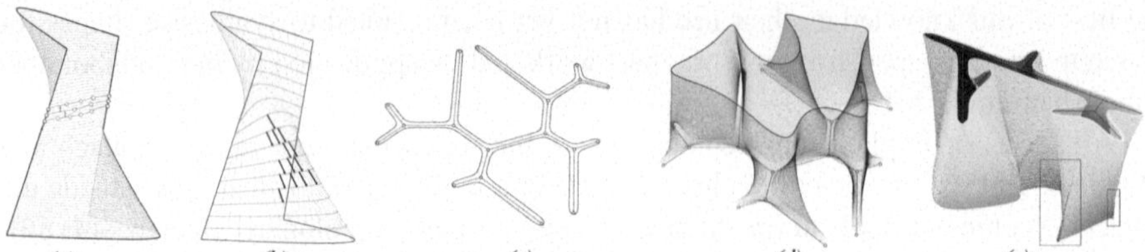

Fig. 1. Processing of mesh representation to generate (a) horizontal and (n) field-aligned task graphs.

are used in medical imaging to recover shapes from a stack of 2D images. This approach has also been recently used to minimise the amount of support structures required for overhanging parts in 3D printing and architecturally, to recreate the columns of Sagrada Familia (Cacace et al. 2017; Monreal 2012).

2.5 Previous Work

The authors have previously articulated and utilised the boundary value formulation to model and realise shapes in 3D Printed concrete (Fig. 1(c, d)). However, such parameterised boundary representations encounter difficulties when the curve expands or shrinks along its normal field and sharp corners and cusps develop or pieces of the boundary intersect (Cohen 1991). This can be seen in some of our previous results (Fig. 1(e)).

3 Processing of Funicular Networks for Concrete Printing

In comparison to previous work (Sect. 2.5), in the current work and paper, we utilise the alternative, initial value formulation and adapt it to model the layered geometries of concrete 3D printing. In particular, we combine them with a skeletal graph representation of 3D funicular networks to process them for physical realisation.

3.1 3D Skeletons and Initial Value Formulation of Front Propagation

It would be possible to use a BRep-based approach to process nodes and bars for 3D printing via the procedural thickening of the skeletal graph and subsequent contouring of the meshes that represent nodes and bars of the network (Fig. 2(a–d)). However, such geometric processing of nodes and bars increase in complexity when processing nodes requiring overhang support and/or with increased topological complexity. This approach holds further complications with regard to assignment of ideal print directions to the node in addition to resolving complex interfaces between the print layers of the nodes and bars (Fig. 2). Furthermore, local processing of print layers does not guarantee good alignment with the funicular force flow over the node.

Our approach, on other hand, capitalises on the rich and easy-to-implement operation set of FRep to alleviate this complexity, easily handle topological changes from

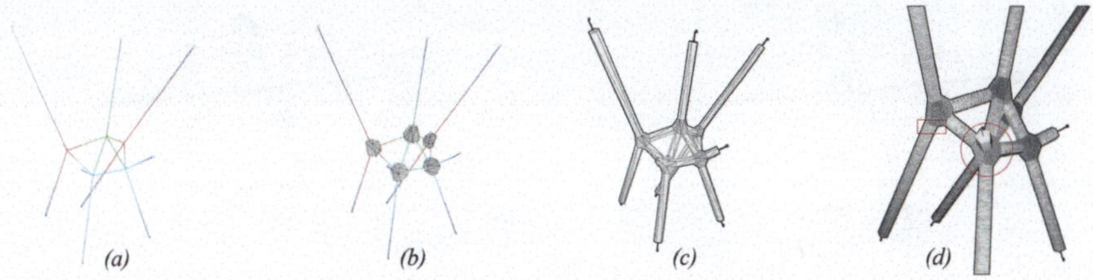

Fig. 2. (a–c) Generating convex hulls around nodes and bars of a graph, (c) contouring the combined mesh. Alignment and interface problems with task graphs.

one print layer to another, naturally specify interior task graphs (printing paths) that add stability to the print and more importantly directly model the task graphs and their evolution in time (Fig. 3).

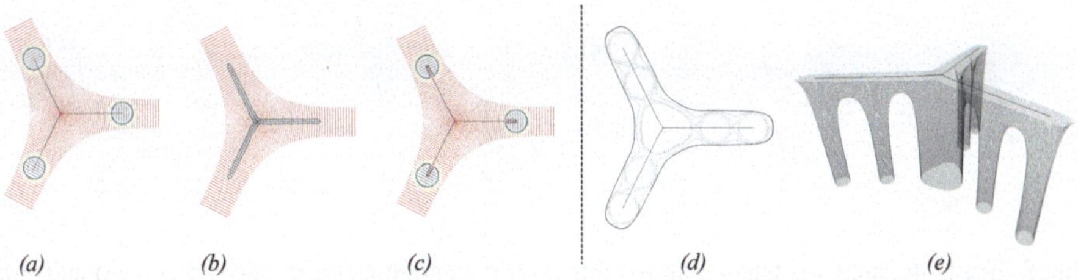

Fig. 3. (a) and (b) two scalar fields. (c) set-theoretic difference operation of (b) on (a). (d) the natural extension of specifying interior task graphs as contours of a scalar field, (e) Natural handling of topological change of many closed curves at the bottom layer to a singular closed curve on the top layer.

3.2 Primitive Field Functions and Operations

There are wide varieties of field functions that have been developed to adapt the use of Function Representation within various application domains. We currently use two simple-to-implement, two-dimensional SDF functions and their booleans, blends and evolution along a normal direction to generate the task graphs. The first field function is the so-called soft-objects field function (Wyvill et al. 1986), which is specified in relation to distance of field points from the nodes of an input graph (Fig. 4(a)). The other field function uses the same function, albeit in relation to the shortest distance from the field point to each of the edges of a 2D input graph (Fig. 4(b)). These simple functions may be replaced by more involved convolution field functions that improve the short-comings of these simple functions (Sherstyuk 1999b; Wyvill and Wyvill 1989).

Fig. 4. (a) Signed distance fields (SDF) in relation to (a) nodes and (b) edges of a skeletal graph.

Booleans

We use standard set-theoretic definitions to specify the Union, Difference and Intersection between two scalar fields (Bloomenthal and Bajaj 1997) (Fig. 5(a–e)).

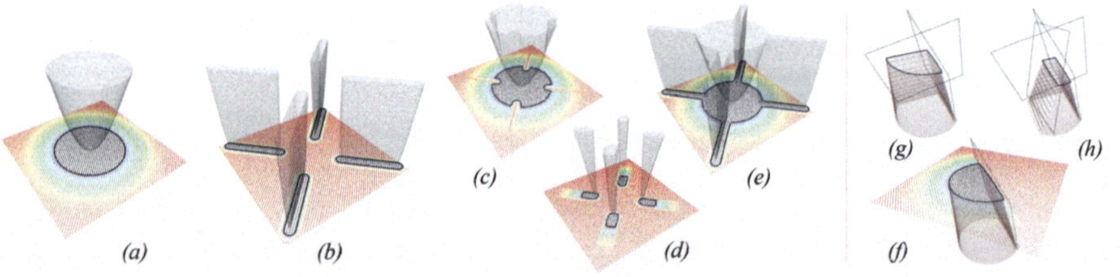

Fig. 5. (c) Difference, (e) Union and (d) Intersection operations on primitive fields (a) and (b). (f) Consequence of trim operation on the scalar field. (h) (g) trim operations can be successively applied.

Trim with Plane

When we use a trim operation using an input plane, we replace the field values of all field points that are on the negative side of the plane (w.r.t its normal) with a value of 1 (Fig. 5(f–h)). In other words, all points on the negative side of the plane are considered outside the object. We use this operation primarily to aid the assembly of printed parts or interface between parts that are consecutively printed (while the first part is still wet).

Blend

We currently use a simple linear additive interpolation function between two given scalar fields to specify a blend (Fig. 6(a and b)). It is common to use the so-called smooth step function to improve the behaviour of the blend at the extremes.

3.3 Processing Nodes

Each of the nodes is printed separately. However, the parts of every node are printed consecutively whilst the previous part is still wet, enabling a material bond across the node. In terms of the processing of the node, this does not pose any additional difficulty.

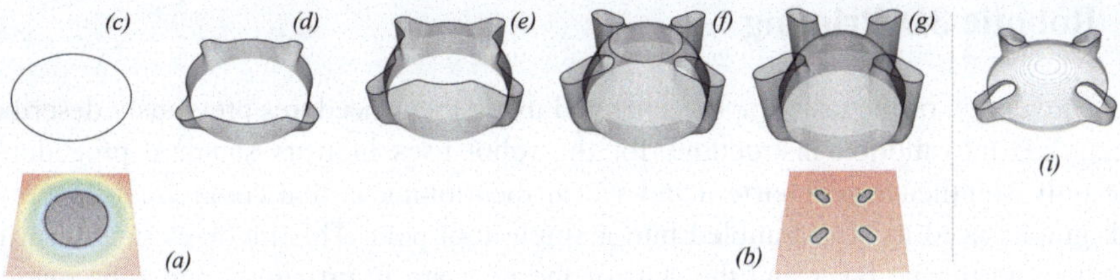

Fig. 6. (c–g) Blend sequence of the zero contour as scalar field (a) blends to scalar field (b). The effect of a faster blend rate on the shape of the geometry described by the stack of zero-contours is shown in (i).

The simplest procedure to produce the task graphs is by translating a circular front along the edge, and simultaneously trimming against the bisector planes defined between pairs of edges (Fig. 7(a)). In other words, in this simple method, the front does not really evolve.

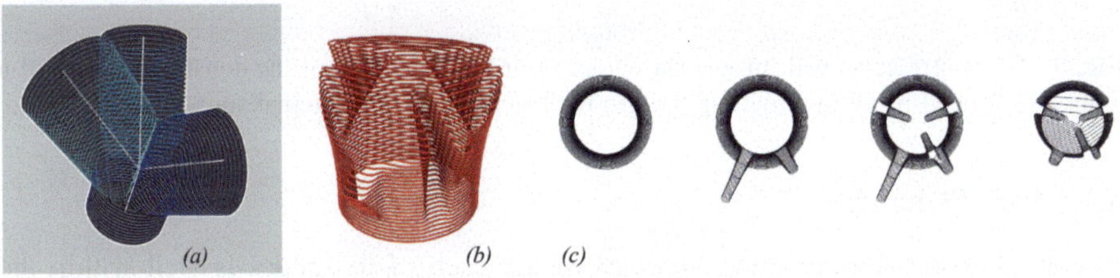

Fig. 7. A nominal procedure of translate and trim to synthesise task graphs for each node segment. (b) Task graphs of one of the four parts meeting at node shown in Fig. 10.

In most practical cases on the other hand, the moving front needs to evolve features to ease steep overhangs and to provide internal support (Fig. 7(b, c)). In such cases, the evolving front utilises the full range of operations: booleans, blends and trims (Fig. 8 (a)). The input primitive fields for these operations are generated by first extracting an oriented sub-graph from edges that meet at a node. This sub-graph is projected onto the ground-plane to get a 2D skeletal graph, which is subsequently used to generate the primitive fields (Fig. 8(b, c)).

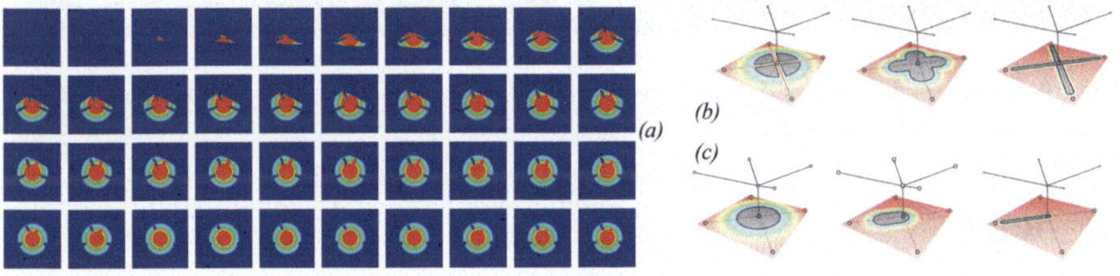

Fig. 8. (a) Secondary composite primitives formed by Boolean operations on primary primitives (b). (b) Primary primitive scalar fields generated as SDFs in relation to the vertices and edges.

4 Robotic 3D Printing

The conversion of the task graphs generated using the procedures previously described (Sect. 3.3) into motion instructions for the robot uses industry-standard procedures. The only significant geometric detail worth mentioning is that consecutive layers of task graphs need to be assembled into a single tool path. This involves first ensuring that the end of one layer and the start of the next are in proximity and subsequently adding a segment that will carry the print head smoothly across layers (Fig. 9).

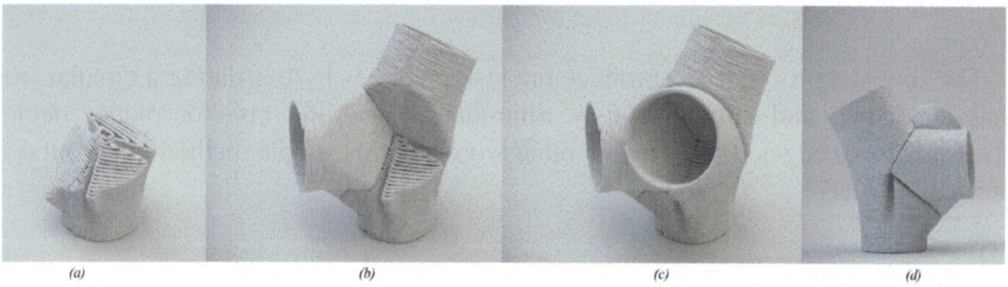

Fig. 9. Computer generated images show (a–c) printing sequence of the constituent parts of a node and (b) end result highlighting the groove feature that adds internal support to the print.

4.1 Path Planning

In current work, the bounding volumes of the each of the nodes is well within the operating sphere of the robot used. Further, the dimensions of the print head and apparatus in relation to the size of the print are also such that the need for extensive collision avoiding path planning is not necessary. However, we simulate and visually inspect the entire print trajectory of the robot along with the print head to ensure there are no collisions, particularly of the print head with the ground (Fig. 10).

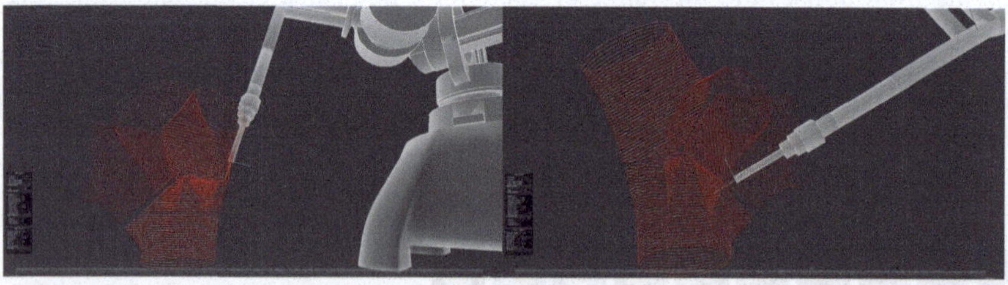

Fig. 10. Print trajectories and visual inspection of potential collision of print head with the already printed parts and/or the ground

4.2 Results

All results use C25 concrete (Fig. 11). The print duration for each of the prints is approximately 30 min and each segment of the nodes typically consists of 90–100

Fig. 11. Photographs showing materially bonded node parts (a, b) and the various types of nodes that vary in valence and angles between incident edges (c).

layers of 4 mm each. The diameter of each of the node segments is approximately 15 cm.

5 Future Work

The current proof-of-concept implementation of FRep for the shape design and its processing for fabrication of robotic 3D Printed concrete demonstrates the novelty and utility of the proposed framework. However, we envisage the following significant avenues of further investigation in the immediate future. Each of the avenues, we believe, are critical to the full-scale realisation of spatial structures: (1) Evolution of the front subject to constraints to ensure the shapes are guaranteed to be stable during print (Bhooshan et al. 2018). Current work circumvents this constraint by using very conservative layer height and overhangs; (2) Automated, structurally informed segmentation of skeletons larger than the print volume. In this regard, the formulation and modelling of equilibrium of rigid-block assemblies would be an important consideration (Frick et al. 2015); (3) Automatic insertion and growth of brackets at nodes that exceed allowed overhang angle; (4) Automatic inclusion of features to enable and ease fixings.

6 Conclusions

The current shape-design and physical results, whilst being early, already demonstrate the following: (1) Adaptation of skeleton-based FRep modelling for shape design for robotic 3D printing of concrete; (2) The subsequent adaptation of FRep to generate and represent tool-paths in robotic 3D printing; (3) First proof-of-concept implementation, opening further possibilities of adapting more sophisticated features from FRep and initial value formulation of front propagation. In particular, the mathematical framework of front propagation that can handle constrained evolution of the front, can be expected to be essential in handling several process and stability related constraints; (4) Potential for translating the proposed method as novel and alternative method of generating geometries of unreinforced, masonry.

In conclusion, the proposed use of a FRep shape representation schema as an alternative to ubiquitous BRep schemes is already providing novel results whilst easing fabrication-related post-production. We believe that this initial work will provide a strong foundation for exploring the design aspects of large-scale concrete printing, the unified representation of design and fabrication related parameters and thus the development of a novel architectural language of concrete extrusion.

References

Bhooshan, S., Van Mele, T., Block, P.: Equilibrium-aware shape design for concrete printing. In: De Rycke, K., Gengnagel, C., Baverel, O., Burry, J., Mueller, C., Nguyen, M.M., Rahm, P., Thomsen, M.R. (eds.) Humanizing Digital Reality, Design Modelling Symposium 2017, pp. 493–508. Springer, Singapore (2017)

Bloomenthal, J., Bajaj, C.: Introduction to Implicit Surfaces. Morgan Kaufmann, San Francisco (1997)

Bloomenthal, J., Shoemake, K.: Convolution surfaces. In: Proceedings of the 18th Annual Conference on Computer Graphics and Interactive Techniques (ACM SIGGRAPH 1991), pp. 251–256. Las Vegas (1991)

Cacace, S., Cristiani, E., Rocchi, L.: A level set based method for fixing overhangs in 3D printing. Appl. Math. Model. **44**, 446–455 (2017)

Cohen, L.D.: On active contour models and balloons. CVGIP Image Underst. **53**, 211–218 (1991)

Desbrun, M., Gascuel, M.-P.: Animating soft substances with implicit surfaces. In: Proceedings of the 22nd Annual Conference on Computer Graphics and Interactive Techniques (ACM SIGGRAPH 1995), pp. 287–290. Los Angeles, CA (1995)

Frick, U., Van Mele, T., Block, P.: Decomposing three-dimensional shapes into self-supporting, discrete-element assemblies. In: Thomsen, M., Tamke, M., Gengnagel, C., Faircloth, B., Scheurer, F. (eds.) Modelling Behaviour, Design Modelling Symposium 2015, pp. 187–201. Springer, Cham (2015)

Hubert, E., Cani, M.-P.: Convolution surfaces based on polygonal curve skeletons. J. Symb. Comput. **47**, 680–699 (2012)

Ji, Z., Liu, L., Wang, Y.: B-Mesh: a modeling system for base meshes of 3D articulated shapes. In: Computer Graphics Forum, pp. 2169–2177. Wiley Online Library (2010)

Kass, M., Witkin, A., Terzopoulos, D.: Snakes: active contour models. Int. J. Comput. Vis. **1**, 321–331 (1988)

Keeter, M.: Hierarchical volumetric object representations for digital fabrication workflows. In: Proceedings of the 40th Annual Conference on Computer Graphics and Interactive Techniques (ACM SIGGRAPH 2013), Poster, Anaheim, CA (2013)

Lu, L., Sharf, A., Zhao, H., Wei, Y., Fan, Q., Chen, X., Savoye, Y., Tu, C., Cohen-Or, D., Chen, B.: Build-to-last: strength to weight 3D printed objects. ACM Trans. Graph. **33**, 97 (2014)

Monreal, A.: T-Norms, T-Conorms, aggregation operators and Gaudí's columns.In: Seising, R., González, V.S. (eds.) Soft Computing in Humanities and Social Sciences, pp. 497–515. Springer, Berlin, Heidelberg (2012)

Pasko, A., Adzhiev, V.: Function-based shape modeling: mathematical framework and specialized language. In: Winkler, F. (ed.) ADG 2002: 4th International Workshop on Automated Deduction in Geometry, Hagenberg Castle, Austria, pp. 132–160. Springer, Heidelberg (2002)

Perugini, P., Andreani, S.: Pier Luigi Nervi's columns: flow of lines and forces. J. Int. Assoc. Shell Spat. Struct. **54**, 137–148 (2013)
Popescu, M., Rippmann, M., Van Mele, T., Philippe, B.: Complex concrete casting: knitting stay-in-place formwork. In: Proceedings of the IASS 2016 Annual International Symposium: Spatial Structures in the 21st Century, Tokyo, p. 1278 (2016)
Rian, I.M., Sassone, M.: Tree-inspired dendriforms and fractal-like branching structures in architecture: a brief historical overview. Front. Archit. Res. **3**, 298–323 (2014)
Rossignac, J.: Blending and offsetting solid models Ph.D. thesis, University of Rochester (1985)
Sethian, J.: Fast marching methods and level set methods for propagating interfaces. In: Comput. Fluid Dyn. Annu. Lect. Ser. 29th, Rhode-Saint-Genese, Belgium (1998)
Sherstyuk, A.: Interactive shape design with convolution surfaces. In: Proceedings of the International Conference on Shape Modeling and Applications, Shape Modeling International 1999, pp. 56–65. Aizu-Wakamatsu, Japan (1999a)
Sherstyuk, A.: Kernel functions in convolution surfaces: a comparative analysis. Vis. Comput. **15**, 171–182 (1999)
West, M.: Fabric-formed concrete structures. In: Proceedings of First International Conference on Concrete and Development, pp. 133–142. Tehran, Iran (2001)
West, M.: The Fabric Formwork Book: Methods for Building New Architectural and Structural Forms in Concrete. Routledge, Abington, OX (2016)
Wyvill, B., Wyvill, G.: Field functions for implicit surfaces. Vis. Comput. **5**, 75–82 (1989)
Wyvill, G., McPheeters, C., Wyvill, B.: Soft objects. In: Kunii, T. (ed.) Advanced Computer Graphics: Proceedings of Computer Graphics Tokyo 1886, pp. 113–128. Springer, Tokyo (1986)
XtreeE (2017). Post in Aix-en-Provence. http://www.xtreee.eu/post-in-aix-en-provence/. Accessed 3 Sept 2018
Zastavni, D.: The structural design of Maillart's Chiasso Shed (1924): a graphic procedure. Struct. Eng. Int. **18**, 247–252 (2008)

Material and Processes

ABB

Thermally Informed Robotic Topologies: Profile-3D-Printing for the Robotic Construction of Concrete Panels, Thermally Tuned Through High Resolution Surface Geometry

Joshua Bard(✉), Dana Cupkova, Newell Washburn, and Garth Zeglin

Carnegie Mellon University, Pittsburgh, PA 15213, USA
{jdbard, cupkova, washburn, garthz}@cmu.edu

Abstract. This paper explores the thermal design and robotic construction of high-performance building components. The complex surface geometry of these components actuate specific thermal behavior in passive building systems through implementing the principles of convection in thermal mass. Our seamless design-to-fabrication workflow uses optimization methods that combine measured thermal data and simulation feedback with advanced modeling and emerging robotic manufacturing techniques. Bridging an understanding of thermal performance, geometry, and manufacturing we suggest direct formal relationships between the behavior of airflow to tool-path planning for a robotic arm. This paper will focus on describing an experimental process we term Profile-3D-Printing that demonstrates a novel approach to the construction of concrete panels with complex surface geometries. This hybrid construction method combines material deposition with tooled post-processing to achieve high-resolution surface definition. The process entails automated delivery of material for selective deposition of panel geometry, and tooled shaping of rough and finish layers for the physical production of computationally generated forms.

Keywords: Additive manufacturing · High-Performance design
Digital concrete · Thermal performance · Robotic fabrication

1 Introduction

Thermally Informed Robotic Topologies is a research project that explores the intersection of high-performance building design and advanced robotic manufacturing. This paper describes Profile-3D-Printing, a novel robotic fabrication process developed by our research team at Carnegie Mellon University. This hybrid additive/subtractive process combines deposition of concrete for rough layup with precision tooling for surface finishing. The motivation for inventing this technique is rooted in previous research by the authors (detailed in the Sects. 1.1, 1.2) regarding the design of geometrically actuated high-performance building components and the robotic manipulation of soft materials for additive manufacturing. In validating the fabrication approach

J. Willmann et al. (Eds.): ROBARCH 2018, *Robotic Fabrication in Architecture, Art and Design 2018*, pp. 113–125, 2019.
https://doi.org/10.1007/978-3-319-92294-2_9

we champion forward design for building performance—in contrast to reverse engineering after initial design—through the design and fabrication of an array of concrete façade panels, thermally tuned to passively reduce heating and cooling loads through precise definition of surface geometry. The robust implementation of our Profile-3d-Printing method could positively impact not just the range of creative expression in architectural form making, but also the ecological footprint of the built environment at large.

1.1 High Performance Building Component Design and Advanced Manufacturing

This project builds on research exploring the seamless design of high-performance, passive building systems that are rooted in the combination of form and matter: topologically complex shapes using engineered materials. The authors have previously demonstrated that these components can be designed as passively activated thermal mass panels, optimized for local climate and tuned to specific heat transfer coefficients through the articulation of surface geometry [1]. We have argued that by using the passive behavior of thermal mass more aggressively it is possible to manipulate the surface geometry [2] of architectural elements to strategically tune the thermal performance of building envelopes to specific climate while reducing heating and cooling loads [3]. However, such surface geometry results in complex form design that requires bespoke mold based manufacturing procedures.

Current building practices struggle with high levels of building component customization, because they heavily rely on mold based production and subtractive milling (Fig. 1). Mold based fabrication methods limit the diversity of construction due to the economics of mold reuse and material waste. Thus, current building practices exclude the greater formal range often required for high-performance building components that are both site and solar-geometry specific. High-performance building design puts pressure on the building industry to develop new methods for the automated manufacturing of custom components.

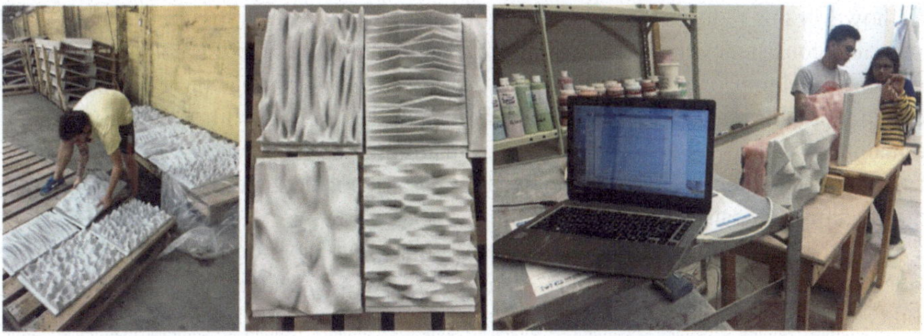

Fig. 1. (left) Prototyping and (right) thermal testing of mold-cast UHPC panels with complex surface geometry.

1.2 Additive Manufacturing for Construction

Additive Manufacturing (AM) techniques afford geometric complexity across scales, the possibility for customization, and the elimination or reduction of costly molds [4]. These technologies applied to construction (Additive Manufacturing for Construction, AMc) hold promise for the manufacture of high-performance building components, but are not yet ready for large scale industrial production. Over the past two decades AMc research has addressed some of the barriers to entry for architectural 3D printing in terms of build volume, production time, and economically viable mix design [5]. Soar and Andreen, early pioneers in AMc argue that if AMc is to have a larger impact on the construction industry, "It must realize greater performance than free-form aesthetics alone. It will enable design freedom combined with greater function, and it must address sustainability head-on" [5]. *Thermally Informed Robotic Topologies* addresses these research challenges with particular emphasis on the creation of novel robotic construction processes and simulation-driven design tools.

Our team at Carnegie Mellon University has developed and tested a hybridized additive manufacturing system for high-performance, modular concrete building components. The workflow is a complete path from modeling and simulation of thermally efficient surface geometries to the robotic deposition and finish tooling of concrete facade panels. The work explores the linkages between the geometric affordances of high-performance panel design and the kinematic freedoms of robotic motion and tooling. This connection between forward design for performance and robotic construction suggests a viable model for addressing the challenges related to the demand for customization of high-performance building components.

2 Methodology

2.1 Mix Design

As part of producing an initial concrete mix for application in a tooled-deposition process, we considered the three factors of workability, strength and aggregate size. It is known that lower water-to-cement ratios (w/c) increase the ultimate strength of concrete [6], however this reduces the workability of concrete. Workability is a measure of yield stress and consistency that is difficult to measure quantitatively, but is usually evaluated in terms of slump or slump spread [7]. For Profile-3D-Printing, the concrete has to remain workable enough for effective and efficient deposition, while being stiff enough to maintain its shape after being troweled. Due to this factor, concrete sand with a large percentage of aggregate fines was selected for easy deposition and smooth incorporation into the concrete panels through troweling.

In order to reduce the w/c ratio while improving workability, admixtures were added to the concrete mixture. One such type of chemical admixture to obtain workability at low w/c ratios are superplasticizers (SP). During the 1980's a copolymer of polymethacrylic acid and polyethylene glycol methacrylate (PEG) formed polycarboxylate ether (PCE) superplasticizers. These became one of the most widely used admixtures due to their ability to lower yield stress at low dosages in low w/c ratio cement [8, 9]. For our mix, a commercial grade superplasticizer was obtained from

BASF. We also included a viscosity modifying agent (VMA), used to both reduce segregation and bleeding along with modifying flow properties [10]. A commercial VMA was also obtained from BASF and added to allow a thicker concrete for more controlled flow and to not allow separation to occur between the aggregate and water.

Type I/II Saylor's cement was obtained and used for testing and ordinary tap water was used. The table below (Fig. 2) shows the mix for a typical optimized batch:

Batch Mix	
Material	**Amount**
Aggregate (kg)	10
Cement (kg)	5
Water (kg)	2
SP (ml)	32
VMA (ml)	45

Fig. 2. (Left) Table: proportions of materials for a batch mixture, (Right) early mix tests and finished panel prototype.

At these proportions a 0.4 w/c ratio was used to tune workability. Both the superplasticizer and viscosity modifier were added in ratios consistent with BASF recommendations, and a delayed addition of SP and VMA was utilized. Addition of 10% of the total water content was followed by one minute of mixing until the rest of the water and additives were added and mixing was completed.

2.2 Component Design and Simulation

Our approach to high-performance passive thermal mass design is based in designing specific mass to surface area ratios for each panel. This requires variation of distributed mass depth across the surface, combined with control of surface smoothness or roughness that results in varied heat absorption and re-radiation rates [1]. Our team's implementation of Profile-3D-Printing serves as a proof of concept in a larger effort to explore a highly customizable design space that links performance based design to large scale additive manufacturing. In contrast to the common design-to-construction pipeline which relies on reverse engineering of design intent to meet performance and fabrication criteria, our project stresses forward design for performance and fabrication. This approach requires the incorporation of computational methods to integrate simulation and construction planning with generative design tools.

The principles for modeling the geometry to produce controlled thermal delay in thermal mass have been described by the authors in previous publications [1, 3, 11]. To summarize this approach, we used genetic algorithms to optimize the key geometric constraints of surface geometry related to area to mass distribution that directly affect thermal performance. Critical factors are the ratio between panel mass and surface area; smoothness of surface features varied through curvature degree of drive geometry; and increase of surface area over identical mass through point-vector displacement constrained to u,v subdivision of base surface. Through multiple rounds of physical testing and thermal simulation we identified a set of modeling principles that rely on curvature degree, scale, and orientation of surface pattern to delay or speed up heat re-radiation. As a result we were able delay heat re-radiation of 35 °C by 75 min within 3 h. [1] This is a significant thermal effect that argues for architecture and manufacturing that is tightly tuned to its thermodynamic design framework, while utilizing logics of computational geometry.

Thermal performance can be generalized through the difference between surface geometry and airflow geometry. Thermal pockets and their size play a significant role in thermal behavior. The visualization of thermal pockets and their relationship to curvature degree is shown in Fig. 3. The link between the modeling of performance based geometries and the possibility for robotic construction was natural in the case of thermally tuned building components with thermal pockets. Panel geometries produced by Profile-3D-Printing are a combination of tool motion and profile knife geometry.

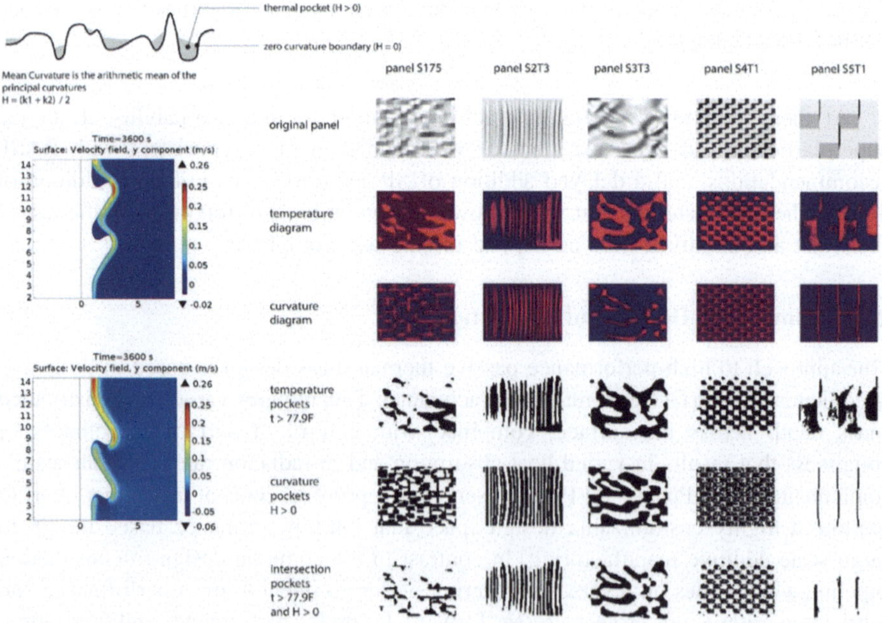

Fig. 3. Surface analysis and simulation of thermal behavior actuated by surface geometry. The difference between the geometry of airflow and surface geometry can be analyzed through mean curvature maps.

Robot tool paths can be directly translated from target surface contours and custom tool profiles can be directly back-engineered from surfaces generated based on thermal principles. While the direct link between thermal simulation and robotic fabrication is not yet fully automated, tool paths and tool profiles that produce surface geometry based in principles of thermal performance were tested extensively.

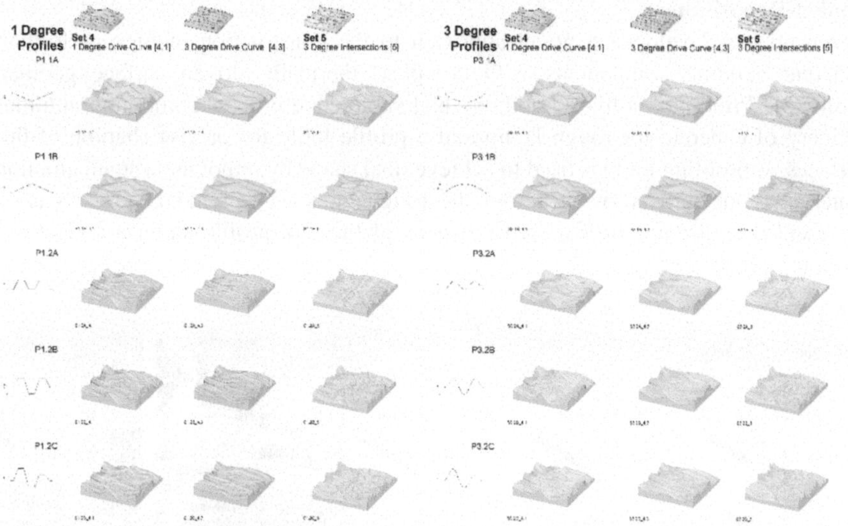

Fig. 4. Matrix of test panels exploring relationship of UV based generation and Profile-3d-Printing fabrication approach.

Based on this rubric we generated a simulation test set to validate the concept of robotic surface figuration guided by performance logic. Design modeling of surface geometry using the UV constraints described above leads directly to toolpath and tool-profile generation—the combination of which determines panel figuration. Our team captured a matrix of recorded simulations achievable with this fabrication approach (Fig. 4). As a set, the simulations explore performative parameters of the system related to size, depth and feature slope. A subset of these simulations were physically fabricated using Profile-3D-Printing and are discussed in the results section below.

2.3 Fabrication Approach

Current AMc approaches are not well-positioned to address the unique set of challenges inherent to constructing high-performance building components with thermally actuated surface geometries. All AMc must overcome the difficulties of building large objects at an economically viable time-scale. This typically entails a course contact zone—combining higher material delivery rates, larger step sizes, and expanded surface contact where finish tooling is involved. Larger contact zones result in a lack of resolution where global form can be approximated, but surface features are not always possible. Thermally actuated surfaces, however, often require a high ratio of surface features to overall surface area. Traditional AM approaches address resolution through fine contact zones, sacrificing build time. Fabrication approaches like Profile-3D-Printing address these competing constraints of high volume output and fine surface feature resolution and could be broadly applicable to high-performance component construction requiring surface detail.

Tooled-Deposition

Our team has developed a novel approach to the construction of high-performance concrete building components with complex, thermally driven surface geometry. Profile-3D-Printing is a hybrid additive and subtractive process combining automated delivery of concrete for rough layup and a profile knife for precise shaping of finish surfaces. A profiling knife is used to achieve final shape by removing a small amount of material as the tool is swept across the rough layup. The removal process can be represented as the union of a set of sweeps of the tool profile along a collection of motion paths (Fig. 5).

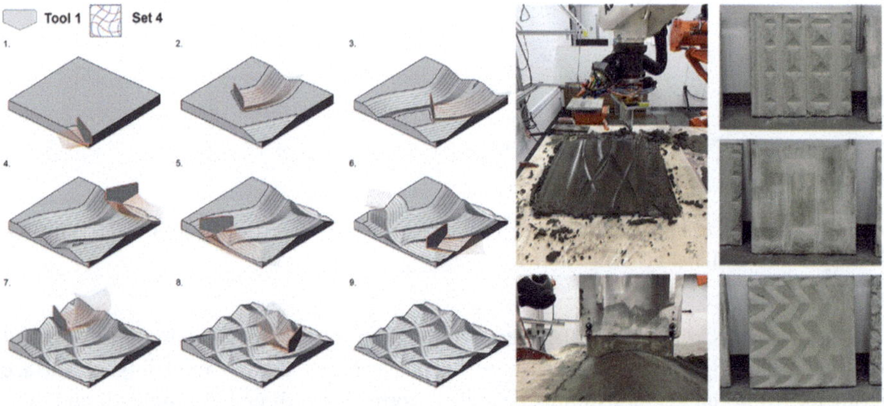

Fig. 5. (left) Trowel toolpath for surface geometry. (right) Sample concrete panel prototypes built with robotic work cell.

This approach is closely related to traditional running moulds and benefits from the high level of surface detail and variation common to these techniques [12]. Unlike traditional running moulds however, Profile-3D-Printing incorporates the kinematic freedom and repeatability of robotic motion planning. The profile can be swept with 6 DoF and multiple sweeps can intersect each other to aggregate over larger areas. For architectural building components, like the facade panels demonstrated by the authors, the process results in seamless construction with near perfect fit to target geometry.

Troweling vs. Profiling

Profile-3D-Printing is similar to but distinct from Contour Crafting, a pioneering technique in the AMc domain combining mechanized extrusion and surface troweling [13]. Both processes apply bulk material along precise computed paths using robotic motion control and subsequently shape it using a profiling tool. However, Contour Crafting exclusively uses a troweling tool in a single deposition-and-shaping pass with a single trajectory, and consequently contacts the surface only once. Profile-3D-Printing uses multiple tooling passes to both decouple the deposition and shaping trajectories and allow iterative refinement of the tooled surface.

The contrast between troweling and profiling plastic materials is stark when considered through the lens of AMc resolution. With troweling approaches similar to Contour Crafting there is a near one-to-one correspondence between extrusion step height and tool contact, where the smoothing action of a straight trowel is used to refine global shape resulting from layered build-up. As a result seams between layers are often visible and cold joints between layers can be an issue [14]. In contrast Profile Printing uses a profiling technique to shape fresh concrete that can be rapidly deposited. For any given tool step, the knife profile can be shaped with smaller features than a trowel. As evidenced by historical molding profiles, profile knives can be fairly intricate. In the case of concrete fabrication the limit for profile detail is constrained by the aggregate size used in the mix design. Thus given tools of equal width, a profile knife can render significantly finer features than a trowel with the same path of travel (Fig. 6). Another affordance of Profile-3D-Printing is that tooling and deposition can be asynchronous. This allows the profile knife to revisit areas to improve finish quality and for greater freedom in motion planning for panel form generation. All of these characteristics combine to make Profile-3D-Printing high-output and high-resolution.

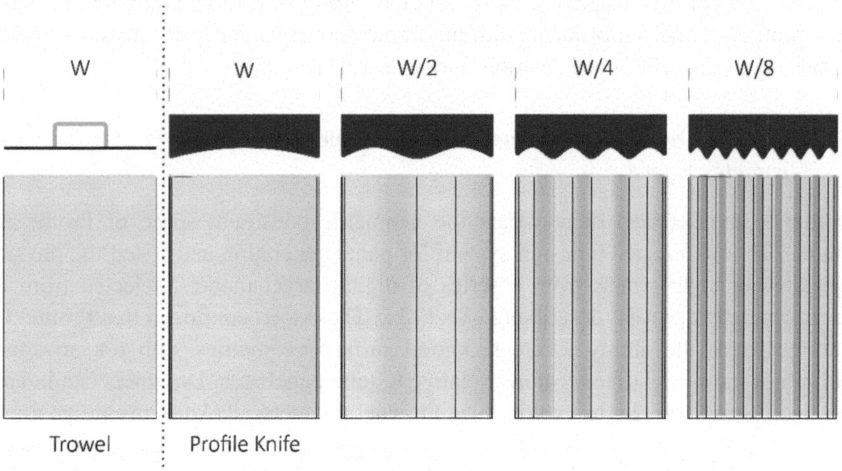

Fig. 6. (left) Profile knives can have significantly more features than a trowel of the identical width.

3 Results

3.1 Workcell Development

Our team has prototyped a robotic Profile-3D-Printing workcell using an ABB IRB 6640 robotic arm. Custom end-of-arm tooling includes an aluminum tool chassis to receive manually interchangeable profiling plates. The 175 mm wide profiling plates are waterjet cut from 1.5 mm thick stainless steel. A custom 3D printed nozzle mounted in front of the profiling knife produces even material distribution across the width of the blade for concrete delivery. Concrete material is delivered from a pressurized canister (5-10 psi) mounted off-arm. Material flow is controlled via an inflatable bladder located in the delivery nozzle. Canister pressure and bladder inflation are actuated using digital outputs from the ABB robot controller. Material flow rate at peak delivery is roughly 15 L/min with a maximum tool traverse speed of 30 mm/s. Depending on overall panel thickness and edge geometry reusable formwork can be used to support boundary walls. Layup without perimeter formwork is possible but requires robotically trimming panel edges after finish shaping.

3.2 Panel Test Series 1: Testing for Process Constraints

After initial setup and verification of the Profile-3D-Printing process our team conducted empirical tests on eight different test panel geometries. The goal of the physical prototyping was to develop heuristics for estimating material behavior and choosing tooling constraints. These constraints impact overall fabrication quality and directly inform the design space of potential surface geometries. Key parameters of the process related to finish quality and panel design include: maximum depth of panel per build (determined by toolZ translation of profile knife), maximum slope of surface features (determined by profile shape and toolX rotation during travel), and maximum rotation of the profiling knife out of plane with the frame perpendicular to the direction of tool motion (determined by toolZ rotation during travel) (Fig. 7).

3.3 Panel Test Series 2: Demonstration of Tooled Deposition Geometry Space

In order to explore and demonstrate the geometric constraint space of Profile-3D-Printing our team created a design system for panel generation and tested the physical fabrication of eight panels from a series of digital target models, selected from the design simulation process described in Sect. 2.2. These tests confirmed that Profile-3D-Printing affords the ability to create varied surface geometries with fine to coarse resolution control of surface features relative to total panel area. Designers can interact with two primary drivers to explore overall panel geometry: 1. knife profile geometry and 2. toolpath controlled through robot motion planning (Fig. 8). The combinatorial results of these two drivers creates a robust space for design and thermal performance.

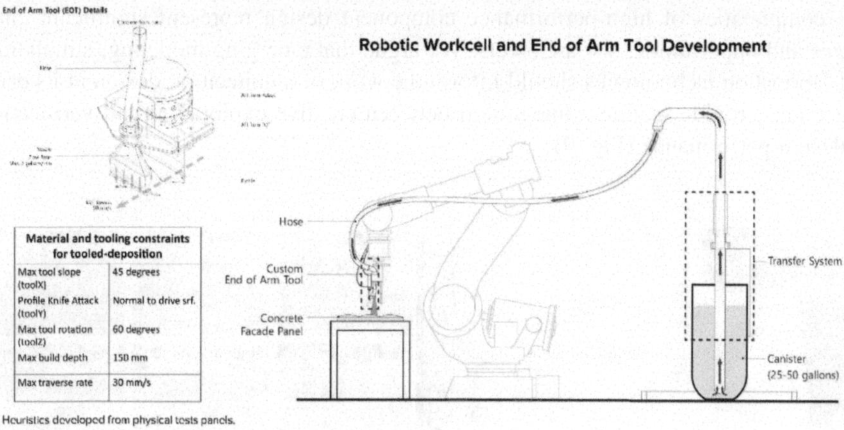

Fig. 7. Heuristic material and tooling constraints garnered from empirical testing of panel creation.

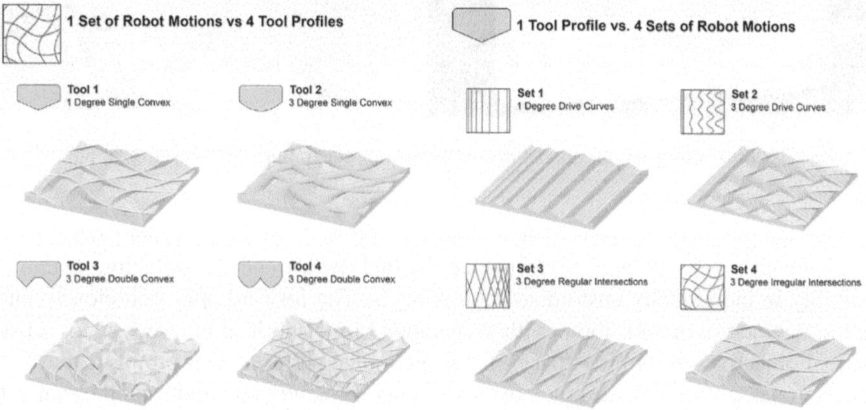

Fig. 8. Large geometric constraint space combining two main parameters of robotic fabrication process: tool profile and toolpath.

4 Conclusion

Thermally Informed Robotic Topologies validated a design to fabrication pipeline enabling forward design for performance and fabrication. Our team tested methods of form generation optimized for thermal performance and robotic fabrication. In doing so, we developed a novel approach to Additive Manufacturing for Construction (AMc) using Profile-3D-Printing. The key benefit of this approach is the combination of high-volume material deposition and multi-resolution feature definition without significant increases in production time. AMc approaches like Profile-3D-Printing are important as high-performance design of architectural building components drives the need for advanced manufacturing techniques with high levels of customization.

The complexities of high-performance component design represent significant challenges and opportunities for architects. We argue that emerging modeling, simulation, and fabrication technologies should inform the work of architectural design at its early stages and provide seamless integration between creative exploration and verification of design performance (Fig. 9).

Fig. 9. Tooling process and finished panel demonstrating Profile-3D-Printing.

Our central hypothesis is that a forward approach to facade panel design can simultaneously achieve high-performance thermal design and cost-effective manufacturability in individually customized components. The forward approach closely integrates geometric optimization with a parameterized fabrication process to satisfy thermal objectives within the space of feasible tooled surfaces. Our prototyped system demonstrates a forward design workflow using geometrically induced thermal rate coefficient values and the novel Profile-3D-Printing robotic fabrication process. Our experiments to date have produced concrete panels with high geometric tolerance and fine surface finish within acceptable fabrication costs. We believe this represents a scalable solution with strong potential for future commercial production.

We are still actively engaged in furthering this research and are currently working to develop a number of key projects components. These include: (a) developing and testing design toolkits that automate the interaction of performance simulation - currently explored through machine learning, (b) fabrication path planning, and tool design with a design friendly user interface; (c) optimizing the Profile-3D-Printing process with new end of arm tooling for better control of material layup and a higher capacity material delivery system; (d) and constructing a full scale wall mockup for thermal testing and process verification.

References

1. Cupkova, D., Promoppatum, P.: Modulating thermal mass behavior through surface figuration. In: Nagakura, T., Tibbits, S., Mueller, C., Ibañez, M. (eds.) Acadia 2017: Disciplines & Disruption, Proceedings of the 37th Annual Conference of the Association for Computer Aided Design in Architecture, pp. 202–211. Cambridge, MA. (2017)
2. Yao, L.-S.: Natural convection along a vertical complex wavy surface. Int. J. Heat Mass Transf. **49**, 281–286 (2006)
3. Cupkova, D., Azel, N.: Mass Regimes: geometric actuation of thermal behavior. Int. J. Architectural Comput. **13**(2), 169–193 (2015). https://doi.org/10.1260/1478-0771.13.2.169
4. Lim, S., Buswell, R.A., Le, T.T., Austin, S.A., Gibb, A.G.F., Thorpe, T.: Developments in construction-scale additive manufacturing processes. Autom. Const. **21**(1), 262–268 (2012). https://doi.org/10.1016/j.autcon.2011.06.010
5. Soar, R., Andreen, D.: The role of additive manufacturing and physiomimetic computational design for digital construction. Architectural Des. **82**(2), 126–135 (2012). https://doi.org/10.1002/ad.1389
6. Mehta, P., Monteiro, P.J.: Concrete: Microstructure, Properties and Materials. Mc-Graw Hill, New York (2006)
7. Marar, K., Eren, Ö.: Effect of cement content and water/cement ratio on fresh concrete properties without admixtures. Int. J. Phys. Sci. **6**(24), 5752–5765 (2011)
8. Marchon, D., Juilland, P., Gallucci, E., Frunz, L., Flatt, R.J.: Molecular and submolecular scale effects of comb-copolymers on tri-calcium silicate reactivity: toward molecular design. J. Am. Ceram. Soc. **100**(3), 817–841 (2017)
9. Yamada, K., Ogawa, S., Hanehara, S.: Controlling of the adsorption and dispersing force of polycarboxylate-type superplasticizer by sulfate ion concentration in aqueous phase. Cem. Concr. Res. **31**(3), 375–383 (2001)
10. Leemann, A., Winnefeld, F.: The effect of viscosity modifying agents on mortar and concrete. Cement Concr. Compos. **29**(5), 341–349 (2007)
11. Cupkova, D., Yao, S. C., & Azel, N.: Morphologically controlled thermal rate of ultra high performance concrete. In: Sabin, J.E., PazGutierrez, M., Santangelo, C. (eds.) Proceedings of the Adaptive Architecture and Programmable Matter Conference - Next Generation Building Skins and Systems from Nano to Macro, San Francisco, CA (2015) https://doi.org/10.1557/opl.2015.569
12. Bard, J., Mankouche, S., Schulte, M.: Morphfaux. In: Brell-Çokcan, S., Braumann, J. (eds.) ROB|ARCH 2012: Robotic Fabrication in Architecture, Art and Design, pp. 138–141. Springer, Vienna (2013). https://doi.org/10.1007/978-3-7091-1465-0_13
13. Khoshnevis, B.: Automated construction by contour crafting - related robotics and information technologies. Autom. Constr. **13**, 5–19 (2004). https://doi.org/10.1016/j.autcon.2003.08.012
14. Wangler, T., Lloret, E., Reiter, L., Hack, N., Gramazio, F., Kohler, M., Bernhard, M., Dillenburger, B., Buchli, J., Roussel, N., Flatt, R.: Digital concrete: opportunities and challenges. RILEM Tech. Lett. **1**, 67 (2016). https://doi.org/10.21809/rilemtechlett.2016.16

Hold Up: Machine Delay in Architectural Design

Zach Cohen(✉)

Massachusetts Institute of Technology, Cambridge, MA 02139, USA
zdacohen@mit.edu

Abstract. This paper introduces an architectural design approach that is founded on working with digital fabrication machines, materials, and time. This approach is called *Machine Delay Fabrication*. Machine Delay Fabrication is contextualized within the lineage of productive delays using two examples from other creative disciplines. It is then contrasted with the "real-time" approach to digital fabrication research and practice. The concept of *machine delay* is outlined along with the three types of machine delay that are pertinent to digital fabrication. A new concrete 3D printing method called *pointillistic time-based deposition*, or *dripping*, is introduced as an example of Machine Delay Fabrication. The setup and variables of dripping are discussed, along with some experimental findings. Finally, dripping is used to demonstrate the constructive and aesthetic possibilities of Machine Delay Fabrication in architectural design.

Keywords: Materiality · Temporality · Robotic deposition
Architectural design · Concrete 3D printing · Productive delay

1 Introduction

Robotic fabrication has reached a turning point in its relatively short history as a discipline: now, it can, and should, depart from its manufacturing origins and establish evaluative criteria beyond utility and efficiency. Recent research has explored the deliberate misuse of machines in order to study robotic fabrication in alternative aesthetic and constructive contexts [4, 11, 13]. This paper builds upon those projects and offers architectural designers a way of working with digital fabrication machines in the contexts of temporality and materiality: Machine Delay Fabrication.

2 The Productive Delay: Stockhausen and Pollock

The productive potential of delay is nothing new for creative disciplines such as music and the visual arts. We will begin by briefly discussing two historical examples of productive delay in order to elucidate the creative differences in Machine Delay Fabrication. We will see that robotic fabrication engages with workflows that develop modes of productive delay not possible in these other disciplines.

The composer Karlheinz Stockhausen (1928–2007) is considered a pioneer of electronic music primarily for his experimentation with tape loops. Stockhausen

J. Willmann et al. (Eds.): ROBARCH 2018, *Robotic Fabrication in Architecture, Art and Design 2018*, pp. 126–138, 2019.
https://doi.org/10.1007/978-3-319-92294-2_10

developed a unique compositional method in which he recorded his soloists with a tape recorder and then, following a delay that he had specified, played the recording back while the instrumentalist continued playing. The soloists were asked to anticipate, remember, or respond to the feedback of the tape loop. The duration of the delay, as well as the duration of the recording, determined how the musician improvised upon the composition and, ultimately, how the improvisation became the piece itself.[1]

Jackson Pollock (1912–1956) also drew upon productive delay as a method of improvisation; however, his process was necessarily solitary. Pollock is famous for innovating "drip painting," in which he laid canvas on the ground and used different techniques to cover it with paint from above. By physically distancing himself from the canvas, Pollock was able to simultaneously witness the paint as it fell and decide where to move next. The delay in movement between artist and paint—the paint always one step behind the artist—gave Pollock a fuller understanding of the fluidity of the material.[2] With this knowledge, Pollock could tune his mix and drip-styles to better control the time that it required for the paint to reach the canvas. For example, if the artist were moving quickly while dripping viscous paints from greater heights, the delay between artist and paint would be longer and sparser lines would land on the canvas.

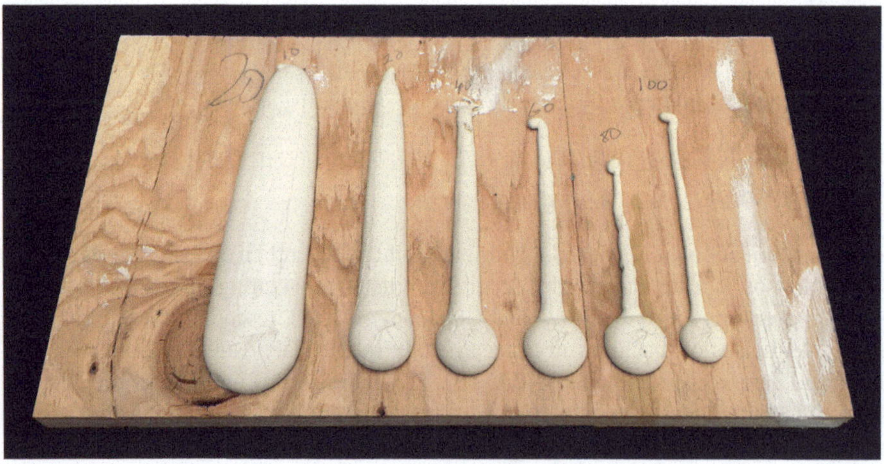

Fig. 1. The line on the right was printed at 100% speed and the line on the left was printed at 10% speed. The bulbs that form at the ends are a result of the material process continuing after the machine process has stopped. The bulb—or delay—is reduced by moving slower, not faster.

[1] Harvey, Jonathan. *The Music of Stockhausen*. Berkeley and Los Angeles, University of California Press (1975). pp. 97–98.

[2] Cernuschi, Claude and Andrzej Herczynksi. "The Subversion of Gravity in Jackson Pollock's Abstractions." Art Bulletin 90 (4). pp. 616–639 (2008).

3 Being Real-Time

Both current architectural design research in robotic fabrication and digital fabrication in general embrace the benefits of real-time techniques without interrogating their disciplinary implications [2, 5, 15]. Architectural designers must question the dominance of real-time techniques in order to assert the uniqueness of their own discipline.

In real-time digital fabrication, researchers equip machines with various sensing mechanisms so that they can immediately respond to, or interact with, the material, the designer, and/or the environment. Digital input, via the sensors, and machine output are so tightly coupled that the process gives the impression of instantaneous interchange—everything appears (to a human) to occur in "real time." However, what we perceive is *only* the appearance of simultaneous cause and effect: not *real* time itself, but a concept of real time. In fact, the electronic signals that enable this processing speed must be carefully sequenced for the real-time feedback loops to function efficiently, or at all—one thing must happen after another and delays, although imperceptible, must and do take place. The alternative approach introduced in this work examines what architectural design possibilities arise when we experiment primarily with the delay rather than with the inputs and outputs. It also examines what design processes emerge when we dwell in, look at, and embrace the delay rather than try to pretend that it is not there.

Real-time digital fabrication techniques are always delayed by the properties of the materials they operate on [10]; machine impulses can move faster than materials can react. Therefore, the digital fabrication machine invariably has to be tuned to the speed of the material process in order to carry out its expected function. When this does not happen, material failures occur. For an architectural designer, this tension between speed and materiality is significant because it means that employing real-time techniques is often, though not always [8], equivalent to attempting to control material rather than to work with it. As a discipline that is necessarily concerned with issues of materiality and uncertainty, architecture must consider whether or not the appropriation of real-time is beneficial for architectural design processes. Further, we should weigh the delegation of sensing to real-time feedback systems against the encapsulation of material knowledge by machines. This research does not advocate for a return to the intuitive methods of Pollock but rather for a way to combine the virtues of the robot, (e.g., repetition) with that of the human designer (e.g., improvisation).

4 The Concept of Machine Delay

Machine delay is the time between machine latency and machine work. In digital fabrication machines, this is the time between digital input and physical output. Normally, designers accept this delay in order for machines to function as expected. In other words, *we* are the ones that are delayed. This paper introduces the *Machine Delay Fabrication* approach in order to create a new designer-machine symbiosis. In this approach, we materialize and manipulate the time taken by digital fabrication machines to do work in order to produce new architectural design possibilities.

Three kinds of machine delay that pertain to digital fabrication have been identified, and each one corresponds to a different scale of time:

- *Integral-machine Delay*: the machine is delayed by its own work, e.g. clock speed or information processing;
- *Interactive-machine Delay*: the machine is delayed by the designer's work, e.g. programmed delays or the time between instructions;
- *Material-machine Delay*: the machine is delayed by the time taken for materials to do their work, e.g. curing, heating, or cooling (Fig. 1)

This paper will focus on an example that combines elements of Interactive-machine Delay and Material-machine Delay.

5 Pointillistic Time-Based Deposition

Pointillistic time-based deposition, or *dripping,* was invented to explore Machine Delay Fabrication. Dripping is enacted by a six-axis robotic arm that has been transformed into a concrete 3D printer (Fig. 3). Small-scale tests were done using a proprietary pneumatic extruder, which utilized the pneumatic outputs on the wrist of the robotic arm. Larger-scale tests were done using a custom printing nozzle and peristaltic pump, which communicated with the robot through a solid state relay. In both instances, digital outputs were controlled in the robot code. Whereas other research in 3D printing concrete is focused on refining machine and material parameters to optimize buildability [3, 9], the dripping system utilized a deliberately minimal system to focus on working *with* the material in order to produce a concrete "drop."

5.1 Delaying the Machine

In the dripping workflow, the designer programs the robot to deposit concrete at specific points in space, and then instructs the robot when to revisit each point and in what order. This instruction is an example of Interactive-machine Delay. Each point, or *drop*, has two delays that manipulate the machine delay: *extrusion time*, which is the duration for which the extruder is active at the drop, and *wait time*, which is the duration for which the extruder waits before moving to the next drop. This is comparable to the "Pause" technique developed by W. Andrew Atwood [1]. However, here, the print is composed *only* of pauses: concrete is laid like bricks.[3]

The time elapsed between each visitation of a drop is the sum of the extrusion times, wait times, and accumulated travel times between drops. This in-between time—or, delay—determines the local structure and aesthetic of the print. For example, a drop that has not been visited for a long duration will conform less to a new deposit than its more recently visited neighbors (Fig. 2). A similar exploration of "revisitations" in concrete 3D printing was conducted in the "Firewall" project by Bekkering Adams [16].

The robotic arm can revisit the exact location of drops in space *and* repeat the exact count between revisitations—it is both spatially *and* temporally precise. Dripping realizes the potential of this precision by turning the robot into a "material clock" that

[3] To see video demonstrations of the dripping system, please visit: https://vimeo.com/270990944 and https://vimeo.com/270990912.

simultaneously materializes and keeps track of time. The dripped thing slowly emerges from the clock and assumes a materiality that is representative of the time that went into its making.

Fig. 2. Longer wait times produce less conformity; longer extrusion times produce larger drops.

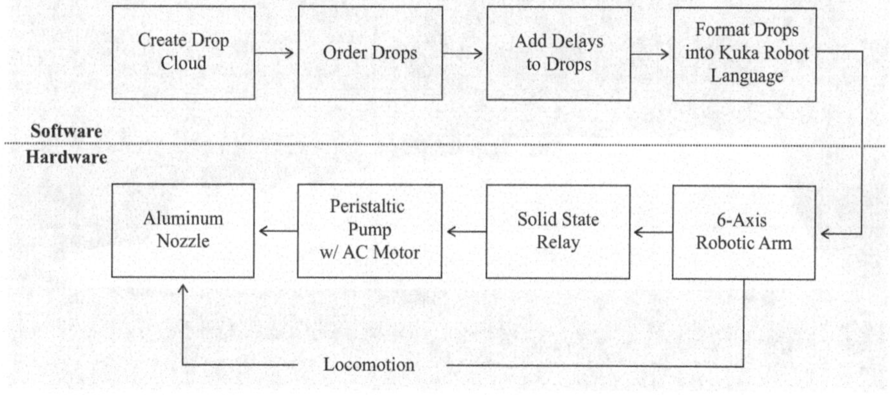

Fig. 3. The software and hardware workflows of the large-scale dripping system.

5.2 Designing the Mix

In the initial experiments (Figs. 4, 5, and 6), white grout was used as the printing material. The material was mixed before being fed into the pneumatic extruder's cartridge. Each test was equal to the cartridge's volumetric capacity (283 g or 10 oz).

Optimizing the water-to-grout ratio proved to be the most challenging part of system development. The objective was to use a mix that would extrude, or leak, even after the pneumatic impulse had desisted, thereby invoking an instance of Material-machine Delay. If the cement was too thick, the mix would only extrude with pressure from the plunger. If it was too thin, the material would slip out and slump into a puddle.

Fig. 4. The cement cured as it was printed, which resulted in a natural material gradient.

Fig. 5. The extruder was not allowed enough wait time between drops and failed to print.

The mix, air pressure, and wait time of each drop had to be calibrated with one another to ensure the continual production of drops. For example, if the mix was too thick and deficient wait time was allotted, the extruder failed to generate enough repeated pressure to deposit material. This failure suggested that the wait times were not only materially productive—e.g., building relationships between drops—but also *necessary* in that they allowed for the extruder to recover amidst its repetition.

Fig. 6. Longer delays cause drops to remain separate from one another.

5.3 Finding Materiality

In the initial experiments that used the pneumatic extruder, all variables were kept constant except for extrusion time and wait time. The constants included course height, order of drops, proximity of drops, mix ratios, air pressure, and nozzle diameter. The tool path was also the same in each experiment: left to right at every course, as well as staggering every other course to create more structural stacks of drops.

In the first tests, excessive wait times (fifteen seconds) between drops resulted in the material curing while it was being printed. Initial courses contained drops; however, less than thirty minutes later the cement became filament-like and distorted the reading of the course altogether (Fig. 4). The resulting materiality was intriguing, but it also cautioned that too much delay could result in *total* machine or material failure.

Reducing the wait time to five seconds between each drop produced more uniformity but did not allow the machine enough time to recover (Fig. 5). As a result, there were less drops on one side of the print than the other. This suggested that the time (approx. three seconds) that it took the robot to traverse back to the starting side after finishing one course added enough recovery. In the penultimate test, the quick curing of the drops prevented them from completely melding into one another upon revisitation. Instead, small holes were formed between successive courses. These voids pointed to one architectural possibility of Machine Delay Fabrication (Fig. 6).

6 The Architectural Design Possibilities of Dripping

Dripping was scaled—in machine, material, and time—in order to speculate on the architectural design possibilities of the initial experimentation in greater detail. The following experiments utilized a specially-tailored concrete mix, along with the aforementioned nozzle and peristaltic pump system. Due to the increase in machine size, the delays (extrusion and wait times) only needed to be increased by five to seven seconds or, depending on the scale and kind of experiment, sometimes not at all. For scale, please note that each test was printed on 19 mm (3/4″) thick plywood.

Two kinds of architectural design possibilities will be discussed in this paper: constructive and aesthetic.

6.1 Constructive

Recent research has uncovered creative alternatives to the inflexibility of conventional slicing-to-layering 3D printing workflows [6, 12]. The pointillistic nature of dripping similarly introduces new ways to build with 3D printers, as well as new forms of architectural components that can be built.

The Wall. Dripping further breaks up the concrete fabrication process—from pours to slices to drops (Fig. 7). Drops can create more structural integrity between layers of 3D printed concrete by increasing the surface area of the connections. Alternatively, cold joints can be strategically distributed to weaken the structure in certain areas. In other words, dripping can allow architectural designers to design failure modes into walls.

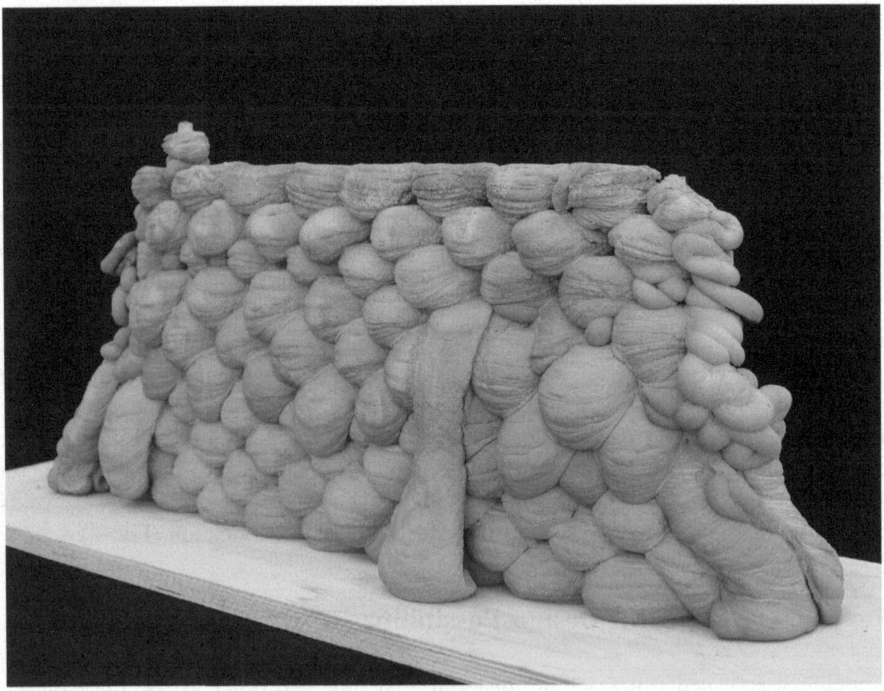

Fig. 7. A standard stack of interlocking eight-second drops with ornamental anomalies.

The Corner. The corner condition poses a unique problem for concrete 3D printing systems (and architectural design in general). If the designer does not programme the machine to slow down or add a radius at corners, the concrete will twist and/or tear. This is because the material cannot change directions as quickly as the machine can.

Dripping allows designers to tune the machine process to be even slower than the material process. The only acceleration at the corners of dripped things is the downward acceleration of the material. Instead of twisting or tearing, dripped corners can melt or bleed (Fig. 8).

Fig. 8. A single wythe of drops with various extrusion times bleeds at the corner.

The Column. If the delays are long enough, drops can stack vertically to form a column (Fig. 9). The assembly of the column can be facilitated by deliberately decreasing the drop height so that the nozzle becomes partially submerged and leaves behind a small depression in each drop. This depression becomes a mechanical joint that ensures that the subsequent drop locks into place and maximizes the surface area of the drop-to-drop connection.

Fig. 9. Drops can stack vertically if they are given time to support themselves.

6.2 Aesthetic

Dripping has a unique architectural grammar that is expressed by pairing the repetition of the robot with the unpredictable amorphousness of the material and improvisatory impulses of the architectural designer. In dripping, and in Machine Delay Fabrication in general, repetition occurs in the process, e.g., in rhythmic extrusion times, but not in the product. For example, the same delay can produces drops that are nearly identical in size, but contain distinct wrinkles, folds, ripples, or marks. This is importantly different from the work of artists who have experimented with 3D printing: Anish Kapoor, for example, is more concerned with creating sculptural products than repeatable and adaptable processes [17].

The repetition in dripping can be tuned to create subtly differentiated architectural patterns. This tuning is similar to the systematization of falling fluid deposition that is described in Klein et al. [7]. Ornaments or apertures can be inserted into dripped patterns by building relationships between the rate of deposition and the rate of curing. For example, longer delays can result in ornamental excess, textural variation, or gaps between neighboring drops (Figs. 7 and 10). Tuning the machine to the material in order to achieve these various aesthetics can be a new form of architectural craft.

Fig. 10. Apertures and wrinkles result from drops resisting conformity.

7 Outlook: Deferring the End

Future development of dripping will continue scaling the system into a construction method. More specifically, dripping will be used to produce both prefabricated and in situ architectural components for old and new buildings. For example, columns could be dripped into dilapidated existent structures as part of their reconstruction. Tests will be carried out to determine the load-bearing capacity of dripped parts, as well as how they can be reinforced and integrated with other construction systems in order to achieve this objective.

Digital fabrication machines are moving from studios and labs to homes and construction sites. As a result, architectural designers need to cultivate ways of interacting with digital fabrication machines that maintain the vitality of their discipline, yet can also evolve to produce the architectural design of the future. Machine Delay Fabrication is a platform for this kind of new design thinking. It aims for a future in which architectural design remains in place, in touch, and, above all, in time.

Acknowledgements. The author thanks: MIT Architecture Researcher Justin Lavallee; MIT Architecture Professors Mark Jarzombek, Sheila Kennedy, Caitlin Mueller, and Mark Goulthorpe; and Joshua C.A. Cohen.

References

1. Atwood, W.A.: Monolithic representations In: Borden, G.P., Meredith, M. (eds.) Matter: Material Processes in Architectural Production, pp. 205–212. Routledge, New York (2012)
2. Batliner, C., et al.: Robot UI. In: Reinhardt, D., Saunders, R., Burry, J. (eds.) Robotic Fabrication Architecture, Art and Design 2016, pp. 377–388. Springer, Switzerland (2016)

3. Bos, F., et al.: Additive manufacturing of concrete in construction: potentials and challenges of 3D concrete printing. Virtual Phys. Prototyping **11**(3), 209–225 (2016)
4. Dörfler, K., Ernst, S., Piškorec, L., Willmann, J., Helm, V., Gramazio, F., Kohler, M.: Remote material deposition: exploration of reciprocal digital and material computational capacities. In: Voyatzaki, M. (ed.) What's the Matter – Materiality and Materialism at the Age of Computation. ENHSA, Barcelona (2014)
5. Dubor, A., et al.: Sensors and workflow evolutions: developing a framework for instant robotic toolpath revision. In: Reinhardt, D., Saunders, R., Burry, J. (eds.) Robotic Fabrication Architecture, Art and Design 2016, pp. 411–425. Springer, Switzerland (2016)
6. Franz, K.: Dual-Axis Precision Deposition System (Plopper). http://www.kaifranz.de/Plopper.html. Accessed 26 Feb 2018
7. Klein, J., et al.: Additive manufacturing of optically transparent glass. 3D Printing Add. Manuf. **2**(3), 92–105 (2015)
8. Johns, R.L.: Augmented materiality: modelling with material indeterminacy. In: Gramazio, F., Kohler, M., Langenberg, S. (eds.) Fabricate: Negotiating Design & Making, pp. 216–223. gta Verlag, Zurich (2014)
9. Le, T.T.: Mix design and fresh properties for high-performance printing concrete. Mater. Struct. **45**, 1221–1232 (2012)
10. Mueller, S.: Interacting with personal fabrication devices, Ph.D. thesis, University of Potsdam (2016)
11. Matsys: Scripted Movement Drawings Series 1. http://matsysdesign.com/2014/07/13/scripted-movement-drawings-series-1/. Accessed 26 Feb 2018
12. Retsin, G., Garcia, M.J.: Discrete computational methods for robotic additive manufacturing. In: Velikov, K., Ahlquist, S., del Campo, M. (eds.) Acadia 2016: Posthuman Frontiers: Data, Designers and Cognitive Machines, Proceedings of the 36th Annual Conference of the Association for Computer Aided Design in Architecture, pp. 332–341. Ann Arbor, MI (2016)
13. Roche, F., et al.: Psychaestenia: pyscho-case studies by New Territories/M4. In: Reinhardt, D., Saunders, R., Burry, J. (eds.) Robotic Fabrication Architecture, Art and Design 2016, pp. 17–30. Springer, Switzerland (2016)
14. Snooks, R., Jahn, G.: Closeness: on the relationship of multi-agent algorithms and robotic fabrication. In: Reinhardt, D., Saunders, R., Burry, J. (eds.) Robotic Fabrication Architecture, Art and Design 2016, pp. 219–230. Springer, Switzerland (2016)
15. Snooks, R., Jahn, G.: Stigmergic accretion. In: Reinhardt, D., Saunders, R., Burry, J. (eds.) Robotic Fabrication Architecture, Art and Design 2016, pp. 399–410. Springer, Switzerland (2016)
16. Teague, L.: 3D print is making an impression on concrete. FRAME, 31 January 2017. https://www.frameweb.com/news/3d-printing-is-making-an-impression-on-concrete. Accessed 30 Apr 2018
17. Valentine, V.L.: Anish Kapoor Stages a Mesmerizing Forest of Concrete. Arts Observer, 24 May 2012. http://www.artsobserver.com/2012/05/24/anish-kapoor-has-staged-a-mesmerizing-forest-of-concrete-at-gladstone-gallery/. Accessed 26 Apr 2018

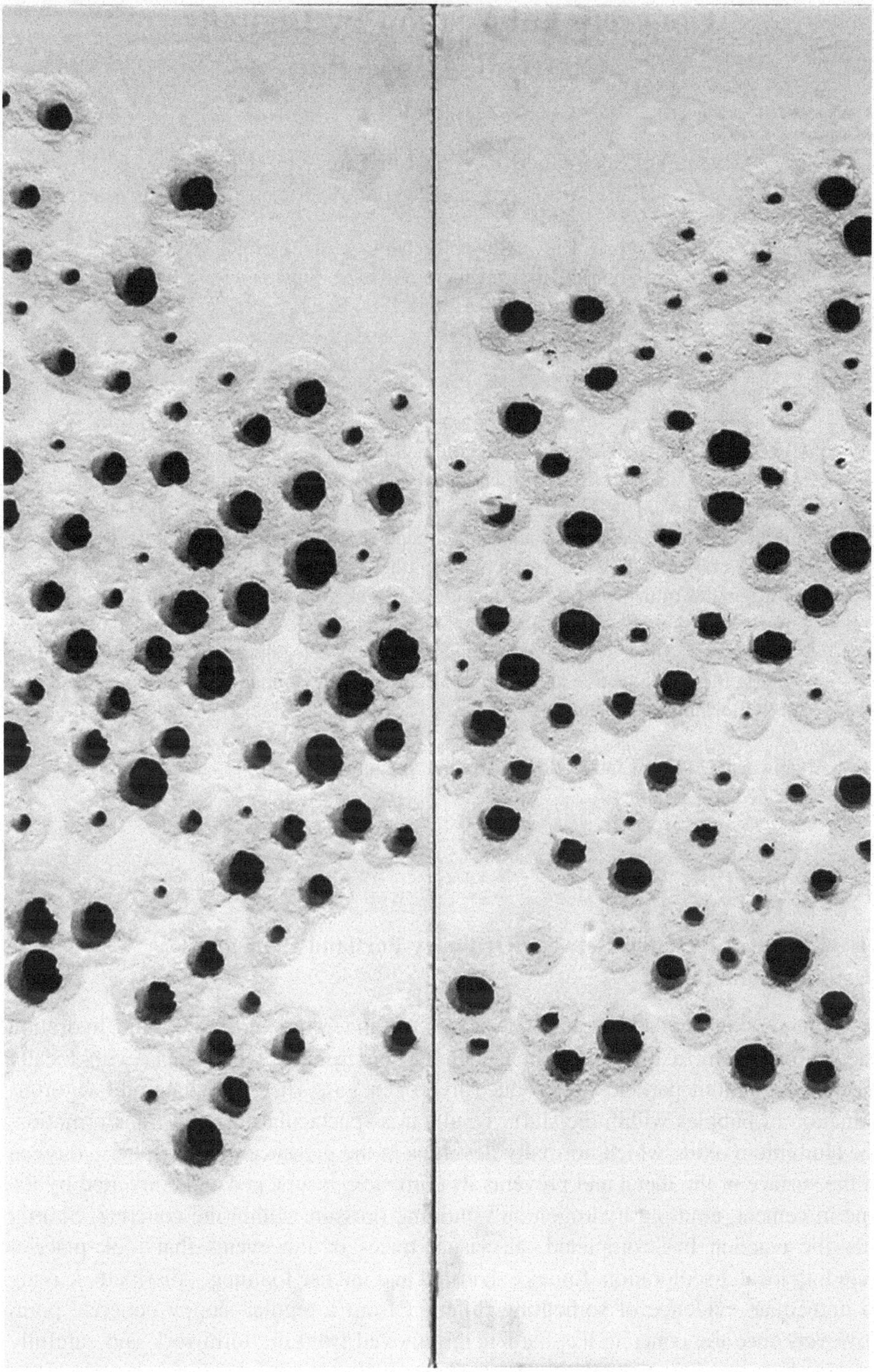

Concrete Fabrication by Digitally Controlled Injection

Ryan Wei Shen Chee(✉), Wei Lin Tan, Wei Hern Goh, Felix Amtsberg, and Stylianos Dritsas

Singapore University of Technology and Design, 8 Somapah Rd, Singapore S487372, Singapore
{ryan_chee, weilin_tan, weihern_goh}@mymail.sutd.edu.sg, {felix_amtsberg, dritsas}@sutd.edu.sg

Abstract. This paper explores the volumetric modification of material properties through precise injection of chemical agents as a new method of concrete fabrication. We inject concrete while in its slurry state with a hydrated aluminium solution, triggering an effervescent reaction. This reaction causes changes in material density which affect the structural integrity and visual characteristics of the end-product once it is set. The process is administered by a purpose-built end-effector attached to an industrial robotic system and guided by a parametric design system. Being able to produce localized differentiation, suggests for an innovative approach to a fabrication technique of texturing, 2D and 3D forming of concrete.

Keywords: Digital fabrication · Additive manufacturing
Architectural robotics

1 Introduction

1.1 Chemical Reaction Between Ordinary Portland Cement and Aluminium

As Ordinary Portland Cement (OCP) cures, it undergoes an exothermic hydration reaction producing a highly alkaline chemical environment that reacts with locally injected aluminium particles to produce hydrogen gas. The immediate and vigorous formation of bubbles within the slurry results in a spectacular material transformation. The aluminium oxide which normally develops in the presence of atmospheric oxygen on the surface of the metal and prevents its corrosion, is attacked and converted by the lime in cement, emitting hydrogen and building pressure within the concrete. Shortly after the reaction has completed, all visual traces of the events that took place – including local discoloration, fumes exhausted and intense foaming – dissipate, leaving no immediate evidence of something different from a regular sloppy concrete pour. However, once the concrete has hardened, removed from the formwork and carefully examined, some quite remarkable properties can be observed which give rise of a spectrum of material effects (Fig. 1).

J. Willmann et al. (Eds.): ROBARCH 2018, *Robotic Fabrication in Architecture, Art and Design 2018*, pp. 139–151, 2019.
https://doi.org/10.1007/978-3-319-92294-2_11

Fig. 1. Left-to-Right: Hydrogen gas seen escaping, concrete of high porosity with some unreacted aluminium, textured concrete is revealed when concrete of high porosity is removed.

1.2 Density Modulation

Hydrogen trapped within the slurry produces volumetric porosity as minute pockets are formed after the concrete has set and all gases have escaped. The presence of micro cavities affects the bulk density characteristics of the composite producing a result which belongs to the family of lightweight, autoclaved aerated or foamed concrete materials. Indicatively, the density of OCP is circa 3,150 kg/m^3 (Lafarge 2015), medium density concrete at circa 2,000 kg/m^3 and thermal conductivity of 1.35 W/mK, while aerated concrete can reach as low densities and thermal conductivities as 300 kg/m^3 and 0.1 W/mK respectively (ISO/DIS 10456 2007). Nevertheless, permeability and often lower load-bearing performance characteristics are also associated with this material innovation dating back in the 1920's (Mathey and Rossiter 1988).

Recent emphasis in sustainable development and the circular economy has resurged interest in aerated concrete for reducing bulk usage of cement and while recycling dross by-products of industrial manufacturing for ceramic materials additives (Studart et al. 2005; Maziah 2011; Kinoshita et al. 2013; Liu et al. 2017). Relevant to this is research in controlled density modulation of material properties for functionally graded materials (FGM), originating from the fields of mechanical engineering and material science (Mahamood and Akinlabi 2017). Examples of influential architectural work include gravity and formwork induced variable porosity in aluminium-based aerated concrete (Cooke 2012) and functionally graded rapid prototyping (Oxman et al. 2011; Duro-Royo et al. 2015). Further development in controlled modulation of material properties at the scale of construction fabrication could lead to architectural innovations such as pre-cast concrete panels that have high structural and thermal performance in the same domain, removing the need for material assemblies such as concrete sandwich walls (Fig. 2).

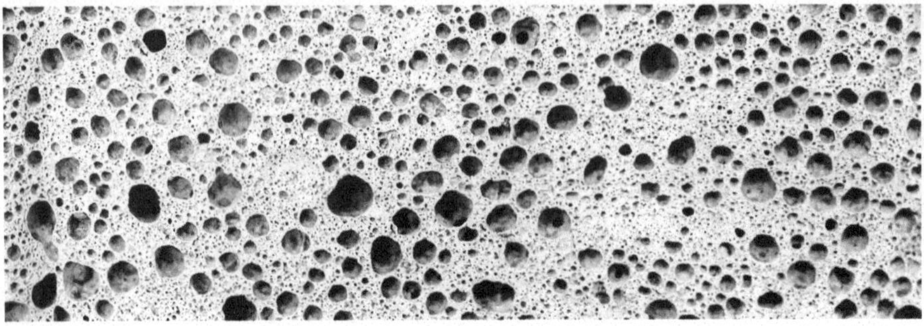

Fig. 2. Detail of porosity characteristics

1.3 Concrete Fabrication

Basic concrete casting techniques involve the use of formwork into which concrete in its slurry state is poured. After a duration of time, the concrete cures into a solidified form, and is subsequently separated from the mold. The fabrication of customised formwork has enabled the manipulation of resultant concrete forms, illustrated by projects such as Ronchamp by Le Corbusier, and Pier Luigi Nervi. Traditionally, the fabrication of such geometries was incredibly tedious and difficult to achieve, however recent innovations in formwork fabrication such as the use of non-rigid tensile fabric as formwork has allowed an easier approach to the creation of non-orthogonal concrete structures that are 3-dimensionally complex and textured (Veenendaal et al. 2011).

2 Background

Our project is situated within the domain of digital design and fabrication. It investigates creative material transformation processes (Lefteri 2007). It is informed by conventional waterjet cutting, the Direct Ink Writing (DIW) 3D printing for FDM (Lewis 2006) and research work in suspended matter within viscous media (Johns et al. 2014). The paper is organized by the following research tasks: (1) Material Studies: investigation of chemical reactions between OPC and additives, (2) Software and Hardware: development of servo-motor controlled syringe injection system, integration of micro-controller dispensing logic with industrial Programmable Logic Control, translation of design geometry to robotic motion, (3) Fabrication and prototyping: design and development of artefacts demonstrating the capabilities of injection fabrication process.

2.1 Material Studies

Initial tests aimed at creating porosity were conducted using such substrates as hydrogen peroxide (Morgulis and Levine 1920) and aluminium powder, known to produce gas bubbles over time and hence material porosity to varying extents. These substrates were directly added and uniformly mixed with slurry concrete to investigate its global effects, with the ability to create and control this porosity as twin goals for these tests. Once set, the concrete samples weighed in less than the control setup but varied in degree of porosity and structural strength. Hydrogen peroxide resulted in a sturdy material with a mixed porosity of different-sized bubbles. However, majority of which formed in the peripheral regions around a nearly-solid concrete core. The addition of yeast as reaction catalyst resulted in a consistent porosity coupled with low structural integrity, possibly due to an over-creation of voids that weakened the cured material.

The aluminium resulted in a similarly-porous material to the latter, with a reasonable degree of structural integrity. The reaction between aluminium and cement creates the formation of bubbles within the slurry. This results in a spectrum of effects: from creating low porosity when minute gas bubbles are trapped internally, dramatically affecting its density producing results affine to aerated or insulated concrete; to producing high levels of chemical erosion and material discontinuities when large volumes of gas reach the surface and escapes. Furthermore, only a small amount of aluminium to cement ratio is required to alter the porosity. This process combined with precise and controlled dispensing capability offers the opportunity for achieving local material differentiation which is profoundly unique compared to conventional additive or subtractive manufacturing processes.

In view of these outcomes and tests showing that small amounts of aluminium powder were able to generate high levels of porosity, the metal was deemed a suitable substrate with a reasonable degree of control for further experimentation (Fig. 3).

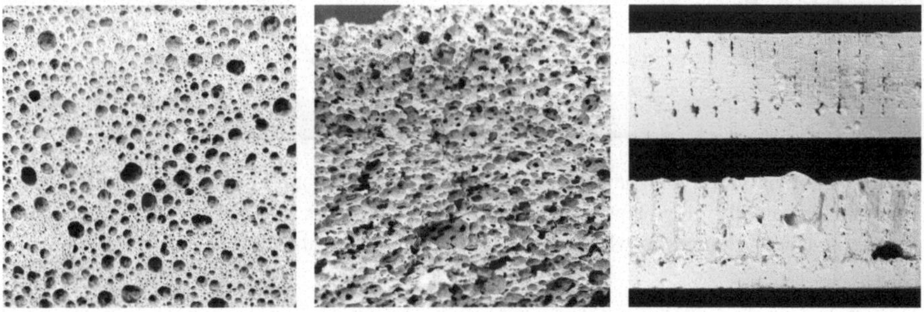

Fig. 3. Left-to-Right: Cement mixed with hydrogen peroxide, cement mixed with hydrogen peroxide and soap, section of localized injection of aluminium mixture.

2.2 Software and Hardware

The instrumentation of our experiments progressively transformed from lab testing using hand tools to the development of a specialized digital design and fabrication process for precise injection. We deployed an ABB IRB1200 series industrial articulated six-axis robot as a programmable and accurate positioning mechanism. Nevertheless, a three-axis cartesian system would have also sufficed as the spatial orientation capabilities of the machine were not required or perhaps not thoroughly explored.

We developed a purpose-built syringe dispenser to control the volumetric injection of the aluminium solution. The end-effector was fabricated using lasercut acrylic sheets, its mechanical components were standard hardware such as threaded rods and metric nuts and miscellaneous bespoke parts such as reduction gears were 3D print. The actuator is powered by a RhinoRMCS high-torque servo motor with integrated encoder and controller electronics. Assembly of these components forms a linear actuator which mechanically depresses and retracts a 10 ml medical syringe's piston. Associated speeds of rotation (RPM) with volume dispensed were measured using the gradations of the syringe to derive the flow-rate formulation (Fig. 4).

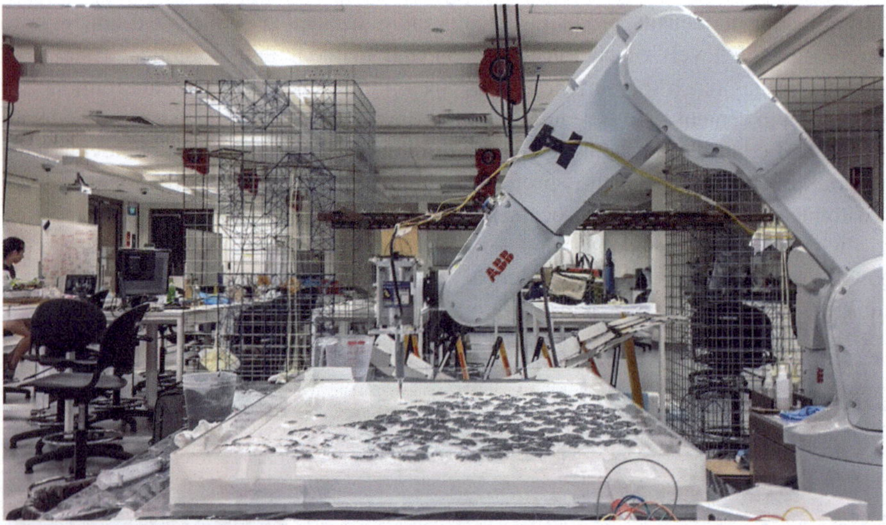

Fig. 4. Robot integrated with an actuator that injects controlled amounts of aluminium. Injected aluminium is initially grey, however once the aluminium has successfully reacted with the cement, the colour changes to white as seen at the left side of the cement mould.

For integrating and coordinating the all mechanical sub-systems, from the robot itself to the injector, we developed a simple Programmable Logic Control (PLC) system. The driver logic was integrated in an ArdBox PLC, a low-cost industrial interface device using Arduino as its microcontroller unit. The benefit of this system is that it can be programmed using standard C/C++ and it mediates 5VDC logic-level signal shifting to the industrial standard of 24VDC required to communicate between the servo motor driver and robot controller. A firmware was developed to convert a digital array of bits from the robot controller to signed speed instructions for the servo drive. As such the dispenser could be controlled by ABB RAPID programming without need for external control and synchronization logic.

Workspace calibration, motion planning, simulation, off-line programming and communications with the robot were performed using the Jeneratiff Digital Design and Fabrication library for the Rhino/Grasshopper visual programming environment. Using parametric modelling techniques enabled rapid experimentation and fine-tuning of critical for the process settings such as motion speed and injection feed-rate and eventually automating the production of prototypes (Fig. 5).

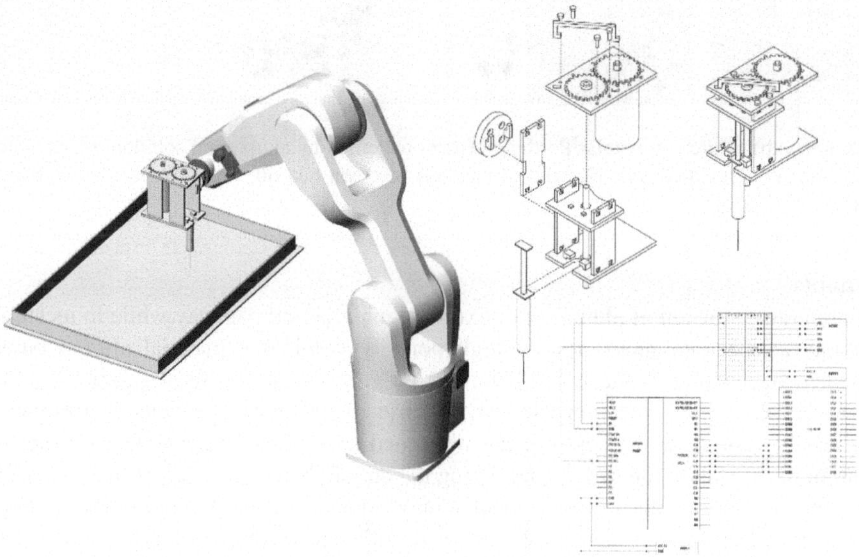

Fig. 5. Left-to-Right: Robot setup, exploded diagram of end-effector and electronic logic

2.3 Fabrication and Prototyping

We designed and fabricated a series of prototypes comprising of flat panels or thicker, narrower blocks of varied dimensions. The goal of these prototypes was to evaluate the potential applications of various process parameters through their small-scale fabrication.

Surface Texturing

The creation of varied surface textures was tested by injecting aluminium of a range of concentrations into a concrete panel of dimensions 400 mm by 500 mm and 15 mm thick. Localisation surfacing texturing was achieved as seen in a point injection of an image, rasterized into a grid of 2-dimensional points. Effects were more localised at points where the aluminium was injected at a later time frame of the fabrication process, allowing the material to have cured to a greater extent. Point grid injections should take place within the window of 20–60 min after the initial pour to ensure best results (Fig. 6).

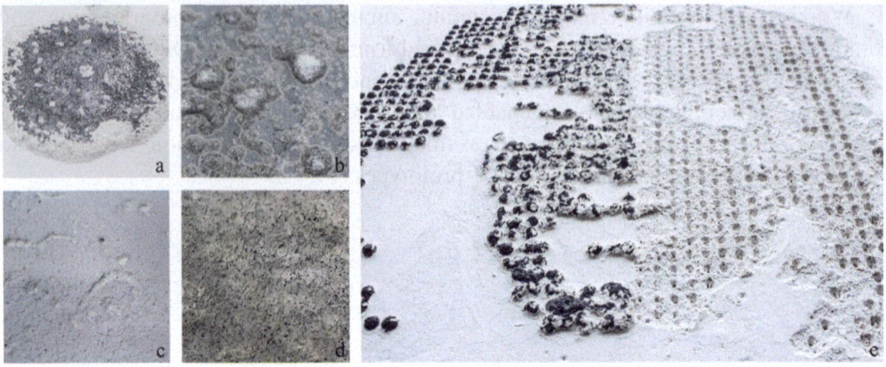

Fig. 6. (a, b) White and regular Portland cement patterned by aluminium solution. (c, d) White and regular Portland cement after the surface has been washed off. (e) Point grid image that is half-washed off.

Cutouts

Continuous extrusion of aluminium mixture along a closed pathway while in its slurry state creates a continuous trail of high porosity within the material after it cures. Weakened concrete in the trail of the needle enables the enclosed geometry to be removed physically. Injection right at the base of the mold is crucial in forming a successful cut-out, as the depth of the weakened area when cured would be from the bottom of the slab to the surface, rendering the cut out being easier to extract. After the concrete has cured the porosity results in weakened material regions that can be removed with a high-pressure water hose. A sufficient level of material discontinuity created by a high level of porosity enabled us to create 2-dimensional, geometric cutouts within a horizontal concrete panel, an outcome similar to CNC milling or water jetting processes.

With the possibility of creating geometric cut-outs, patterns were generated in a panel of similar dimensions 400 mm by 500 mm and 15 mm thick to test the customizability of cut outs with different shapes and sizes. For this process a mixture of aluminium powder (1–3 μm) and water at a concentration of 29.3 mol/dm^3 was used. With the success of introducing different geometries into the cut-out process, therein lies the potential of casting without custom molds, and achieving customizable geometric patterns via the injection process only (Fig. 7).

We designed and fabricated a large prototype comprised of three white cement panels with dimensions of 1,650 mm by 850 mm and 35 mm thick in total. The objective of the prototype was to evaluate the process parameters beyond the previously created small and medium size prototypes. The design is comprised of 335 perforations derived from an algorithmic process blending imagery data from multiple sources. The combined weighted sums of overlapped raster image intensities were converted to an abstract perforation pattern artwork.

The range of diameters used, span between 5 mm and 30 mm, with minimum distance between circle perimeters at least 5 mm. The constraint was derived empirically, where below this nominal tolerance the concrete became too brittle for further

Fig. 7. (a) Circular cut-out sample motion (b) Cut-out of vernacular ventilation brick panel (c) Segmentation by continuous variable injection experiment

processing. The syringe needle of 0.75 mm internal diameter was ground flat and depth-set at the very bottom of the acrylic mould. Linear motion speed was 75 mm/sec and injection flow rate at 0.375 ml/sec. Those settings ensured full penetration of the solution with enough material available to reach the top surface.

To dislodge the non-fully disconnected concrete plugs from the panels we used a power washing system for ejection. The hole diameters were surprisingly consistent, with average error of ± 1.2 mm or approximately between 0.2 mm to 3 mm with respect to largest and smallest programmed hole sizes. In addition, due to upward direction of escaping gasses, the interior wall finish of perforations was rough including a nominal draft angle similar to waterjet cutting edge-tapers but with inverse orientation. The prototype panels weight 16 kg each, required approximately 20 min of the injection process, at least 24 h of curing and 5 min of high pressure water blasting.

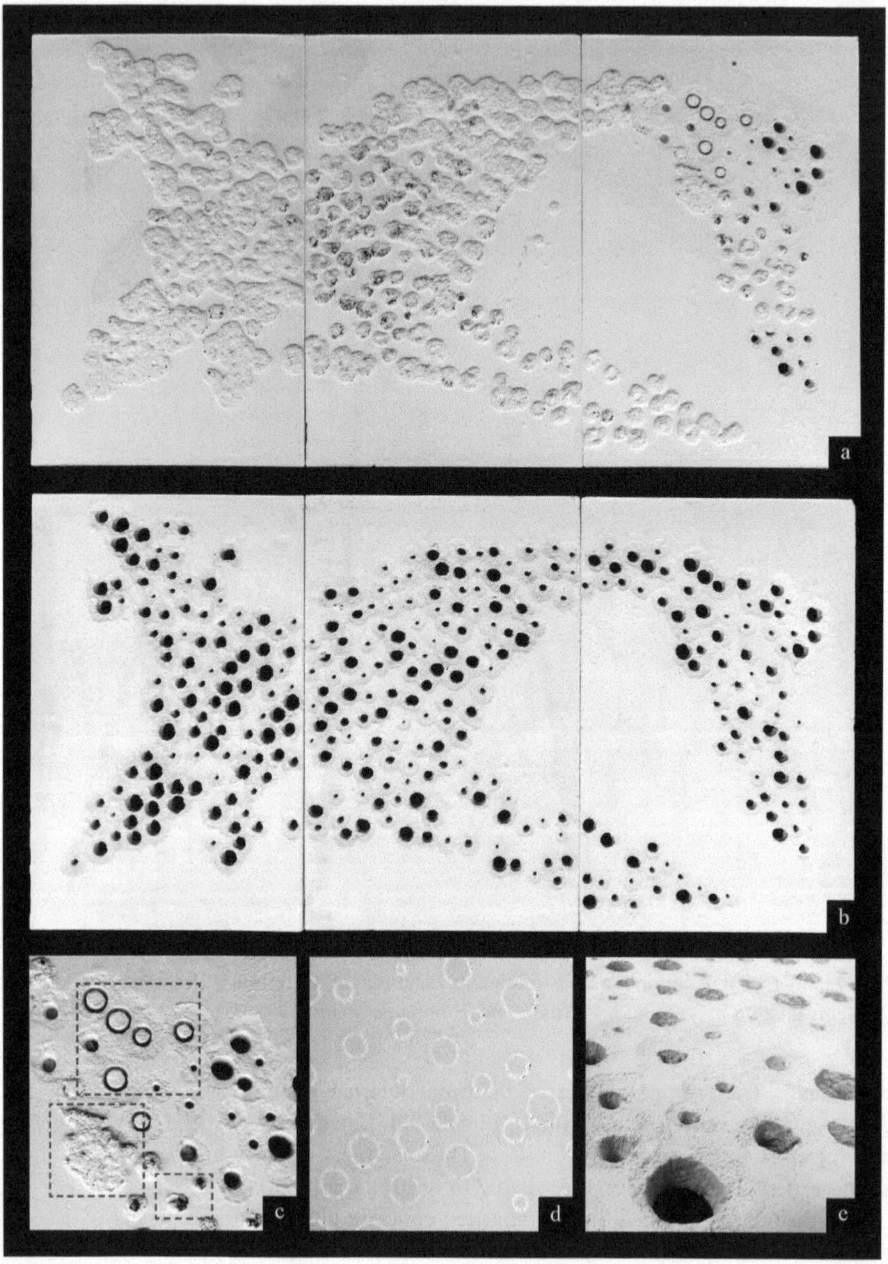

Fig. 8. (a) Top view of panels after the cement has cured (b) Top view of panels after water blasting, (c) Top detail view of aluminium not reaching the mold's bottom hence some areas were not fully cut. Grey areas indicate excess unreacted aluminium whereas white areas indicate concrete of high porosity. (d) Bottom detail view of highly smooth surface finish due to acrylic mold. (e) Top detail view of the textured perforations with approximately square draft angle.

Each panel used in total 30 ml of equal aluminium-to-water solution by weight. The excess aluminium solution that reached the top and pooled around the perforation produced a highly patterned, unforeseen and beautiful result which combines both the cutting and texturing features of the process (Fig. 8).

3D Sculpting

Varying the vertical height during continuous extrusion of the aluminium mixture leads to the transformation of the entire material surface, similar to a manner of sculpting. The extrusion was injected into a block-shaped mold of vertically thicker dimensions 100 mm by 200 mm by 30 mm to allow a greater vertical range of variability. For this process a mixture of aluminium powder (1–3 μm) and water at a concentration of 29.3 mol/dm^3 was used. Using a lower motor speed led to incomplete trails as the robot arm moved along the line (trail on the right). The width of eruptions across each line varied from 20–25 mm due to the increasing amount of aluminium injected as the motor speed increases. Beyond a motor speed of 15, the width of the eruption does not increase as the motor speed increases. However, the width of the cavity trail after post processing was generally the same at 5 mm (Fig. 9).

Fig. 9. 3D Sculpting on a concrete brick

3 Conclusion

Prototypes created from experiments thus far point at the potential of material injections as a concrete fabrication system for a wide range of applications. An alternative to existing 3D printing technologies which build up material differentiation in fine resolutions, we offer a case study in influencing material densities using controlled conditions that trigger inherent material properties and their reactions. By controlling the depth, motion and injection flow rate, we can produce direct cuts and even surface contours similar to those created by milling machines. At this phase, we have not exploited the capability towards a concrete functional objective, such as modulation of structural or environmental performance, but focused on developing experience and understanding the process parameters towards controlling end results. This requires additional work in Design of Experiment models to enable prediction of composite properties based on parameters such as concentration of aluminium mixture affecting the porosity of concrete, the amount of aluminium injected in relation to the depth of the concrete component affecting the degree of the resultant effervescent reaction, and spacing between the toolpath affecting the resolution of concrete texture.

This multi-material process allows for an integrated building process that could change the current concrete pre-fabrication workflow. In conclusion, our wet injection method suggests for an alternative approach to concrete fabrication for some applications potentially may be a faster and more cost-effective, or just purely aesthetically and creatively interesting. Through this process, the technique of material injections can be seen as a fabrication system for differentiated materials that is at the intersection of additive and subtractive manufacturing.

Acknowledgments. This research work was supported by Singapore University of Technology and Design, Office of Education and the Digital Manufacturing and Design Centre.

Reference

Building materials and products, Hygrothermal properties, Tabulated design values and procedures for determining declared and design thermal values (2007). <ISO/DIS> 10456. Accessed 12 Dec 2017

Cooke, T.G.: Lightweight Concrete: Investigations into the production of variable density cellular materials, Master's thesis, Massachusetts Institute of Technology (2012)

Duro-Royo, J., Mogas-Soldevila, L., Oxman, N.: Flow-based fabrication: an integrated computational workflow for design and digital additive manufacturing of multifunctional heterogeneously structured objects. Comput. Aided Des. J. Special Issue on Geometric and Physical Modeling for Additive Manufacturing (2015)

Johns, R.L., Kilian, A., Foley, N.: Design approaches through augmented materiality and embodied computation. In: McGee, W., Ponce de Leon, M. (eds.) Robotic Fabrication in Architecture, Art and Design 2014, pp. 319–332. Springer International Publishing Switzerland (2014)

Kinoshita, H., Swift, P., Utton, C., Carro-Mateo, B., Marchand, G., Collier, N., Milestone, N.: Corrosion of aluminium metal in OPC- and CAC-based cement matrices. Cem. Concr. Res. **50**, 11–18 (2013)

Lafarge North America Inc., Lafarge Portland Cement, Safety Data Sheet, Version 2.0 (2015)

Liu, Y., Leong, B.S., Hu, Z.T.: Yang., E.H.: Autoclaved aerated concrete incorporating waste aluminum dust as foaming agent. Constr. Build. Mater. **148**, 140–147 (2017)

Lefteri, C.: Making It: Manufacturing Techniques for Product Design. Laurence King Publishing (2007)

Lewis, A.J.: Direct ink writing of 3D functional materials. Adv. Funct. Mater. **16**, 2193–2204 (2006)

Mathey, G.R., Rossiter, Jr, J.W.: A Review of Autoclaved Aerated Concrete Products, U.S. Department of Commerce, National Bureau of Standards, National Engineering Laboratory, Center for Building Technology, NBSIR 87-3670 (1988)

Maziah, M.: Development of Foamed Concrete: Enabling and Supporting Design, Ph.D. thesis, University of Dundee (2011)

Mahamood, R.M., Akinlabi, E.T.: Functionally Graded Materials. Springer, New York (2017)

Morgulis, S., Levine, V.E.: Decomposition of Hydrogen Peroxide by organic compounds and its bearing on the catalase reaction (1920)

Oxman, N., Keating, S., Tsai, E.: Functionally graded rapid prototyping, innovative developments in virtual and physical prototyping. In: Proceedings of the 5th International Conference on Advanced Research in Virtual and Rapid Prototyping (2011)

Studart, A.r., Innocentini, M.d.m., Oliveira, I.r., Pandolfelli, V.c.: Reaction of aluminum powder with water in cement-containing refractory castables. J. Eur. Ceram. Soc. 25(13) (2005)

Veenendaal, D., West, M., Block, P.: History and overview of fabric formwork: using fabrics for concrete casting. Struct. Concr. **12**(3), 164–177 (2011)

Towards the Development of Fabrication Machine Species for Filament Materials

Maria Yablonina(✉) and Achim Menges

Institute for Computational Design and Construction,
University of Stuttgart, Keplerstr. 11, 70174 Stuttgart, Germany
{maria.yablonina, achim.menges}@icd.uni-stuttgart.de

Abstract. The research presented in this paper explores the concept of deploying collaborative heterogeneous robot systems where machines are working together towards a common fabrication goal. Augmenting or replacing existing industrial-robot fabrication processes with task-specific architectural construction machines has the potential to expand the design space of digital fabrication methods beyond the limitations of previously existing strategies. The proposed system implies the development of a library of hardware solutions as well as a digital control tool to enable successful execution of fabrication tasks.

This research is focusing on heterogeneous mobile robotic fabrication strategies specific to filament materials. Deploying smaller robots for manipulation of lightweight thread-like materials allows building significantly larger structures. Multiple task-specific machines developed in this research are designed to carry, manipulate, anchor and pass filament materials in an on-site architectural environment of interior space. This paper presents the current state of the catalogue of robot species developed in this research as well as the experiments and demonstrators performed to evaluate them. Ultimately this research aims to create a larger toolbox of hardware and software tools and methods for heterogeneous teams of custom single-task fabrication and construction robots.

Keywords: Heterogeneous robot teams · Mobile robots · Filament material
Task-Specific robotics · Robotic ecosystems

1 Introduction

Advancements in digital fabrication in the past few decades have been transformative to both the geometrical complexity of architectural objects one can produce and their properties, perpetually expanding the design space. Shifting from the Industrial Revolution paradigm in construction towards a customized approach to unique fabrication processes largely became possible through deploying computer controlled tools for construction. At the forefront of this development are research initiatives that have been appropriating industrial robots to architecture through augmenting them with custom tools and software. However, these machines introduce rudimentary limitations, inherited from the assembly line processes they were initially designed for.

J. Willmann et al. (Eds.): ROBARCH 2018, *Robotic Fabrication in Architecture, Art and Design 2018*, pp. 152–166, 2019.
https://doi.org/10.1007/978-3-319-92294-2_12

The research presented in this paper aims to demonstrate the potential of further expanding the design space of architectural form and materiality through augmenting or replacing off-the-shelf industrial machines with a catalogue of task-specific mobile robot species (see Fig. 1). The current catalogue was explicitly developed for fabrication with filament materials as their unique scale and weight affordances intuitively suggest a low payload high reach fabrication system, diametrical to the conventional industrial arm. Building upon the existing knowledge in the field, this research aims to apply robotic filament winding techniques in fabrication scenarios beyond the constraints of a production hall. Comparatively effortless deployability of small mobile machines suggests an opportunity for using existing interior architectural environments and their features as anchoring areas and frameworks for temporary filament architecture capable of adapting and relocating based on programmatic requirements of the space.

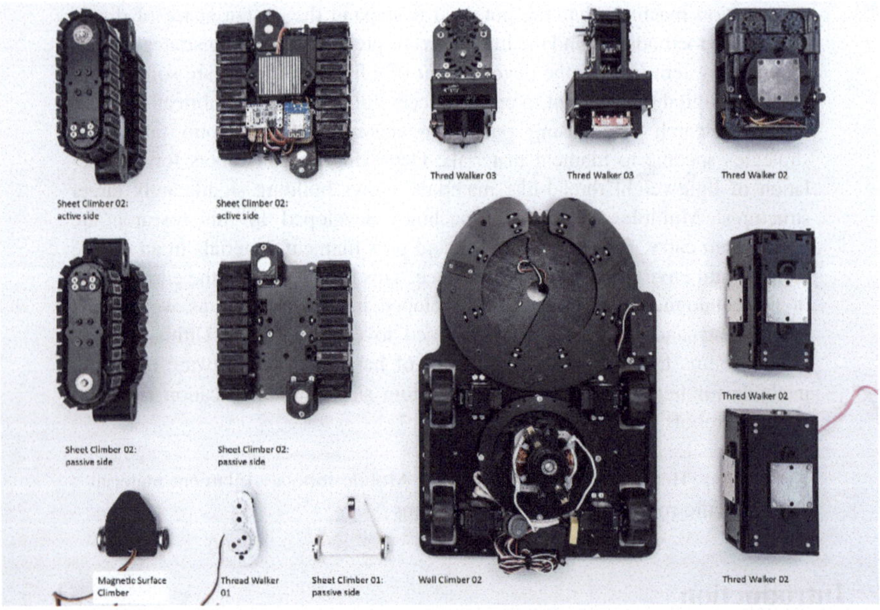

Fig. 1. The catalogue of filament-specific mobile robotic species

2 Background

The essential distinction between manufacturing and construction processes, expressed in the physical relationship between the product and the robot, requires significant rethinking of an automation process towards a system where the product that is the building is stationary, and robots must change their location [1]. Construction robots that emerged as a series of off-site task automation processes in the early 1980s have gradually proceeded to a construction site to become robotic on-site factories towards mid-1990s, combining tasks into an integrated process. However, recent developments in management approach, digital workflow and integration made it possible for construction companies today to re-visit the concept of Single-Task Construction Robot, creating multi-machine systems where automation operations are not directly and rigidly connected, providing workshop-like flexibility, which is becoming more desirable when more individuality of a product is demanded [2].

Currently deployed Single-Task Construction Robots in the industry primarily aim to supplement existing work tasks in conventional and at best slightly altered construction environments [2]. However, a diametrical approach can be seen in the academic and research community. Growing availability and variety of robotic tools along with an increase in ability to handle computational complexity has produced a significant leap in methods and materials one can apply in the context of architectural construction, thus increasing the design space of architectural form.

Filament materials, such as fiber composites that have been a promising technology in construction since the late 1950s [3, 4] are now gaining new value, being freed from the limitations of conventional manufacturing methods through the implementation of industrial robots and complementing fabrication processes like coreless filament winding [5] and integrated formwork [6]. While these industrial machines allow for high inbuilt flexibility through reprogramming [7], they still introduce severe limitations in workspace scale and mobility, residual from the static manipulator - moving product logic of an assembly line (see Fig. 2a). A way to avoid these limitations is to produce smaller modules for assembly of larger architectural elements [8] (see Fig. 2b) or to augment industrial machines with other types of robots designed for large-scale material manipulation and transportation [9]. This approach allows creating scenarios where multiple machines operate collaboratively performing tasks they are best suited for.

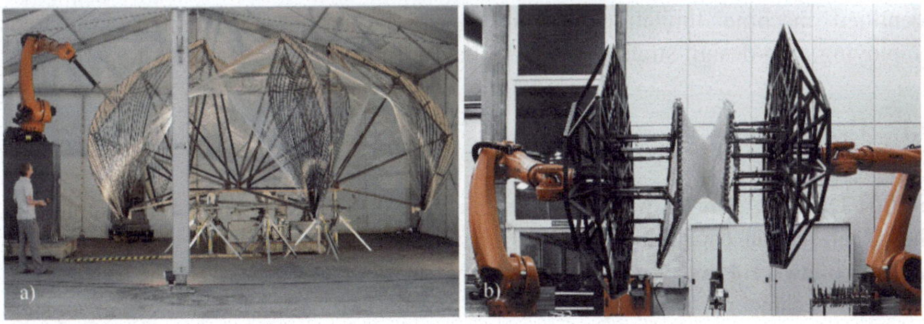

Fig. 2. a) Fabrication process of the ICD/ITKE Research Pavilion 2012 (image copyright: ICD/ITKE, University of Stuttgart). b) Fabrication process of the ICD/ITKE Research Pavilion 2013-14 (image copyright: ICD/ITKE, University of Stuttgart)

Collaboration of multiple machines in a task space has been an area of interest in robotics and computer science since late 1980's [10]. Today one can observe a growing interest in applications of mobile robotic swarm concepts to construction. Projects like TERMES [11] deploy multiple mobile robots for assembly with a designed material, clearly demonstrating the potentials of swarm-based and multi-robot methods in architecture. In the Aerial Construction research conducted by Gramazio Kohler Research group of the Eidgenössische Technische Hochschule in Zurich [12], a team of aerial machines is deployed to create tensile structures out of rope. Through the choreographed movement of aerial robots in relationship to each other, connections between threads are achieved, making the integrity of a tensile structure possible, clearly demonstrating the potential of creating new configurations of material arrangements through machine-machine collaboration, beyond merely increasing the efficiency of an existing process.

Single-Task Mobile Robots allow to distribute the increasing complexity of fabrication goals across families of simpler machines, utilizing a concept of a heterogeneous robot team [13] in an architectural context as is demonstrated in the Minibuilders project by IAAC research group [14], where task-specific machines perform their functions in a hierarchy of steps, each building upon the result of the previous machine's construction sequence.

3 Methods

The presented research aims to create a library of tools, locomotion systems and integration devices for custom filament material-based fabrication robots completed with matching software for object design and robot control. Development of the library is an on-going process, and this paper demonstrates its current stage.

Evolution of the tools in the library, as well as the decision-making process on which machines or systems are to be added, is approached in an experimental framework of iteratively completing on-site demonstrator objects. Site and programmatic requirements of the objects become input for new machine iterations, while identified machine limitations inform the properties of the demonstrator, thus employing a co-design strategy for the robot and the object created by it.

3.1 Demonstrator 01

Setup. Demonstrator 01 was conceived as an experimental setup for validating predicted fabrication capabilities of collaborating mobile robotic units with a minimum number of two robots. While only one type of robot was deployed for this demonstrator, it became a preliminary exploration of mobile robotic filament winding system prior to the introduction of a heterogeneous robot team which is described in the next chapters. For this demonstrator a custom robot capable of locomotion along vertical surfaces of a typical interior space and filament material manipulation [15] (see Fig. 3) was designed. This Wall Climbing robot uses vacuum strategy to adhere to the surface [16], and thus requires a continuous high voltage power supply which results in a tethered configuration.

A furniture-scale object and a complementing winding sequence were designed incorporating the limitations of the tethered locomotion system, such as maximum travel distance and the requirement for entangle-free path planning strategy. The

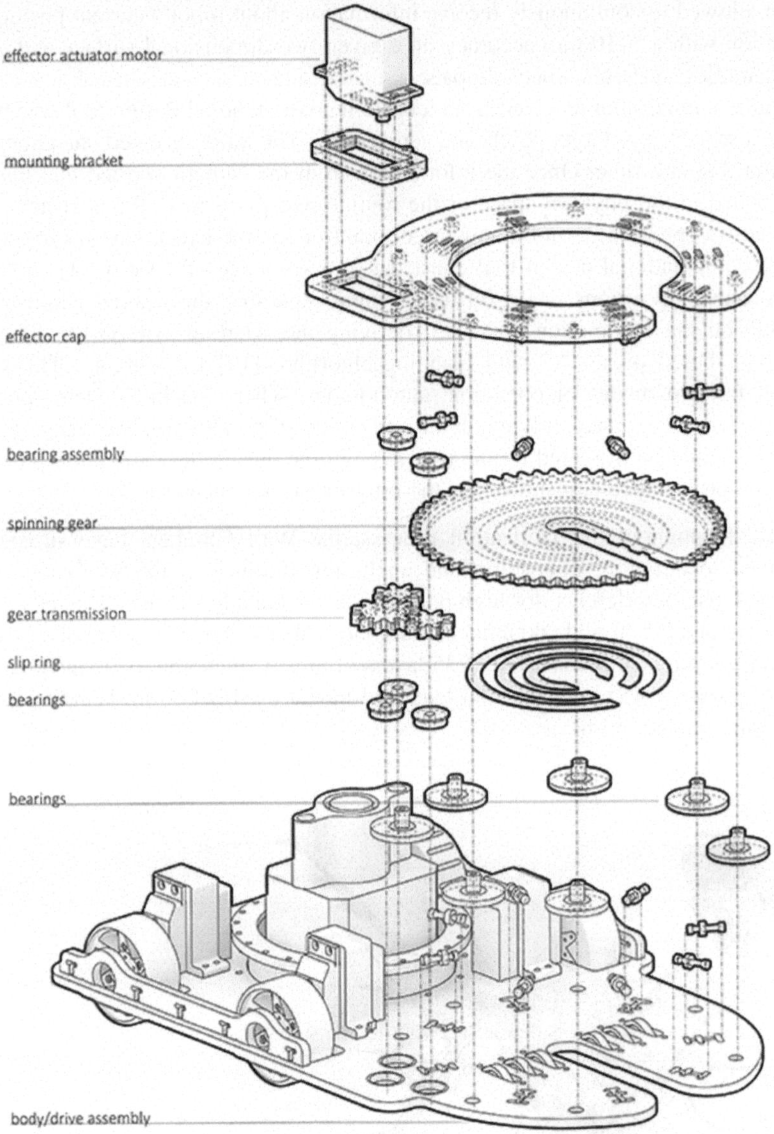

Fig. 3. Wall climber robot's custom material manipulator

winding syntax was developed based on the coreless filament winding technique [17] where a series of layers are applied iteratively with an increasing offset to create a hyperbolic surface.

Localization and Control. In order to accommodate autonomous navigation in the fabrication environment, each Wall Climber was augmented with an external camera and a reacTIVision marker [18] attached to the body of the robot. This perception system allowed to continuously receive information about robot's current position and orientation with a 5–10 mm accuracy on a given two-dimensional surface and correct for locomotion system's wheel slippage.

Path planning input received from a custom computational design tool was broken up into a sequence of target points and action flags. The robot received one target point or action flag at a time. Once the information from the camera verified that the point was reached in the physical space or the routine was performed, the next item in the sequence was processed, and a new target point or routine execution was triggered.

The computational design tool consisted of a sequence of 3 steps: (1) computing optimal anchor positions based on user's input regarding the desired geometry and available on-site locomotion areas; (2) deriving the winding syntax; (3) generating robot path based on an A* path planning algorithm [19] for winding. The system allowed for two modes of operation: autonomous, where machines were controlled entirely from the software, relying on the perception system for feedback, and a manual mode that could be triggered by the user at any time during the fabrication process to troubleshoot, or make adjustments to the on-going task execution.

Results. Throughout the fabrication process, the Wall Climbers have successfully performed locomotion, interaction and anchoring routines in the semi-autonomous mode with user interference for high precision operation adjustments. The result was a 2,5-m-long and 0,5-m-wide doubly curved hollow fiber structure capable of supporting a human (see Fig. 4). It consisted of 35 layers of thread, anchored to 26 anchors. Total count of passes is 455, and the total length of thread used is 800 m. Winding process took approximately 50 h.

Fig. 4. Demonstrator 01 result

Demonstrator 01 has proven the feasibility of using custom wall climbing robotic units for fabrication of on-site architectural objects and has provided ground for further expanding the robotic system towards a heterogeneous robot team. Limitations discovered in the process of completing the structure are to be taken into consideration in future development: travel distance of a tethered locomotion unit, geometric restrictions of a 90° passing routine, imprecision of the chosen localization method and tensioning system for the material.

3.2 Demonstrator 02

Setup. Demonstrator 02 meant to further explore the scalability of the collaborative mobile robot strategy as well as to examine the potential of a heterogeneous robot team building upon the Demonstrator 01. The goal was to increase the span between the anchoring surfaces creating a structure that would span between two parallel walls positioned at a distance from each other. To achieve this, a new fabrication robot type was developed, extending the capabilities of the passing routine: The Thread Walker. This robot was designed to move along a fiber or a rope spanning the overall distance of the fabrication space. It was equipped with an electromagnetic cartridge-carrying effector that would allow passing of the material between the two fabrication surfaces. The Thread Walker picked up the filament cartridge from one of the Wall Climbers, then crossed the distance between the anchor walls, passed the cartridge to the other Wall Climber, which in turn attached the filament to one of the anchors (see Fig. 5) and then passed the cartridge back.

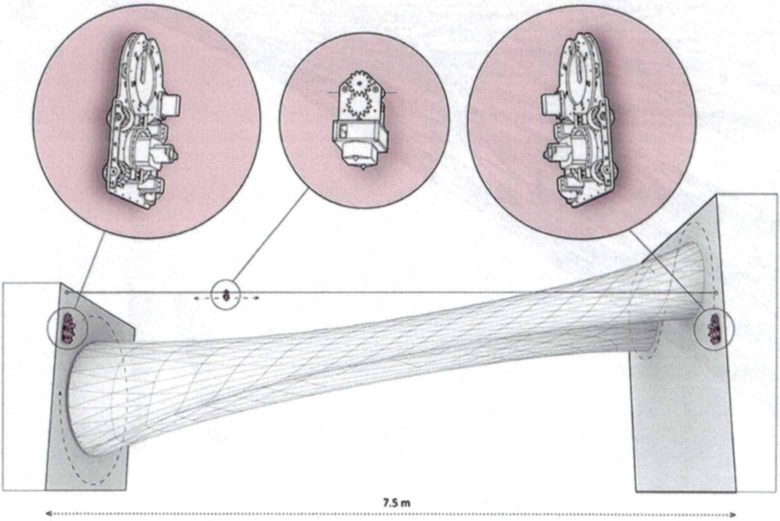

Fig. 5. Demonstrator 02 setup schematics

Localization and Control. Like in Demonstrator 01, the Wall Climbing robots were controlled autonomously using an external perception system and visual markers. For this iteration, the perception hardware was upgraded to higher resolution cameras, and the visual marker strategy was changed from Fiducial to ArUco markers [20] which provided higher tracking accuracy and speed. The Thread Walker was using a more straightforward positioning strategy, as its movement is one-dimensional. The machine was equipped with a pair of optical interrupters that were coupled with pre-calibrated positioning plates installed directly on the transportation thread. During the Thread Walker movement, the optical interrupter would flag that the passing position was reached and the robot would stop and wait for the signal from the Wall Climber to release or engage the cartridge-holding effector. One full cycle of anchoring, passing and anchoring a single thread took approximately 3–5 min (depending on anchor location) in autonomous mode and 10–13 min in manual mode. The critical time constraint for one cycle was a result of speed limitation of the Thread Walking locomotion system, as well as Idle Time[1] for wall climbing robots waiting for the anchoring routines to be completed by its robot counterparts.

Results. The result of Demonstrator 02 was a 7.5-m-long structure made of polyamide thread spanning between two surfaces, fabricated on site in an exhibition space (see Fig. 6) over the course of 10 days prior to the exhibition opening.

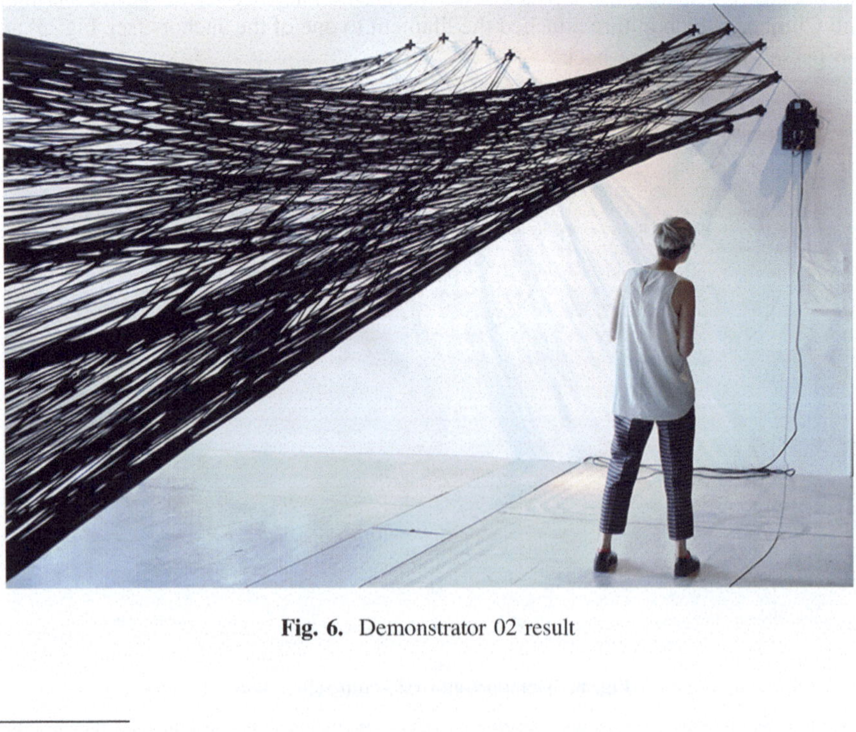

Fig. 6. Demonstrator 02 result

[1] Idle Time - The unproductive standstill of a machine from end of completion to the beginning of the processing of the next material [2].

Unexpected direct sunlight on-site conditions have impacted the perception system's precision and reliability which resulted in only part of the structure being fabricated autonomously during no-daylight hours, while the rest of the process had to be performed with manual user control. Filament tension in the system was relying on the mechanical friction inside of the material cartridge, however depending on the anchor position that was not always sufficient for providing enough force over the large distance, thus an actuated tensioning system is to be implemented in the future iterations.

3.3 Demonstrator 03

Setup. For Demonstrator 03 the nature of the on-site fabrication logic has been changed significantly to provide an experimental setup for evaluation of a tether-less locomotion system. Building upon the knowledge acquired from preceding demonstrators, this iteration addressed issues of idle time and tethered locomotion. The anchoring surfaces of this demonstrator were not an on-site architectural feature, but rather designed objects that could be easily moved and installed at various sites.
The setup consisted of two one by two meter suspended frames with transparent sheets of polycarbonate completed with pre-installed anchors. The frames were positioned opposite each other at a two-meter distance. The thickness of the sheet material accommodated for deployment of a custom magnet-based sheet locomotion robot performing the anchoring routine on each of the two surfaces: The Sheet Climber. Its locomotion system consists of two parts: actuated wheeled unit and its passive counterpart, both equipped with strategically placed magnets that ensure that the sheet material is always clamped in-between the aligned units. This system allows for two-dimensional navigation along the surface of any inclination and curvature.

Two transportation cables were span between the surfaces prior to the fabrication process, each occupied by a Thread Walker, performing passing routine previously developed for Demonstrator 02 (see Fig. 7). Matching the number of thread walkers to the amount of anchoring robots allowed to significantly reduce idle time of the overall fabrication process.

As magnetic locomotion method does not require power for surface adhesion, the robot was designed to be powered from a battery, which eliminated the need for an actuated anchoring effector, replacing it with a pre-choreographed movement of the robot around an anchor to attach filament. Tether-less locomotion units' ability to freely maneuverer between anchors provided an opportunity to execute more complex winding syntax, including planning filament interactions with previously placed layers. Thus the design criteria for the demonstrator was to leverage this feature of the new robot. The outcome of the fabrication process was described by two sets of arrays of straight non-intersecting filaments span between the two surfaces. Two colors of filament were used to highlight the winding logic of a fabrication process with two bobbins used simultaneously by four robots.

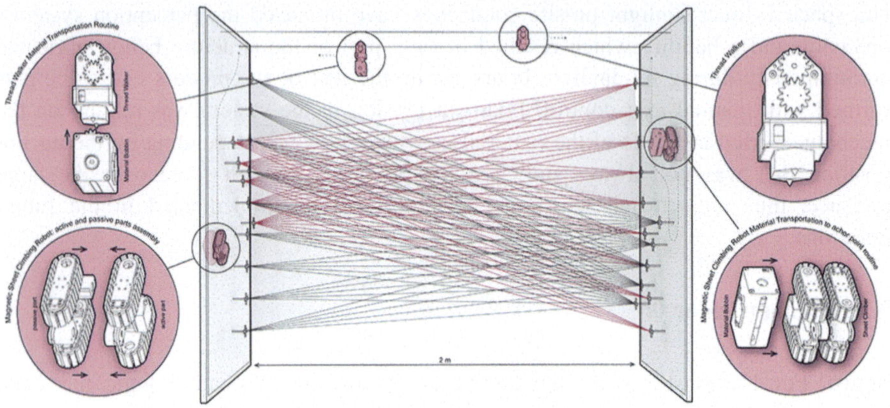

Fig. 7. Demonstrator 03 setup schematics

Additionally, upgrades were made to the tensioning system through implementation of an actuated tensioner and sensor inside the material cartridge. The actuator was installed on the material spool and was rotating it based on the sensor data whenever the tension was too high or too low.

Localization and Control. For Demonstrator 03 significant upgrades were made to the localization system. The camera system was replaced with an off the shelf infrared laser-based external machine vision system and active markers placed on the back side of the anchoring robot, ensuring that they were not obstructed by the structure. This system provided a more reliable and precise (up to 2–4 mm tolerance) position tracking throughout the fabrication process.

Results. The result of Demonstrator 03 was a sculptural object suspended in the center of an exhibition space, providing the viewers with an opportunity to explore it from any vantage point (see Fig. 8).

The structure was fabricated semi-autonomously by robots over the course of 5 days prior to the exhibition opening. Robots were exhibited as part of the installation. Improvements made to the localization system allowed to increase the amount of time the robots were running autonomously to 70%, occasionally requiring user interventions for passing routines between anchor and thread robots.

For the future stages of this research, the Sheet Climber could provide a promising strategy for locomotion along full-scale sheet construction materials, such as windows, façade, and interior finishing panels. To achieve this, sufficient upgrades to the locomotion system hardware are to be considered, including more powerful magnets and motors.

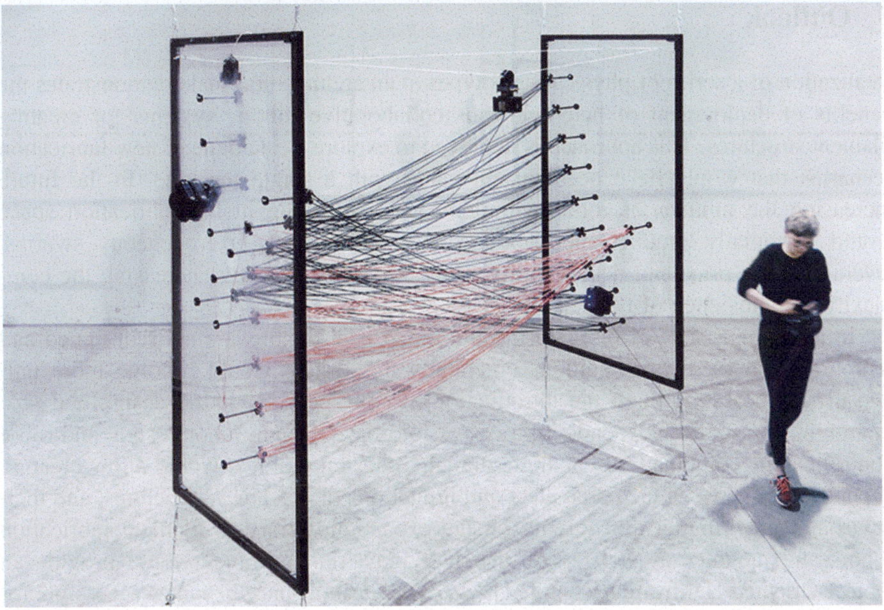

Fig. 8. Demonstrator 03 fabrication process

4 Results and Reflection

The presented series of experiments demonstrate the potential of using heterogeneous robotic systems for achieving complex tasks in fabrication scenarios. However, several aspects of the system require further development to match the efficiency of existing off-the shelf machines.

The control system that is currently centralized implies limitations on the computational capabilities of the setup, reducing the possible complexity of tasks as well as the number of units deployed in one fabrication space. Distributing the control system across multiple onboard computing units for each robot would allow for a combined strategy of local and global control, where global fabrication tasks are dispatched by the central server while local tasks such as pathfinding are performed locally on each machine. This development requires upgrades on the hardware side of the perception and sensing systems. While perception systems used in the experiments have proven to be increasingly successful, each introduces its unique limitations, which become more present with growing complexity of desired structure geometry. A promising path towards a robust localization would be an augmentation of existing perception strategies with a variety of additional sensors, introducing a level of redundancy that would guarantee successful localization in situations of camera/marker obstruction.

5 Outlook

Realization of a series of physical prototypes at an architectural scale demonstrates the benefits of deployment of heterogeneous collaborative robotic systems for creating filament structures. This approach is deployed to explore a catalogue of new fabrication scenarios that would have been unachievable with a single machine. In the future increasing the number of robots simultaneously operating in the fabrication space would potentially require approaching these systems as heterogeneous swarms, leveraging low price and deployability of mobile robots towards increasing the complexity and efficiency of tasks.

In the current state of research machines and control software are redesigned and reprogrammed for every iteration. Potentially the system would become more universally applicable through the introduction of concepts of modular design and programming. Currently independent processes for object design, robot design, and robot control can be combined into a single digital co-design tool that would allow creating dependencies between the parameters and limitations of machines, structures, and their environment. A digital tool relying on the expanding library of filament fabrication robots and modules as well as off-the-shelf industrial machines would provide an Object-Oriented Hardware system [21] where the design space becomes accessible for exploration by a larger pool of users.

The proposed system together with other work of research in robotic applications in architecture demonstrates the possibility of building a hierarchical multi-species robotic ecology capable of navigating in the future construction site environments and performing fabrication tasks cooperatively. Introducing modularity and a level of control and sensing redundancy would allow to move away from material-specific machines towards a broader ecosystem of tools that can be deployed for construction. Through adding or removing hardware and software modules, one can significantly adapt the fabrication process to changing requirements.

Heterogeneous robot teams introduce the discourse of robotic ecosystem, moving towards observing fabrication strategies as "relationships between organisms and their environment" [22]. Ultimately, the research aims towards multi-material applications that can expand the possibilities for computationally designed and digitally fabricated architecture across the spectrum of scales and construction methods.

References

1. Cousineau, L., Miura, N.: Construction Robots: The Search for New Building Technology in Japan. American Society of Civil Engineers (ASCE), Reston, VA (1998)
2. Bock, T., Linner, T.: Construction Robots: Elementary Technologies and Single-Task Construction Robots, vol. 3. Cambridge University Press, New York (2016)
3. Phillips, S.: Plastics. In: Colomina, B., Brennan, A., Kim, J. (eds.) Cold War Hothouses: Inventing Postwar Culture, From Cockpit to Playboy. Princeton Architectural Press, New York (2004)
4. Knippers, J., Menges, A.: Fibres rethought - towards novel constructional articulation. Detail Rev. Archit. **15**, 21–23 (2015)

5. Prado, M., Dörstelmann, M., Schwinn, T., Menges, A., Knippers, J.: Coreless filament winding: robotically fabricated fiber composite building components. In: McGee, W., Ponce de Leon, M. (eds.) Robotic Fabrication in Architecture, Art and Design 2014, pp. 275–289. Springer, Switzerland (2014)
6. Vasey, L., Baharlou, E., Dörstelmann, M., Koslowski, V., Prado, M., Schieber, G., Menges, A., Knippers, J.: Behavioral design and adaptive robotic fabrication of a fiber composite compression shell with pneumatic formwork. In: Combs, L., Perry, C. (eds.) Acadia 2015: Computational Ecologies: Design in the Anthropocene, Proceedings of the 35th Annual Conference of the Association for Computer Aided Design in Architecture, pp. 297–309. Cincinnati, OH (2015)
7. Bock, T., Linner, T.: Robot-Oriented Design: Design and Management Tools for the Deployment of Automation and Robotics in Construction. Cambridge University Press, New York (2015)
8. Doerstelmann, M., Knippers, J., Menges, A., Parascho, S., Prado, M., Schwinn, T.: ICD/ITKE research pavilion 2013–14: modular coreless filament winding based on beetle elytra. Arch. Des. **85**(5), 54–59 (2015)
9. Felbrich, B., Früh, N., Prado, M., Saffarian, S., Solly, J., Vasey, L., Knippers, J., Menges, A.: Multi-machine fabrication: an integrative design process utilizing an autonomous UAV and industrial robots for the fabrication of long-span composite structures. In: Nagakura, T., Tibbits, S., Mueller, C., Ibañez, M. (eds.) Acadia 2017: Disciplines & Disruption, Proceedings of the 37th Annual Conference of the Association for Computer Aided Design in Architecture, pp. 248–259. Cambridge, MA. (2017)
10. Tan, Y., Zhong-yang, Z.: Research advance in swarm robotics. Defence Technol. **9**(1), 18–39 (2013)
11. Petersen, K., Nagpal, R., Werfel, J.: TERMES: an autonomous robotic system for three-dimensional collective construction. In: Durrant-Whyte, H., Roy, N., Abbeel, P. (eds.) Robotics: Science and Systems VII, pp. 177–184, MIT Press, Cambridge (2012)
12. Mirjan, A., Gramazio, F., Kohler, M.: Building with flying robots. In: Gramazio, F., Kohler, M., Langenberg, S. (eds.) Fabricate: Negotiating Design & Making, pp. 267–271. UCL Press, London (2014)
13. Stranieri A., Ferrante E., Turgut A.E., Trianni V., Pinciroli C., Birattari M., Dorigo M.: Self-organized flocking with a heterogeneous mobile robot swarm. In: Lenaerts, T., Giacobini, M., Bersini, H., Bourgine, P., Dorigo, M., Doursat, R. (eds.) Advances in Artificial Life, ECAL 2011, pp. 789–796. MIT Press, Cambridge (2011)
14. Jokic, S., Novikov, P., Maggs, S., Sadaan, D., Jin, S., Nan, C.: Robotic positioning device for three-dimensional printing, ArXiv, Spain. https://arxiv.org/ftp/arxiv/papers/1406/1406.3400.pdf. Accessed 25 July 2018
15. Yablonina, M., Prado, M., Baharlou, E., Schwinn, T., Menges, A.: Mobile robotic fabrication system for filament structures. In: Sheil, B., Menges, A., Glynn, R., Skavara, M. (eds.) Fabricate: Rethinking Design and Construction, pp. 202–209. UCL Press, London (2017)
16. Dethe, R.D., Jaju, S.B.: Developments in wall climbing robots: a review. Int. J. Eng. Res. Gen. Sci. **2**(3) (2014)
17. Knippers, J., La Magna, R., Menges, A., Reichert, S., Schwinn, T., Weimar, F.: ICD/ITKE research pavilion 2012 – coreless filament winding on the morphological principles of an arthropod exoskeleton. Architectural Des. **85**(5), 48–53 (2015)
18. Kaltenbrunner, M., Bencina, R.: reacTIVision: a computer-vision framework for table-based tangible interaction. In: Proceedings of the 1st International Conference on Tangible and Imbedded Interaction, TEI 2007, pp. 69–74. Baton Rouge, LA (2007)

19. Hart, P.E., Nilsson, N.J., Raphael, B.: A formal basis for the heuristic determination of minimum cost paths. IEEE Trans. Syst. Sci. Cybern. SSC **4**(2), 100–107 (1968)
20. Garrido-Jurado, S., Muñoz-Salinas, R., Madrid-Cuevas, F.J., Marín-Jiménez, M.J.: Automatic generation and detection of highly reliable fiducial markers under occlusion. Pattern Recognit. **47**(6), 2280–2292 (2014)
21. Peek, N.: Making Machines that Make: Object-Oriented Hardware Meets Object-Oriented Software. Ph.D. thesis, Massachusetts Institute of Technology (2016)
22. Hensel, M., Menges, A.: Morpho-ecologies. Architectural Association, London (2006)

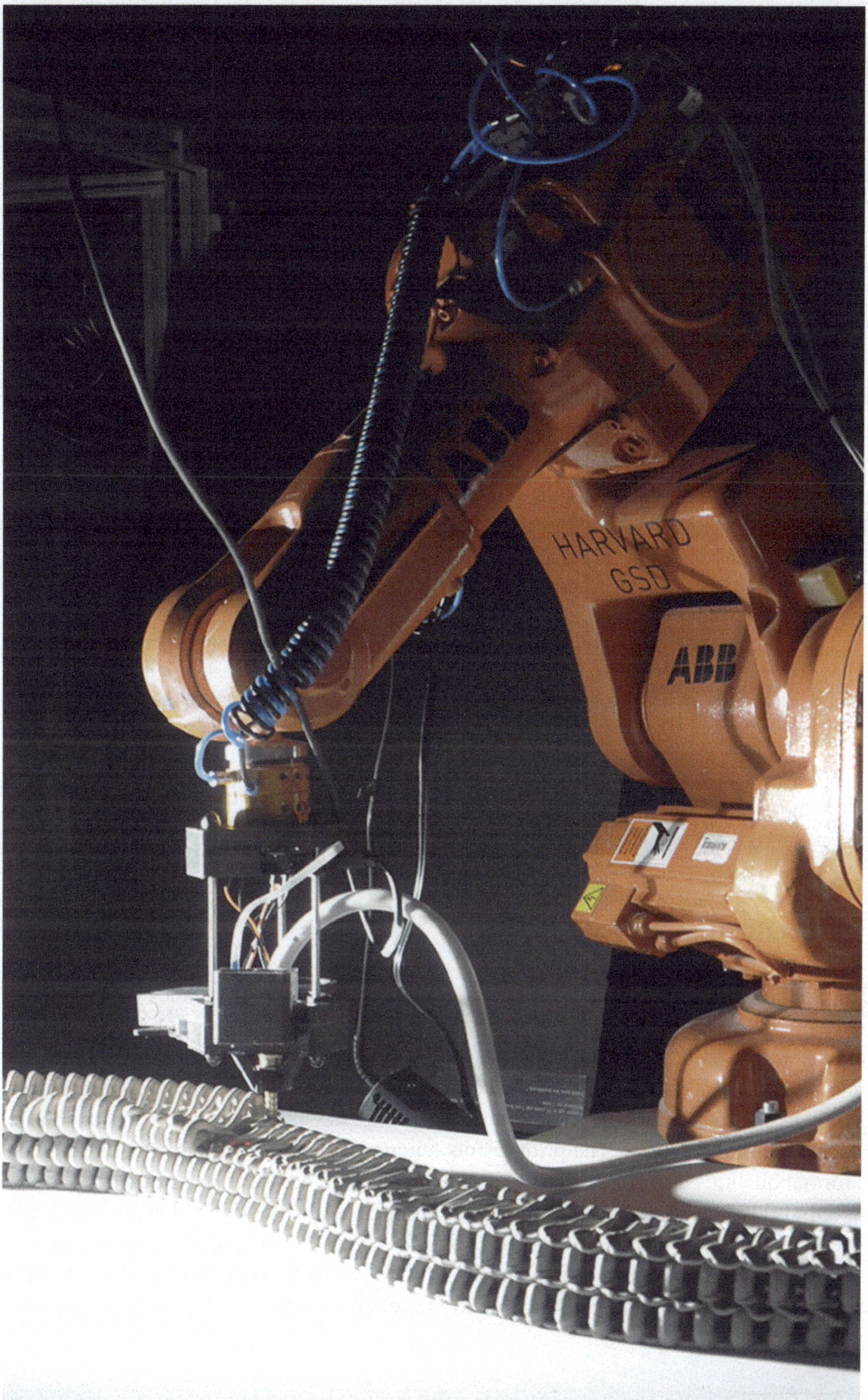
HARVARD
GSD
ABB

Spatial Print Trajectory

Controlling Material Behavior with Print Speed, Feed Rate, and Complex Print Path

Sulaiman AlOthman, Hyeonji Claire Im(✉), Francisco Jung, and Martin Bechthold

Graduate School of Design, Harvard University, Cambridge, USA
hyeonjiim@gsd.harvard.edu

Abstract. Current digital clay fabrication techniques comply with the innate material behavior of clay by extruding in two-dimensional layers. This method inevitably uses an excess amount of material and is a time-consuming process that does not take advantage of the viscous properties of clay. However, by utilizing spatial print trajectories with embedded print parameters (e.g. print speed and extrusion rate), the extrusion behavior of the material can be controlled via simulating actions like anchor, drag, and pull of the clay at the nozzle tip. The aforementioned spatial print trajectory can then form a voxel that can be heterogeneously controlled in order to quickly form self-supporting complex geometries with different density, macro-porosity, and structural rigidity. The print path can also be scaled up to exploit the potential of digital fabrication at the construction scale.

Keywords: Digital soil fabrication · Spatial print trajectory · Material behavior
Print speed · Print path · Extrusion rate · In-situ printing
Self-supporting lattices · Return-loop

1 Introduction

Construction scale 3D printing has been developed for over two decades (Penja 1997), yet major obstacles remain to be overcome. Printing solid concrete or ceramic components is typically based on extruding the material in continuous layers over the previously deposited material. The process continues to be slow and lacks precision. Openings cannot be properly configured since the wet concrete or clay is unable to bridge unsupported spans. The print results remain limited in their performative as well as formal qualities. Polymer printing, on the other hand, has demonstrated the ability to not only bridge unsupported gaps, but even produce mesh-like forms that accelerate the overall build speed in construction (Hack and Lauer 2014). Polymer mesh printing creates truss-like, triangulated geometries which are challenging to print with viscous materials like concrete and ceramic due to the relative softness of the material, and the time needed for it to harden.

Authors by alphabetical order. Authors contributed equally to the preparation of this manuscript.

J. Willmann et al. (Eds.): ROBARCH 2018, *Robotic Fabrication in Architecture, Art and Design 2018*, pp. 167–180, 2019.
https://doi.org/10.1007/978-3-319-92294-2_13

This research aims to address the challenge of 3D printing complex geometries that are non-conforming to the innate material behavior of clay, with the expectations of having aspects of the work translate into a 3D printing method cognizant of the specific behaviors of the material being printed to exploit the inherent behaviors in pursuit of novel form. The work focuses on extrusion based clay printing, which is a widely adopted principle modeled after polymer printing that uses fused filament fabrication. Resulting data on a series of experimental studies that test a computational workflow was recorded to compensate for the anticipated deformations of unsupported clay extrusion, which culminated into the ability to deposit wet clay over unsupported spans of up to 20 mm. The work presented is potentially far reaching. First, it is possible to 3D print mesh-like geometries on the construction scale out of ceramic materials. Doing so shortens print time and enhances material efficiency. Second, the work presented opens up opportunities for 3D printing of both porous and solid conditions through the careful choreography of nozzle movements. The resulting porous material patterns can potentially allow for a modification of insulation and mechanical properties through a controlled variation of solid to void ratios. The principles – here developed through the microcosm of printing clay – can apply to other materials. Of particular promise appears to be printing lattice-systems with high-tensile strength fiber concrete. The ability to do so with concrete opens up applications for on-site printing but not for clay due to the need to fire the material. Towards the end of the paper, possible applications are sketched out. Concrete lattices can produce visually porous forms, but can also serve as a substructure for the quick application of spray-on concrete.

2 Background Review

Institutions like the IAAC and digital fabrication manufacturers like WASP are utilizing robotic fabrication to extrude and deposit clay at a constant speed along a linear print path. Both projects, Pylos (IAAC) and Big Delta (WASP), show potential applications of clay to a large-scale construction in a time and cost efficient manner. Even though much research has been done on the 3D printing of clay, controlling porosity of the end product entirely through the 3D printing process remains challenging. For construction scale printing the generation of insulating (hence porous) and solid (hence load bearing) areas could be pivotal. Integrating porosities has been shown primarily at a millimeter scale, with ceramic extrusion diameters between 0.25 and 0.41 mm and voids in the same order of magnitude (Minas et al. 2016). A master's thesis produced porous patterns while avoiding unsupported areas, but did not supply any thermal performance evaluation (Hinz 2016). The ratio of solid to void largely determines the thermal conductivity of the cellular solid (Gibson 1999). The research claims that, in order to increase thermal resistance, maximizing the pathway of thermal energy through the mass should minimize heat flux through the solid. Lattices with fewer connecting areas and unsupported bridge features can be expected to perform better as thermal insulators.

Extrusion of concrete for 3D printing has been widely demonstrated. The transfer of knowledge from the clay-based work presented here to concrete, however, requires that the concrete quickly strengthens in order to bridge unsupported areas. Work on large scale concrete extrusions demonstrates the technical feasibility of this assumption (Guerrini and Roberta 2014). Extruding unsupported concrete forms then opens opportunities such as the ones shown in the Mesh-Mould project at the ETH Zurich where a non-concrete mesh was sprayed with concrete as a mold and in turn became the reinforcement. This will be discussed later in the paper.

3 Overview of the Experiment

The process of developing a method for 3D printing clay without support involved an experimental approach focused on three aspects: materiality, controllability, and scalability for potential application in large scale construction. A key parameter was the choice of the most suitable clay body, and several different types were tested. The viscosity of the wet clay body relates closely to the required extrusion speed at the nozzle, and both relate to the pressure supplied in the clay reservoir. In the case where the clay mixture was too liquid, the extrusion rate was easily controlled but the clay trajectory was unpredictable, and the intended geometry was difficult to achieve. Through multiple tests and evaluation processes the controllability and plasticity of the clay mixture was optimized, resulting in the selection of a cone 6 porcelain clay for all experiments. This type of clay is stiff enough to allow for the printing of self-supporting lattice systems utilizing novel spatial print trajectories. The ramifications of scaling the print path, heating the printed geometry, and mixing fibers for tensile reinforcement were also studied, with a particular interest in larger scale meshes and porosity for use in the construction context.

3.1 Experimental Setup

The DeltaWASP 20 40 with LDM (Liquid Deposit Modeling) extruder was used to carry out the experiments. A stationary 3 L clay container was pressurized with 5–7 bars of pressure. Clay flowed through a flexible ½″ diameter hose to a nozzle mounted on the print head. Nozzles with 9.50 mm, 7.37 mm and 3.20 mm diameter openings were tested. The work provided the knowledge on how the variables of air pressure, nozzle diameter, cartridge volume, as well as clay composition relate to the ability to create predictable unsupported printed geometries. The research also investigated the influence of alcohol and additives in the clay body, as well as the application of heat through a heat gun in order to preempt sagging of unsupported areas immediately after deposition. Clay, water, alcohol, and nylon fiber were mixed at the ratio of 1 kg clay to 70 g of water, 2 spray shots of alcohol, and 5 g of fibers. The inclusion of the fibers was found to greatly improve stiffness of the freshly extruded clay.

3.2 Procedures

Clay is malleable in the wet state, but takes longer to harden than polymers, metal or concrete. The time delay involved in the drying of clay to the green state allows it to be manipulated as the clay can be printed by first anchoring and then pulling the material along through a spatial drag-movement of the nozzle. Folding the clay extrusion upon itself induces a self-supporting loop that can withstand the dead-load of more loops that are printed on top. The self-supporting loop can be arrayed like voxels in a series of return-loops to form a complex geometry. The positioning of these return-loops can vary to adjust the macro-geometry, porosity, density and structural rigidity of the overall intended structure. The aforementioned processes require a specific mix of clay, print speed, extrusion rate and a predetermined print path. The mix of clay inherently affects the pressure needed to push the clay and also the plasticity of the material being extruded. In order to create the effect of anchoring and stretching the clay, the aforementioned clay body helps to avoid cracks during drying.

The geometry of the loop was parameterized into curve divisions. The angle between the curves divisions are recorded, and the difference between two consecutive angles is calculated. The latter value relates closely to the print speed parameter, reflecting a high print speed at low differences between two angles, and a low speed at high differences. The speed changes throughout with a maximum speed of 10 (mm/s) and a minimum speed of 4 (mm/s) in the looping areas. Next, a supplementary modular print path is needed to produce successive layers, cross-link previously produced forms, and achieve a stable, taller overall component. The secondary print path expands and fortifies the surface area above the vertical return-loop structure by rotating the subsequent return-loop to its side, thereby increasing the points of contact for anchoring and creating lateral supports. The goal was to achieve a double curving lattice structure to prove the print path's capability of printing complex geometries in less time, with less material and with self-sustaining structural capacity.

3.3 Impact of Spatial Print Paths on the Anchor, Drag and Pull Maneuver

The spatial print path is one of the key factors in reproducing the anchor, drag and pull maneuver needed to generate spatial clay lattices. After testing several different heights of the drag path within the loop structure it was observed that the height of the print path has a minimal effect on the actual height of the loop structure produced. Whether the height of the drag path was 25 mm, 30 mm, or 40 mm (Fig. 1), the height of the drag path had almost no influence on the resulting print.

3.4 Impact of Incremental Print Speed/Extrusion Rate on the Anchor, Drag and Pull Maneuver

The anchor portion of the print motion is primarily dependent on the speed and contact point of the nozzle. Congestion occurs when the nozzle is slowly extruding very close to the point of contact - here a surplus of material builds up. This induces a tapering effect of the material when quickly extruded vertically or horizontally from the anchor point (Fig. 2). The tapering of material is the drag portion of the print maneuver and innately has structural implications that allow the material to stretch and span to the next anchor point without disconnecting or drooping. Following anchoring a swift pulling motion with the nozzle creates a return-loop—double loop—and a lateral connection between the loops (Fig. 2). As lateral connections are made the points of

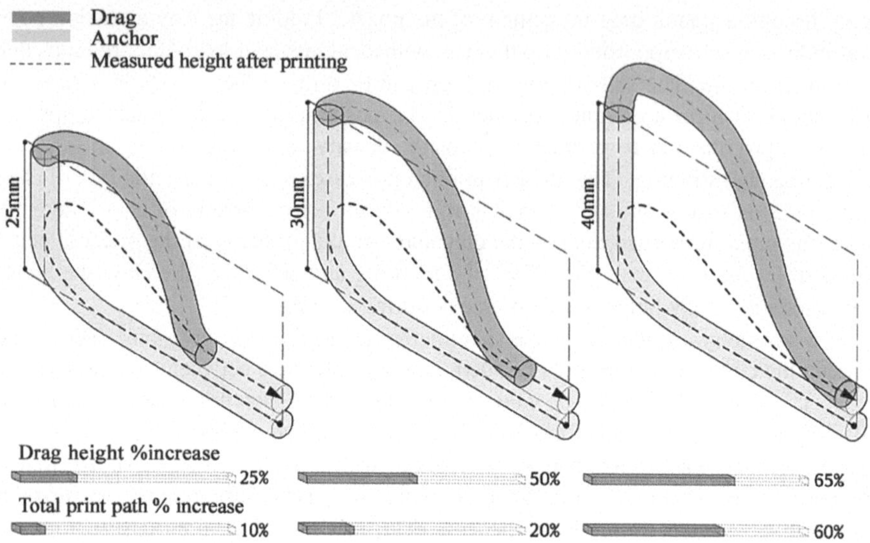

Fig. 1. Increase in drag height (25 mm, 30 mm and 40 mm) vs. height of loop extruded

contact are thickened for the next floor of return-loops because they share the same anchor points. This creates increased surface area that permits a denser anchor for the return-loops above. By creating a network of loops and bridges, the structure evenly distributes the weight, vertical and lateral forces making it stronger.

3.5 Modular Return-Loop Prototypes

Three types of return loops were prototyped, each based on the angle between the loops: 90°, 45° and 0°. Each type incorporated an incremental speed across an array of return loops for both the anchor and the drag paths. The 90° version had a lower range

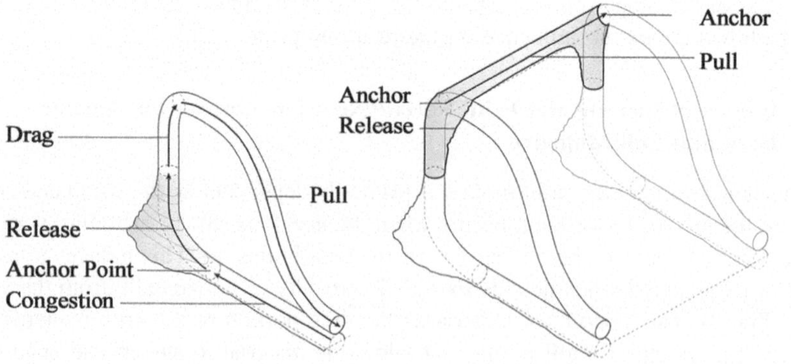

Fig. 2. The nozzle releases the congestion of surplus material and pressure at the anchor point (L). Material can be extruded horizontally from the anchor point to bridge a connection with another anchor point (R).

of extrusion speed since more clay was required at the anchor, and the straight path enabled the clay to remain in its position as more clay was deposited. This type featured a higher resulting path, but with less surface area for the next level to anchor. On the other hand, the 45° and 0° return loops were printed with a larger range of incremental speed as their anchor paths were not straight and thus had less support when deposited. These two types featured more surface area for anchoring consecutive levels and also acted as a bracing element for the vertical return loop (90° loop).

4 Results

4.1 Print Time, Density, and Structural Integrity

Each modular return-loop prototype has corresponding total print times and densities. The 0° return-loop required 6 levels with a total print time of 1 h and 52 min. It is the densest prototype with 0% porosity across the shape (Fig. 3). The resulting element was extremely durable and capable of withstanding dead loads or gravity and itself without visible deformation. The 45° return-loop required 6 levels with a total print time of 47 min making it the next densest prototype with 18% visibility (Fig. 3). The slanted nature of the print path induced a loop structure that leaned on itself in order to be structurally stable. Thus, the structural rigidity was very similar to the 0° return-loop since there were no free standing elements. Crevices that are formed in between each of the slanting loops created porosity and led to reduced overall density. The 90°

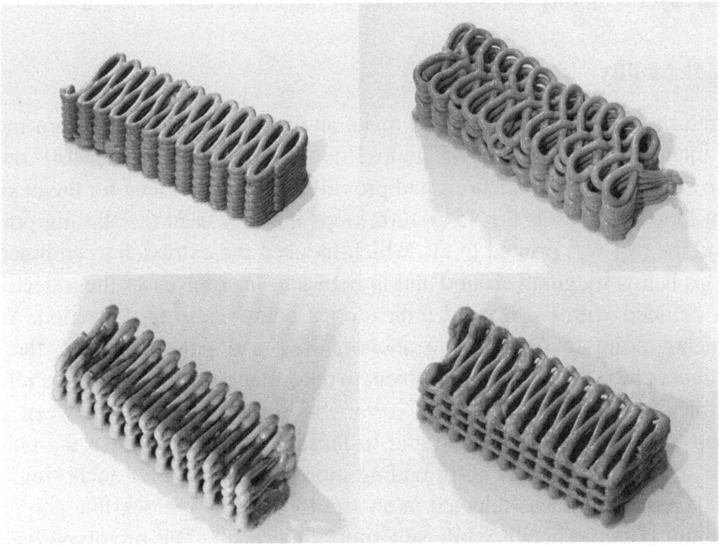

Fig. 3. The 0° return-loop: 6 levels, 1 h and 52 min print time, 0% porosity (Top L), The 45° return-loop: 6 floors, 47 min, 18% porosity (Top R), The 90° return-loop: 4 floors, 6 min print time, 35% porosity (Bottom L), The 90°//0° return-loop: 6 floors, 25 min print time, 25% porosity (Bottom R)

return-loop required 4 floors with a total print time of 16 min which made it the least dense prototype with the highest porosity at 35% (Fig. 3). The tradeoff for this print path was that the loops were freestanding and relied on the anchor points to hold itself up. Any sort of force applied laterally shattered the structure in its entirety. Therefore, a hybrid formation of either the 90°//45° or 90°//0° was necessary to achieve structural rigidity and height with efficient use of materials. The 90°//45° did not yield promising results because the slants of the 45° return-loop hugged and toppled the 90° return-loop below. The 90°//0° return-loop, on the other hand, provided the necessary reinforcement at the anchor points and secured the 90° return-loop below with lateral support. The 90°//0° return-loop required 6 levels with a total print time of 25 min which made it a slightly denser prototype with 25% porosity (Fig. 3). The structure was able to accommodate dead and live weight.

Heating each layer immediately after printing for about a minute at 450 °F was necessary to ensure sufficient anchoring for the consecutive layer without deforming the layer underneath. Otherwise, any weight that the loop encountered at the green state either deformed or broke the print. Heating the printed geometry at higher temperatures above 650 °F rapidly evaporated the water in wet clay, which caused the printed layer to crack. Furthermore, the temperature and time of heating each layer directly impacted the material bonding between layers, which in turn compromised the overall structure of the resulting print. Anything above 8 s and 650 F would make the printed layer insusceptible to bonding with the wet layer of clay printed on top. The printed layers would wedge apart from each other as the clay dried, no longer providing a connection to distribute the dead load. However, with the prescribed amount of heat over time, the layers bonded even better than non-heated layers due to its ability to maintain form and the printed layer's affinity for the moisture in the consecutive print above.

4.2 Predictability

The 45° return loop was introduced to make an alternating 90°//45° return-loop structure. The intention was to lay the 45° return loop structure on top of the 90° return loop in order to create lateral connections and provide more surface area for the next layer of 90° return loops. However, the 45° return loops innately used the starting point as the anchor and the rest was printed in air, which induced the extrusion to embrace the 90° return-loops below by going around and in between. In some cases the trajectory of the extrusion collided with the printed return-loops below causing deflection. Without a proper anchor point to execute the anchor, drag and pull maneuver, the material behavior cannot be controlled and is subject to deformation by deflection. In reference to the aforementioned observations, a 0° return loop was necessary between each 90° return loop to keep the latter stable and in tension, and to minimize the tendency to disconnect at the anchor point by providing more surface area for anchoring.

The 90° return loop was selected in an aim to generate a block-like geometry with the maximum possible height while self-supporting itself. The prototype had reached 28 levels after the clay tank was emptied. Unlike the 45° and 0° return loops, the 90° return loop had a small surface area for anchoring succeeding level, which undermined the rigidity of the overall geometry when printing several levels. In addition, it was necessary to shift each level on the longitudinal axis to accommodate for the slight slant caused by the drag action. Heating each printed level led to shrinkage and cracking, which meant a smaller surface area for anchoring and the connection between the anchor and the top of the loop structure started to crack apart (Fig. 4). This disconnection from the designated anchoring surface or the highest stress point further undermined the overall structural rigidity of the prototype. Moderate vertical force or any measure of lateral force dismantled the 90° return loop structure. Lateral supports and extended surface area for anchoring was necessary.

Fig. 4. Uncontrolled deformation due to deflection and insufficient anchoring (L), 90° return loop structure is offset every level. (M), and Cracks due to shrinkage after heating (R)

4.3 Discrepancy in Digital Print Path and the Actual Print

The combination of 90° and 0° return loops was adopted to create a series of levels in order to form a double curved volume. The process began with dividing the modeled surface into voxels and a constant width for each level. Then the return-loop was modeled within the defined boundaries of each voxel, allowing the articulation of the surface through a voxelized return loop system. After printing the first level of 90° return loops, a heating process was introduced to solidify the clay. This facilitated the anchoring process of the consecutive level and also maintained its height despite the small compression force from anchoring the next level. The actual loop height was not implemented within the generative print path, and thus it was necessary to obtain the measurement of each level's height after printing and heating it. Then, each succeeding level was modified based on the resulting deflection after printing and heating to anchor properly. It was essential that the nozzle tip did not touch the top of the anchoring surface or else the contact disrupted the printed level or the friction led to shifting of the already printed levels out of the printable boundary. Without considering the deflection caused by the material plasticity, 14 levels of a combination of 90° and 0° return loops articulated the overall double curved surface (Fig. 5). Nevertheless, 14 levels only reached 74% of the target height recorded with a substantial decrease in print path length, which is almost half of that when printed with a 2D layering technique (Fig. 5). It was concluded that the articulation of the material system to form a double curved

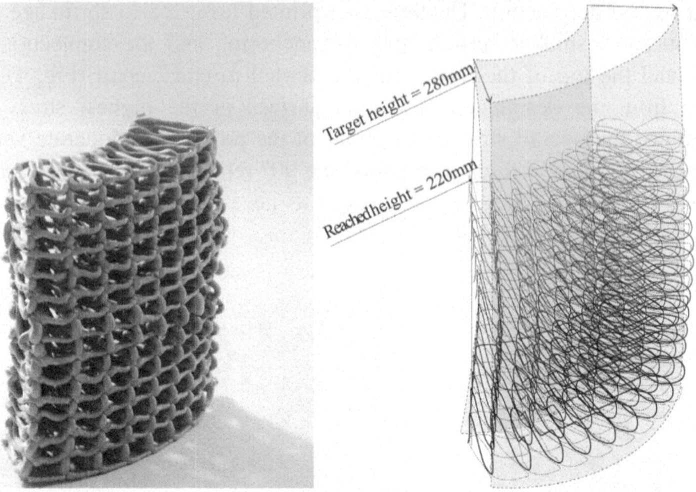

Fig. 5. Printed physical height (R) vs. Projected digital height

surface helped minimize the amount of clay required for printing and thus enabled a significant reduction in print time. In addition, the material system achieved a 25% visibility within the articulated surface, allowing light and air to penetrate through.

4.4 Nozzle Diameter Scalability

Several limitations were seen after using the 9.5 mm diameter nozzle, including the printer's bed size being 20 × 20 cm and the limited volume of the clay tank. Thus, a scaled down version of an extruding nozzle (3.2 mm diameter) was selected for printing the next prototypes. The switch in nozzle diameter brought up the need to understand the relationship between the extruder's nozzle diameter and the resulted print height, which in turn relates to the deflection of the 3D printed loop. It was recorded that when the height of the print path of a loop geometry with a 9.5 mm nozzle ranged between 30–40 mm, the resulting height after printing ranged between 24–26 mm, reflected a deflection of 6–14 mm. On the other hand, scaling down the loop size by the ratio of a diameter change (from 9.5 to 3.2 mm, thus by a factor of 0.33) with a similar morphology yielded a resulting height of 17 mm with a deflection that ranged only between 1.5–2 mm. Scaling overall geometry and nozzle diameter by a factor of 0.33 scaled deflections by a factor of 0.25 to 0.14 – a highly non-linear correlation.

The deflections were recorded after printing. In order to build higher prototypes with more levels it was important to calculate the deflection through structural analysis, and using the material properties of wet clay. Once deflections can be anticipated through analysis the print path can be corrected in order to achieve an outcome that is closer to what has been envisaged. The structural analysis tests were performed using Karamba, a structural analysis plug-in in Grasshopper 3D. Under linear static analysis for loops with heights of 40 mm and 19 mm, and cross sections of 9.5 mm and 3.2 mm accordingly, it was found that the deflections decreased by a factor of 6, confirming approximately what had been experimentally observed. This result not only helped in the process of printing each level at the correct height, but it also showed that knowing the relationship between the nozzle's cross section and the height of the loop could contribute to creating more precise geometries with many levels considering (Figs. 6 and 7).

Fig. 6. Scalability Test, 7.37 mm vs. 3.20 mm nozzle.

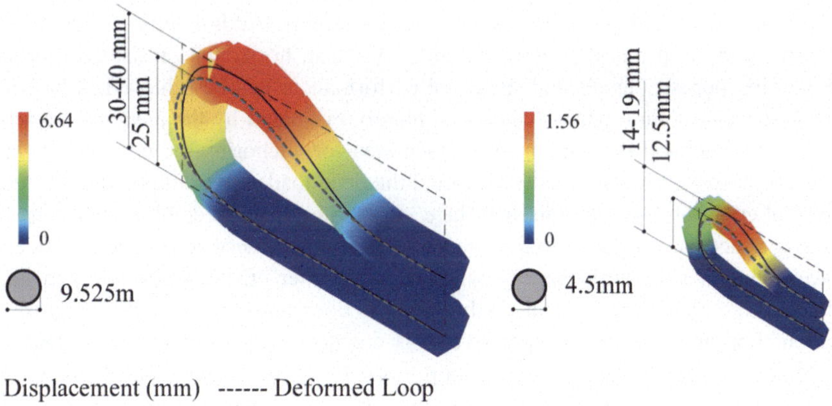

Fig. 7. Lateral supports for return-loop structure.

5 Conclusions and Outlook

The experiments have shown that the process of depositing wet clay over unsupported spans can form mesh-like structures that optimize print time, material use and structural performance. The process is controlled by the amount of clay deposited from the print head, which is configured by the print speed parameter. Reducing the print speed exhibits more material deposition with a thicker clay extrusion. This provides a good structural support for the subsequent model layers to build on with less deformation. On the other hand, bridging between thick clay extrusions required less material deposition that was achieved by increasing the print speed to an appropriate value without disconnecting the clay extrusion or letting it sag. These processes were embedded in an integrated computational workflow that ensured alternation between two values of material deposition to continuously print mesh-like structure with any given form.

We have shown that a double curved mesh-like form can be printed with the developed computational workflow. A feedback system would be desirable in order to correct the location of each successive level of the printed model. Manual measurements of the deflections showed a range of difference between 1 and 4 mm. Deformations varied between each level of the printed form. These measurements were averaged, and the value was fed back into the computational workflow to correct the location for the next level. The disparity of the deformation values necessitated collecting data for each single level of the printed model, which slowed down the printing process. Hence, integrating a digitally driven sensor environment in a closed loop system will allow for simultaneous feed of deformation values for each support condition while printing to ensure proper model correction.

The material deposition of the clay printing process can be manipulated in several ways in order to introduce variation in the porosity across the mesh-like structure. For example, increasing the material deposition at the vertical support of the print path translates into a slower print speed and the print will have less porosity and more structural support. On the other hand, a decrease in material disposition would increase the porosity level and result in less structural support. Additionally, by combining different return-loop morphologies including, vertical, horizontal or slanted morphology, varying porosity levels and structural performances can be achieved. It has to be noted that the structural performance was merely evaluated by the angle of the return loops, which dictated the amount of surface area for anchoring the consecutive layer above and how well it was able to withstand the dead load. For example, the 90° return loop resulted in the least amount of surface area and bracing needed for anchoring, and thus less structural support. Fired prototypes were also observed to have a noticeable improvement in structural integrity without any further cracking. However, a large-scale structure cannot be fired in a kiln and thus does not pertain to the scope of this research. For future work, compressive tests are necessary to precisely evaluate the structural performance of each return loop morphology in its green and fired state. In addition, we can also control the porosity by combining both strategies: the amount of material deposition and print path morphology. This intrinsic behavior of the printing process opens the opportunities to integrate mechanical and thermal properties into the system, which can be varied across any given geometry by controlling the ratios of solid to void through the aforementioned strategies.

This research constitutes a strategy for extrusion-based clay printing to rapidly construct mesh-like forms with controlled mechanical performances. The process enables a viscous material—clay—to be deposited over unsupported spans to structurally connect extrusions. The amount of material deposition and print speed can

Fig. 8. Speculative use of spatially printed concrete for remote site.

Fig. 9. Further development: Double curved surface out of the 90°//0° return-loop.

potentially be tailored to accommodate the material behavior of other, similarly viscous materials, in particular ultra-high strength fiber concrete. It is conceivable that truss-like forms could be printed with ultra-high strength fiber concrete (UHSFC), such that these could act as formwork and reinforcement for spray on concrete much like in the mesh-mould project discussed earlier. Given the good tensile and bending strength of UHSFC, it is conceivable that these systems could provide an alternative method for on-site concrete printing, or even be used to print regolith for extraterrestrial constructions (Figs. 8 and 9).

Acknowledgments. This research was partially funded by the Kuwait Foundation for the Advancement of Sciences (KFAS) under project code "CB18-65EA-01", and the authors would like to sincerely thank Jose Luis Garcia del Castillo Lopez (Harvard University GSD) and Kathy King (Harvard University Ceramics) for the valuable feedbacks to this arduous research process.

References

Gibson, L.J.: Cellular Solids: Structure and Properties. Cambridge University Press, Cambridge (1999)

Hack, N., Lauer, W.V.: Mesh-Mould: robotically fabricated spatial meshes as reinforced concrete formwork. Archit. Des. **84**(3), 44–53 (2014). https://doi.org/10.1002/ad.1753

Minas, C., Carnelli, D., Tervoort, E., Studart, A.R.: 3D printing of emulsions and foams into hierarchical porous ceramics. Adv. Mater. **28**, 9993–9999 (2016)

Penja, J.: Exploratory investigation of solid freeform construction. Autom. Constr. **5**, 427–437 (1997)

Hinz, K.: Brick geometries: 5-axis additive manufacturing for architecture. Master's thesis. Harvard Graduate School of Design (2016)

Guerrini, G.L., Roberta, A.: Process for the production and form preservation of an extruded hollow product made of cementitious material. EU Patent EP1957245, 20 August 2014

PYLOS (n.d.). http://pylos.iaac.net/main.html#main. Accessed 17 Nov 2017

Wang, B.: 40 foot tall 3D Printer for printing houses. https://www.nextbigfuture.com/2015/09/40-foot-tall-3d-printer-for-printing.html. Accessed 1 Jun 2018

An Additive and Subtractive Process for Manufacturing with Natural Composites

Stylianos Dritsas[1,2(✉)], Yadunund Vijay[1], Marina Dimopoulou[1], Naresh Sanadiya[1], and Javier G. Fernandez[1]

[1] Singapore University of Technology and Design, Singapore 487372, Singapore
dritsas@sutd.edu.sg
[2] Architecture and Sustainable Design, 8 Somapah Road, Singapore 487372, Singapore

Abstract. We present research work on a manufacturing process deploying natural composite materials. The objective of the project is to create a sustainable manufacturing process integrating materials, hardware, software and fabrication logic from the ground up. We deploy a bioinspired natural composite comprised by renewable, widely available, biodegradable and low-cost natural components. Material properties closely resemble those of high-density foams or low-density timbers and it is produced without any petrochemical or harmful solvents associated with adverse environmental effects. We designed a mobile material deposition system using the Direct Ink Writing method, with work envelope of over 3 m vertically and indefinite horizontal range, comprised of industrial robotic hardware and purpose-built mechanical mobile platform. We performed testing in characterizing material properties with and without the introduction of the printing process, tightly integrated material behavior with manufacturing and developed design software for direct transition from design to production. To address scaling, we approached the fabrication process from the perspective of fusing the best principles from both additive and subtractive manufacturing, offering geometric freedom and material efficiency of additive manufacturing while targeting production and quality efficiencies of subtractive and forming processes. We believe this process has the potential of significant impact on general manufacturing as well as the building industry.

Keywords: Robotics · Additive manufacturing · Natural composites

1 Introduction

While great leaps in the development of additive manufacturing technologies have and actively take place in nearly every domain of design, two fundamental challenges persist, namely addressing the problem of transition from small to large scale as well as sustainable integration of additive manufacturing materials in ecological cycles. Even though our research work is fundamentally motivated by investigating the implications of prioritizing sustainability in additive manufacturing, using environmentally benign bioinspired materials, we also demonstrate a method for achieving large-scale artifacts with unexpected physical and mechanical properties.

J. Willmann et al. (Eds.): ROBARCH 2018, *Robotic Fabrication in Architecture, Art and Design 2018*, pp. 181–191, 2019.
https://doi.org/10.1007/978-3-319-92294-2_14

Scaling additive manufacturing is complex: Firstly, increase in physical size cannot be achieved in reasonable timeframes by deposition of microscopic layers of material. Production time scales at a cubic asymptotic rate to the size of the product at worst. Secondly, products exhibit multiple scales both physically and conceptually. If their finest details dictate deposition and production rate, an approach that enables multiple resolutions is required to avoid stalling the entire process. Finally, material processes do not scale linearly without significant effort. Indicatively, the surface finish deteriorates by scaling the layer thickness; process parameters such as temperature, cooling and drying are coupled with heat transfer and dimensional range of fabrication; mechanical performance is related with the scale at which material is fused.

Our approach on the challenge of scale proposes a hybrid fabrication process combining concepts from additive and subtractive manufacturing (Fig. 1). We deposit material and perform shape modifications, such as conforming dimensions, smoothing surfaces and eventually aim at adding features at lower scales than the fixed-size deposition range. The objective for such approach is to retain the geometric freedom and material efficiency of additive manufacturing processes but increase production speed of to achieve large-scale at reasonable rates. The dimension of time is more critical parameter than geometry or space in approaching the notion of scalability.

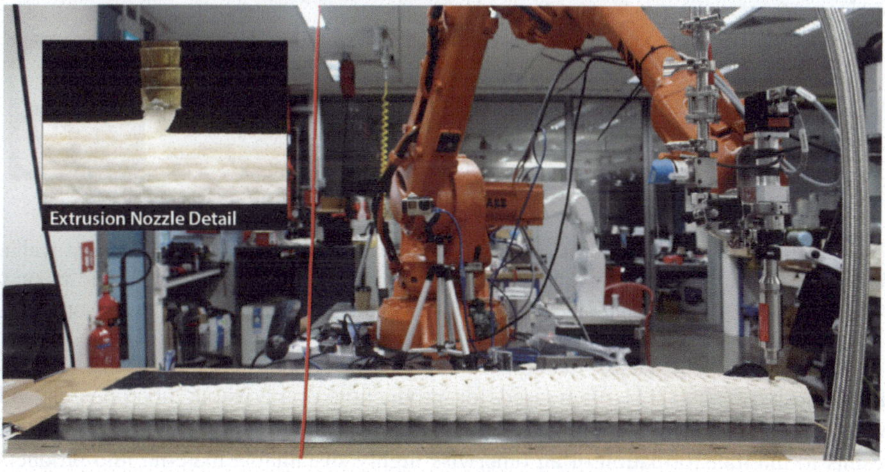

Fig. 1. A hybrid additive manufacturing system based on the direct ink writing method comprised of an industrial robotic arm, material supply and precision dispensing components.

2 Material Design

Popular materials for 3D printing, especially those used for rapid prototyping [1, 2], are derived from petroleum products such as thermoset and thermoplastic polymers. Not all plastics however are equal from a sustainability point of view, as thermoplastics such as ABS can be recycled and PLA is compostable in controlled environments [3]. Inorganics such as plaster [4], clay [5, 6] or sand [7] as well as organics such as wax [8], if unmodified, may be considered environmentally friendly due to modest embodied energy. Metal alloys and glasses can be considered eco-efficient for their high recyclability content. Biological composites offer a profoundly different value proposition when the entire life cycle impact is considered.

The material used here originates in prior research work on bioinspired composites [9, 10]. Its composition is derived from the cell wall of a family of fungus-like microorganism comprised of a cellulose-chitin organic composite. Its components, first and second most abundant natural polymers on earth, are sourced in industrial grades or from waste by-products of timber and food processing. The physical and mechanical properties of the bio-composite such as bulk density and stiffness, are circa 0.4 kg/m^3 and 0.2 GPa respectively, which are within the range of high-density polyurethane foams and low-density timbers. Its appearance, when wood particles are used, resembles fiber board products while in pure form it resembles compressed paper (Fig. 2).

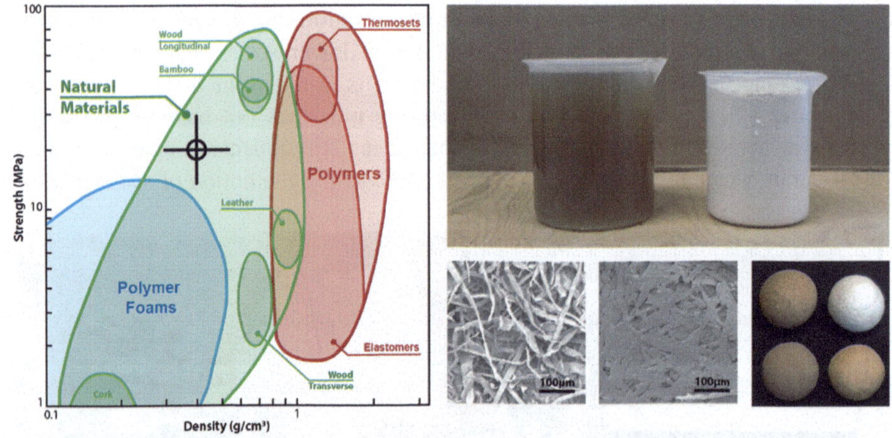

Fig. 2. Left: Ashby plot of the material's physical and mechanical properties situated within range of common natural and synthetic polymers. Right: Material ingredients before mixing, cellulose and wood flour samples and electron microscope photography of fibrous composite.

Cellulose is notorious for being a difficult to dissolve and bond, requiring solvents and adhesive resins [11] often used for production of plywood, fiber boards and additive manufacturing wood-plastic composites [12]. Introduction of those chemicals, results to the transformation of an otherwise highly sustainable material into products which cannot be easily recycled or disposed [13]. Unlike general wood-plastic composites, fully biological materials present a sustainable alternative. Relevant material science work includes study for wood bio-composites fused by glutinous [14] or ligneous [15] organic adhesives from agricultural sources and additive manufacturing with organic composites such as starch [16], protein-based adhesives from animal byproducts [17], and gelatinous bio-composites [18] with chitinous matrices.

Key challenges of this project are in the concurrent design and development of the material along with its manufacturing process. Variations of ingredients and admixtures produce significant changes in material properties associated with process parameters such as flow control, due to viscosity changes, dimensional stability, due to shrinkage under evaporative hardening, phase separation, due to grain size distribution and even mold growth, due to contamination.

3 Fabrication System

While our fabrication process resembles Fused Deposition Modeling, due to its progressive linear bead extrusion, it is also affine to the Direct Ink Writing method [19, 20]. Unlike FDM where linear filament is fed from a spool, our process deploys a viscous colloid transported hydraulically by a pump. As the material cures at ambient conditions, the process does not require temperature control for thermoformed dispensing and fusion. However, convective air flow accelerates drying and various modes such as robot mount air supply or external fans have been used.

The system is comprised of several integrated hardware and software sub-systems: (a) Spatial Positioning, (b) Material Transport, (c) Hardware-Software Interfaces and (d) Design-to-Fabrication CAD/CAM. The positioning system is based on an industrial articulated robot, namely a commercial six-axis robot with 20 kg payload and 1650 mm horizontal reach. The machine is mounted on a hydraulic scissor-lift mobile platform. The platform's vertical travel, up to 1600 mm from the ground, allows for a combined maximum vertical reach of 3700 mm while horizontally, with some calibration effort after relocation, it can be expanded indefinitely (Fig. 3).

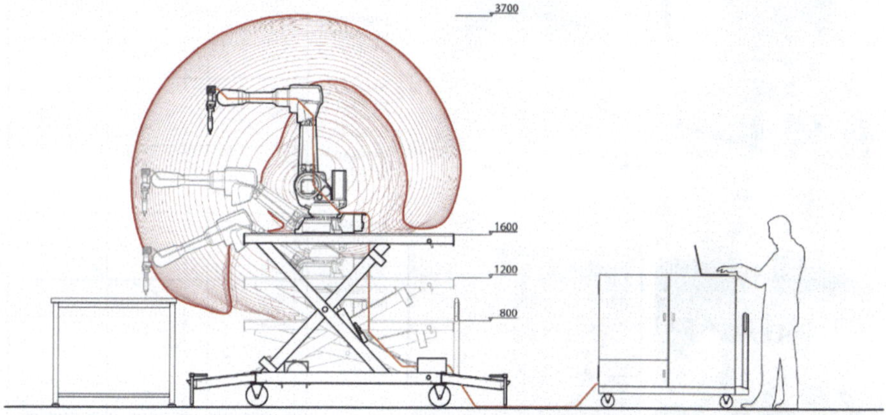

Fig. 3. The digital fabrication system is comprised of an industrial robot mounted on a hydraulic scissor-lift mobile platform. The controller and material supply are separate mobile units.

The material supply system is composed of two units, namely a material unloading pump with 50 Lt capacity and a precision dispensing unit mounted on the robot. The unloading system offers the ability to transport viscous materials without pulsation or shear artifacts present in reciprocating pumps. In addition, the dispensing unit deploys an auger screw cavity transport design, which unlike solenoid valves systems, allows for precise flow control enabling drip and tail prevention. The nozzle's inner diameter can be at maximum 12 mm while flow rate can reach up to 3.5 ml/s. For shaping operations, we employ a PTFE coated nozzle jacket and for subtractive operations we deploy a pneumatic die grinder mounted at 90° about the extrusion nozzle.

Integration between the material supply and positioning system is done via Programmable Logic Control. The firmware design is kept to a minimum, namely digital-to-analog conversion for dispensing flow control and digital switching for peripherals. Control logic is pushed upstream on the design-to-production software based on more modern programming paradigms. The digital fabrication library [21], within the Rhinoceros/Grasshopper parametric environment, is used for kinematic simulation, machine code generation and bidirectional communications.

4 Process Parameters

The most challenging aspect of the process is achieving dimensionally predictable and repeatable results. To define the various controllable process parameters such as bead size, nozzle offset, feed rate and motion speed, during production and after printed artifacts are cured, is a complex multi-dimensional problem which required significant effort in mathematical modeling. Through a series of empirical measurements, a numerical approximation model [22] was fitted and verified as per its capabilities of predicting the process outcomes (Fig. 4).

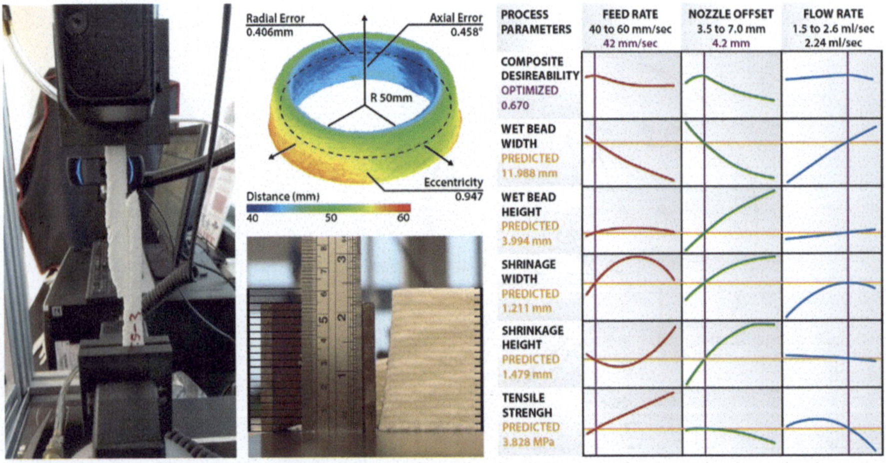

Fig. 4. Left: Material fusion evaluation using universal testing equipment. Middle: Dimensional control testing using optical measurements. Right: Multi-parameter optimization model results.

Process parameters currently used are conservative, with bead sizes of 7 mm, flow rate of 2.2 ml/s and feed rate of 42 mm/s. In comparison to widely available low-cost desktop 3D printers, this amounts to approximately three orders of magnitude faster build rates. The nozzle offset used is 3 mm which due to compaction results into 12 mm wall thickness. Interlayer dependency affects typically three layers below the current except for the bottom and top three which due to print-bed interface and absence of material above, as at the bottom adhesion prevents shrinkage at the same rate as interim layers, while at the top, larger exposed surface area accelerates drying. The success of the model is evident by the nearly uniform layer height achieved vertically.

5 Process Fusion

For generating machine pathing from design geometry, we developed several segmentation and motion planning algorithms. Initially implementations followed a process very similar to FDM, where a solid is decomposed into layers, exterior contours are traced to achieve best surface finish and interior is fractionally infilled with a structural support patterns. However, as bead sizes used here are substantial in size, the exterior surface is bound to rough layered finish. In addition, while the material is highly resilient when cured, in its wet state it exhibits viscoelastic mechanical characteristics which prevent large spanning and cantilevering. We thus had to fundamentally rethink the process logic to respond to material opportunities and limitations.

Fig. 5. Top: Raw and smoothed samples. Bottom: Surface finishing and smoothing passes.

The interior structure of the solid is first constructed at high build rates, preferably with maximum porosity to accelerate drying. Then exploiting the self-adhesive characteristics of the material, we first coat the support and then shape it to smooth finish. The process fuses concepts from additive and subtractive manufacturing by decoupling roughing and finishing guided by material affordances while targeting increased production speed and product quality. In recent iterations of the still under development process, we begun performing deposition and forming in overlapped mode. This allows to take advantage of the window between wet and dry state. Benefits of such approach are: the ability shape geometries, externally and internally, that could not be machined after the built is complete, due to physical access constraints; improve dimensional accuracy and/or relax extrusion constraints, as the secondary process can enforce surface boundaries by performing conforming actions (Fig. 5).

6 Geometric Modeling

The first family of algorithms for support structures is based on dimensional reduction from orientable solid to closed contours and then to scan-lines. Scan-lines stacks seen from section, are patterned by top-down injection of cavities and incremental expansion into protruding arches. This fractal heuristic, inspired by false-arches, produces vault-like structures aligned with the compressive logic of the material (Fig. 5). Unfortunately, due to challenges that emerge from abrupt flow rate modulation required to achieve stripping, we moved to a method that aims to retain fixed flow rate when possible while minimizing backtracking (Fig. 6).

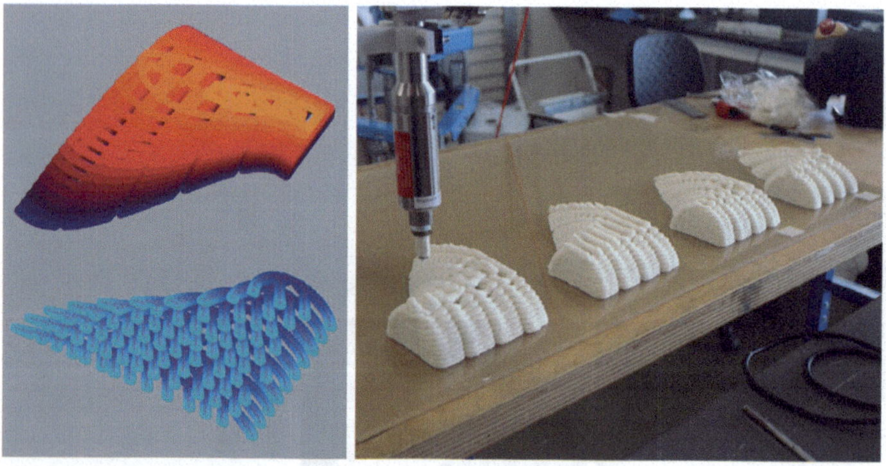

Fig. 6. Left: Computer simulation of continuous sweep line (red) and vault-like support structure (blue). Right: Support structure prototypes based on various continuous sweep algorithms.

The second family of algorithms is based on a parallel sweep-line method as each layer is scan-converted into a serpentine path with minimal state transitions. Modifications of this approach catering to material and fabrication process include interpolated edge-smoothing and fusion of the scan-line boundaries for achieving consistent exterior finish and patterned inter-scan-line overlapping for stiffening and control of shrinkage. This approach enables fractional volume filling of typically under 50% rate which improves air circulation for faster drying.

7 Fabrication and Evaluation

Numerous parts have been printed during iterative improvement of the process. Evaluation models include linear and circular primitives to allow measuring dimensional accuracy, layer compaction, shrinkage uniformity, tensile and bending performance. Even if the material is still medium in viscosity, resembling toothpaste in consistency, it is possible to build single walls of up to 250 mm high. To demonstrate the efficiency of the process and versatility of the material we produced an airfoil measuring 1,200 by 300 by 100 mm. The blade is comprised of two printed halves which required 1.5 h each. After drying for 24 h, they were fused and coated using a thin layer of the material and the surface was finished using sanding. The blade weights approximately 5 kg, it is comprised of one material which can be 3D printed, used as adhesive and processed using conventional wood working techniques (Fig. 7).

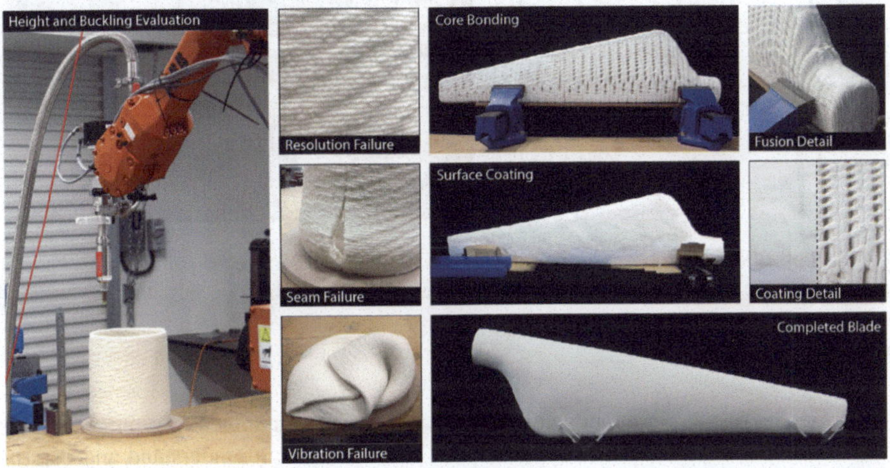

Fig. 7. Left: Vase design evaluating the buckling height of single wall extrusion. Right: Wind blade design evaluating mixed-mode fabrication combining 3D printing and woodworking.

To investigate the ability to create larger scale objects we are developing a prototype for a composite column installation design (Fig. 8). The geometry is expressed as an implicit surface and segmented into layers. As of now, we have not embedded in the machine planning code the ability to continuously vary the process parameters such as motion speed and flow rate. This means that differential drying is a major challenge as extended printing time may result to departure from the pre-set geometry used for motion planning. The pre-set model currently is statically generated from the original design geometry and its dry steady-state after its shrinkage prediction, from the mathematical approximate model, has been incorporated to stretch the original model. For instance, to achieve 100 mm vertical sections, the model must be scaled vertically to 150 mm to account for drying and forced layer compaction for density and stiffness modulation.

Nevertheless, we are investigating the option of either ultra-lightweight single wall segmentation with conservative height-step of 200 mm or composite double wall profiles which target 500 mm high segments. The composite profile parametric model incorporates a double wall design with interior webbing for increased stiffness, micro-patterning the exterior surface and it is produces a continuous path per layer to optimize production time.

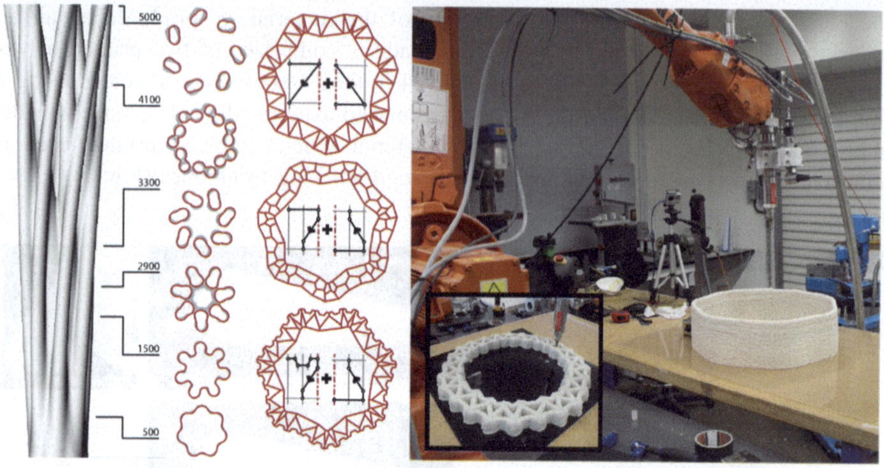

Fig. 8. Left: Elevation of composite column model. Middle: Parametric pathing with various web and surface texture options. Right: Prototypes of under construction installation.

8 Conclusions

We presented results on the development of a composite manufacturing process using natural composite materials. While there is a lot more work to understand and control every parameter of the material and fabrication process, there is good evidence that it can produce sustainable artifacts at literally large scales. We do not know the limits of such scaling but it would be naïve to assume that no assembly will be required to span the orders of magnitude and systemic complexity covered by architectural artifacts. In this respect, it makes sense to target minimizing the production time to leverage scaling. Our process suggests that fusing the best aspects of additive and subtractive manufacturing processes may be a reasonable way forward. As this is a fundamental technology research work, little time has been devoted to specific targeted applications. However, there are foreseeable uses based on material properties, in both product and interior design, such as furniture and finishes, industrial such as from packaging and naval to aerospace, and architecture, such as insulated lightweight partitions, acoustical panels or even free-form precast molds. Success in digital manufacturing with two of the most globally abundant and locally available biological material in the world, positive steps towards this direction were presented here, may have significant impact upstream in design as well as downstream in general manufacturing and overall a sustainable future.

References

1. 3D Systems Inc.: 3D Printing Materials. https://www.3dsystems.com/materials/. Accessed 10 Feb 2018
2. Materialise: Materials http://www.materialise.com/en/manufacturing/materials/. Accessed 10 Feb 2016

3. Martin, O., Averous, L.: Poly(lactic acid): plasticization and properties of biodegradable multiphase systems. Polymer **42**, 6209–6219 (2001)
4. Bard, J., Mankouche, S., Schulte, M.: Morphaux, recovering architectural plaster by developing custom robotic tools. In: Brell-Çokcan, S., Braumann, J. (eds.) ROB|ARCH 2012: Robotic Fabrication in Architecture, Art and Design, pp. 138–141. Springer, Vienna (2013)
5. Friedman, J., Kim, H., Mesa, O.: Experiments in additive clay depositions. In: McGee, W., Ponce de Leon, M. (eds.) Robotic Fabrication in Architecture, Art and Design 2014, pp. 261–271. Springer International Publishing Switzerland (2014)
6. Dunn, K., Wozniak O'Connor, D., Nemela, M., Ulacco, G.: Free form clay deposition in custom generated molds, producing sustainable fabrication processes. In: Reinhardt, D., Saunders, R., Burry, J. (eds.) Robotic Fabrication Architecture, Art and Design 2016, pp. 317–325. Springer International Publishing Switzerland (2016)
7. Gramazio, F., Kohler, M.: Procedural Landscapes, Architecture and Digital Fabrication. http://www.dfab.arch.ethz.ch/web/d/lehre/211.html. Accessed 10 Feb 2018
8. Gardiner, B.J., Janssen, R.S.: FreeFab development of a construction-scale robotic formwork 3D printer. In: McGee, W., Ponce de Leon, M. (eds.) Robotic Fabrication in Architecture, Art and Design 2014, pp. 131–144. Springer International Publishing Switzerland (2014)
9. Fernandez, G.J., Mills, A.C., Samitier, J.: Complex micro-structured 3D surfaces using Chitosan Biopolymer. Small **5**(5), 614–620 (2009)
10. Fernandez, J.G., Ingber, D.E.: Manufacturing of large-scale objects using biodegradable chitosan bioplastic. Macromol. Mater. Eng. **299**(8), 932–938 (2014)
11. Lu, Z.J., Wu, Q., McNabb Jr., S.H.: Chemical coupling in wood fiber and polymer composites: a review of coupling agents and treatments. Wood Fiber Sci. **32–1**, 88–104 (2000)
12. 3D Systems Inc.: Tree-D Printing in Wood: https://www.3dsystems.com/blog/foc/freedom-of-creation-develops-tree-d-printing. Accessed 12 Nov 2016
13. Franke, R., Roffael, E.: Recycling of particle and fiberboards (MDF). Holz Roh Werkst. **56** (1), 79–82 (1998)
14. Lei, H., Pizzi, A., Navarette, P., Rigolet, S., Redl, S., Wagner, A.: Gluten protein adhesives for wood panels. J. Adhes. Sci. Technol. **24**, 1583–1596 (2010)
15. Pizzi, A.: Recent developments in eco-efficient bio-based adhesives for wood bonding: opportunities and issues. J. Adhesion Sci. Technol. **8**, 829–846 (2006)
16. Lam, C.X.F., Mo, X.M., Teoh, S.H., Hutmacher, D.W.: Scaffold development using 3D printing with a starch-based polymer. Mater. Sci. Eng. **20**, 49–56 (2002)
17. Tan, R., Sia, C.K., Tee, Y.K., Koh, K., Dritsas, S.: Developing composite wood for 3D-printing. In: Janssen, P., Loh, P., Raonic, A., Schnabel, M.A. (eds.) CAADRIA 2017: Protocols, Flows and Glitches, Proceedings of the 22nd International Conference of the Association for Computer-Aided Architectural Design Research in Asia, pp. 831–840. Suzhou (2017)
18. Mogas-Soldevilla, L., Duro-Royo, J., Oxman, N.: Water-based fabrication. 3D Printing Add. Manuf. **1**, 141–151 (2014)
19. Stuecker, N.J., Miller, E.J., Ferrizz, E.R., Mudd, E.J., Cesarano, J.: Advanced support structures for enhanced catalytic activity. Ind. Eng. Chem. Res. **43**(1), 51–55 (2004)
20. Lewis, A.J.: Direct ink writing of 3D functional materials. Adv. Funct. Mater. **16**, 2193–2204 (2006)
21. Dritsas, S.: An advanced parametric modelling library for architectural and engineering design. In: Chien, S.F., Choo, S., Schnabel, M.A., Roudavski, S. (eds.) CAADRIA 2016: Living Systems and Micro-Utopias: Towards Continuous Designing, Proceedings of the 21st International Conference of the Association for Computer-Aided Architectural Design Research in Asia, pp. 611–620. Melbourne (2016)
22. Vijay, Y., Sanandiya, N., Dritsas, S., Fernandez, G.J.: Control process settings for large-scale additive manufacturing with natural composites. In: Proceedings of ASME (2018). in press

Hard + Soft: Robotic Needle Felting for Nonwoven Textiles

Wes McGee(✉), Tsz Yan Ng, and Asa Peller

University of Michigan, Ann Arbor, MI 48109, USA
wesmcgee@umich.edu

Abstract. This project explores the development of an additive manufacturing technique for nonwoven textiles. Nonwoven textiles, based on natural materials, synthetic polymers, or blends of the two, have numerous performative aspects, including excellent acoustic absorption, thermal insulation, and tactile characteristics. Felt is a typical example of a nonwoven material, and can be manufactured by both wet or dry processes. One example of a dry process involves needle felting, whereby fibers of the textile are meshed and entangled when punched together. This process binds the material together seamlessly without the addition of sewn thread or adhesives. Needle felting can range in scale from hand craft techniques with a single needle to large scale web processing. Integration into a robotic process not only enables precision and speed in manufacturing but also extends needle felting as a three-dimensional process, allowing for local differentiation of stiffness and other properties across a homogeneous solid. Through a customized digital workflow, formal and material properties can be varied at local level within a component. By developing a fully integrated design to production methodology for influencing these properties, this research opens a wide range of potentials for nonwoven textiles in architectural applications. The research involves three areas of development; the process tooling for robotic felting, the digital workflow that enables the formal and material properties to be specified computationally and embedded into the machine code, and prototypes of architectural elements such as acoustic panels and furniture demonstrating different techniques and processes.

Keywords: Robotic needle felting · Nonwoven textile
Additive manufacturing · CNC manufacturing

1 Introduction

Felt is a nonwoven textile produced by pressing or matting fibers together. The fibers can be natural (wool, bamboo), synthetic (polyester), or blended (viscose). The natural type, commonly wool, is considered one of the world's oldest known textiles, and has a wide variety of uses including furniture, clothing, rugs, musical instruments, etc. (Pietsch and Fuchs 2016) In industrial applications, technical felt is often used in gaskets for vibration dampening or for moisture control. It is fire-retardant and self-extinguishing. (Dent 2009) It absorbs sound (Ballagh 1996; Berardi 2016) and can be molded or impregnated with gelatin or resin for stiffness. (McKracken 1856) Due to the highly interlocked nature of the fibers, felt can be cut easily without the edges fraying.

J. Willmann et al. (Eds.): ROBARCH 2018, *Robotic Fabrication in Architecture, Art and Design 2018*, pp. 192–204, 2019.
https://doi.org/10.1007/978-3-319-92294-2_15

Natural felt has thermal, mildew-resistant, and water-resistant qualities that make it ideal for architectural applications such as interior surfaces as padding or as a buffer between high-moisture areas. (Corscadden et al. 2015) These applications and characteristics highlight the versatile properties of felt and encompass an even longer history of human making in its manipulation. From traditional handcrafting techniques passed on from nomadic tribes such as 'wet-felting' - to contemporary industrial processes of steam pressing for mass-production, felt can be considered one of the most adaptive and transmutable materials available to designers.

It is within the extensive trajectory of felt that this project is positioned. This research seeks to expand the processing of felt beyond traditional textile fabrication methods like cutting and sewing. Typically, felt is patterned, cut into parts, and joined, using equipment similar to a sewing machine or layered together with adhesive. The project seeks to link an industrial process for manufacturing felt, namely needle felting, with computational design techniques and robotic automation. Needle felting, traditionally done by hand with a barbed needle, is a technique whereby fibers of the textile are meshed and entangled when punched together. (Pietsch and Fuchs 2016) This process is seamless and does not require the addition of thread or adhesives. Robotic needle felting extends this into a three-dimensional process, enabling the capacity to work with complex geometric forms while developing variability in the characteristics of felt. Stiffness and deformation become manipulatable aspects of a composition, and layers are bonded together mechanically to produce a homogenous whole.

Additive manufacturing (AM), commonly known as 3D printing, has revolutionized the design to production workflow in a wide range of disciplines. While AM processes have been developed for a wide range of materials, from ceramics to plastics to metals, there have been very few investigations into their applications for textiles. While notable examples exist, such as weaving and 3D knitting, (Ahlquist 2016) these processes impose limitations on the resulting thickness and fiber density. Given the unique capacity of felt to be seamlessly "added" into a cohesive solid, it presents a unique opportunity to investigate the potentials of an AM approach to fabricating geometrically complex components. By developing a digitally controlled methodology for influencing the properties of nonwoven textile, this research opens a wide range of potentials for architectural design applications.

2 Robotic Felting

Needle felting, as a process, straddles two extremes. At one end are traditional craft-based techniques that involve hand punching, and the other, industrial-based manufacturing with mechanized processes. Hand felting is labor and time-intensive but offers nimble control over the material in ways that industrial web manufacturing (continuous roll) is not capable of. This manual technique allows punching from any direction with different depth for each punch, and the amount of force and speed applied to the punching can be variable, responding to feedback from the material. The advantages of mechanized felting, as done for the mass production of web materials, are speed, automation, and consistency of the rate and the force of punching. The disadvantages are that it only punches in one axis, is limited to producing a layer of consistent thickness, and the force and speed for punching is typically not varied across the course of a production run.

Since the industrial revolution, industrial processing for felt manufacturing has remained relatively consistent. (Dent 2009) The novelty of robotic felting combines the advantages of hand felting with an automated process, whereby the designer can control the direction and speed of punching synchronously with the motion of the machine to layer material variably across a substrate. This research specifically focuses on the development of the felting tool as an automated manufacturing process for dealing with complex geometries as well as various techniques to create an inventory of specific performance characteristics (Fig. 1).

Fig. 1. Robotic end-effector V1.0.

3 Research, Development, and Implementation

The development of the robotic felting process involved three interrelated aspects, including materiality, process tooling, and geometric design parameters. Within each are a series of variables that affect the calibration of the other two to achieve the desired effect. These challenges are highlighted below.

3.1 Nonwoven Textiles

Nonwoven textiles can be natural (wool, bamboo, et al.), synthetic (polyester, aramid, et al.), or a blend of both. The characteristics of the textile can vary based on the thickness, density, and composition of the fibers in terms of length and shape. Pressed felt is a quasi - isotropic material (with random directionality of fibers within the planar direction of the sheet), higher density and strength of the fiber is generally proportional to the stiffness of the textile. Sheets can range from highly elastic to rigid. The fiber length also affects the mechanical properties; felt made from short fibers has less tensile strength and stiffness as they become disentangled more easily (Fig. 2) (Ghosh 1994). Less dense felts are more elastic given the ability of the fibers to slide past one another. The complexity of working with nonwoven materials involves understanding not only the material specification but also how it was produced, whether by steam pressing, needle felting, or hydro-entangling. There is a significant interaction between the properties of a given felt and the parameters of the joining process.

3.2 Process Tooling

The development of the robotic end-effector for the felting process included a number of fundamental goals. Virtually all additive manufacturing processes rely on some type of material discretization. The fundamental unit can range from the scale of a brick (Bonwetsch 2006) to a continuous fiber (Vasey 2015), with new methods constantly under development. Given the availability of nonwoven textile as a rolled sheet material, a continuous tape was identified as the unit of material. As a precedent, the automated tape lamination process, used widely in the aerospace industry for composite layup manufacturing, was adapted (Taylor 1980). Previous work by the authors (Seyedahmadian et al. 2015) explored this process in the use of thermoplastic matrix composite tows to produce differentiated surface geometries. In this application, the process has the additional capacity to tailor material properties along a continuous stripe of the felt tape.

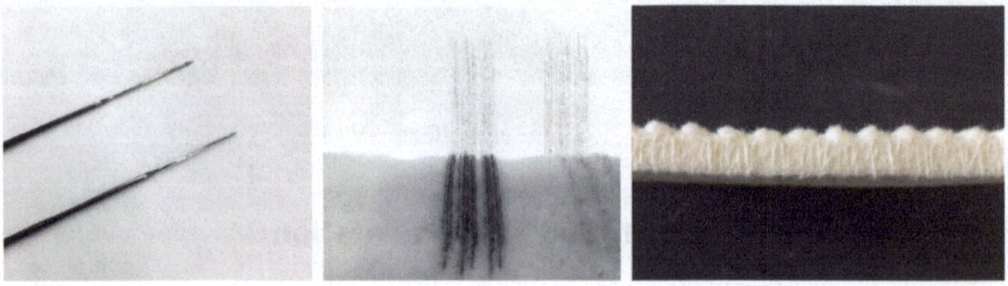

Fig. 2. Felting needle (left), Fiber meshing (center), Multi-layer felting (right)

The felting process can be further divided into two functions, which can be controlled independently. The needle punching is based on a reciprocating mechanism, identical to small scale felting and sewing machines. There are a wide variety of barbed needles to choose from and they come in different sectional shapes, from triangular to 5-pointed star, to twisted, to reverse - each catering to different purposes from general binding to surface texturing. Reverse needles have barbs that pull the fibers in the opposite direction of punching. To achieve sufficient fiber meshing, the rate of punching should be sufficiently high, and the needle stroke should engage as many barbs as possible; in this case it is set to 16 mm (Hearle and Purdy 1971).

While this can be compounded by increasing the count of needles in the travel direction, it is also important that the dwell time of the needle in the punched hole must be sufficiently short, as there is no allowance for the continuous travel of the robot (this would add a significant complexity). The end-effector utilizes a synchronous servo to allow for a user -driven rate of punching which responds in real-time ($\sim$ 12 ms latency) to the travel speed of the robot. The maximum dwell time depends on material and substrate stiffness, and experience shows that punch rates below 1 cycle/mm risk breakage of the needle. The maximum punch rate is limited to avoid over punching the material, effectively severing the fibers and cutting the textile. At the targeted travel speeds of 10–30 mm/s, punch rates range between 10 and 50 cycles per second. This high rate requires careful engineering of the needle carrier to be light and extremely stiff; any resonant vibration will create motor control challenges, though these can be addressed by increasing the rotor moment of inertia. In addition, the system must be able to achieve these rates while punching through multiple layers of material.

The second function is adapted directly from automated composite tape placement processes and provides the ability to cut and restart the feed motion of the tape. This consists of a pneumatic cutter and a synchronous servo-stepper feed motor. This motor must perform two discrete functions. The first is the initial loading of material before the start of each path. The second is to feed the material synchronously with the robot motion. This also adds the ability to over or under feed material, which can produce narrowing effects in the taped path (effectively stretching the material). Felt tapes have low in-plane compression stiffness, which required significant development to feed reliably (Fig. 3).

Fig. 3. Felted strips on substrate, highlighting repeatability of cut and restart to follow subtle curvature in the shingle prototype.

The process control is handled by an industrial software PLC (programmable logic controller), TwinCAT 3, which communicates via the EtherCAT protocol with a Kuka KR120HA industrial robot. The PLC handles the motor trajectory planning and synchronization with the robot TCP (tool center point) motion. To better facilitate design exploration, control of process parameters such as punch rate and material feed are driven via the machine code, as opposed to being completely offloaded to the PLC. This is accomplished using the SuperMatterTools (SMT) (McGee, 15) plugin for Rhino, which allows offline programming and simulation for robotic and CNC machines. SMT provides a user configurable interface to allow precise control over the process, which in this case includes three primary parameters. The punching rate (cycles/mm), start and end of punching along the path, and cut point relative to the end of the path can all be set independently. SMT also provides an application programming interface to allow integration with Rhino via python, C#, and grasshopper components. In this research a custom grasshopper component is used to set the previously described functions variably across the parametrically driven geometry. For instance, the needle punching action can be started and stopped (pulsed) along a path and/or the rate of punching can be geometrically informed. The ability to control process parameters at a geometrically local level is an important consideration for the techniques described below (Fig. 4).

3.3 Design Parameters

In exploring different techniques for robotic felting, the investigation focused on three key techniques that address surfacing conditions as well as structural stiffness. Each is derived based on (1) the quality of the mechanical bonding from needle felting, in conjunction with (2) surface geometry being felted on, and (3) the substrate necessary for the punching process. As discussed above, mechanical bonding can vary depending on the type and length of the fibers. Shorter fibers such as those in industrial (or engineered) wool felt are less able to develop a strong bond. While the fibers may be visibly punched through, they are too short to permanently entangle. In terms of the surface geometry being felted on, the design space is constrained due to the tool geometry and/or robot's range of motion. Primarily this relates to the ability of the end-effector to follow concave geometries. In this case the limitation is a minimum radius of 500 mm in the travel direction. Additional geometric constraints relate to the design of the "foot" which guides the needles and applies pressure to the material. The substrate for the punching process requires a thickness of approximately 20 mm to clear the needle extension from the bottom surface of the foot. Depending on whether the felted surface is to be left in place or lifted off after felting, a variety of substrates can be used. These include expanded polystyrene (EPS) foam (cheap, readily shaped), thicker felt sheets (durable but limited to developable surfaces), or stiff beds made of nylon brushes (ideal but expensive, available as flat surfaces only).

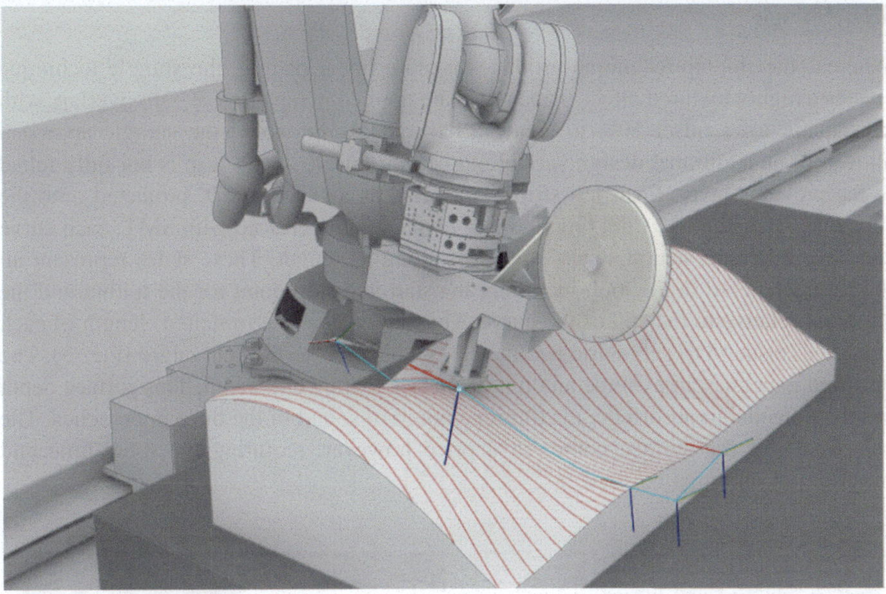

Fig. 4. SuperMatterTools render of robotic felting process

4 Case Studies

As an application of the developed techniques, several experimental prototypes were produced. The tests expose potentials for further study around topics like highly customized acoustic treatments and new surfacing strategies toward textile applications in furniture and architectural interiors.

4.1 Shiplap

As a direct test of the automated tape placement process, this technique utilizes continuous rolls of material which are layered using a variable offset across a surface. The needle distribution is modified to leave approximately 25% on one side of the tape unfelted, creating an effect similar to a shiplap. For instance, for a 38 mm strip, the next layer was offset by 19 mm so that half of the strip would overlap the previous layer, and of that half, only 25% is felted (75% of the total width). The range of overlap takes into account a certain amount of lateral shrinkage that occurs in the punching process. This technique is relatively fast as most of the time is spent felting at full speed (approximately 30 mm/s with the current setup) (Fig. 5). Of the three primary techniques studied, the shiplap technique was the most successful and durable as a surface, as well as allowing a high degree of formal variation.

4.2 Shingle

Whereas the shiplap technique produces a more linear pattern, the shingle technique creates a highly textured effect. Tape strips are applied in a pattern of zero overlap, with subsequent rows offset 50% to create a half-lap. Additionally, the shingle approach allows for an additional design variable to be introduced, as the strip is not fully felted to the previous layer. Each felt strip is modeled as a curve in 3D, projected onto the substrate. By using a custom Grasshopper component interfaced with SMT, each curve can have different "action states" specified along its length. These states represent all the parameters fed to the tool, including the start and stop point for the felting and the cut point relative to the end of the curve. This allows the free (unfelted) length of each shingle to respond to geometric parameters, in this case surface curvature (Fig. 5). The resulting panel has an extreme amount of surface texture and depth. This surface depth has the potential to absorb the maximum amount of sound of the three approaches. The complexity of the surface comes at a price however, requiring the most time and material per unit area covered.

Fig. 5. (left) Shiplap technique with variable overlap, (right) Shingle technique.

4.3 Quilt

The quilting approach forgoes the use of the taping process and uses only the needle punching action to pattern and secure the surface. The felt sheet is precut and patterned as a developable surface to fit the 3D substrate and stacked with additional layers of polymer felt and polymer batting to create an approximately 20 mm thick pad. Needles are removed from the carriage to produce a narrow band which simultaneously compresses and binds the layers to the substrate. The result is similar to typical upholstery techniques but is accomplished from only one side of the material, as opposed to both sides in most sewing and tailored fiber placement processes which require a bobbin (Fig. 6). While this process can cover the most surface in the least amount of time compared to the previous techniques described, the overall binding to the substrate is less effective. After extensive handling (especially pushing onto the pillowed surface), the quilted lines begin to lift as the binding releases from the pressed internal batting. For interior acoustic panels that are out of reach of everyday contact, this technique is more could be appropriate.

4.4 Stiffen

The final technique explored was a hybrid process based on the shiplap approach above combined with a thermally induced residual stress. By layering the felt with a specialized nonwoven polyester known as Fosshape, a thermally formed composite can be created. Fosshape has a high shrinkage when raised to 150C. When felted with non-thermally active felts, this produces considerable surface deformation in the previously flat part. As a composite panel, highly variable curvatures can be designed based on the amount of overlap (thickness) and direction and the felt layers (Fig. 6, right) Additionally, as the thermoplastic layer shrinks, it has the effect of constricting onto the interlayer fibers created during the punching process, behaving like Z reinforcement in woven laminates, and significantly improves the bond between the felted layers.

Fig. 6. (left) Quilt technique, (right) laminated with Fosshape showing surface deformation

Based on the potential of the stiffen technique to produce doubly curved surface geometries from a flat production method, a series of tests were conducted to characterize the basic controlling parameters of the process. Five test panels were produced with the only variation being the overlap of subsequent paths of tape. To clearly quantify the deformation, the layup was unbalanced, with one layer of Fosshape and one layer of wool felt strips. The test panels were then heated in an oven at 150C for approximately 20 min and then allowed to cool (Fig. 7). The curved panels were then scanned with a 2D laser scanner positioned by an industrial robot. The minimum radius of curvature of the panel vs. the overlap percentage is shown in Fig. 8. As the overlap increases from 50% to 75%, the overall thickness and density of the panel increases, providing more resistance to the stresses caused by the shrinking thermoplastic, resulting in a larger radius of curvature. Future work will explore more complex layups which result in highly anisotropic behaviors.

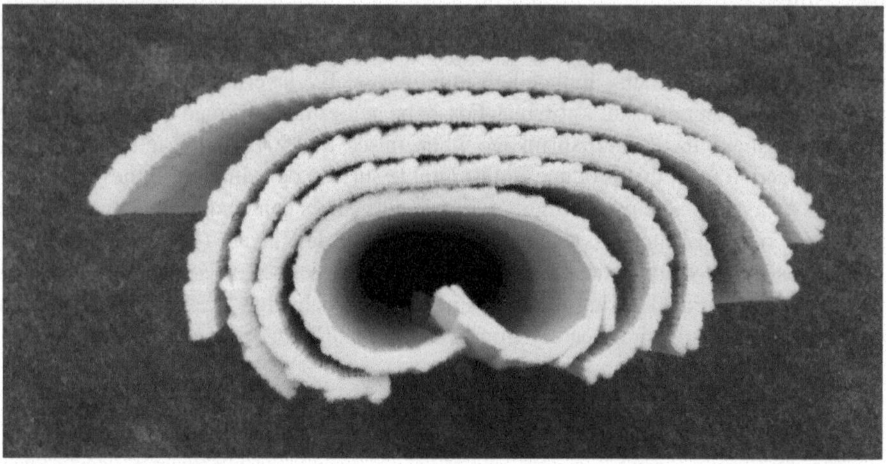

Fig. 7. Variable overlap panels after thermal treatment

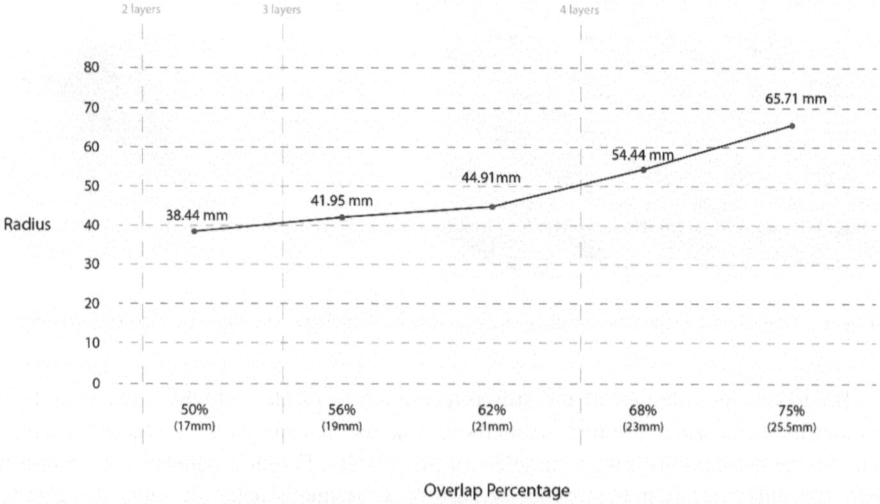

Fig. 8. Radius of curvature vs. overlap percentage

5 Conclusions

Robotic technologies continue to extend the capabilities of architects, designers, and artists to manipulate a wide range of materials. The underlying goal of this research has been focused on advancing the capabilities of nonwoven textiles like natural felt beyond two-dimensional applications toward the production of highly customized geometries with variable material properties. The developed process tooling has been refined to a level that is now capable of continuous production, and future work will focus on the development of improved interlayer bonding as well as methodologies for extending the process to a broader range of material properties and surface geometries, including anisotropic composites and variable hardness surfaces. The research thus far, proposes multiple trajectories. The first aims toward acoustic mitigation applications that take advantage of the material properties of nonwoven textiles. Additional testing will further characterize the anisotropy of the felted layers, with the goal of developing a computational design tool for predicting the curvature produced by the thermal treatment process. This can be combined with a range of synthetic felts, like aramid felt, with the goal of producing semi-structural components. As nonwoven textiles contribute over 50% of the global textile market, there is significant potential in the exploration of new applications.

Acknowledgments. This work was generously supported by the 2018 Taubman College Research Through Making Programme, as well as the University of Michigan Office of Research. Research Assistant Rachael Henry supported the work and Jared Monce, Drew Bradford, and Carlos Pompeo provided production assistance.

References

Pietsch, K., Fuchs, H., Cherif, C.: Textile Materials for Lightweight Constructions: Technologies – Methods, Materials, Properties. Springer, Berlin (2016)

Dent, A.: Felt Technology in Fashioning Felt, pp. 92–135. Smithsonian Institute, New York (2009)

Ballagh, K.O.: Acoustical properties of wool. Appl. Acoust. **48**(2), 101–120 (1996)

Berardi, U., Iannace, G., Di Gabriele, M.: Characterization of sheep wool panels for room acoustic applications. In: Proceedings of the 22nd International Congress on Acoustics: Acoustics for the 21st Century, Buenos Aires (2016) (2016)

McKracken, J. Improved Process of Stiffening Hat Bodies. US Patent No. 15,664 (1856)

Corscadden, K.W., Briggs, J.N., Stillesba, D.K.: Sheep's wool insulation: a sustainable alternative use for a renewable resource? Resour. Conserv. Recycl. **86**, 9–15 (2015)

Ahlquist, S.: Integrating differentiated knit logics and pre-stress in textile hybrid structures. In: Thomsen, M., Tamke, M., Gengnagel, C., Faircloth, B., Scheurer, F. (eds.) Modelling Behaviour, pp. 1–14. Springer, Cham (2015)

Ammayapan, L., Moses, J., Shunmugam, V.: An Overview of the Production of Nonwoven Fabric from Woolen Materials. https://www.researchgate.net/publication/296769658. Accessed 05 Oct 2018

Ghosh, S., Dever, M., Thomas, H., Tewksbury, C.: Effects of selected fiber properties and needle punch density on thermally-treated nonwoven fabrics. Indian J. Fibre Text. Res. **19** (1994)

Bonwetsch, T., Kobel, D., Gramazio, F., Kohler, M.: The Informed Wall: applying additive digital fabrication techniques on architecture. In: Luhan, G.A., Anzalone, P., Cabrinha, M., Clarke, C. (eds.) Acadia 2006: Synthetic Landscapes, Proceedings of the 25th Annual Conference of the Association for Computer Aided Design in Architecture, pp. 489–495. Louisville (2006)

Vasey, L., Baharlou, E., Dörstelmann, M., Koslowski, V., Prado, M., Schieber, G., Menges, A., Knippers, J.: Behavioral design and adaptive fabrication of a fiber composite compression shell with pneumatic formwork: computational ecologies. In: Proceedings of the 2015 ACADIA Conference (2015)

Taylor, P.: The development and use of a tape laying machine. In: Symposium on Fabrication Techniques for Advanced Reinforced Plastics. IPC Science and Technology Press, Salford (1980)

Seyedahmadian, A., Torghabehi, O., McGee, W.: Developing a computational approach towards a performance-based design and robotic fabrication of fibrous skin structures. In: Proceedings of the International Association for Shell and Spatial Structures Symposium (IASS): Future Visions, pp. 1–12. Amsterdam (2015)

Hearle, J.W.S., Purdy, A.T.: The structure of needle punched fabric. Fibre Sci. Technol. **4**, 81–100 (1971)

McGee, W., Pigram, D.: Formation embedded design: a method for the integration of fabrication constraints into architectural design. In: Taron, J.M. (ed.): Acadia 2011: Integration through Computation, Proceedings of the 31st Annual Conference of the Association for Computer Aided Design in Architecture, pp. 122–131. Calgary/Banff (2011)

Construction and Structure

1800mm
1000mm
770mm
74°
103°

SCRIM – Sparse Concrete Reinforcement in Meshworks

Phil Ayres[1(✉)], Wilson Ricardo Leal da Silva[2], Paul Nicholas[1], Thomas Juul Andersen[2], and Johannes Portielje Rauff Greisen[2]

[1] Centre for Information Technology and Architecture, The Royal Danish Academy of Fine Arts, School of Architecture, 1435 Copenhagen, Denmark
phil.ayres@kadk.dk

[2] The Danish Technological Institute, Gregersensvej 2630, Taastrup, Denmark

Abstract. This paper introduces a novel hybrid construction concept, namely Sparse Concrete Reinforcement In Meshworks (SCRIM), that intersects robot-based 3D Concrete Printing (3DCP) and textile reinforcement meshes to produce lightweight elements. In contrast to existing 3DCP approaches, which often stack material vertically, the SCRIM approach permits full exploitation of 6-axis robotic control by utilising supportive meshes to define 3D surfaces onto which concrete is selectively deposited at various orientation angles. Also, instead of fully encapsulating the textile in a cementitious matrix using formworks or spraying concrete, SCRIM relies on sparsely depositing concrete to achieve structural, tectonic and aesthetic design goals, minimising material use. The motivation behind this novel concept is to fully engage the 3D control capabilities of conventional robotics in concrete use, offering an enriched spatial potential extending beyond extruded geometries prevalent in 3DCP, and diversifying the existing spectrum of digital construction approaches. The SCRIM concept is demonstrated through a small-scale proof-of-concept and a larger-scale experiment, described in this paper. Based on the results, we draw a critical review on the limitations and potentials of the approach.

Keywords: 3D concrete printing · Textile reinforcement · Robotic fabrication

1 Introduction

1.1 Context and Motivation

Concrete continues to be the most used material within the construction industry, with use predicted to increase dramatically over the coming years [1]. Despite its relatively low impact compared to many other construction materials, the extent of global consumption has made concrete use a primary contributor to CO_2 production [2]. Coupling increased scrutiny of environmental credentials with the prospect of escalating use creates a high-impact arena for research efforts directed at improvements, advancements and innovation within all concrete fabrication aspects.

J. Willmann et al. (Eds.): ROBARCH 2018, *Robotic Fabrication in Architecture, Art and Design 2018*, pp. 207–220, 2019.
https://doi.org/10.1007/978-3-319-92294-2_16

Over the last 40 yrs., the associated fields of digital practice and digital fabrication have matured as potent sources of innovation - advancing, evolving or introducing novel design and construction practices. Within the context of concrete fabrication, the literature evidences modern digital approaches having contributed to: (a) Formwork strategies supporting free-form geometry, e.g. single use moulds [3]; adaptive formwork [4, 5]; (b) Stay-in-place formworks [6]; fabric formworks [7]; (c) Formwork-free strategies, e.g. 3D Concrete Printing [8, 9] and robot-assisted generative manufacturing [10]; (d) Tailored reinforcement strategies, e.g. MeshMould and integrated thin-shell roof [11, 12]; (e) Adapted slip-forming approach, e.g. Smart Dynamic Casting [13].

Yet, whilst the targeting of concrete fabrication through the optic of digital-design research has enriched and diversified the landscape of approaches, the transfer to high-impact and disruptive changes within industry remains nascent; and 3DCP is a particular case. Despite 3DCP garnering substantial research interest over the last decade, and having been demonstrated at various scales, key challenges remain that hinder its broader acceptance. A recent [accepted] publication by Buswell et al. [14] offers examples of application and identifies the spectrum of research prospects.

1.2 Scope

This paper presents an approach to 3DCP that takes point of departure in layer deposition based on extrusion, but critically intersects the method with aspects of textile/fabric formwork and tailored reinforcement techniques. This hybrid approach offers advantages that overcome key challenges facing layer extrusion, namely restrictions on build orientation and incorporation of reinforcement. The approach also articulates a distinct architectural expression. To start, we contextualise our approach to relevant technologies and research work. Next, we describe our methodology, experimental setups and results. Then, we examine key implications and identify principle contributions and limits of the approach, particularly those related to imprecision due to dynamics of the meshwork. Finally, we outline future development work that could contribute to improve our approach.

2 State-of-the-Art

The SCRIM approach combines 3DCP and integrated reinforcement technologies. As such, we dedicate this section to a review on 3DCP, while highlighting the latest digital fabrication process with integrated reinforcement strategies.

In a nutshell, 3DCP technology uses computer-controlled placement of material to build a concrete component without formwork. As such, it enables the production of complex geometries in a fully-automated setup, while featuring material savings, faster production time for complex products, and new architectural insights and design strategies. The fundamentals of the process comprise (a) mixing, (b) pumping, (c) controlling and (d) extrusion. These basic steps are combined to configure a 3DCP setup, and the construction application lends itself in categories such as off-site factory-based production of components and on-site, large-scale automated construction. Notice, however, that the existing 3DCP demonstrations (see examples in [14]) are followed by time-consuming processes and/or costly full-scale testing, obstructing general uptake by industry. Besides the absence of technical standards or specification, the main obstacles preventing 3DCP from moving ahead to the market include: the process robustness and reinforcement strategies. The SCRIM approach represents an innovation towards the latter barrier as well as an attempt to overcome limitations in relation to build orientation, which is primarily vertical.

As for integrated reinforcement strategies, MeshMould [11] is one of the latest technologies that provides means to robotically fabricate spatial meshes that compose reinforced concrete without formwork. Likewise, a generative approach for manufacturing of complex concrete structures without formwork and using robotic spray technology is proposed in [10]. In essence, SCRIM could be interpreted as a variation of the above mentioned approaches, since they rely on a spatial reinforcement as the support base for concrete. Nonetheless, in the SCRIM approach we selectively deposit concrete by 3D printing rather than spraying and/or pouring and vibrating it. Besides these technologies, the formwork system technology for shell construction developed for the NEST HiLo [12] represents an innovative solution that uses integrated fabric formwork for modern constructions. This technology offers a degree of control over the shape such that it can be easily optimised for improved structural behaviour and other criteria compared to traditional geometries [12]. Similar to SCRIM and MeshMould, it is the textile material serving as basis for supporting concrete, except a cable-net falsework provides additional support.

3 Methodology and Experimental Setups

The SCRIM approach combines selected aspects of the methods outlined in the previous section. In principle, we employ commercially available Carbon-Fibre Reinforced Polymer (CFRP) meshes to define a target geometry that acts as a combined stay-in-place-formwork and reinforcement. The textile basis of the mesh requires the use of boundary restraints and adequate tension to resist applied loads (mostly self-weight) when robotically printing cementitious material. The material is selectively deposited to achieve specific design goals. In this light, we invert the conventional understanding of 'adding' tensile capabilities to concrete through the introduction of continuous (rebar) or discrete (fibre) ductile reinforcement; rather, we sparsely 'add' compressive capabilities to the CFRP meshwork. In this section, we provide details of the general setup and describe two experiments showcasing the SCRIM approach.

3.1 Process Setup

The installations from the High-Tech Concrete Lab, at The Danish Technological Institute, served as basis for the two SCRIM experiments. The 3DCP setup comprises a 6-axis industrial robot (Fanuc R-2000iC/165F), a progressive cavity pump (NETZSCH with flow rate up to 100 dm^3/h), a 3.0 m long steel-wire concrete rubber hose (Ø32 mm), and custom-design nozzles produced with ABS plastic and metal parts. Additional details are described in [15]. For this setup, we set a robot work volume of 2000 × 1000 × 2000 mm (length x width x height). In Experiment 1, we used a rectangular nozzle with cross-section of 40 × 10 mm; in Experiment 2 we used a round nozzle with Ø26 mm.

The design-to-production framework relies on a custom algorithm developed in Rhino/Grasshopper that translates toolpath designs into Gcode files. This algorithm generates point coordinates and associated orientation vectors; the information required for the end-effector to follow a specified path. The generated files are then sent to the robot using RoboDK, which computes optimal toolpath through inverse-kinematics.

3.2 Cementitious Material Setup

A batch of 36 dm^3 was used in each experiment. In Experiment 1, the mix design comprises a binder system with CEM I 52.5 N and fly ash mixed with fine sand (max. particle size: 1.0 mm), water (water/binder ratio: 0.38), a retarder admixture (0.5% by wt. of cement) to maintain a long open time, and a high-range water reducer admixture (0.1% by wt. of binder). All materials were kept the same in Experiment 2, except for the binder system - now composed of CEM I 52.5R (White Cement) and limestone filler (water/binder ratio: 0.36) as well as a Viscosity-Modifying Agent (0.1–0.05% by wt. of cement). A paste composed of limestone filler, red pigment and water was added to modify the concrete colour.

3.3 Setup for Experiment 1 - Testing the SCRIM Concept

Experiment 1 aimed to test the production viability of the SCRIM concept. A carbon-fibre reinforced polymer textile with open-mesh grid of 30x30 mm was used to construct two meshworks with curved geometry – a tapered half-cone and a ¼ pipe as shown in Fig. 1. The ¼ pipe (with 500 mm radius) was divided in two sections (P_1 and P_2). In section P_1, the mesh was tied back to a support surface with an 18 mm offset. While in P_2, two layers of mesh were overlapped with a translational offset to double the mesh-grid density, i.e. the final grid is 15x15 mm. The two layers were intermittently tied to prevent separation. The mesh spanned to the boundary of the formwork frame, with tension wires used to pull the mesh surface into a closer cca. of the target surface. Next, clamping rails were applied to both the top and bottom edges of the mesh. To systematically test the adhesion of deposited material in non-horizontal orientations, a linear 'square-wave' toolpath with increasing vertical travel was defined and repeated for each section.

The same mesh was used to prepare a tapered half-cone with 920 mm edge length and respective diameters 200 mm/100 mm. The mesh was free-spanning with boundary restraints cut from 15 mm shuttering plywood. The mesh-grid configuration is similar to that used on section P_2. The continuous toolpath for this target was defined as cca. sinusoidal pattern with 3.5 'waves' bilaterally mirrored.

Fig. 1. Meshwork setups for Experiment 1.

3.4 Setup for Experiment 2: Scaling up

Experiment 2 aimed at scaling up the SCRIM concept. The goals for this experiment were to: (a) address a larger-scale mesh target, (b) increase spatial complexity by combining two CFRP meshes, (c) demonstrate the use of applied material as a permanent junction mechanism between discrete meshes, and d) test the effect of an intersecting toolpath on deposition and adhesion performance.

In this experiment, a CFRP mesh with grid size of 14 × 7 mm was used in combination with a Ø26 mm nozzle. The target meshwork comprised two CFRP planes (1800 × 1000 mm each) intersecting with an angle of 103° and a 74° inclination from horizontal, giving a 2nd-order rotational symmetry in plan and bilateral symmetry in elevation. One of the CFRP meshes was cut along the intersection line and re-joined with ties. The assembly is shown in Fig. 2. The CFRP meshes were held top and bottom using aluminium Keder rails (40 × 10 mm), with aluminium plates holding the rails at the prescribed intersection angle.

This assembly was attached to a support superstructure to tension the mesh and transfer loads during construction and curing. The superstructure was made up of Ø34 mm steel tubes held with Kee-klamp connections and screwed into the build base. With the exception of cross-bracing and 'outrigger' supports, its configuration approximated the boundaries of the mesh planes to minimise obstacles for the robot.

Fig. 2. Meshwork setup for Experiment 2.

The toolpath for this experiment was developed from a knit-like pattern to test intersection conditions. The total toolpath length was 38.0 m, yielding a concrete consumption of cca. 20.2 dm^3. This pattern was projected onto the 'open' face of the mesh assembly to define a toolpath, modified at the boundary conditions to ensure continuity. This strategy prevented start-stop in the pumping process.

4 Results

4.1 Experiment 1

The open cell size in section P1 was similar to the nozzle width (40 mm); thus, a slight amount of material fell through the mesh. This falling effect in P1 was more pronounced in the vertical parts of the mesh (Fig. 3a). This effect indicates that the fresh concrete mix cannot cope with the bending moment and shear stresses in the open cell. Alternatives to tackle this effect include the use of fibers in the mix design, and production of a stiffer mixture – with the drawback that these yield greater challenges to 3DCP; namely, stiff mixtures are prone to filament tearing and/or splitting during extrusion [16]. Also, the use of a denser mesh (as in P2) provides better support – minimizing the falling effect as shown in Fig. 3b. Notice, however, that due to the deviation of meshwork from the assembly in P2, this resulted in a mismatch between the CAD model and toolpath; specifically, the curvature in the model was greater than that of the assembly, hampering the maintenance of a regular tool offset during printing. Thus, in certain areas, the extruded material was excessively pushed through the mesh as can be seen in Fig. 3c. In the cone section, a greater distance between the nozzle and mesh was noticed (Fig. 4a); as a result, the material-mesh connection was apparently weaker than that observed in the deposition areas in P1 and P2, though the push effect was eliminated (Fig. 4b, c).

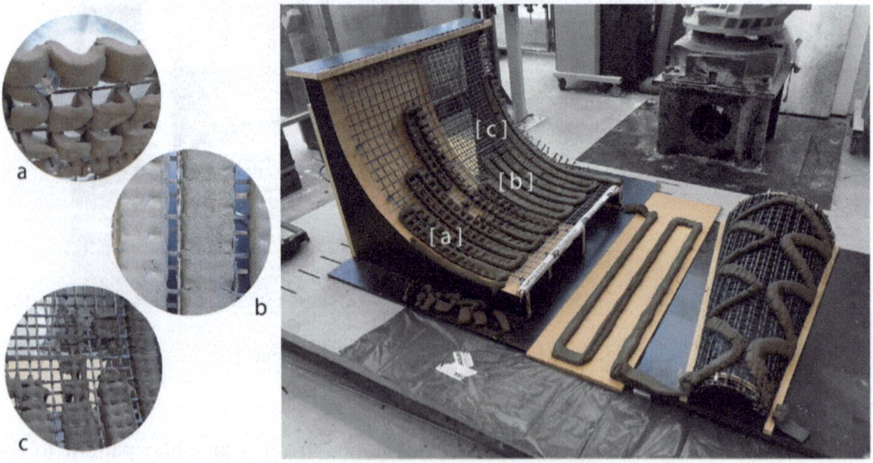

Fig. 3. SCRIM results of ¼ pipe in Experiment 1.

In general, Experiment 1 showed that, despite localised challenges, there was a good integration of the mesh within the concrete and that printing in vertical conditions is plausible.

Fig. 4. SCRIM results of tapered cone in Experiment 1. Released free-standing assembly (right), with details of deposition process and results (left).

4.2 Experiment 2

The 3DCP started at the top of the mesh with a nozzle travel speed of 30 mm/s. Good adherence and encapsulation was observed on the first two rows (Fig. 5a), but, as further material was added, the distance between nozzle and mesh increased due to mesh deflection under the applied self-weight. This reduced the filament adhesion (and sometimes direct contact) to the mesh. After cca. 20 kg of deposition, material sheared down the 74° mesh inclination. In a second attempt, we reduced the pattern length to 16 m (Fig. 6), corresponding to 18 kg material. The nozzle travel speed was increased to 60 mm/s, thus decreasing the linear material deposition rate by half. In this case, we achieved markedly better material adhesion and encapsulation to the mesh, with only localised sections of material collapse due to mesh dynamics, as detailed on Fig. 5b.

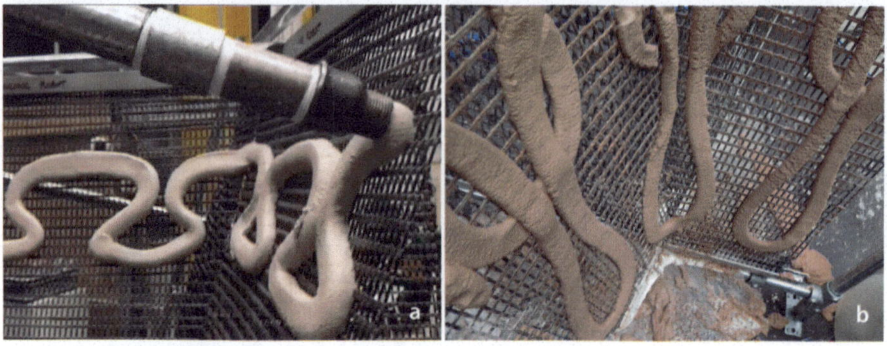

Fig. 5. SCRIM results in Experiment 2. (a) First deposition attempt; (b) second deposition attempt

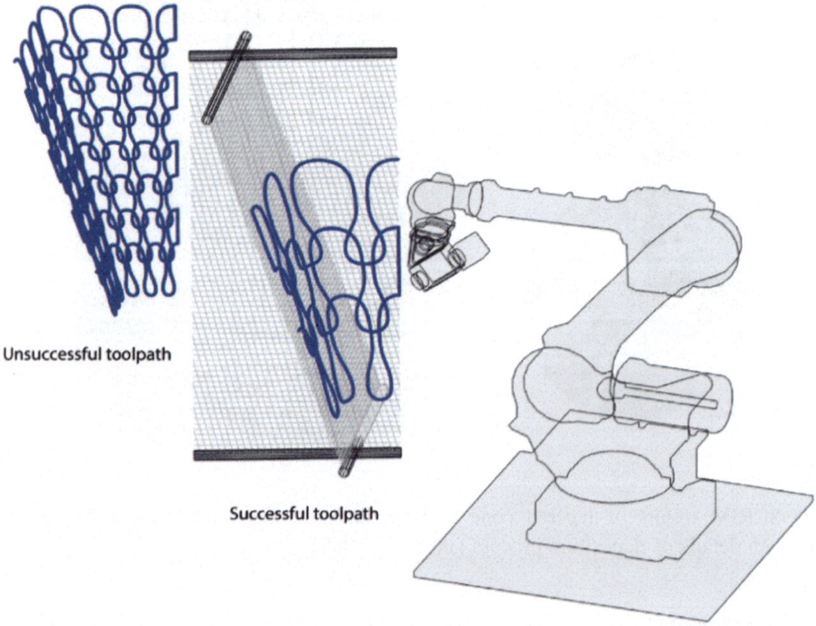

Fig. 6. Toolpaths for Experiment 2.

In general, Experiment 2 showed that heavy deposition of material in partially supported meshes causes sufficient deflection to impact consistency in "layer height". Also, the use of a slow travel speed and dense toolpath intersections contributed to failure in adhesion. By reducing the amount of deposited material and increasing travel speeds, we succeeded in demonstrating: (a) increased complexity of the toolpath, (b) deposited material acting as a bonding mechanism between discrete meshes, and (c) adhesion of material even with toolpath intersection in non-horizontal conditions.

5 Discussion and Further Work

Having demonstrated the plausibility of the SCRIM concept through two experiments, we now outline key contributions, implications and limits of the approach as well as prospective developments.

5.1 Contributions and Implications

The SCRIM concept offers three main contributions: (a) fully exploiting 3D spatial capabilities of conventional robotics, allowing greater freedom in build orientation that includes near-vertical material deposition; (b) resolution of the challenge to integrate reinforcement by inverting the conventional concept of 'adding' reinforcement to concrete, to printing on CFRP meshes that act as reinforcement; and (c) decoupling the printing process from a time-critical dependency on very-early age properties of previously extruded layers. These contributions imply that the SCRIM approach (a) supports greater geometric freedom in design targets, extending beyond extruded geometries prevalent in existing 3DCP approaches; (b) supports intricate optimised geometries that selectively place material to achieve design aims, further reducing material use; (c) introduces a novel tectonic language that diversifies the existing spectrum of digital construction approaches.

5.2 Limits

Limits of the Method. A principle challenge in the SCRIM approach is that the meshworks require tensioning and exhibit dynamics when load (mostly from the material self-weight) is applied during printing. This can result in layer height irregularities, with consequences on mesh encapsulation and adequate adhesion as witnessed in the first print attempt of Experiment 2. Although SCRIM offers greater freedom in build orientation and the geometry of design targets, the mesh deformation places additional emphasis on tool path verification, which can be more complex in comparison to 2D layered approaches.

Limits in Exploration. To date, the SCRIM experiments have focused on developing the production processes and exploring deposition of cementitious material in non-horizontal conditions; production of designed components is yet to be explored, and, by extension, a study of their cured performance and tectonics in the context of larger assemblies.

5.3 Further Work

In the short-term, incremental advances shall focus on applying the SCRIM concept to the production of indicative building elements with determined performance demands, so that measures of success can be established through testing. Lightweight building elements fulfilling a range of roles can be envisaged, such as load bearing partitions (Fig. 7) or suspended ceiling elements, with each element being individually graded in porosity to accommodate multi-criteria design demands and being produced either

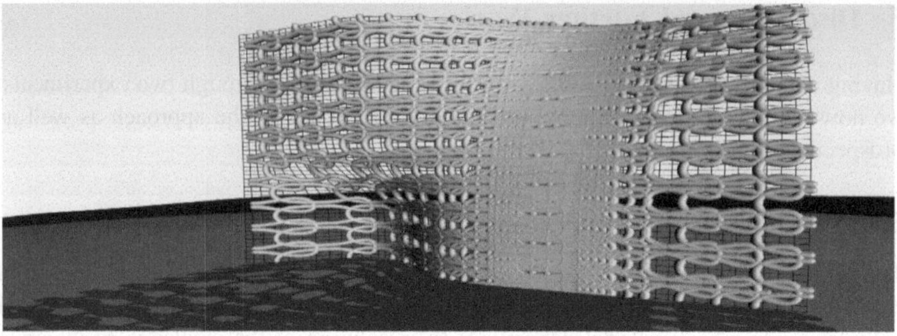

Fig. 7. Speculative visualisation of an envisaged building element - a graded internal partition.

through off-site factory-based production, or by on-site automated construction. A further potential use could be as prefabricated stiffened reinforcement elements, marking an interim step towards massive concrete casting. An interim casting process utilising knitted textiles in combination with cement pastes to produce stay-in-place formwork has been recently demonstrated [17]. The use of SCRIM could mark a hybrid approach combining reinforcement and stay-in-place formwork. Partial, or localised, massive casting would extend this idea further, with compelling architectural potential.

In the longer-term, further development shall focus on refining and advancing the potentials of SCRIM by means of the following:

(a) **3DCP material:** investigating (a1) the effect of 3D printed concrete rheology on the encapsulation between mixture and mesh, (a2) the use of accelerators to control the concrete setting time on demand, and (a3) the effect of fibres on the early age mechanical properties of 3D printed concrete. The goal in this investigation would be to optimise the mix design to achieve an optimal material deposition on CFRP meshes, preventing the falling and/or sliding effect noticed in our experiments.

(b) **CFRP mesh design and production:** investigating the production of bespoke meshes will allow for refined tailoring of structural performance (during and post-printing), provide optimised cell sizing in areas of deposition and support customised incorporation of other architectural elements.

(c) **Adaptive robotic control:** two complementary approaches can be considered: (c1) scanning and registering of constructed meshes to provide 'as-built' data and generate the deposition path; and (c2) online toolpath adaptation through real-time scanning to enable optimised layer height in response to mesh dynamics.

(d) **Integrated design environment:** simulation and analysis of fabrication, in concert with post-fabrication performance, will be used to inform and optimise deposition pattern in relation to design goals. Strategies such as stress-line additive manufacturing (SLAM) [18] could be productively combined with the current aesthetically oriented patterning approaches.

6 Conclusion

This paper has introduced a novel hybrid construction concept that intersects robot-based 3D Concrete Printing and textile reinforcement meshes to produce lightweight elements, namely, *Sparse Concrete Reinforcement In Meshworks*. We have described the process setup and results of two experiments that demonstrate how the SCRIM concept offers advantages that tackle key challenges facing conventional 3DCP based on vertical layer extrusion, i.e. restrictions on build orientation and incorporation of reinforcement. In addition, SCRIM supports subsequent processes of element addition and embellishment as well as articulating a distinct architectural expression in contrast to the approaches reviewed in the literature. We have examined key implications of SCRIM, identified principle contributions, stated limits of the approach, outlined future development work that could contribute to further improvements during and post-production and suggested two application scenarios that we envision as uses for the method.

Acknowledgements. Danish Agency for Science and Higher Education for funding the project "3D Printet Byggeri" at the Danish Technological Institute. The authors gratefully acknowledge the assistance of Stian Vestly Holte, Kit Wai Chan, Alma Bangsgaard Svendsen and Suna Ezra Petersen, enrolled at the Master of Architecture programme *CITAstudio: Computation in Architecture*, KADK. The authors also wish to thank the anonymous reviewers for their comments and suggestions to improve the quality of the paper.

References

1. Wray, P., Scrivener, K.: Straight talk with Karen Scrivener on cements, CO_2 and sustainable development. Am. Ceramic Soc. Bull. **91**, 47–50 (2012)
2. Ritchie, H., Roser, M.: CO_2 and other Greenhouse Gas Emissions (2017). https://ourworldindata.org/co2-and-other-greenhouse-gas-emissions. Accessed 28 Feb 2018
3. Pottmann, H., Wallner, J.: Geometry and Freeform Architecture (2010). http://www.geometrie.tugraz.at/wallner/matharch.pdf. Accessed 28 Feb 2018
4. Hickert, S.: Evaluation of free-form concrete architecture, moulding systems and their technical potentials. J. Facade Des. Eng. **3**(3–4), 273–288 (2015)
5. Costanzi, C.B.: 3D Printing Concrete Onto Flexible Surfaces. Delft University of Technology (2016). https://repository.tudelft.nl/islandora/object/uuid%3A84d36c2e-8969-4432-b1a5-c9c02e6304f6. Accessed 08 May 2018
6. Nelson, M.S., Fam, A.Z., Busel, J.P., Bakis, C.E., Nanni, A., Bank, L.C., Henderson, M., Hanus, J.: Fiber-reinforced polymer stay-in-place structural forms for concrete bridge decks: state-of-the-art review. ACI Struct. J. **111**(5), 1069–1079 (2014)
7. Veenendaal, D., West, M., Block, P.: History and overview of fabric formwork: using fabrics for concrete casting. Struct. Concrete **12**(3), 164–177 (2011)
8. Pegna, J.: Exploratory investigation of solid freeform construction. Autom. Constr. **5**, 427–437 (1997)
9. Khoshnevis, B.: Automated construction by contour crafting related robotics and information technologies. Autom. Constr. **13**, 5–19 (2004)

10. Neudecker, S., Bruns, C., Gerbers, R., Heyn, J., Dietrich, F., Dröder, K., Raatz, A., Kloft, H.: A new robotic spray technology for generative manufacturing of complex concrete structures without formwork. In: Procedia CIRP (2016)
11. Hack, N., Lauer, W.V.: Mesh-Mould: robotically fabricated spatial meshes as reinforced concrete formwork. Architectural Des. **84**(3), 44–53 (2014)
12. Block, P., Schlueter, A., Veenendaal, D., Bakker, J., Begle, M., Hischier, I., Hofer, J., Jayathissa, P., Maxwell, I., Mendez Echenagucia, T., Nagy, Z., Pigram, D., Svetozarevic, B., Torsing, R., Verbeek, J., Willmann, A., Lydon, G.P.: NEST HiLo: investigating lightweight construction and adaptive energy systems. J. Build. Eng. **12**, 332–341 (2017)
13. Lloret, E., Shahab, A.R., Linus, M., Flatt, R.J., Gramazio, F., Kohler, M., Langenberg, S.: Complex concrete structures: merging existing casting techniques with digital fabrication. Comput. Aided Des. **60**, 40–49 (2015)
14. Buswell, R., Leal da Silva, W.R., Jones, S.Z., Dirrenberger, J.: 3D printing using concrete extrusion: a roadmap for research. Cement and Concrete Research (2018, accepted for publication)
15. Leal da Silva, W.R.: 3D printed constructions - engineering perspective (Presentation Slides). In: 3D Printet Byggeri Conference, Denmark (2017). https://www.dti.dk/_/media/67675_3DPrintetByggeri_Konference_TeknologiskInstitute.pdf. Accessed 28 Feb 2018
16. Bos, F., Wolfs, R., Ahmed, Z., Salet, T.: Additive manufacturing of concrete in construction: potentials and challenges of 3D concrete printing. Virtual Phys. Prototyping **11**(3), 209–225 (2016)
17. Popescu, M., Reiter, L., Liew, A., Van Mele, T., Flatt, R.J., Block, P.: Building in concrete with an ultra-lightweight knitted stay-in-place formwork: prototype of a concrete shell bridge. Structures **14**, 322–332 (2018)
18. Tam, K.M.M., Coleman, J.R., Fine, N.W., Mueller, C.T.: Robotics-enabled stress line additive manufacturing. In: Reinhardt, D., Saunders, R., Burry, J. (eds.) Robotic Fabrication Architecture, Art and Design 2016, pp. 351–361. Springer International Publishing Switzerland (2016)

KUKA
makita

Versatile Robotic Wood Processing Based on Analysis of Parts Processing of Japanese Traditional Wooden Buildings

Hiroki Takabayashi[1(✉)], Keita Kado[2], and Gakuhito Hirasawa[2]

[1] Building Research Institute, 1 Tachihara, Tsukuba, Ibaraki, Japan
tkbys@kenken.go.jp
[2] Graduate School of Engineering, Chiba University, 1-33 Yayoi-cho, Inage-ku, Chiba, Japan
kado@chiba-u.jp, hirasawa@faculty.chiba-u.jp

Abstract. Processing parts of wooden buildings by robots and computerized numerical control (CNC) machines has been attempted in recent years. However, the shapes of the processed parts are constrained by the machines and attached tools. Modified parts cannot be applied to traditional wooden buildings, as the shapes of the parts are an aspect of the tradition. This paper introduces a method of versatile processing of wooden building parts using an articulated robot, based on the analysis of part shape and carpentry tools used in Japanese wooden building construction. The goal is to process the original part shape of the construction method as is, without optimizing it for the robot and machining. We analyze Japanese carpentry tools and processing methods and propose a method to flexibly process wooden building parts using a circular saw, square chisel, vibration chisel, and router. Then, using robots employing these tools, we process parts of a five-storied pagoda of a Japanese traditional wooden building. We consider flexible processing of the wooden building parts.

Keywords: Japanese traditional wooden architecture · Part processing
Industrial robots

1 Introduction

Since ancient times, wooden buildings have been widely used in Japan for temples, shrines, and residences. Traditional Japanese wooden buildings are extremely complicated to construct, requiring advanced skills. These buildings also have historical and cultural value, and registration of culturally important properties continues to increase each year [1]. However, *Miya* carpenters, i.e., skilled workers who maintain and repair these buildings, are decreasing each year [2], and it is expected that demand for workers will exceed the supply in the near future.

A conventional construction method exists for Japanese wooden houses. The elements of this method are common with those of traditional wooden buildings, such as the use of *Tsugite-Shiguchi*, which are joining methods for parts, and the requirement of carpenters to handcraft techniques. Because of the decrease in skilled craftsmen and

J. Willmann et al. (Eds.): ROBARCH 2018, *Robotic Fabrication in Architecture, Art and Design 2018*, pp. 221–231, 2019.
https://doi.org/10.1007/978-3-319-92294-2_17

rationalization of production, precuts are becoming mainstream, which are parts processed using machines in factories. However, the shape of a part processed using existing pre-cut systems are generally simplified, compared to that prepared by carpenters. Thus, precuts cannot be utilized for construction or repair of traditional wooden buildings, where the part shapes are considered part of the tradition.

On the other hand, articulated robots (hereinafter, simply called "robots"), which are widely used in manufacturing industries, such as the automobile industry, enable flexible and versatile movements that are similar to those of a human arm. Thus, we can expect to create parts with shapes that are challenging for existing processing machines and machine tools by utilizing a robot.

This paper reports the versatile processing of wooden building parts using a robot, based on analysis of part shapes and carpentry tools employed in Japanese wooden building construction. The goal is to realize robot processing without excessive modification of the part shape used in the existing wooden construction method, with parts processing being equivalent to handwork, including realization of the fitting shapes of joint parts utilizing joining methods, such as *Tsugite-Shiguchi*.

2 Related Work

Recently, the use of computerized numerical control (CNC) machines for processing parts in wooden building production has become more frequent. For example, the Hans Hundegger AG [3] machine is highly productive and used worldwide, including Japan. CNC machines primarily machine parts by the rotation of tool blades, such as those of circular saws and drills. Thus, there are restrictions on the shapes that can be machined; for example, uncut parts remain in the inside corners. Conventional Japanese wooden building employs *Tsugite-Shiguchi,* similar to traditional wooden buildings. However, shapes of parts processed by machines are streamlined and rounded according to the processing machine, yielding a different shape than that obtained via carpentry (Fig. 1). A recent example is the new Tamedia office building, designed by Ban and constructed using a CNC machine [4]. This building consists of a wooden frame with columns and beams, having a structural form in which an elliptical beam penetrates a column. This architecture constitutes a new structural form unique to wooden buildings, with CNC milling machines being used to process the parts.

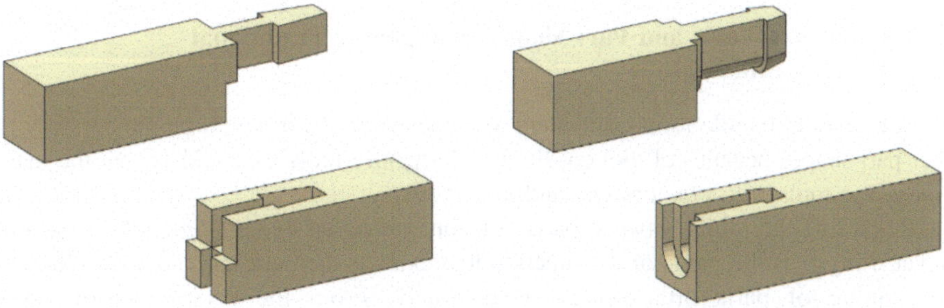

Fig. 1. Original shapes of *Kohikake-Kamatsugi* (left) and machined shapes (right)

Robotic work on wooden building parts processing has also been reported. At the 2011 ICD/ITKE Research Pavilion [5], panels with finger joints on polygonal thin plywood were processed through robot milling. However, this pavilion consists of many panels, and each panel shape and finger joint position is different. Thus, it is necessary to confirm whether each panel can be processed at the design stage. In this work, using shape generation parameters that consider constraints on robot and tool processing, a method is developed to simultaneously confirm the geometric consistency and production possibility of a shape during shape generation. Dank and Freissling [6] have regarded a robot with a router as the ultimate CNC machine, producing a framed pavilion by combining parts having joining shapes that can be realized via milling. In their work, visual programming languages were used throughout the project, and each phase was described parametrically and correlated. The part shape and toolpath were simultaneously generated in the detailed design of the joint part, especially by defining the mill diameter, the robot mechanism, and the fixation (position and angle) of the workpiece as shape generation parameters. Robot processing for pavilion production using curved cross-laminated timber (CLT) panels has also been reported [7]. In that approach, the dovetail-based joint shape and the groove in which the joint part fits are cut by milling or using a circular saw. An effort to attach a band saw to a robot and process a free-form surface obtained from a workpiece shape has also been performed [8].

Thus, various studies have been conducted on the application of CNC machines and robots to wooden building parts processing. In those works, as an attempt to rationally realize new architectural designs, the part shape design and tool path generation are derived through a correlated parametric design method. However, the part shape is mostly designed through machining using these machines or a machining system customized to produce a specific design.

This paper targets robot processing that does not alter the original part shape used in the wooden construction method. For example, Japanese traditional wooden building parts cannot be produced using existing machining technology; however, such production may be realized using the new parts processing method presented in this work, which incorporates robots and CNC machines. The ability to process parts of Japanese traditional wooden buildings, which are complex on the international scale, indicates the applicability of robot processing to other wooden building construction methods.

3 Robot Tools and Toolpath Generation

3.1 Carpentry Tools and Part Shapes in Japanese Traditional Wooden Buildings

If a robot could handle tools equivalent to carpentry tools, it would be unnecessary to alter part shapes because of tool constraints. Carpentry tools vary widely, but the main processing tools are saws, chisels, and planes [9].

Regarding the joints between parts, a technique called *Tsugite-Shiguchi* is used in Japan. *Tsugite* is the name of the junction that extends the part, and *Shiguchi* refers to the joining of parts orthogonally or obliquely. From the perspective of stress transmission, workability, design beauty, etc., there are many types of joint connections, and the shapes are often complex [10].

The processing procedure employed by a carpenter is shown in Fig. 2, where the wood is first cut and incised, mainly with a saw. Then, a chisel is used for further refinement, particularly of parts that are difficult to cut with a saw. Using these two types of tools, various shapes can be processed [11]. A *Tsugite-Shiguchi* part has a broad three-dimensional shape with a combination of planes. Other parts have curved surfaces, such as the wooden framework called *Tokyou,* the warped roof parts called *Kayaoi,* and round columns. The curved surfaces of the outer shapes of such parts are often finished with a plane after rough cutting with a saw or chisel.

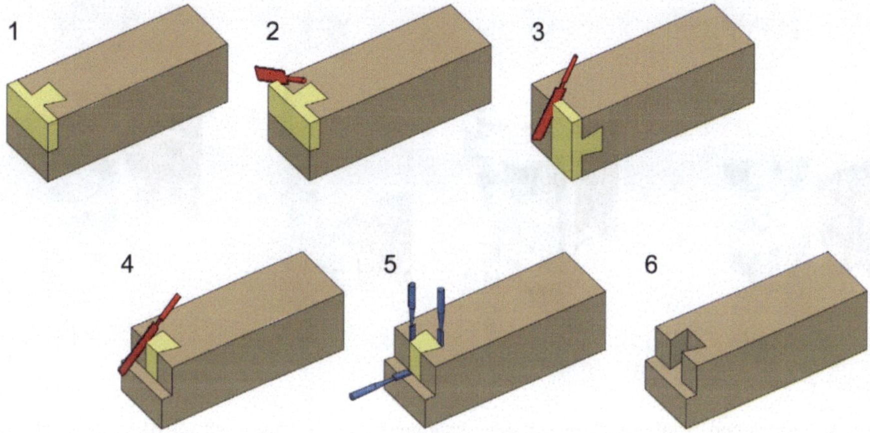

Fig. 2. Carpenter's processing procedure for *Koshikake-Aritsugi*

3.2 Robot Tools

A strategy to robotically process parts while alternating between multiple tools is required, as wooden building parts cannot be processed satisfactorily with a single tool. However, various shapes can be processed, even with a small number of tools. Using multiple tools unnecessarily increases the time and work required for tool exchange, while also complicating the entire parts processing system, including the software.

Circular saws, jigsaws, and reciprocating saws (saver saws) have been developed as electric tools to replace handsaws. A jigsaw or reciprocating saw is capable of some curvilinear processing. Here, however, we use a circular saw that can efficiently cut in a straight line, the original role of the saw. A power square chisel that can be used for creating a square hole, such as a mortise, has been developed. The square chisel is a tool combining a cone drill bit and a chisel blade comprised of a square cross section covering the cone. Accurate and efficient drilling with no rounded corners is possible. On the other hand, it is impossible to process sharp interior angles and curved surfaces with a square chisel alone. Internal corners are processed by a vibrating chisel (electric wood carving machine). It has a mechanism through which the blade moves up and down rapidly and is essentially identical to a hand chisel. It is suitable for relatively delicate processing and can remove internal corners that remain after processing with a square chisel. For machining curved surfaces, milling is suitable. Depending on the shape, round holes may be drilled.

Based on the above considerations, in this study, we developed four types of tools for the robot, i.e., a circular saw, square chisel, vibrating chisel, and router [12]. Figure 3 shows the developed tools and examples of processing with each tool. In the developed approach, parts are processed by manually exchanging these tools.

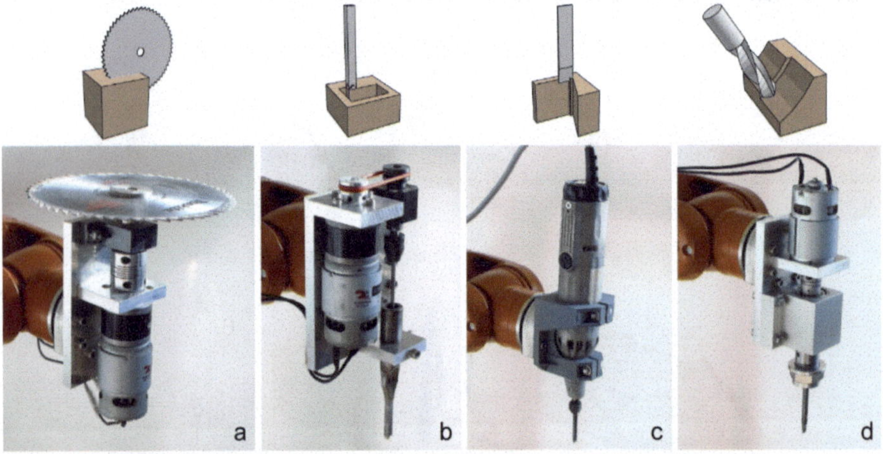

Fig. 3. Developed tools and processing examples: (a) circular saw, (b) square chisel, (c) vibration chisel, and (d) router

3.3 Parts Processing System

Robot motion during processing is generated according to the path of the tool. We previously proposed [13] a method to calculate the toolpath from the part geometry for each tool. A computer-aided-manufacturing (CAM) programme that implements this method and calculates the path of each tool from the three-dimensional part shape described in stereolithography (STL) format was developed. Geometric elements, such as normal vectors and edges, required for toolpath calculation and the calculation procedures differ for each tool. In this CAM programme, the operator selects a tool or geometric element for calculation from a three-dimensional model of a part and inputs it through a dialog window. After confirming the validity of the toolpath by simulation, an operation instruction is given to the robot through a command transmission programme (commander). The robot controller uses a package that transmits and receives data to and from external devices through the network via the transmission control protocol (TCP)/Internet protocol (IP). In the experiments conducted in this study, which are described below, the KR 6 R 700 (KUKA, Augsburg, Germany) robot was used, with Ethernet KRL (KUKA, Augsburg, Germany). The commander also provides signals to the operator, such as those indicating that workpiece rearrangement is required.

4 Experiments: Roof Corner of Five-Storied Pagoda of Japanese Traditional Wooden Building

Advanced skills are required for processing traditional wooden building parts. In particular, the roof includes a wooden framework called *Tokyou* and a roof frame, and the parts thereof are intertwined with the corner parts in a complex manner. Thus, many complex part shapes are involved. Confirmation that these parts can be processed using the developed system would indicate the system's applicability to various wooden building parts. Therefore, we conducted production tests on the roof corners of traditional wooden buildings. In addition, parts of the main structure, excluding design-like sculptures found in *Kibana*, were examined.

The parts processing subject in this test was the roof corner of a five-storied pagoda of a traditional wooden building. The production part is shown in Fig. 4. This three-dimensional model was created from the digital archive research [14] of traditional wooden buildings that we are continuously performing in our laboratory, and was modeled precisely based on the part-fitting shapes. Originally, we intended to conduct a

Fig. 4. Three-dimensional model of five-storied pagoda with production part indicated

full-scale test; however, a small robot (KUKA KR 6 R 700) was employed in this test, so it was conducted at a 1/5 scale. The workpiece was cypress wood. Lumber with different cross-sectional dimensions was prepared and selected according to the part dimensions. The workpiece was affixed to the workbench with a vise.

Toolpaths were generated by the developed CAM programme for each of the parts constituting the roof corner and machined by a robot with tool exchange. Figure 5(a)

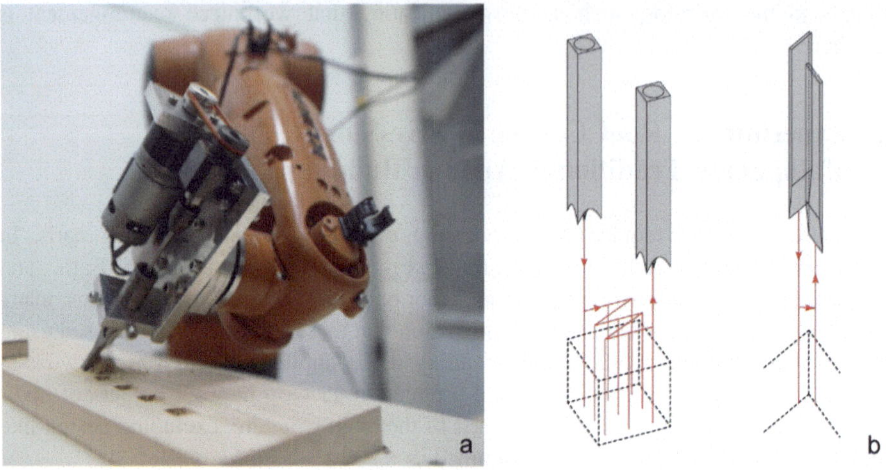

Fig. 5. Robot processing with (a) a square chisel and (b) two types of chisel motion

shows *Sumigi* (hip rafter) processing. Thus, it is possible to process a hole with a square cross section in which the *Taruki* (hip jack) is inserted using a square chisel. Several square holes with different sizes were required to be processed in this test. We

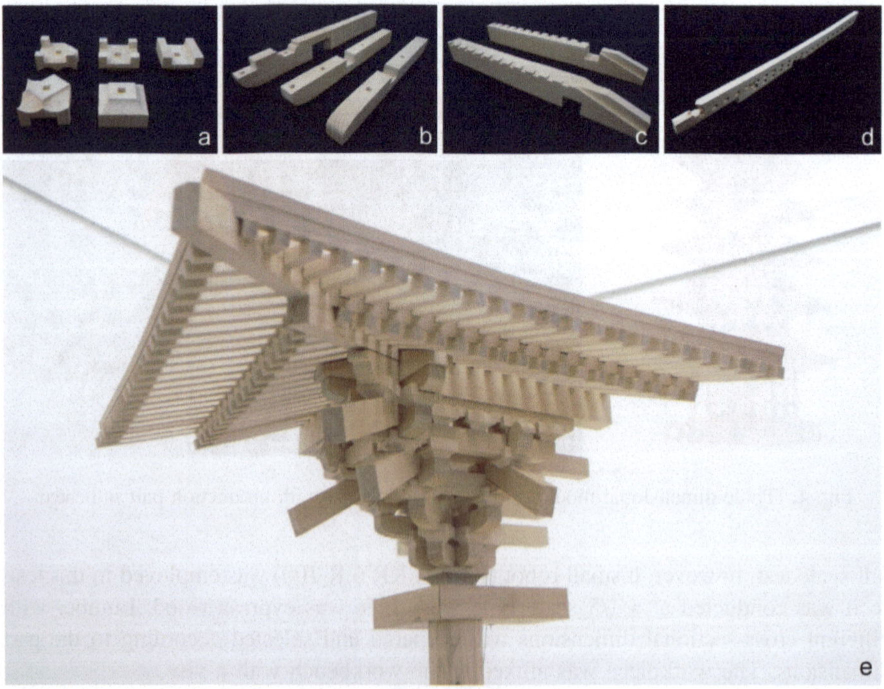

Fig. 6. Processed parts of (a) *Masu,* (b) *Hijiki,* (c) *Gagyou,* (d) *Sumigi,* and (e) model comprising assembled processed parts

prioritized fine processing and used chisels smaller than the hole size. A square chisel produced holes of varying sizes by making multiple passes in a grid pattern. The vibrating chisel used a flat blade with a narrow width. Acute inner corners were machined by cutting twice along the two faces composing the corner (Fig. 5(b)).

Figures 6(a)–(d) show some of the processed parts. There were approximately 80 types of parts and a total of 160 parts. Although the part shapes were diverse, part processing was possible by applying the same workflow without requiring special treatment, such as preparing jigs for each shape. Figure 6(e) shows the model with the assembled parts. The characteristic features of the traditional wooden building were apparent, such as the wooden framework, called *Mitesaki* in the corner and the subtle *Taruki* and *Kayaoi* curves of the warped roof. Thus, the parts with various shapes can be processed using the proposed system without altering the details of the shape, such as the joint structure.

5 Discussion

In the processing test, parts of traditional Japanese wooden buildings could be processed without modifying their shape. Square hole processing by square and vibration chisels was effective, resulting in parts that were equivalent to a carpenter's handwork. Rotary cutting tools can be handled with five axes, but these chisels cannot be handled unless there are six or more axes. Tools that effectively utilize high degrees of freedom can overcome the weaknesses of existing machining. Also, the software utilized by the system is not customized for traditional wooden buildings. The ability to process parts of traditional wooden buildings with complex shapes shows the applicability to other designs and construction methods. Furthermore, it was confirmed that individual work, such as marking and jigs for processing, is not required for each part shape, which is a major difference from handwork. This shows that regardless of the part shape being the same or different, processing is not affected.

Meanwhile, the processing marks of the rotary cutting tool and the roughness of the bottom resulting from the structure of the square chisel remained on the finished surface. The former is believed to be improved somewhat by increasing the mechanical rigidity of the tools and robots. The latter usually disappears after mating the pieces, which is also true for square chisels used in handwork. Moreover, depending on the shape of the part, the workpiece must be rearranged, such that it reverses during processing. Within the scope of this paper, humans perform these using simple fixtures, such as a vice; however, for CNC machines, like Hunderegger machines, the movement of the workpiece and the change in posture are almost fully automated. The same is true for tool replacement. The amount of human effort and cost of that effort can be reduced, and the optimization of the production system for processing parts of various shapes is a future task.

6 Conclusion

This study proposed a technique of flexibly processing wooden building parts using robots and demonstrated the processing of wooden frame parts for a five-storied pagoda of a Japanese traditional wooden building. From an analysis of carpentry tools and processing methods, the tools needed to process various part shapes were selected (i.e., circular saw, square chisel, vibration chisel, and router). In addition, a toolpath generation mechanism that does not depend on a specific design or design method was implemented, which can machine various shapes. The results of the processing test indicated that the parts processed by the robot were comparable to the handwork of carpenters. However, manual tasks, such as creating toolpaths using CAM, exchanging tools, and rearranging workpieces during processing, continue to exist, and solving these issues will truly realize versatile robotic wood processing.

The ability of robots to fabricate parts comparable to those produced by humans also implies that parts produced by each can be combined. Overall, human carpenters are skillful in expressive and delicate processing like sculpture, whereas robots with sensors can perform effective processing, exploiting variations in shapes and physical properties found in natural materials. Increased collaboration between humans and robots will facilitate more opportunities to combine parts created by each. These developments will not only eliminate the shortage of craftsmen, but create new wooden buildings that make full use of traditional wood knowledge and designs.

Acknowledgments. This work was supported by a JSPS Grant-in-Aid for JSPS Research Fellow grant number 16J02060.

References

1. Statistics Bureau of Japan Homepage, Historical Statistics of Japan Chapter 26 Culture and Leisure, Cultural Properties 26-16. http://www.stat.go.jp/english/data/chouki/26.htm. Accessed 2 Mar 2018
2. Statistics Bureau of Japan Homepage, Population Census. http://www.stat.go.jp/english/data/kokusei/index.htm. Accessed 2 Mar 2018
3. Hans Hundegger AG Homepage. https://www.hundegger.de. Accessed 2 Mar 2018
4. Ban, S.: Tamedia Shin Honsha (Tamedia New Office Building). In: Shinkenchiku-sha (ed.) Shinkenchiku 92(3), pp. 39–49. Shinkenchiku-sha, Tokyo (2017)
5. Menges, A.: Morphospaces of robotic fabrication. In: Brell-Çokcan, S., Braumann, J. (eds.) ROB|ARCH 2012: Robotic Fabrication in Architecture, Art & Design, pp. 28–47. Springer, Vienna (2013). https://doi.org/10.1007/978-3-7091-1465-0_3
6. Dank, R., Freissling, C.: The framed pavilion.In: Brell-Çokcan, S., Braumann, J. (eds.) ROB|ARCH 2012: Robotic Fabrication in Architecture, Art and Design, pp. 238–447. Springer, Vienna (2013). https://doi.org/10.1007/978-3-7091-1465-0_28
7. Robeller, C., Nabaei, S., Weinand, Y.: Design and fabrication of robot-manufactured joints for a curved-folded thin-shell structure made from CLT. In: McGee, W., Ponce de Leon, M. (eds.) Robotic Fabrication in Architecture, Art and Design 2014, pp. 67–81. Springer International Publishing Switzerland (2014). https://doi.org/10.1007/978-3-319-04663-1_5

8. Johns, R.L., Foley, N.: Bandsawn bands feature-based design and fabrication of nested freeform surfaces in wood. In: McGee, W., Ponce de Leon, M. (eds.) Robotic Fabrication in Architecture, Art and Design 2014, pp. 17–32. Springer International Publishing Switzerland (2014). https://doi.org/10.1007/978-3-319-04663-1_2
9. Takenaka Carpentry Tools Museum, Jousetsu tenji zuroku (Permanent exhibition catalog). Kobe (2009)
10. Sumiyoshi, T., Matsui, G.: Mokuzou no tsugite to shiguchi (Joining method of wooden architecture). Kajima Institute Publishing, Tokyo (1989)
11. Fujisawa, K., Tadokoro, H.: Mokuzou kenchiku no kidori to sumitsuke (Cutting and marking method of wooden architecture). Inoueshoin, Tokyo (2001)
12. Takabayashi, H., Motoike, R., Aida, K., Kado, K., Hirasawa, G.: Study on parts processing of the traditional wooden construction method using articulated robot. AIJ J. Technol. Des. **22**(50), 331–334 (2016). https://doi.org/10.3130/aijt.22.331
13. Takabayashi, H., Motoike, R., Kado, K., Hirasawa, G.: Study on tool path generation for wood processing using articulated robot. AIJ J. Technol. Des. **22**(51), 813–816 (2016). https://doi.org/10.3130/aijt.22.813
14. Kado, K., Hirasawa, G.: Part classes and their instances on digital archive of traditional wooden architecture. J. Archit. Plan. (Trans. AIJ) **76**(662), 877–886 (2011). https://doi.org/10.3130/aija.76.877

Form Finding of Nexorades Using the Translations Method

Tristan Gobin[1,2,3(✉)], Romain Mesnil[1], Cyril Douthe[1], Pierre Margerit[1], Nicolas Ducoulombier[1], Leo Demont[1], Hocine Delmi[1], and Jean-François Caron[1]

[1] Laboratoire Navier, UMR 8205, École des Ponts, IFSTTAR, UPE, Champs-sur-Marne, France
t.gobin@hal-robotics.com

[2] Université Paris-Est, Ecole Nationale Supérieure D'architecture de Paris-Malaquais, Laboratoire Géométrie Structure Architecture, Paris, France

[3] HAL Robotics Ltd, London, UK

Abstract. The aim of this paper is to discuss the dialectic form-finding of a complex timber structure based on an innovative structural system: shell-nexorade hybrids. Nexorades, also known as reciprocal frames are elegant structures that suffer from a relatively poor structural behavior due to in-plane shear and bending of the members. Introducing plates as bracing elements significantly improves their performance, but increases the manufacturing complexity and sets high tolerance constraints. We present the fabrication and assembly of a 50 m^2 timber pavilion with 6-axis robotized milling. The use of a mobile robot and fixed machining stations is explored to allow for maximal flexibility of iterations between design and fabrication.

Keywords: Complex timber structure · Non-standard structure
Reciprocal frame · Nexorade · Free-form architecture · Robotic construction

1 Introduction

The last decades have seen the emergence of complexly shape structures in architectural design. The convergence between modelling and manufacturing tools allows architects and engineers to explore new formal possibilities. Timber structures are no exception to this trend of free-form architecture, as they already rely on digital fabrication, like 5 axis CNC milling [10–12].

Connections are particularly difficult to handle in timber structures, because of complex geometrical arrangement of the members and of structural systems that rely on the bending stiffness of the joints. As such, timber structures often rely on heavy steel connectors, like the 500 hexavalent steel nodes constructed for the roof of the Crossrail station [4]. *Nexorades*, also known as *reciprocal frames*, aim at simplifying the connection details by having each member supported along the span of another member. The connections are thus two-valent. The computational morphogenesis of this structural system has been pioneered by Baverel and Nooshin [1], and it has been recently applied in the context of robotic construction [14].

J. Willmann et al. (Eds.): ROBARCH 2018, *Robotic Fabrication in Architecture, Art and Design 2018*, pp. 232–241, 2019.
https://doi.org/10.1007/978-3-319-92294-2_18

The counterpart for the ease of assembly is the bending dominant behavior due to the T-shaped junctions. The optimal shape for nexorades are not funicular because an axial force in one element is converted in bending and shear in the attached members: this makes them unusual structures for engineers [3]. Using plates as bracing components is a promising direction to improve their efficiency, the resulting structural system is called *shell-nexorade hybrid* because it is based on the geometry of a nexorade and the plates are expected to yield a shell-like behavior [7]. Their practical application are not studied yet, especially the trade-off between structural performance and ease of manufacturing. The geometrical complexity of shell-nexorade hybrids make robotic fabrication a natural path to explore. The aim of this paper is thus to discuss robotic fabrication of shell-nexorade hybrids through the study of a 50 m^2 pavilion. The need for an integrated design and fabrication approach is stressed in this paper, as it allows an efficient collaboration between different actors of the construction project.

2 Geometry of Shell-Nexorade Hybrids

The form finding of nexorades is here based on the method of translations, where all the members are translated to create eccentricities [1]. It is shown in [7] that the eccentricities and engagement lengths depend linearly on the translations parameters. The shape-generation method is thus based on the creation of a linear subspace, a well-known topic in computer graphics, which has been applied to the modelling of polyhedral meshes recently [8]. The design exploration is eased by this approach, and local fitting problem are expressed with the least square method and solved instantly.

The proposed design space is related to *parallel transformations*, which preserve facet planarity [8]. Applying our method to a mesh with planar facets yields a nexorade with nearly planar panels. Such a mesh can be obtained by solving a fitting problem with the marionette technique [6]. This is shown on the left of in Fig. 1: the reference geometry is shown in light gray and the final mesh with planar quads is shown in black.

After reciprocation, some eccentricities are necessarily introduced. The right of Fig. 1 shows a close-up of a fan: the thick black lines correspond to the beam neutral axis, while the gray lines correspond to planar panels. One has thus to deal with the relative eccentricities between members (white dots), but also with the eccentricities between beams and panels (orange thick dots). This geometrical complexity is investigated in the followings with the design and fabrication of a prototype, shown in Fig. 1.

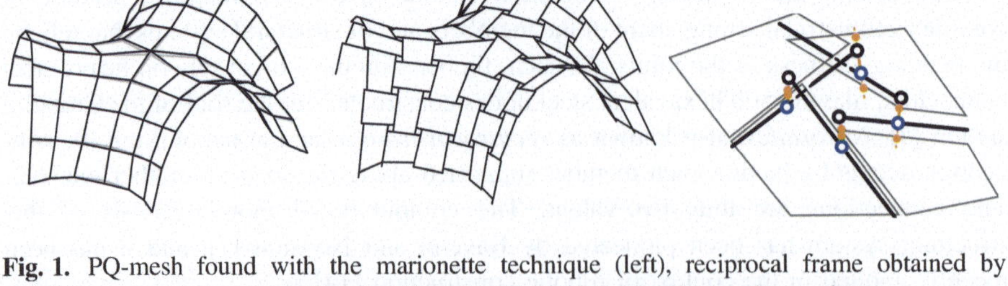

Fig. 1. PQ-mesh found with the marionette technique (left), reciprocal frame obtained by translation of the edges (middle): the opposite edges of the quads remain co-planar.

3 Design Workflow and Fabrication: Integrated Approach

The case-study presented in the following is based on several constraints. No lifting equipment was available, which limited the possibilities to have large prefabricated components, like envisioned in [14]: all the beams were to be assembled on-site manually. The structure was also to be checked through classical calculations, derived from Eurocode 5 and technical agreements. Screwed connections are suited for these matters, but they require guides during assembly. We choose to use tenons and mortises to place the beams correctly. Additionally, pilot holes have to be made in the beams to avoid timber splitting and to facilitate the assembly. The ease of on-site assembly leads thus to a significant complexity of the resulting beams.

3.1 Integrated Design Workflow

The geometrical complexity of a free-form pavilion requires the creation of a computational workflow with maximal automation. This is achieved here with custom C# libraries for form finding and interaction with Karamba [9] and HAL through Grasshopper. The workflow is illustrated in Fig. 2. For the pavilion studied here, the main constraints are the utilization in the connection details and the manufacturing of the beams. The designers can leverage different design parameters (in bold in Fig. 2): the reference geometry, defined as a NURBS, the parameters of form-finding for nexorade, and fabrication parameters, like the position of the working space for the robots.

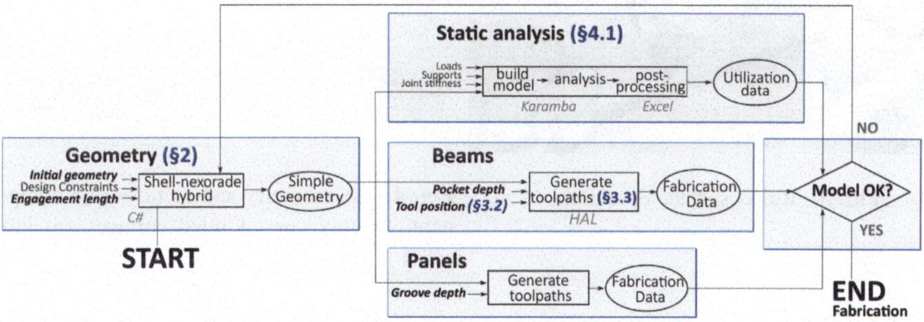

Fig. 2. ROD [2] Computational workflow

Design iterations are first performed on global geometrical parameters, which influence greatly the structural response and fabrication. Once a satisfying geometry is found, the detailed path-planning begins, eventually with local adjustments of the geometry or fabrication parameters (change of engagement length of one fan, or change of groove depth for one beam, etc.). This hierarchy in the design iterations allows for a quick exploration of the design space based on the experience of the project team.

A full automation of the path-planning optimization is possible within the chosen framework and should be the topic of future work.

3.2 Designing for Flexibility and Modularity

Five axis machines are well implanted in industries and especially in timber fabrication: the programming environment is now robust, and 5-axis machines are also more rigid than 6-axis robots and usually more precise for comparable speeds. These advantages make them as 'classical' tools for digital fabrication [11]. However, the cost, flexibility and modularity which were required for the present project drove us to use 6 axis robots for most of the operations. The scenarios and processes have indeed evolved with the design and technical constraints of the pavilion construction, and some operations could be performed more efficiently without swapping tools, for example on fixed stations. Using a gripper to transport the component to different areas of the workshop for specific tasks solves this constraint and allowed for a truly modular robotic concept (Figs. 3 and 4).

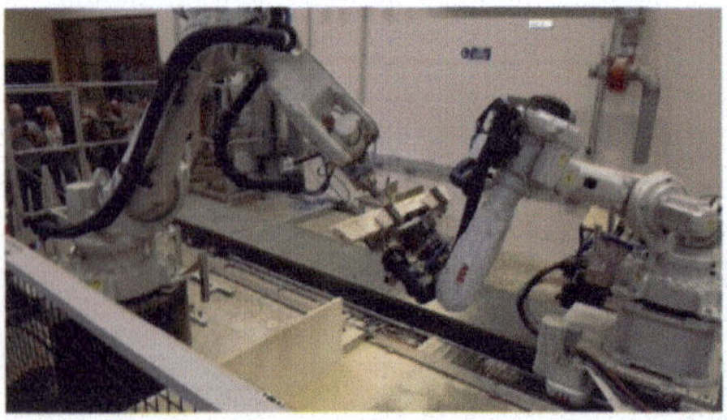

Fig. 3. Ruled surface operations

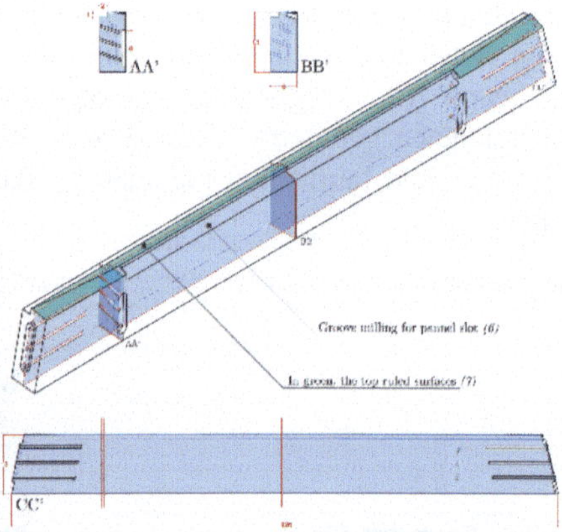

Fig. 4. Axonometry and sections of a component. Doubly curved ruled surfaces are indicated in green.

The cell layout is constituted of two robots, shown in Fig. 5. One is mounted on a 9 meters track and referred as *gripper robot*. The other is fixed and referred as *fixed robot*. Those two robots can work together or independently. All end-effectors are constituted by automated tool changer and can mount specific tools, which allows us to easily swap the tools between the two robots if needed. For the project we use a milling head mounted on *fixed robot* and a pneumatical gripper mounted on *gripper robot*. Various stationary tools are dispatched along a side of the track where *gripper robot* has access.

It is more complicated to program robots due to the non-linearity of geometrical constraints (accessibility, collisions, large rotations of wiring and cables). However, new robotic framework as *HAL | robot programming & control* [13] allow to integrate easier robots or CNC machine in the design environment, as shown in Fig. 2.

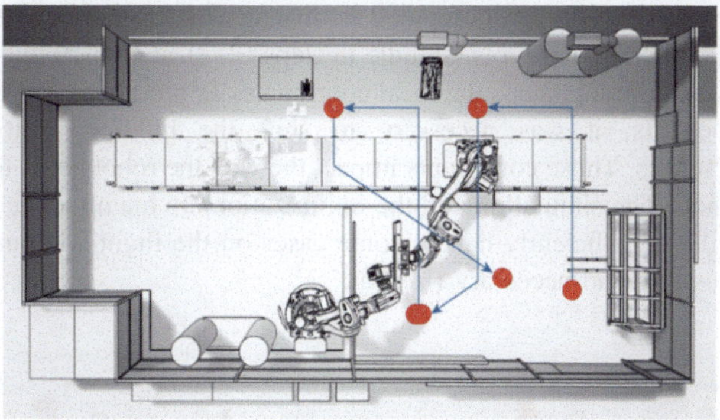

Fig. 5. Workflow in the robotic cell and corresponding machining steps: fixed feeder (1), stationary circular saw (2), milling (3–7), fixed wood rooter (8).

This way to design a small modular production environment allows to explore ROD [2] methods related to small series. With the evolution of digital tools, design gains make it possible to spend less time reconfiguring a production space for which robot are convenient at. One could say that it will be less efficient for reasons like a reduce stiffness that require to lower milling speed. However, this is an advantage only for large series and robot use fits well to transform the timber industry's practices in particular, where production often returns to that of small series. Besides, recent improvement in the controllers and dynamic models allows a better active compensation and fine tuning and thus to increase the rigidity of the axes. We use such features during our production.

3.3 Operations and Component Geometry Fitting

The manufacture of a component consists of several steps for which the gripper is always under stress. The use of two robots allows greater freedom in the machining of the components because it is possible to use different positions with *gripper robot* to maximize accessibility for *milling robot*. Because of the robot's mobility and the robustness of the clamping, the component can thus navigate between the different workshops without compromising machining accuracy (±1 mm).

In detail, eight independent operations are identified in the fabrication of the beams. *(1)* At **picking step**, the *gripping robot* get the rough material on a dispenser. This step allows the alignment of the beam to the local coordinate system of the robot end-effector. *(2)* Directly after, **the shaping,** where each element is cut to an approximate size, is done on a stationary circular saw. Then the *gripping robot* is moving the component to the common work space for a collaborative machining sequence. Several operations occur as follow: *(3)* **the two tenons milling**, *(4)* **the two mortises milling**, *(5)* **the ruled surface milling**, *(6)* **the groove milling** and finally *(7)* **the bottom-surface milling**. After those operations, the *gripping robot* move for *(8)* **drilling operations** to the wood rooter. Finally, the component is bringing back to the dispenser.

All geometrical aspects are computed to manage the distance to avoid collision between the clamp and other tools (spindle in steps *(2–7)*, stationary tool frames in *(2)* and *(8)*) since the gripping keeps the local coordinates as precise as possible. For some specific components, it was necessary to shift the holding position to avoid unmachinable spaces. Those constraints impact the way the robot holds the component and its dimension. The simulation and the optimization are mainly done automatically but the possibility to efficiently handle some cases on the flight without breaking the workflow is possible and necessary (Fig. 6).

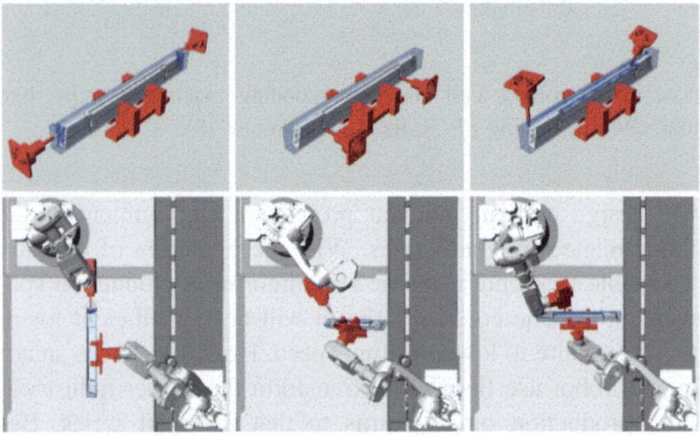

Fig. 6. Operations (3), (4) and (5). First row is tool collision detection and second row toolpath and robot simulation and reachability checking.

4 Discussion

4.1 Structural Benefits of Shell-Nexorade Hybrids

The design workflow includes an automated structural calculation. The connection details are designed according to the specifications of a technical agreement, while the overall structural calculation is performed with Karamba. This allows to quickly assess the structural reliability of the design. Nexorades are less efficient than conventional gridshells, because of low nodal valency: the plates introduce in-plane stiffness and solve this issue. The hybrid structure is up to 10 times stiffer than the unbraced nexorade, with a 30% mass increase. It is also more efficient than a quadrilateral gridshell with rigid joints based on the reference mesh used as an input for the edge-translation technique. Another benefit is the reduction of the forces in beams and connections details shown here for a non-symmetrical wind load on Fig. 7: bending moments, and thus utilization are divided by eight on average with the introduction of plates as bracing elements. For this load case, and assuming a joint rotational stiffness of 100 kNm/rad, the displacement of the unbraced nexorade and gridshell are 75 mm and 26 mm respectively, while the constructed structure has a maximal displacement of 4.4 mm.

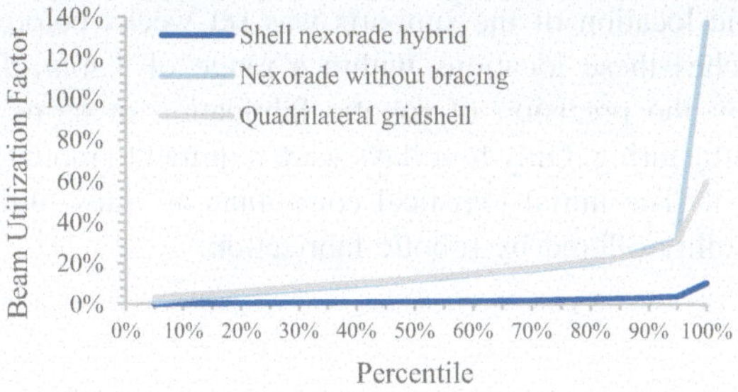

Fig. 7. Beam utilization factor for a wind load for the shell-nexorade hybrid shown in this paper, the structure with the same geometry but without panels, and a gridshell based on the Planar-quad mesh on the Left of Fig. 1.

This illustrates the fact that shell nexorade hybrids are more efficient than quadrilateral gridshells, which are themselves more efficient than nexorades. The relative performance between shell-nexorade hybrids and quadrilateral gridshells under non-symmetrical load is similar to the one observed between fully-braced gridshells and quadrilateral gridshells, whose bracing relies on the joint and member bending stiffness [5]. It can thus be inferred that shell-nexorade hybrids can be as efficient as triangular gridshells or ribbed shells. The construction of shell-nexorade hybrids does not however require the construction of complex connectors between the members.

4.2 On-Site Assembly

The structure consists of 102 beams and 48 plywood panels, manufactured off-site and assembled *in-situ*, so that the respect of fabrication tolerances is crucial for the assembly process. The beams and panels weigh less than 5 kg each and can be assembled by two people without difficulty. The assembly sequence starts from the construction of a tripod, which requires temporary supports, as seen in Fig. 8. The structure is then built by cantilevering beams from the tripod.

Fig. 8. Different stages of construction: tripod with temporary supports (left), cantilevering structure (middle) and completed structure (right)

The topographic location of the supports was set weeks before the construction. The structure matches these locations within a range of 1 mm. This illustrates the potential offered by the precision of robotic fabrication in terms of sequencing of assembly and manufacturing. Only four days were required to mount the structure, with no lifting equipment. The initial logistical constraints are thus fulfilled thanks to the precision and flexibility offered by robotic fabrication.

5 Conclusion

This paper presented the design workflow and fabrication process of an innovative structural system. Shell-nexorade hybrids combine the ease of assembly of nexorades and the strength of thin shell structures. The addition of plates, which are mechanically attached to the beams of a classical nexorade significantly increases the structural stiffness. This attachment is guaranteed at the cost of higher manufacturing complexity, which requires the use of robotic fabrication. A workflow integrating fabrication constraints related to geometry, robot reachability, and structural analysis was necessary to efficiently iterate and improve the design. The prototype presented in this paper demonstrates the potential of this design approach, as it sustained decennial wind load and snow loads with no visible damage since its completion in October 2017. The project discussed in this paper also shows that robotic fabrication can be adapted to the specific requirements of a project with a complex geometry.

The authors believe that the concepts developed here can be deployed at larger scales, due for example of the compliance of the structure and screwed connection details with building structural standards. Spans in the range of 20 meters can be envisioned if one uses industrial manufacturing tools and other materials, for example by using cross-laminated timber instead of plywood for bracing.

References

1. Baverel, O.: Nexorades: a family of interwoven space structures. Ph.D. thesis, University of Surrey (2000)
2. Bock T.-A.: Robot-oriented design. In: Proceedings of the 5th International Symposium on Robotics in Construction, ISRC, pp. 135–144. Tokyo (1988)
3. Brocato, M.: Reciprocal frames: Kinematical determinacy. Int. J. Space Struct. **26**(4), 343–358 (2011)
4. Harris, R.: Engineered timber structures in the UK. In: 3rd Forum International Bois Construction, pp. 1–12, Beaune (2013)
5. Mesnil, R., Douthe, C., Baverel, O., Léger, B.: Linear buckling of quadrangular and kagome gridshells: a comparative assessment. Eng. Struct. **132**(1), 337–348 (2017)
6. Mesnil, R., Douthe, C., Baverel, O., Léger, B.: Marionette Meshes: covering free-form architecture with planar quadrilateral facets. Int. J. Space Struct. **32**(3–4), 184–198 (2017)
7. Mesnil, R., Douthe C., Baverel O., and Gobin T.: Form-finding of reciprocal frames with the method of translations. Automation in Construction, accepted (2018)
8. Poranne, R., Chen, R., Gotsman, C.: On linear spaces of polyhedral meshes. IEEE Trans. Visual Comput. Graphics **21**(5), 652–662 (2015)

9. Preisinger, C.: Linking structure and parametric geometry. Architectural Des. **83**(2), 110–113 (2013)
10. Robeller, C., Weinand, Y.: Fabrication-aware design of timber folded plate shells with double through tenon joints. In: Reinhardt, D., Saunders, R., Burry, J. (eds.) Robotic Fabrication in Architecture, Art and Design 2016, pp. 166–177. Springer International Publishing, Switzerland (2016)
11. Scheurer, F.: Materialising complexity. Architectural Des. **80**(4), 86–93 (2010)
12. Scheurer, F., Stehling, H., Tschümperlin, F., Antemann, M.: Design for assembly—digital prefabrication of complex timber structures. In: Proceedings of the International Association for Shell and Spatial Structures Symposium (IASS): Beyond the Limits of Man, pp. 1–7. Wroclaw (2013)
13. Schwartz, T.: HAL. In: Brell-Çokcan, S., Braumann, J. (eds.) ROB|ARCH 2012: Robotic Fabrication in Architecture, Art and Design, pp. 92–101. Springer, Vienna (2013)
14. Willmann, J., Knauss, M., Bonwetsch, T., Apolinarska, A.A., Gramazio, F., Kohler, M.: Robotic timber construction—expanding additive fabrication to new dimensions. Autom. Constr. **61**, 16–23 (2016)

KUKA
KUKA

Sub-Additive 3D Printing of Optimized Double Curved Concrete Lattice Structures

Christopher A. Battaglia(✉), Martin Fields Miller, and Sasa Zivkovic

Cornell University, Ithaca, NY 14850, USA
cab523@cornell.edu

Abstract. The research presented in this paper investigates architectural-scale concrete 3D printing for the fabrication of rapidly-constructed, structurally-optimized concrete lattice structures. *Sub-Additive Manufacturing* utilizes a three-dimensional tool path for deposition of material over a mechanically-shaped substructure of reusable aggregate. This process expedites the production of doubly-curved concrete form by replacing traditional formwork casting or horizontal corbeling with spatial concrete arching. Creating robust non-zero Gaussian curvature in concrete, this method increases speed over typical concrete fabrication practices. Utilizing robotics to integrate a streamlined workflow from digital design to physical fabrication, *Sub-Additive* leverages digital workflows to produce structurally, materially, and spatially optimized building components while dramatically reducing waste material. Addressing digital form finding and optimization, material behaviors (both concrete and supportive aggregate), nozzle design and novel utilization of robotic fabrication, this paper introduces a series of key concepts for *Sub-Additive Manufacturing*, radically advancing concrete 3D printing at full scale.

Keywords: Concrete 3D printing · Structural optimization
Concrete lattice structures

1 Intro and Context

Advancements in additive manufacturing technology have provided the tools to rapidly produce unique forms, shifting the focus from mass production to mass customization [1]. As a viscous material contingent upon long curing times, concrete has historically relied on wasteful, discarded molds for cast-in-place constructions. In contrast, current architectural research takes concepts from small scale PLA 3D printing and applies similar objectives to building materials such as terracotta and concrete: recent developments in additive manufacturing make it possible to construct building components directly through material deposition and therefore avoid custom formwork. Research such as Contour Crafting by Behrokh Khoshnevis of USC [2], TU Eindhoven's 3DCP programme [3], or XtreeE [4], explores concrete 3D printing in two-dimensional layers and, when stacked, these layers create a three-dimensional form [5, 6]. To create a three-dimensional object that is not merely a pure extrusion of a planar geometry, the method of corbeling is used to offset the next deposited layer no more than half of the printing width of the previously extruded line [5, 6]. When an offset more than half of

J. Willmann et al. (Eds.): ROBARCH 2018, *Robotic Fabrication in Architecture, Art and Design 2018*, pp. 242–255, 2019.
https://doi.org/10.1007/978-3-319-92294-2_19

the printing width is required, additional support is added to the fabrication process to stabilized the geometry [7]. A time consuming process which also limits the range of the degrees of curvature in the object, corbeling without the addition of structural reinforcement or a concrete cast interior core cannot handle load perpendicular to the printed direction, causing potential delamination and structural shearing of the building component.

By replacing corbeling as a means of creating curvature, *Sub-Additive Manufacturing* reduces fabrication time and creates structural continuity in compressive structures. Finite element solvers are deployed to determine stress from digital thin-shell form [8] and translate structural information into three-dimensional toolpaths. Precise material deposition increases structural and material efficiency in construction components, and reduces the amount of extrusion layers needed to produce objects of greater curvature. By utilizing a reconfigurable aggregate surface to print upon [9], the supporting structure offers an adaptable and mass-customizable surface without the waste of traditional formwork. This paper focuses on outlining *Sub-Additive*'s parameters and limitations in the production of curvature, the calibration of minimum thicknesses for structural continuity, the application of three-dimensional reinforcement, the manipulation of extrusion layers, direct deployment of finite element solutions in shell geometries, and center of gravity optimization for balance in situ, all tested through a typology of full-scale structural concrete arches.

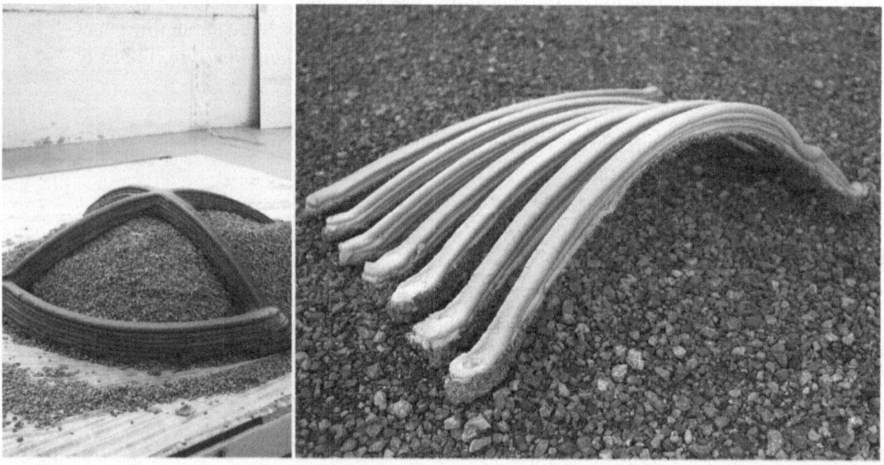

Fig. 1. *Sub-Additive* spatial printing, gravel subsurface, angles of repose from 5 to 40°

2 Methodology

2.1 General Principle and Curvature Conventions

The practice of land-forming to cast shell structures becomes an inspiration in the *Sub-Additive* process: the Philips pavilion at the Brussels Expo '58 was one of the first modern examples to panelize a large double curved surface through land-forming [10].

Casting on a shaped hyperbolic sand form, multiple panels were produced as sections of the surface, transferred to the site, and post tensioned to create the structure. Additional seminal references of sand pile formwork include the work of Heinz Isler and Mutsuro Sasaki, - who both used this method to construct double curvature shell structures [8, 9].

The Sub-Additive 3D Printing Concept. To explore spatial concrete printing, a concrete material was developed which has the ability to support additional layers deposited upon itself while still wet and, when fully cured, handles compressive forces with a minimum unreinforced thickness of about 5 cm. The concrete mixture includes locally available aggregates and portland cement compounds. One batch contains 11.5 kg portland cement, 19.5 kg fine sand, 25.2 kg Mortar Mix Type S, 110 g Thermo-Lube, 35 g Superplasticizer #5, 10 g Nylon fibers, and 11.5 kg of water. When analyzed for its compressive strength, at around 62 N/mm^2, it is about three times stronger than regular grade concrete. The mix has an open time of approximately 1 h and can be extruded for a maximum layer height of 10 mm, to an overall print height of 100 mm before allowing additional time for the material to set. While the concrete is only stable after 2 days of curing and fully cured at 26 days, an interchangeable bed system has been developed to allow for continuous printing. 1-5 mm gravel was chosen as a supporting aggregate for its jagged geometry which offers a relatively high angle of repose. The loose mound of granules on the print bed can easily be reshaped, using the printer's gantry or the robotic end-effector Fig. 2. The printing surface is no longer a flat datum, but rather a manipulable surface, customized to the element which needs to be fabricated in coordination with the computer model. With the introduction of a reconfigurable and recyclable substrate, double-curved surfaces can be created in a fraction of the time required by traditional casting methods.

Fig. 2. Digital carving of gravel sublayer, printed panel

First investigations into *Sub-Additive* 3D printing using gravel support were performed on Daedalus Fig. 2, a large format open-source gantry printer [11] before calibrating to a KUKA KR200. Using Daedalus, concrete material is loaded into a hopper on the X-axis gantry, while a Nema 34 stepper motor drives an auger placed inside the nozzle to extrude the material. The print nozzle remains constantly parallel to the cartesian x-y plane, which results in differentiated layer thicknesses on steeper inclines. In contrast to the gantry system, the KUKA KR200's 6-axis range of mobility allows for a greater degree of curvature and constant extrusion thickness on the double curvature. With the ability to orient the print nozzle normal to the curvature of the surface, higher degrees of precision within the component are achievable. Using a progressive cavity pump outside the robotic work cell, material is loaded into the pump's hopper and transferred to the extrusion nozzle via a 2 cm diameter high pressure hose.

2.2 Testing Parameters of Sub-Additive Printing

To understand the baseline parameters in *Sub-Additive* printing, multiple tests were conducted across various scales. The information acquired was needed to inform the design of architectural components that fully leverage the advantages of the *Sub-Additive* concept.

Testing Angles of Repose. Using the gantry printer platform, a series of catenary arches were first printed to determine the angle of repose achievable in the formation of the gravel substrate Fig. 1. Single extrusion lines of 9 mm layers were printed over various catenary surfaces from a controlled datum of 0°, at 5° increments, until reaching 45°. Each arch was printed to a thickness of 54 mm. A maximum slope degree of 45° was determined for gravel stability, however, the concrete prints were only stable to a maximum of 40°. At a 40° surface incline, the material can only be printed perpendicular to the slope. If the extrusion line travels parallel across the slope, the concrete material rolls off the constructed mound. At an angle of 35° in slope, multi-directional printing becomes reliable.

In coordination with this test, the behavior of the gravel mound during the printing process was analyzed, looking for settling of the mound or shifting of gravel aggregate. The gravel chosen is sharp and angular in geometry, adding surface area to the individual aggregate and creating more points of contact between individual pieces of gravel. During the extrusion of the first print layer, the concrete starts to hold the mound in place, resisting forces that would result in the mound settling. Gravel surfaces that exhibit degrees of curvature over 40° did see evidence of shifting. Staying within the boundary of 35° of curvature, the gravel substrate is able to keep its geometry, and also resist the forces of setting due to the weight of the fabrication process.

To analyze global curvature accuracy of the completed concrete component, 3D scanning of the finished print was used to compare the spatial surface geometry to the computer model. Using a FARO 150 LIDAR scanner with 1 mm precision, a point cloud of the 3D printed geometry was generated and upon analysis, no significant deviations in curvature accrued. Although minimal movement was exhibited in the gravel aggregate when the first layer of concrete material came in contact with the aggregate, this did not affect the overall curvature of the geometry.

Minimum Printing Thickness. Single line arches were printed with an extrusion thickness of 9 mm per layer, at increments of 1 layer each time until reaching a thickness of 54 mm. The first layer is crucial in attaching itself to the aggregate subsurface, but its structural continuity is not enough to hold together the extrusion line. The gravel aggregate does get absorbed into the initial layer, with the print nozzle occasionally dipping into the aggregate subsurface. Upon reaching 36 mm in overall print depth (4 x 9 mm layers) single line extrusions became stable, effectively supporting their self-weight. At 54 mm (6 x 9 mm layers) the sections were stable enough to be considered for larger architectural components.

Layer Thickness Manipulation in 3D Printing. To increase adaptability in future prints, the research team developed and tested sectional variation of structural parameters within thin-shell lattice components. Through the use of G-code manipulation and a scalar extrusion rate in relation to the height above the previous extruded layer, the gantry platform can execute single line printing geometries that taper from 54 mm to 9 mm over the course of 1 m Fig. 3. By manipulating thickness at the scale of the layer, global geometries can taper in section to increase structural efficiency and deposit material where structurally necessary.

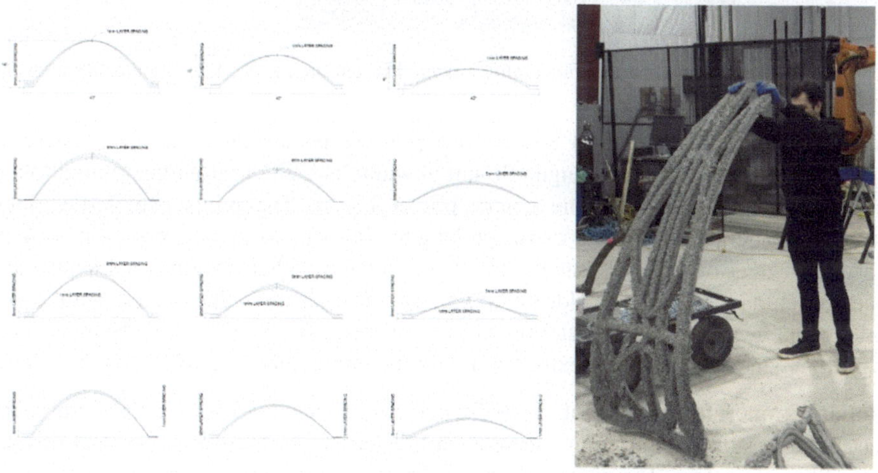

Fig. 3. Layer thickness manipulation diagrams, 1.75 m pointed arch

Structural Reinforcement in Spatial 3D Prints. In the case of *Sub-Additive 3D Printing*, reinforcement needs to conform to complex toolpaths and reside within the thin 54 mm sectional profile. CNC rebar/wire bending is one method for laying rebar into the concrete during fabrication. Flexible forms of steel reinforcement or fiber reinforcement can be robotically integrated in the component fabrication process. Adding steel or synthetic fibers into the print mixture itself increases structural strength within the layer but does not address layer-to-layer discontinuity. The research team manually tested different possible reinforcement materials such as flexible steel chain link, mesh, long strand pins, and various fibers to add additional strength to the printed components and combat issues of delamination between layers Fig. 4. Initial testing results show the feasibility of inserting a secondary material during the fabrication process between the layers to enhance fusion. Other future reinforcement techniques are currently in planning and will be discussed in the future investigation section of this paper.

Fig. 4. Integration of structural reinforcement, 150 cm arch tests, 1.25 m catenary arch

Arch Geometry Testing. To test various printing parameters, a series of shallow arches measuring 150 cm in length, 30 cm in width, and 25 cm from the ground to the apex, were first printed on the gantry printer Fig. 4. The prints explored various structural pattern densities and extrusion lengths. In total, six arches were printed with varying degrees of material usage to find the optimal ratio between material deposition and structural capacity. Exploring a variety of toolpaths, the arches with short orthogonal geometries were determined to have lesser structural continuity than those with long smooth curvatures extruded across the entire curvature of the mound.

Robotic Concrete Printing of Arch Structures. Using a KUKA KR200, the team tested the feasibility of printing concrete while orienting the extrusion nozzle normal to the surface curvature. The pump allows for the use of a thicker concrete and a decrease in material viscosity when compared to the gravity-fed auger system of the gantry 3D printer. Similar in approach but varying in material usage and process, the project Filligree Robotics by Martin Tamke explores 3D printing normal to a curvilinear surface though clay by first scanning an unknown surface and then applying the technique of over-forming [9]. In order to test robotic concrete printing normal to the surface, a 90 cm gravel mound with double curvature was triangulated along the length of the surface, as has been successful in the gantry test examples. With an extrusion diameter of approximately 2 cm at 9 mm in layer thickness, the 6-axis robot allowed for consistent layer height along the entire surface. The increased precision enables the exploration of shells with greater degrees of curvature and tighter toolpath densities for more intricate ornamentation and higher structural efficiency.

3 Structural Optimization and Workflow Strategy

3.1 Tool Path, G-Code, and KRL

Through the use of digital modeling tools and structural analysis software, thin shell geometries are transformed into three dimensional tool paths used for both the KUKA and gantry printer platforms. Using a combination of finite element structural analysis solvers such as Karamba and Millipede to determine the shell geometry and extract lines of principal stress, the *Sub-Additive* process advances typical gantry 3D printing techniques by incorporating the variable Z vectors within the G-code to follow a three-dimensional curve. The gantry printer utilizes traditional G-code whereas the KUKA robotic platform used KRL code. Grasshopper plug-ins such as Silkworm or KUKA| prc convert curve geometries into manipulated G-code or KRL code. By calibrating the extrusion parameter as well as entry and exit velocities along the curve, layer thicknesses can be manipulated in coordination with overall curvature.

3.2 Structural Optimization

To ensure a smooth digital workflow, a form finding exploration was conducted that computationally explores a series of catenary and pointed arch structures. Leveraging concrete as a highly compressive material combined with the linear deposition inherent to 3D printing, principal stress lines were used to generate toolpaths [12], creating lightweight lattice double curvature shells Fig. 5. In consideration of movement and installation of large-scale components, further optimization of components explores manipulation and adaptation of material placement for a more ideal center of gravity

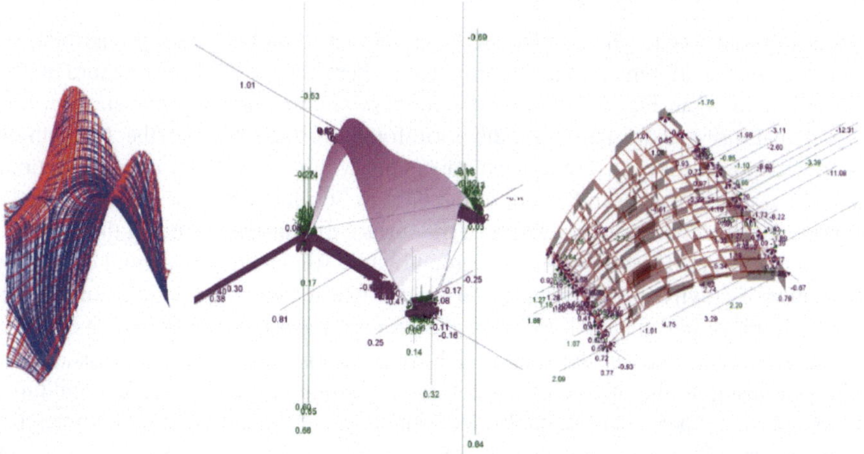

Fig. 5. Structural optimization analysis using Karamba, extraction of principal lines of stress

when erecting the shell forms into their final position. Evolutionary solvers are deployed to rapidly test and determine material distribution for ease of erection and balance in situ. Each arch shell was optimized for the multiple structural stages it endures throughout the process of fabrication and assembly. Each side of the arch was optimized to maximize curvature without exceeding the maximum print angle of 35°.

4 Results

4.1 Concrete 3D Printing of Arch Structures (Gantry Printing)

1.25 m Catenary Arch. To determine parameters of the *Sub-Additive* printing system, the research project investigates a series of structural and ornamental applications. The typology of the arch enables a close focus to explore the relationship between form, structural efficiency, and double curvature. In an initial test, a single curvature catenary arch was divided into two halves, taking into account appropriate angles of repose Fig. 4. In previous prints, curvatures were printed convexly, whereas in the 1.25 m arch, the two components were printed on concave gravel sub-surfaces. By tooling the gravel in this manner, the surface uses 30% less gravel by volume and reverses the relationship of printed surface to gravel texture.

1.75 m Pointed Arch. Continuing to move up in scale, the team printed a pointed arch with an apex of 1.75 m. The pointed arch has the advantage of shallower surface angles relative to the normal cartesian plane and creates a planar contact surface between the two architectural components. In this arch, double curvature was introduced within the geometry of the surface Fig. 3. Through the double curvature, the arch gains stability, strength, and additional visual depth. A resulting discovery upon extraction was that the curvature on each side of the arch allowed the piece to essentially "roll" along its edges, making the processes of erecting the two pieces to their final position noticeably easier. This center of gravity manipulation was then further explored as explained in the form finding section of this paper.

2.75 m Pointed Arch. The final pointed arch fabricated on the gantry printer platform combines all the aforementioned testing parameters into a 2.75 m structure with a 1.25 m interior span Fig. 6. Utilizing the concave double curved gravel surface, lines of principal stress extracted from the form-found surface became the toolpath for material deposition. Each component tapers in width as well as overall sectional thickness transforming from 81 mm at the base to 36 mm at the apex. The density of the print increases where the thickness of the surface diminishes, reducing the weight of the arch the higher it resides above the ground while still creating structural continuity. The center of gravity in each component is such that it causes the double curvature to essentially pinch and flatten 1.75 m off the ground. Each component weighs roughly 150 kg and was printed in the span of 1 h. Imprecision caused by the extrusion nozzle's relationship to the surface resulted in some shifting of the material and buildup of excess concrete where multiple toolpaths terminated. The future investigations section of this paper will discuss strategies for increasing the overall precision of 3D printed *Sub-Additive* assemblies.

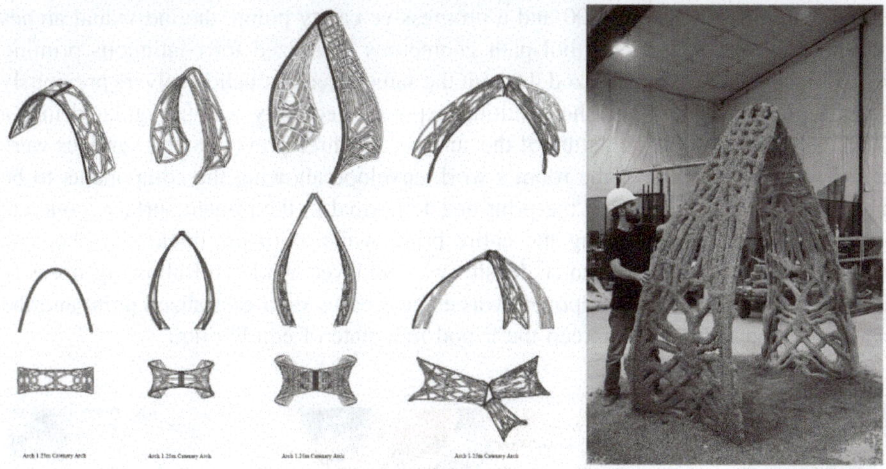

Fig. 6. Sub-Additive arch structure typology, 2.75 m pointed arch

4.2 Concrete 3D Printing of Tri-Arch Structures (Robotic Platform)

Spatializing the arch, a spanning structure of three doubly curved geometries - each different in length, width, curvature, and structural pattern - was designed around an optimized central point. The resulting tripod Fig. 7, composed of various arching components utilizes the Galapagos evolutionary solver to determine its overall form. Each individual shell component has an independent center point used in composing the tripod, keeping the differentiating length of each arch within its goal of optimizing their relationship to the origin. The connection detail at the top creates a center opening, so the arches only touch along their side and at two points along the ground.

Fig. 7. Robotically printed tri-arch structure

Utilizing the KUKA KR200 and a progressive cavity pump, the individual arches currently require closed loop tool path geometries optimized for continuous printing Fig. 8. Each surface was analyzed through the same force evaluation solvers previously stated within this paper and the resulting exported geometry was triangulated into a facets within the double curvature of the surface. All three gravel mound surfaces were tooled within the radius of the robot's work envelope allowing the components to be fabricated in succession. With the print nozzle normal to the printed surface, concrete layers became consistent along the entire print, while extrusion thicknesses became more refined and thinner. Greater density was achieved while still allowing porosity through the surface. Each component was erected based on a centralized point and the enacting compressive forces keep the tripod in a state of equilibrium.

Fig. 8. Robotic concrete printing of tri-arch with progressive cavity pump

4.3 Sub-additive Process Advantages

Advantages of the *Sub-Additive* construction method are evident in the areas of formwork, structure, speed, and geometry. By using the desired printing platform to create both the support surface and extrusion process, speed and efficiency are achieved through the execution of multiple processes within a singular fabrication platform. This then is evident in the resulting component, two different types of textures. The top layer of extruded concrete, and the granulated texture of the gravel subsurface, recalling the multiple processes used in the components fabrication. By utilizing a re-usable medium in the creation of the sub-surface, material waste is significantly reduced when compared to the use of custom formwork in typical casting methods. The process of printing volumetrically in three-dimensional space allows concrete prints with complex curvatures to be fabricated with increased structural efficiency. The continuity achieved in the printing of a single line, in contrast to the stepover exhibited in corbeling, allows forces to move through elements without having to transfer over multiple cold joints as a result of the 3D printing process. The individual lines of fabricated concrete work in tandem with the direction of force loads, optimizing material deposition for design elements. Using less material than casting, the architecture becomes both structurally efficient and lightweight, allowing for various degrees of porosity.

5 Future Investigations

Further development of the *Sub-Additive* technique requires investigations into material, geometry, and post-processing. The concept of this fabrication strategy can be applied to a multitude of material combinations in various printing mediums and subsurface aggregates. Sub-surfaces such as sand, salt, recycled plastic, crushed concrete, or crushed glass, have varying degrees of precision in the creation of the surface and display different visual qualities in fabrication. A manipulable wet sand mold has the advantage of creating a more detailed subsurface, increasing the accuracy of the edge conditions of the print.

Furthermore, the implementation of robotic learning strategies, paired with high-resolution 3D scanning technology [13] will allow for a more efficient shaping of and response to aggregate sub-surfaces. The *Robotic Construction Lab (RCL)* at Cornell University is currently collaborating with building industry partners on implementation and initial research of smart sub-surface shaping.

Integrating reinforcement is paramount for scaling-up architectural investigations and expanding the current limitations of the process. The *Robotic Construction Lab (RCL)* at Cornell University currently investigates fiber reinforcement of *Sub-Additive* thin-shell lattice structures. Considering the material makeup of the concrete mixture, advancements in natural short and long fibers can drastically increase the tensile strength of the component, eliminating the need for steel reinforcement. Collaborating with fiber scientists, civil engineers, and mechanical engineers, the addition of post-processed natural fibrous materials has the potential to significantly increase the strength of the concrete upon curing and also reduce the ratio of layer failure during the printing process. Long synthetic or natural fibers would then act as tensile structural reinforcement when robotically extruded along with the concrete material. The flexible continuous fiber strands can conform to the three dimensional toolpaths in coordination with the concrete extrusion.

Finally, similar to the Compound Fabrication project developed at MIT [14], robotic CNC mill post-processing of printed concrete components will enable precise joinery and an additional level of overall accuracy. Using a process of 3D scanning and stone milling, the completed geometries can be refined surficially and along the edges. Joinery pockets can be accurately placed for the insertion of connectors used in concrete precast construction. Any minor issues in resolution resultant from the 3D printing process can be overcome through the deployment of CNC milling for final finishing and refinement. Different than solid stone carving, the material removed during the milling process is relatively minimal as the concrete has already been accurately placed during the printing process and could be reintegrated as aggregate for either substrate or in future prints.

6 Conclusion

Additive manufacturing processes continue to be an evolving field of research across scales, progressing towards spatial fabrication. Projects such as Liquid Printing developed at MIT [15] or WirePrint developed at Cornell [16] explore new methods for rapid spatial 3D printing. As opposed to traditional horizontally layered printing methods, *Sub-Additive Printing* facilitates the spatial deposition of concrete material, taking full advantage of a new fabrication technology. While small scale 3D printing has quickly proliferated, execution of mass-customized geometries at an architectural scale requires the development of efficient and robust techniques and processes to achieve large scale implementation. Minimizing waste materials and expanding potentials for integration of computationally derivative smart geometries, *Sub-Additive Manufacturing* creates a system for the rapid production of full scale building components. By operating with a palette of readily available standard construction materials, this system relies upon the implementation of advanced robotics, computational technologies, novel processes and techniques to create an elegantly simple means of fabricating complex form. Through the utilization of three dimensional toolpaths, force evaluation solvers, and complex curvature, the *Sub-Additive* research project has the ability and potential to rapidly produce structurally efficient, lightweight thin shell lattice building components. As exhibited in the arch typology investigations, this method of robotic digital fabrication expands the discourse of spatial mass customization to include concrete 3D printing at a one-to-one architectural scale.

Acknowledgements. The Cornell Robotic Construction Laboratory (RCL) has received generous support from: AAP College of Architecture, Art, and Planning: The Department of Architecture at Cornell University, and HY-Flex Corporation. This work is also supported by the academic student researchers at the Cornell Robotic Construction Laboratory.

References

1. Carpo, M.: The Alphabet and the Algorithm. MIT Press, Cambridge (2011)
2. Khoshnevis, B.: Automated construction by contour crafting—related robotics and information technologies. Autom. Constr. **13**(1), 5–19 (2004)
3. Bos, F., Wolfs, R., Ahmed, Z., Salet, T.: Additive manufacturing of concrete in construction: potentials and challenges of 3D concrete printing. Virtual Phys. Prototyping **11**(3), 209–225 (2016)
4. Gosselin, C., Romain Duballet, P., Roux, N.G., Dirrenberger, J., Morel, P.: Large-scale 3D printing of ultra-high performance concrete–a new processing route for architects and builders. Mater. Des. **100**, 102–109 (2016)
5. Hwang, D., Khoshnevis B.: Concrete wall fabrication by contour crafting. In: 21st International Symposium on Automation and Robotics in Construction (ISARC 2004), Jeju, South Korea (2004)
6. Lim, S., Buswell, R.A., Le, T.T., Austin, S.A., Gibb, A.G.F., Thorpe, T.: Developments in construction-scale additive manufacturing processes. Autom. Constr. **21**, 262–268 (2012)

7. Retsin, G., Garcia, M.J.: Discrete computational methods for robotic additive manufacturing. In: Velikov, K., Ahlquist, S., del Campo, M. (eds.) Acadia 2016: Posthuman Frontiers: Data, Designers and Cognitive Machines, Proceedings of the 36th Annual Conference of the Association for Computer Aided Design in Architecture, pp. 332–341. Ann Arbor, MI (2016)
8. Sasaki, M.: Flux Structure. TOTO (2005)
9. Isler, H.: New shape for shells. In: Bulletin of the International Association for Shell Structures, vol. 8, pp. 123–130 (1961)
10. Zephir, A.: Le Corbusier: Philips Pavilion, Brussels, 1958. A Treasury of World's Fair Art and Architecture (2005). Accessed 1 Jul 2018
11. Zivkovic, S., Battaglia, C.: Open source factory: democratizing large-scale fabrication systems. In: Nagakura, T., Tibbits, S., Mueller, C., Ibañez, M. (eds.) Acadia 2017: Disciplines & Disruption, Proceedings of the 37th Annual Conference of the Association for Computer Aided Design in Architecture, pp. 660–669. Cambridge, MA (2017)
12. Tamke, M., Evers, H.L., Nørgaard, E.C., Leinweber, S., Hansen, F.T.: Filigree robotics. In: Velikov, K., Ahlquist, S., del Campo, M. (eds.) Acadia 2016: Posthuman Frontiers: Data, Designers and Cognitive Machines, Proceedings of the 36th Annual Conference of the Association for Computer Aided Design in Architecture, pp. 280–289. Ann Arbor, MI (2016)
13. Keating, S., Oxman, N.: Compound fabrication: A multi-functional robotic platform for digital design and fabrication. Robot. Comput. Integr. Manuf. **29**(6), 439–448 (2013)
14. Hajash, K., Sparrman, B., Guberan, C., Laucks, J., Tibbits, S.: Large-scale rapid liquid printing. 3D Printing Add. Manuf. **4**(3), 123–132 (2017)
15. Mueller, S., Im, S., Gurevich, S., Teibrich, A., Pfisterer, L., Guimbretière, F., Baudisch, P.: WirePrint: 3D printed previews for fast prototyping. In: Proceedings of the 27th ACM Symposium on User Interface Software and Technology, UIST, pp. 273–280. Honolulu, HI (2014)
16. Tam, K.-M.M., Coleman, J.R., Fine, N.W., Mueller, C.T.: Stress line additive manufacturing (SLAM) for 2.5-D Shells. In: Proceedings of the International Association for Shell and Spatial Structures Symposium (IASS): Future Visions, Amsterdam (2015)

Investigations on Potentials of Robotic Band-Saw Cutting in Complex Wood Structures

Hua Chai and Philip F. Yuan(✉)

Tongji University, 1239 Siping Road, Shanghai, China
Philipyuan007@tongji.edu.cn

Abstract. In the field of wood manufacturing, CNC milling seems to be the only way to deal with geometrically complex wood components. Robotic band-saw, the combination of robotic fabrication and band-saw cutting provides a feasible solutions for producing curved wood beams without the immense time consumption of CNC milling method. This paper explores potentials of robotic band-saw cutting technology in the construction of complex wood structures. The research processes of two large-scale research pavilions, from structure design and optimization, Glulam production, robotic fabrication to site assembly, are presented in this paper to explore an integrated workflow of complex wood structures construction based on robotic band-saw. As the two pavilions show, robotic band-saw technology has the ability to perform high efficiency and accuracy both in the plane and space curved wood elements, and shows the potential to take over part of the work of CNC in wood construction.

Keywords: Robotic band-saw · Complex wood structure · Curved beam
Ruled surface

1 Introduction

CNC milling seems to be the only way to deal with geometrically complex wood structures in the field of wood manufacturing. Built projects like Centre Pompidou Metz and the Nine Bridges Golf Club, both of which were designed by Shigeru Ban, were fabricated through milling with the indispensable technical support of Design-toproduction [1]. CNC milling method not only consumes a lot of time, but also produces lots of material wastes. In recent years, with the increasing demand for complex wood structures, new possibilities of wood fabrication were explored by employing industrial robots in the fabrication process, which has shown great potentials in a variety of projects [2]. Among them, the combination of robotic fabrication and band-saw is one of the feasible solutions to produce geometrically complex wood elements without the immense time consumption of CNC milling method. The material efficiency of robotic band-saw was researched and demonstrated for the first time by researchers from Greyshed and Princeton University who used robotic band-saw to cut a series of curved strips from an irregular flitch [3]. Researchers from RMIT University has further studied the robustness of this new craft on many technical issues in regard

J. Willmann et al. (Eds.): ROBARCH 2018, *Robotic Fabrication in Architecture, Art and Design 2018*, pp. 256–269, 2019.
https://doi.org/10.1007/978-3-319-92294-2_20

to speed, accuracy and material finish in the mass-customization of ruled surface production [4]. Robotic band-saw also appeared in an ongoing AA School programme to prototype complex wood structures [5].

While many advantages have been shown in previous studies, there are still great potentials in robotic band-saw that haven't gotten enough attention. The fabrication of curved beams, especially space curved beams, is among the most meaningful abilities of robotic band-saw cutting technology when it comes to complex wood structure. The research presented in this paper was devoted to exploring curved wood beams cutting method with robotic band-saw for complex wood structures. Both the fabrication of plane curved and space curved wood beams were tested through the design and fabrication of two large-scale research pavilions. The design possibilities of complex wood structures brought by robotic band-saw cutting were also shown in the integrated working process from tool development, structural performance-based design, Glulam production to robotic fabrication. This method has the potential to take over part of the work of CNC in wood construction.

2 Methods and Tools

2.1 Tools

Band-saws are employed in wood industry in many areas from primary log conversion to furniture manufacturing. Compared with circular saws, band-saws have advantages such as lower kerf waste and noise levels [6]. More importantly, as the width of the band-saw blade is much smaller, it's possible to cut ruled surfaces with a band saw by continuously rotating (either material or blade) while cutting. While in conventional band-saw cutting process workers need complex jigs to stabilize the cutting path when feeding wood through a band-saw, robots can make this process flexible and precise. Robotic band-saw works in a similar way with hot-wire cutting, creating ruled surfaces through the moving of hot-wire/blades. However, as band-saw blade, which cannot be simplified into a line, should always be parallel to the tangential direction of target surface throughout the cutting process, robotic band-saw cutting is more complicated than hot-wire cutting.

Fig. 1. Robotic band-saw

In this research, a modified 14-in. band-saw was installed on KUKA robots to conduct ruled surfaces cutting (see Fig. 1). Some spare parts of the band-saw were removed to reduce weight and increase the cutting range. With a Motor of 1500 W, the saw is possible to cut materials with a cross-section of 35 cm × 30 cm. Bimetallic blades, which have higher stiffness and durability than high-carbon steel ones, were employed all over the research process.

2.2 Research Method

The scope of this research is to figure out the capacity of robotic band-saw in irregular wood beams cutting, including both plane curved and space curved beams. For practical considerations, Glulam was employed as testing materials. The method proposed here is to cut the target irregular beam from a raw beam, a plane curved one with uniform rectangular section which can be produced with existing Glulam production technology.

The limitations imposed by the fabrication method on the design is that the surfaces of beams needs to be designed as ruled surfaces. Except that, the beams are free to be designed into a wide variety of forms, from plane curved ones to space curved ones, with section from uniform ones to ones that varied with other parameters, such as the inner forces.

The volume of raw beam, which is expected to accommodate the target irregular beam, need to be minimized to reduce material wastes. The minimal volume of the raw beam can be found with the help of Genetic Algorithm, after which the files of the raw beams will be sent to factories to produce the Glulam.

The robotic fabrication process starts from modeling the tools and simulating the cutting process in computer. After raw beam being fixed in the reachable range of the saw, the position of the beam will be measured by the robot and modeled into computer. By placing the model of target beam inside the model of the raw beam, the robot can know precisely where needs to be cut. The robot simulation, path planning and programming process can be carried out in Rhino and Grasshopper with KUKA|prc. The robotic cutting process is quite simple and efficient with the output G-code.

This research was carried out through the design and fabrication of two large-scale wood grid shells, the first concerned mainly on the cutting of plane curved beams while the second focused on space curved ones.

3 Wood Pavilion 1: Design and Fabrication of Complex Wood Structure with Plane Curved Beams

The first project is a large-scale Unstrained Grid Shell made of Glulam, which tried to test the capacity of robotic band-saw in mass-customization of large-scale plane curved wood elements [7].

3.1 Fabrication-Oriented Structural Design

The grid shell was designed to be a funnel-shape column structure. Viewed from the top, it is an oblique quadrilateral grid system. The structure consisted of 16 beams which were all plane curves with length varied from 5.8 m to 7.5 m. The initial cross-sections of the beams were designed to be 100 mm × 100 mm. By introducing structural performance-based design, the sections of the beams were optimized to be varied according to structural simulation results. Grasshopper plug-in Millipede was employed in this stage to conduct FEM (finite element method) analysis of the structure under gravity. The thickness of the beams was kept unchanged during the optimization while the height of the beams varied between 120 mm and 200 mm. In this way, each beam remained plane curved, with 2 surfaces flat and parallel. After the optimization, the two irregular surfaces of each beam were all transformed into ruled surfaces so as to be fabricated with robotic band-saw technology. As the thickness was quite small compared to the length, this transformation can hardly be visually perceived. For assembly considerations, the 16 beams of the structure were divided into four layers according to assembly order, different layers were connected with traditional mortise-tenon joints instead of metal connectors (see Fig. 2).

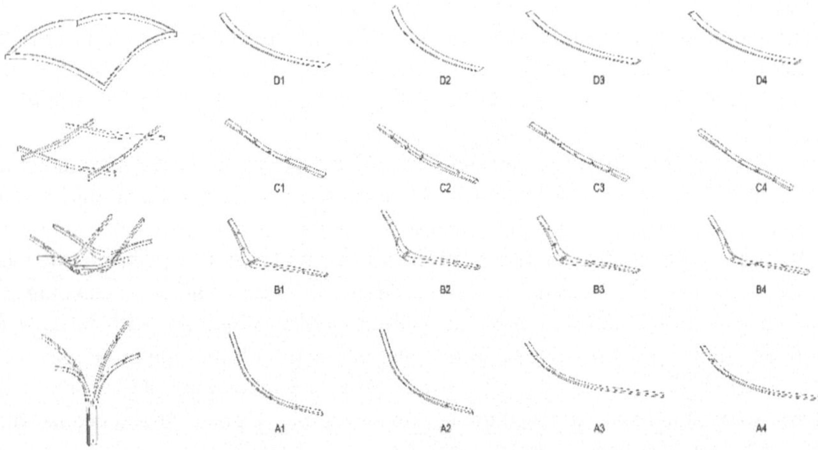

Fig. 2. 16 plane curved beams were divided into four layers

3.2 Raw Beams Production

The beams were oriented onto a horizontal plane to extract the outer contour. After offsetting the contour 1 cm outwardly to make tolerance, the maximum width of the contour was taken as the cross-sectional width of the raw beam. The raw beams were produced in the factory with Douglas fir Glulam, one of the most commonly used materials in the construction of wood structures (see Fig. 3). The production process was guided by CNC templates which was used to inform the curvature of timber beams. In spite of this, the accuracy of the beams was difficult to be guaranteed in the manual production process, 1 cm more tolerance was given in this stage.

Fig. 3. The production process of raw beams

3.3 Robotic Fabrication

The band-saw end-effectorwas installed on a gantry robot system, a triple-axis gantry with a hanging KUKA KR120 robot. The range of the gantry robot system is 11 m × 6 m × 3.6 m, which provides great possibilities for large-scale fabrication.

The raw beam was firstly fixed to two movable and height-adjustable tables with woodworking clips, and the part to be cut was cantilevering outside the tables. After measuring the exact position of the beam, the model was generated in Rhino and Grasshopper, and then robot tool paths were designed and simulated with KUKA|prc (see Fig. 4).

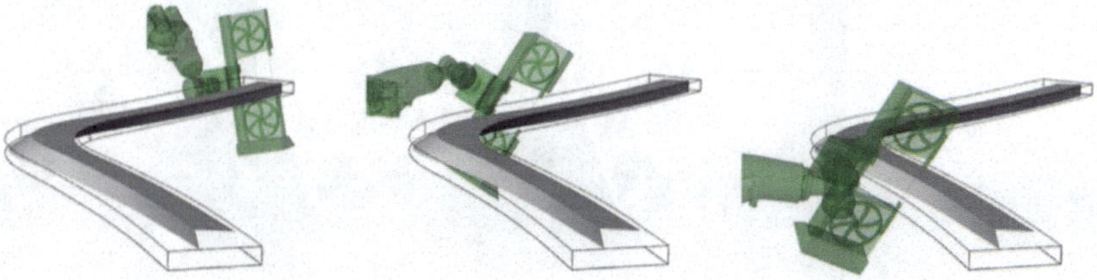

Fig. 4. The simulation of robotic cutting with KUKA|prc

Each beam was cut out through four cuts, respectively top and bottom surfaces, and two ends surfaces. The overall cutting speed was about 5–8 m/h (see Fig. 5). The time cost by each beam can be restricted within 3 h. Although two or three workers were needed to carry and fix the beams, only one robot operator was needed during cutting process. As for the fabrication of mortise-tenon joints, a 24000 rpm spindle was equipped on the same gantry robot to mill the slots after the band-saw cutting process.

3.4 Assembly On-Site

The wood structure was installed on a steel base. By placing the beams of different layers in place in the right order, the site assemble process was completed by five workers in two days, including the setup of steel base, installation and removal of

Fig. 5. Robotic band-saw cutting

Fig. 6. On-site assembly

scaffoldings (see Fig. 6). The mortise-tenon joint made things a lot easier, and the accuracy of robotic band-saw cutting effectively ensured the successful installation.

The wood pavilion appears as a funnel-shaped structure with a height of 7 m, the maximum cantilevered span of which is up to 4.5 m (see Fig. 7).

3.5 Research Evaluation

This project explored the entire process of robotic band-saw in complex grid shell construction from structural optimization, Glulam production, and robotic fabrication to site assembly. The pavilion preliminarily validated the effectiveness and robustness of robotic band-saw technology in plane curved beams cutting.

Fig. 7. The final pavilion

4 Wood Pavilion 2: Design and Fabrication of Complex Wood Structure with Space Curved Beams

Currently, while plane curved Glulam could be easily produced in the factory, there were still no efficient solutions for space curved GLT production. In this context, the second project is a geometrically complex strained grid shell, the space curved edge beam of which provides an opportunity to explore the method and capabilities of robotic band-saw in space curved Glulam cutting.

4.1 Structure Design

Strained grid shell has been one of the most efficient structure system to cover large spans by lightweight construction [8]. This pavilion is a self-support Enneper surface-shaped strained grid shell, consisting of three parts: continuous laths, rotatable joints and rigid edge beam. The edge beam which resisted the bending action of the laths, was curved three-dimensionally. Since this project intended to challenge robotic band-saw with complex beams cutting, the edge beam was also structurally optimized to improve its complexity. The analysis model of the grid shell was built in Karamba, a FEM plugin for Grasshopper which could simulate the structural performance under different loads and boundary conditions, and output structural parameter values corresponding to the input geometry. The cross-sections of edge beam were optimized according to the output structural information (see Fig. 8), material utilization efficiency in this case. The initial cross section of edge beam for analyzing was 80 mm × 100 mm. After the optimization, the cross-sectional dimensions of edge beam varied with material utilization efficiency between 60 mm × 90 mm and 90 mm × 120 mm. The optimization process create a direct connection between form and inner forces of the grid shell.

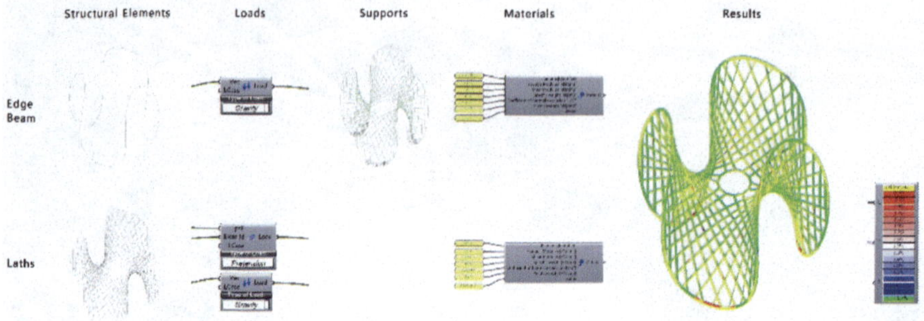

Fig. 8. Structural optimization with Karamba

4.2 Raw Beams Production

The edge beam was divided into 12 segments according to the machine range (see Sect. 4.3). As the beams were space curved, it was more complicate to find the form of raw beams with minimal volume. Genetic Algorithm was employed in this stage to find the smallest volume that can accommodate the target beams by rotating the beams on XZ and YZ planes to changing the projecting angles onto XY plane. After finding the form of raw beam for each segment, the same method was used to production the Glulam raw beams in factory as the first project. Spruce, another common material for wood construction was used this time. As the structure is centrosymmetric, the 12 segments actually had only two different shapes, and all the segments shared the same raw beam mold.

4.3 Digital Fabrication

The resulting complexities of the grid shell were fully addressed in the fabrication and erection process. Different digital fabrication methods were explored in the fabrication of different parts. The laths were straightened onto a plane, and subdivided into segments which can fit into the plywood boards of 1220 mm × 2440 mm, and milled with CNC machine.

The fabrication of space curved edge beam segments was the main concern of this stage. The band-saw was mounted on a robust KUKA R2700 this time to cut the beam segments from the raw beams (see Fig. 9). The robot programing and simulation process were almost the same with the first project (see Fig. 10). As there were substantial spatial torsions in most surfaces that need to be cut, the tables for fixing the material were not feasible in this case. Triangle wooden frames, which is more flexible to be customized for each beam, were developed as supports. In this way, each beam only needed to be turn around once during the cutting process. Each segment was cut from the raw one through 6 cuts (see Fig. 11). During the fabrication process, the speed of the robot was continuously changing according to the thickness to be cut. The material was softer than that of the first project. As the thickness of the raw beams increased relatively, the overall cutting speed actually hasn't changed much.

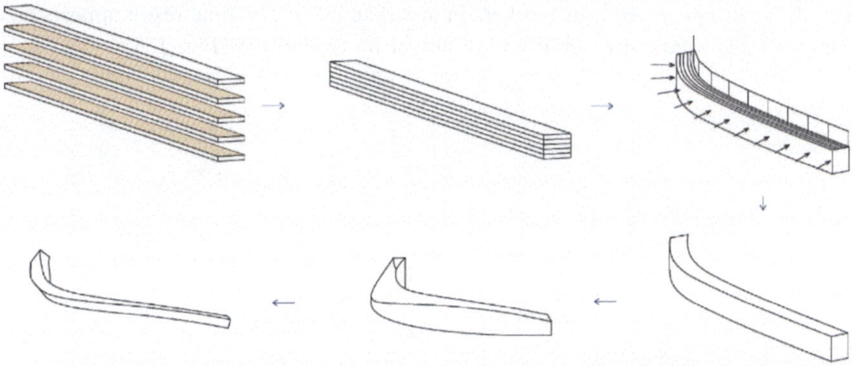

Fig. 9. The diagram of raw beams production and robotic cutting

Fig. 10. Robotic cutting simulation in KUKA|prc

Fig. 11. Robotic cutting process

4.4 Assembly On-Site

As the edge beams and laths are all fully informed, the on-site assembly process of the large scale structure was quite simple and efficient. First of all 12 edge beam segments were connected with bolts and erected in place; then laths were assembled one by one in way similar to weaving (see Fig. 12). The entire assembly work was done in a joint effort of one designer and four workers in less than 20 h. The final result appears as an Enneper surface-shape grid shell with a height up to 6 m (see Fig. 13).

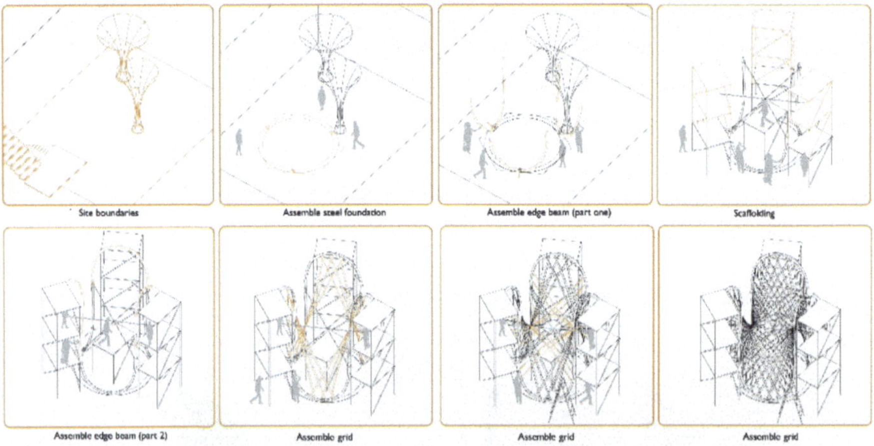

Fig. 12. The diagram of on-site assembly

Fig. 13. The second pavilion

4.5 Research Evaluation

The final structure achieves the goal of self-support. Although both the laths grid and edge beam are quite soft, the combination of these different systems forms a much more rigid structure. This project follows the same process as the first one and extends the application scope of robotic band-saw to space curved wood beams. The final pavilion shows the feasibility and efficiency of robotic band-saw cutting in more complicate structures.

5 Conclusion and Discussion

As the projects show, the robotic band-saw performs high efficiency and accuracy not only in plane curved surfaces, but also high quality three-dimensional ruled surfaces through the continuous rotation of the blades. While shown in this paper its capacity of the mass-customization of full scale geometrically complex wood components, robotic band-saw also have the ability to further adapt to the requirement of industrial mass production. Due to the reason that band-saw has the smallest possible kerf of any mechanical wood cutting method, the process is timesaving compared with the immense material consumption of a CNC milling process. At the same time, the resulting surface of robotic band-saw cutting is qualitatively smoother and more continuous than that of CNC milling. After the applications above, we managed to reduce the tolerances within 3 mm, and increase the machine speed to around 5 m/h (15 cm thick GLT). In further researches, the speed and tolerance can be compared with CNC milling quantitatively. It is expected that robotic band-saw could take over part of the work of CNC in wood industry.

Although this technique has great advantages in material efficiency, there are still some deficiencies to be improved. It is undeniable that there is still a waste of material due to the volume difference between the raw beams and the target ones. The waste may be minimized through the optimization of gluing technology by employing more precise CNC template to guide the material distribution, which can minimize the volume difference between the raw and target beams. In addition, some part of the cutting process could be more automated, such as speed controlling, material measuring.

DigitalFUTURE
Shanghai

Acknowledgments. This research is funded by National Natural Science Foundation of China (Grant No. 51578378), National Key R&D Programme of China (Grant No. 2016YFC0702104), Sino-German Center (Grant No. GZ1162), and Shanghai Science and Technology Committee (Grant No. 16dz1206502, Grant No. 16dz2250500, and Grant No. 17dz1203405).

References

1. Scheurer, F.: Materialising complexity. Archit. Des. **80**(4), 86–93 (2010)
2. Menges, A., Schwinn, T., Krieg, O.D. (eds.): Advancing Wood Architecture: A Computational Approach. Routledge, Oxford (2016)
3. Johns, R.L., Foley, N.: Bandsawn bands: feature-based design and fabrication of nested freeform surfaces in wood. In: McGee, W., Ponce de Leon, M. (eds.) Robotic Fabrication in Architecture, Art and Design 2014, pp. 17–32. Springer International Publishing, Switzerland (2014)
4. Williams, N., Cherrey, J.: In: Reinhardt, D., Saunders, R., Burry, J. (eds.) Robotic Fabrication Architecture, Art and Design 2016, pp. 294–303. Springer International Publishing Switzerland (2016)
5. AA School Robotic Fabrications programme website. http://roboticfabrications.aaschool.ac.uk/project/2017/. Accessed 27 Feb 2018
6. Ulsoy, A.G., Mote, C.D., Szymni, R.: Principal developments in band saw vibration and stability research. Holz als Roh- und Werkstoff **36**(7), 273–280 (1978)
7. Yuan, P.F., Chai, H.: Robotic wood tectonics.In: Sheil, B., Menges, A., Glynn, R., Skavara, M. (eds.) Fabricate: Rethinking Design and Construction, pp. 44–49. UCL Press, London (2017)
8. Chilton, J., Tang, G.: Timber Gridshells: Architecture, Structure and Craft. Taylor & Francis, New York (2016)

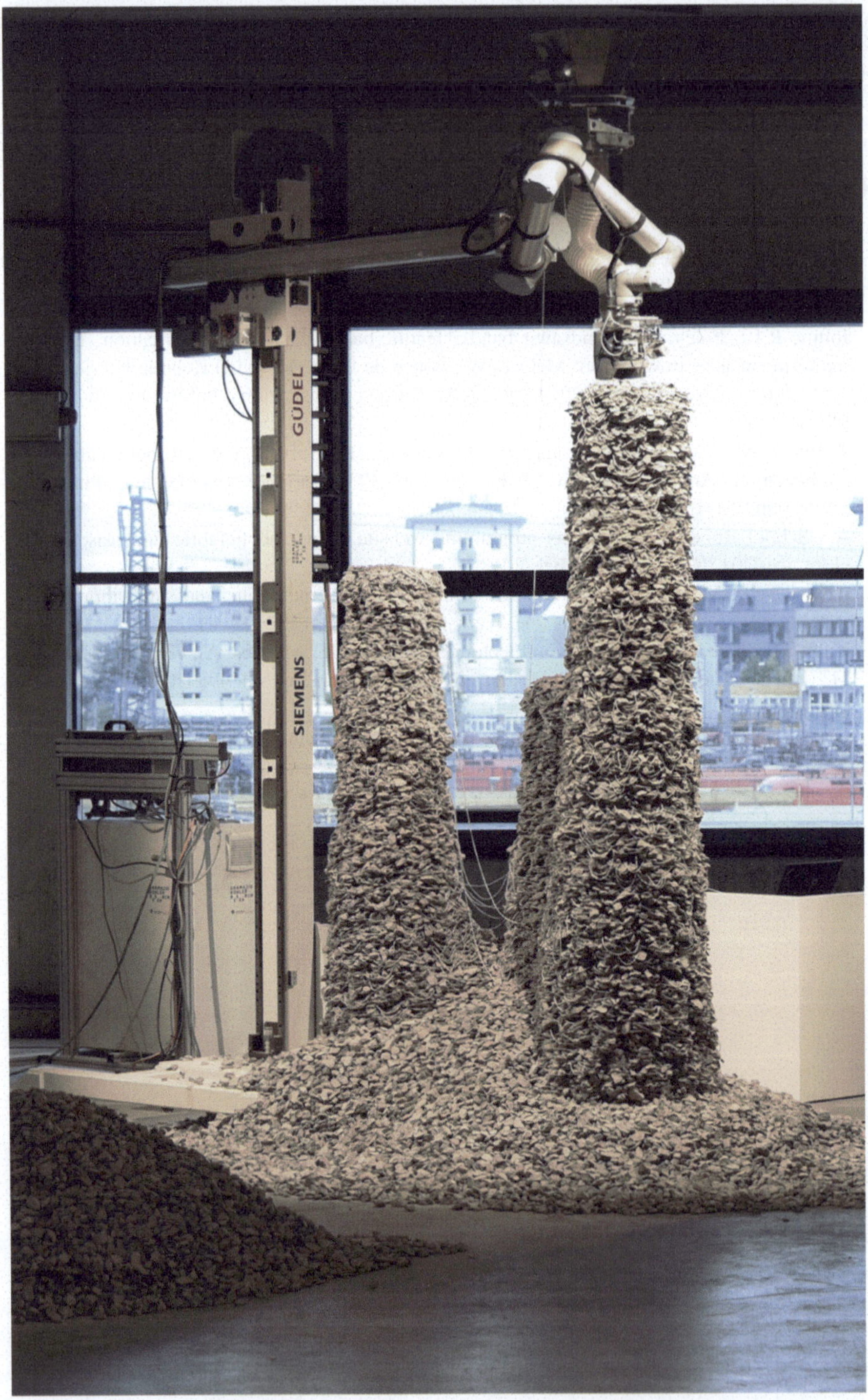
GÜDEL
SIEMENS

Direct Deposition of Jammed Architectural Structures

Petrus Aejmelaeus-Lindström(✉), Gergana Rusenova, Ammar Mirjan, Fabio Gramazio, and Matthias Kohler

Gramazio Kohler Research, Chair of Architecture and Digital Fabrication, ETH Zurich, Zurich, Switzerland
{apetrus, rusenova, mirjan, gramazio, kohler}@arch.ethz.ch

Abstract. The research presented in this paper investigates novel techniques and tools for robotic fabrication of fibre reinforced granular structures built without any type of formwork. Combining granular jamming with strategically placed, continuous reinforcement, allows for precise fabrication of Jammed Architectural Structures (JAS) out of crushed rock and string that are fully recyclable. By further combining the material system with computational design and robotic fabrication enable to place and compact materials directly where needed, without formwork, allows building bespoke architectural structures in an additive, fully reversible and waste-free manner.

The paper describes techniques and challenges for implementing a robotic direct deposition of JAS for fabrication of building-scale and loadbearing artefacts, and highlights the prospects of in-situ fabrication that is exemplified by two one-story tall prototypes. The robot direct depositions the material system in space, without formwork: first, a robot places string in loop-based pattern; second, it places small piles of aggregates inside the string loops, on top of the previously aggregated material; third, it carefully packs the piles and activates the string trough tension.

Keywords: Jammed Architectural Structures · Robotic fabrication
Loadbearing structures · Granular material · Sustainable construction
Direct deposition

1 Introduction

Jammed Architectural Structures (JAS) investigate a fully reversible construction system to create load-bearing freeform structures from granular bulk materials. JAS build upon the physical phenomena by which granular material changes from a loose, liquid state to a solid state and back again, known as the jamming transition [1, 2], and in case of JAS by changes in confinement. Instead of cementing particles to create a solid material such as concrete, the aggregates solidify when the total volume reduces and the particles cannot anymore rearrange. An early example of such jamming in architecture are gabion cages, where lose rocks are jammed by external confinement provided by cages [3]. More recently, a series of construction experiments demonstrate

J. Willmann et al. (Eds.): ROBARCH 2018, *Robotic Fabrication in Architecture, Art and Design 2018*, pp. 270–281, 2019.
https://doi.org/10.1007/978-3-319-92294-2_21

novel methods to apply jamming to build stable structures. This includes wrapping of aggregates in a non-porous membrane and creating a negative pressure [4], or by purposely designing aggregates to interlock and jam because of self-confinement [5–8]. Jammed materials based on aggregates designed to self-confine, solidify immediately without any external confinement, which makes them suitable for fabrication with and without formwork [7, 8].

Combining aggregates with fibres that provide the self-confinement necessary for the jamming transition allow for fabrication of jammed structures out of bulk materials. This can be done by mixing fibres with aggregates and incrementally pack it inside formwork. The fibre-gravel composite takes the shape of the formwork immediately at demoulding. The structures can be casted in segments, while reusing the formwork, allowing for fabrication of structures that are larger than the formwork [9].

A process similar to powder based 3D printing enables controlled fabrication of freeform jammed structures by layering robotically laid string with manually placed aggregates inside a fabrication container [9–13]. An exhibition project at the 2015 Chicago Architecture Biennial demonstrates this principle with a 4-m-tall structure standing on four slender legs [9, 10]. The structure was robotically in-situ fabricated and fully recycled at the end of the exhibition. Pulling out the string network separates the composite back to string and aggregates. A drawback of this construction method is the dependency on a fabrication container, making it more difficult for scaled up applicability and realisation of larger structures. A container free fabrication method for JAS would eliminate the upscaling difficulties and allow for larger geometric flexibility. In this paper, we present methods and techniques to robotically direct deposit load-bearing JAS without any formwork or container.

2 Three Methods of Direct Deposition Fabrication

There are three different methods allowing container free fabrication. First, a closed slip-mould, which is moving upwards during fabrication of the aggregates and string composite. Second, an open slip-mould moving in the vertical plane during the fabrication, repeated in layers to building in height. Third, direct deposition of aggregates and string without slip-mould.

2.1 Closed Mould Direct Deposition

A robot lays out string in a circle-based pattern inside a slip mould (Fig. 1a), covering the whole opening of the mould. Following this, an operator manually pours and distributes a one particle thick layer of aggregates, followed by packing the composite with a handheld concrete compactor (alternatively with vibration). The slip-mould moves upwards to allow for the fabrication of the next layer. Each layer uses the same string pattern and aggregate placement. The composite expands during the packing process, which makes it necessary to use a tapered slip mould (5°) to reduce friction between the slip-mould and the composite.

The slip mould directs the packing energy downwards and insures an even compression, resulting in structures with strong surfaces, free from loose particles. When

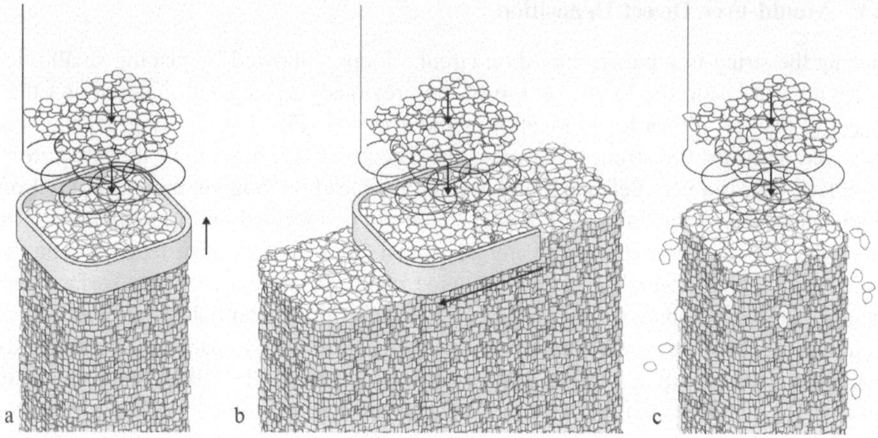

Fig. 1. (a) Closed mould direct deposition. The mould is moving upwards. (b) Open mould direct deposition. The mould is moving horizontal building up layer by layer. (c) Mould-free direct deposition.

the mould is moved upwards a small amount of aggregates falls-off (<10% of the total amount of aggregates). Rotation and tilting of the mould allow fabrication of a range of column-like structures that are dependent on the cross-section of the slip-mould.

2.2 Open Mould Direct Deposition

The robot lays string and places gravel inside a horseshoe shaped slip-mould, followed by manual packing (Fig. 1b). The slip-mould moves away from the composite in the vertical plane, creating an aperture between mould and the already aggregated material. The deposition of string and aggregates takes place in the aperture and the mould moves to the next position. The procedure repeats until the whole cross-section of the intended geometry is completed. The slip-mould moves to one layer above the original position and repeats the process. The string pattern and the gravel placement changes for each slip-mould position and adapts to its position as well as to the overlap between the mould and the previously fabricated structure.

Packing the composite in the open slip-mould with the same force as the closed slip-mould, results in the expansion of the composite being pushed out of the opening, risking deforming and damaging the previously fabricated structure. Packing with less force avoids the damaging material expansion, but results in less precise and strong surfaces. The process is efficient in terms of aggregate placement (<10% of the aggregates falls off). Rotation, tilting and repositioning in the horizontal plane, allows for fabrication of wall-like elements that can be straight as well as curved and branched. The wall thickness depends on the size and shape of the slip-mould.

2.3 Mould-Free Direct Deposition

Placing the string in a pattern based on circular loops, followed by placing small piles of aggregates inside the loops, on top of the previously aggregated material and then carefully packed allows for fabrication without mould (Fig. 1c). Packing expands the piles and activates the string trough tension. The process repeats until the structure is complete. This is a very delicate fabrication process leading to larger amounts of fall-off aggregates during fabrication (20–40%), which can be recycled during fabrication. The process allows for geometries with any horizontal cross-section, only depending on the underlying support, which allows for a wide range of vertical or flatter surface curvatures. Thus, demonstrates mould-free direct deposition potential to fabricate architectural scale structures with more geometrical flexibility compared to the methods described earlier with a mechanically simpler method, and is therefore especially suitable for digital design and robotic fabrication methods.

3 Robotic Fabrication and Computational Design

Precision is an important factor to fabricate tensile reinforced JAS with mould free direct deposition. It is important to maintain an exact position and pattern of the string, precise position and amount of aggregates, constant layer thickness and overall height in space as well as a homogenous packing rate for predictable load-bearing behaviour. Inconsistencies in booth the design information and the fabrication steps leads to imprecise geometry, structural failures and deformation, and therefore requires the mould-free direct deposition digital fabrication tools and computational design tools.

The digital fabrication setup consists of a 6-axis collaborative robotic arm (Universal Robot UR 10, 10 kg payload) mounted on a two-axes gantry system, which is used to reposition the robotic arm. The robot is equipped with a multi-functional end-effector consisting of a string laying device, a gravel measuring and dispensing device and a pneumatic compactor (Fig. 2.1). The string-laying device feeds string with motor driven rolls synchronized with the speed of the robot (0.15 m/s). The robot lays a 2 mm polyester reinforced textile string (chosen for its flexibility) according to a digital blueprint (Fig. 3a). If the underlying layer is not precise, the string pattern will locally fail, and require a second string layer, which is a time consuming approach to handle tolerances. To avoid this, the process aims to maintain the quality of the underlying layers. A self-adopting aggregate placement method allows for compensation when the build-up of aggregates is too fast or too slow and coupled with a precise packing method, it ensures that the top-surface is flat and at the correct height.

Crushed 16–32 mm porphyry aggregates (chosen for its non-directional structure and material strength) are measured and dispensed with a system consisting of a soft nozzle mounted on the end-effector, connect with a polyurethane hose (120 mm diam.) to a hopper (1800 cm^3) mounted on the gantry system (Fig. 2.2). The opening of the hopper ends in a measuring chamber (35 cm^3) coupled with a pneumatic shutter. Two electric vibrators mounted on the sides of the hopper un-jam the aggregates, which flow into the chamber. Turning of the variations re-jam the aggregates in the hopper and

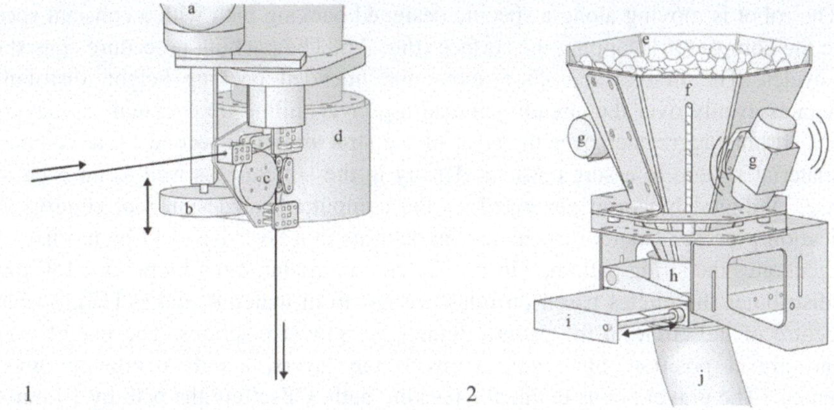

Fig. 2. 1. Combined end-effectors (a) Robot arm, (b) Pneumatic packing device, (c) String laying device, (d) Soft nozzle for gravel placement. 2. Gravel measuring and dispensing device, (e) 1800 cm^3 hopper, (f) Slot showing gravel level, (g) Electric vibrators, (h) 35 cm^3 measuring chamber, (i) Pneumatic shutter, (j) Polyurethane hose.

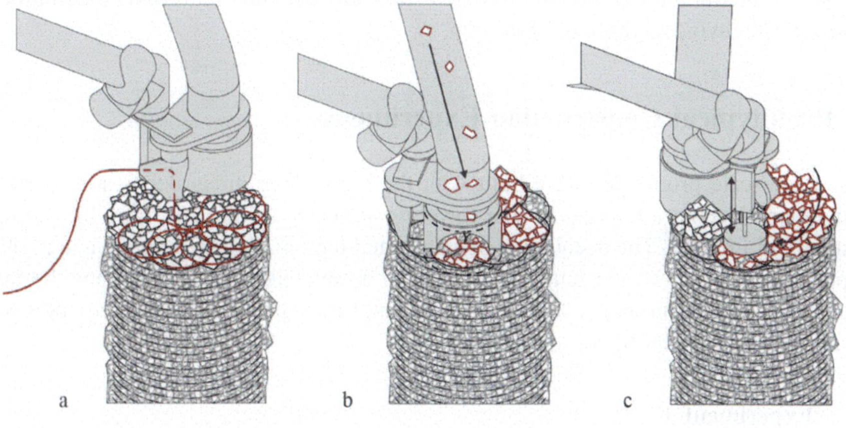

Fig. 3. (a) The string is laid in a pattern based on circular loops, (b) Placement of aggregates inside the circles in small piles (c) The aggregate piles are packed. Packing expands the piles, which tension and thereby activate the string.

allows the shutter in the bottom of the chamber to open and dispense the aggregates. The chamber and shutter system is necessary to assure a constant aggregate amount.

The robot moves to the dispensing position, pushes the soft nozzle down and seals against the underlying surface, where it dispenses the gravel, before moving to the next position (Fig. 3b). Distribution and compacting of the loose aggregates are done with a pneumatic cylinder (80 N force, 10 Hz and 60 mm amp.) equipped with a round aluminium plate (thickness 20 mm, 100 mm diam.) and mounted on the end-effector in the other end.

The robot is moving along a specific designed packing path with a constant speed while the compactor is tapping the surface (Fig. 3c). The packing procedure consist of two cycles. The first cycle, offset above the intended packing height, distributes aggregates evenly over the already jammed top-layer, filling-up eventual cavities and pushes surplus aggregates over the edge of the structure. The second cycle compacts the material system to assure constant density in the structure as well as an even top surface. A digital blueprint generated by the computational design tool controls the fabrication process. The tool calculates the patterns in a layer based fashion. First, the tool generates the string pattern, which is based on circular string loops (d = 150 mm) and distributes the circles based on rules derived from material studies [15]. Second, the centre of the circle in the pattern defines the gravel positions. The use of every second gravel position, alternating between the layers, avoids overlaying gravel placement. The gravel points define the packing path. Offsetting the path by 10 mm in height creates the first packing cycle and the final packing cycle constitutes of the path in original height. The computational design tool accounts for repositioning of the robotic arm, by dividing the fabrication pattern into segments that fit within the reach of the robotic arm. It alternates these divisions, to avoid vertical seems between the fabrication segments. Similarly, the start and end loop of the string pattern is alternated through the pattern. By doubling these loops, the fabrication process becomes more robust. Furthermore, the gravel placement and the packing order alters during the fabrication to avoid asymmetric build-up.

4 Prototypical Construction Experiments

Two large-scale prototypes were built with the mould-free direct deposition method: Experiment 1 being a tall and slender towering structure for the 2017 Ars Electronica Festival in Linz [15]. The installation demonstrated the in-situ fabrication live in public. Experiment 2 is a wall element combined with a concrete slab that explores the fabrication of larger continuous surfaces (>3 m^2) and their behaviour under compression from heavy horizontal building elements.

4.1 Experiment 1

The objective of Experiment 1 is fabrication of a one-storey high structure with mould-less direct deposition fabrication, exploring vertical production within achievable height and reach of the fabrication setup (1700 × 1100 × 2700 mm L × W × H) while maintaining minimal volume and thereby fabrication time. The experiments further explores possibilities to recover from fabrication errors.

The experiment consists of fabrication and evaluation of an artefact with a double curved base (700 mm tall) with slender and conical towers (2700, 1800, and 2100 mm tall including the base, with top radiuses measuring 190, 175, and 175 mm). The fabrication of the base is layer-based, and split in two segments adopted to the reach of the robotic arm. The gantry-system repositions the robot once every layer. The first position of the robot is at the maximum extension of the vertical axis where the robot fabricates the first segment of the layer. The gantry-system moves the robot to the second position, 700 mm along the horizontal axis, to fabricate the second part of the layer. The process repeats in an alternating way, fabricating two layer segments on top of each other in between every repositioning. Every 20 layers, the gantry-system repositions the robot 200 mm upwards. The axis system moves in a low speed due to security reasons and accounts for 15% of the total fabrication time. Thus, it is desirable to reduce the amount of positions along the axis system. The fabrication can be continuous for each column, which dramatically reduces the repositions of the robot. Due to collision risks in-between the structure, the robot arm and the end-effector is the fabrication of the columns done in chunks ranging between 25 and 50 layers (250–500 mm). The aggregate fall-off (40%) lays around the base structure to give the impression of a sculpture growing out of a pile of loose gravel.

The fabrication is stable for the base of the structure as well as the first parts of the columns. At layer 110 (1100 mm height) the fabrication of the middle column becomes unstable, due to asymmetric geometry with small local surface radiuses (<140 mm) and few (<6) gravel placement points. This leads to uneven builds-up with local failures and erosion at the top surface. Attaching corrugated board at the erosion points and manually placing aggregates restores the top surface and stops the erosion. After three new layers removes the temporary support. When the surface radius decreases below 200 mm, the failure frequency increases and requires temporary support every 20 layers (200 mm). Below this radius, the layer thickness varies and thereby needs controlling and compensation, by adjustments to the fabrication pattern up or downwards in 5 mm increments.

Fabrication of the 1.08 m^3 structure takes 40 h and requires 6000 m string and 5000 gravel poses (1.75 m^3). The base builds up in a homogenous, stable way with a 25 mm offset outwards compared to the string pattern. The first tower measures 2680 mm compared to the supposed 2700 (270 layers fabricated), the mid tower measures 1875 instead of 1800 mm (180 layers fabricated) and the last tower measures 2070 instead of the 2100 (210 layers fabricated). The varying heights of the columns depends on the uneven layer build-up occurring when the surface radius decreases below 200 mm, due to the discrete nature of the gravel placement and the few (<6) gravel positions. Any change in gravel positions affects dramatically the total amount of placed gravel. This, especially in combination with asymmetric pattern results in a higher degree of surface failures obstructing the process from recovering fabrication errors.

The experiment demonstrates for the first time successful automated robotic fabrication (string laying, gravel placement and packing) of one-story tall, vertical and geometrically articulated structure with mould-free direct deposition. The process autocorrects small errors during fabrication of larger surface curvatures (radius <200 mm), meanwhile smaller surface curvatures and slenderness requires occasionally temporary support (Fig. 4).

4.2 Experiment 2

Experiment 2 tests the feasibility of in-situ (outdoors, protected from direct precipitation) fabrication of one-storey, loadbearing wall element. It takes advantage of the full reach of the fabrication setup and aims to explore production of continuous

Fig. 4. (a) Experiment 1 at the 2017 Ars Electronica Festival in Linz. The fall-off aggregates lays around the base of the structure, suggesting that the robotically laminated sculpture grows out of a pile of loose gravel [14]. (b) Close-up of a fabrication failure after use of temporary support.

Fig. 5. (a) Experiment 2. Loaded with 2700 kg. (b) Close-up between layer 50 and 100 showing the surface quality. The difference between fabrications with wet aggregates (bottom) compared to dry aggregates (top) is clearly visible.

surfaces with respect to surface quality, recovery from fabrication errors as well as the loadbearing behaviour of the structure over a 20-day time span.

The experiment consists of a tapered, curved 2700 mm tall wall, loaded with a 2700 kg stack of concrete elements. The walls curvature increases the structural depth. The ends of the wall are shaped as half-circles with varying radiuses ranging from 400 mm at the baste to 200 mm at the top, derivate from “Experiment 1”. The fabrication process and sequence is the same as described for the base of "Experiment 1". The aggregate fall-off (30%) is recycled during the fabrication. The finished structure is loaded with the concrete slabs for 20 days. In the end, removing the slab allows for deconstruction by pulling the continuous string.

Due to rainy weather was the first 70–75 layers (700–750 mm) fabricated with humid gravel. The humidity reduces the friction in-between the aggregates, unjamming the aggregates inside the hopper. This obstructs the measuring chamber and shutter system, leading to inconsistent aggregate placement and a low surface quality and faster build-up in the centre of the structures. This requires compensation every 10–15 layer by placing aggregates only at the external gravel points, avoiding the internal gravel points. Fabrication faults due to humid gravel are occasionally damaging the top surface, leading to erosions and requires temporary support (>30 layers in-between the repairs). Above layer 75 is the structure fabricated with dry aggregates.

The completed structure is loaded with a 2700 kg concrete slab with a forklift and connected to a safety chain. The weight is equivalent to a compressive force of 0.05 MPa (2.5% of the maximum force used in uniaxial load to explore loadbearing properties of the material system [15]). Due to asymmetric positioning of the load, the structure bends over time (10° rotation after 20 days) (Fig. 6a). By removing the weight, the structure straightens up and returns almost to its original shape.

Fig. 6. (a) 10° rotation after 20 days. (b) By rewinding the string, the materials are restored to their original state.

Fabrication of the 2 m^3 structure takes 60 h and requires 9200 m string and 7500 gravel depositions (2.65 m^3). The structure measures 2720 mm compared to the supposed 2700. The limited reach of the setup restricts the fabrication 268 layers. The first 70–75 layers fabricated with humid gravel, results in surface deviations of ±25 mm. The string pattern is loose and clearly visible on the surface. Above layer 75 is the surface homogenous with 25 mm offset of the string pattern. The ends of the wall are more precise and more stable to fabricate compared to the towers with same radius in

"Experiment 1". The build-up of the structure is as well more continuous when the layers consists of a higher gravel position count.

The experiment demonstrates the first robotic in situ outdoors fabrication of loadbearing jammed structures with mould-free direct deposition. The process is more stable with larger surface curvatures (>200 mm) in combination with larger cross-sections compared to the towers in "Experiment 1". It shows that the fabrication method as well as the material system is sensitive to humidity. Furthermore, it demonstrates that it is possible to regain precision and surface quality after significant (Fig. 5b), continues surface failures.

5 Conclusions and Future Challenges

This paper outlines a novel, robotic construction method to create load-bearing jammed structures without the need of formwork. The integration of computational design and robotic fabrication enable to place and compact the bulk material directly where needed to aggregate bespoke architectural structures in an additive, fully reversible and waste-free manner. As such, this mould-less direct deposition expands the design space of digital granular construction and fosters the creation of large-scale JAS.

Following the examination of the research parameters for the individual components of direct deposition fabrication, the paper identifies the architectural implications of integrating these components to realise full-scale prototypical structures. This includes the observation of creeping behaviour under asymmetric load and geometric limitations when building slender structures. A key challenge is the outdoor construction in wet or humid conditions, altering the jamming conditions between aggregates and weakening the surface quality of the structure.

To become a competitive construction system, the different fabrications steps, including string laying, gravel placement and compacting require optimisation to improve the speed and robustness of construction. The string, gravel and packing parameters needs to be further analysed to gain precision in the build-up. Additionally, the range of the robotic system is limited by the size of the machine, operating in constrained working space that limiting the application in architecture to small structures. Switching the process to a mobile robotic system [16], however, allows reaching points in space otherwise not accessible to stationary robots. As such, a mobile robotic fabrication system assembling readily available, cheap and fully recyclable building material, indicates the potential for sustainable construction beyond the small artefact; that is, at the full scale of architecture.

Acknowledgments. The research project "Design and Robotic Fabrication of Jammed Architectural Structures" is a collaboration between Gramazio Kohler Research (Institute of Technology in Architecture, ETH Zurich) and Prof. Hans J. Herrmann's group of computational physics (Institute for Building Materials, ETH Zurich). The authors would like to thank the researchers and technicians that have contributed to the ongoing research: Pavel Iliev, Dr. Falk Wittel, Dr. Romana Rust, Marco Palma, Stephané de Weck, Jesús Medina Ibáñez, Michael Lyrenmann, Philippe Fleischmann and Dominik Werne. The research is supported by the ETH Zurich Foundation.

References

1. Song, C., Wang, P., Makse, M.A.: A phase diagram for jammed matter. Nature **453**, 629–632 (2008)
2. Liu, A., Nagel, S.: Jamming and Rheology: Constrained Dynamics on Microscopic and Macroscopic Scales. Taylor & Francis, London, New York (2001)
3. Rammeli, A: Le diverse et artificiose machine del capitano Agostino Ramelli: nellequali si contengono varij et industriosi movimenti, degni digrandissima speculatione, per cavarne beneficio infinito in ogni sorte d'operatione: composte in lingua Italiana et Francese, Paris, pp. 334–338 (1588)
4. Huijben, F.: Vacuumatic formwork: a novel granular manufacturing technique for producing topology-optimised structures in concrete. Granul. Matter **18**(2), 1–23 (2016)
5. Jaeger, H.M.: Celebrating soft matter's 10th anniversary: toward jamming by design. Soft Matter **11**, 12–27 (2015)
6. Murphy, K., Roth, L., Peterman, D., Jaeger, H.: Granular construction: aleatory construction based on jamming: stability through self-confinement. Archit. Des. **87**, 74–81 (2017)
7. Dierichs, K., Menges, A.: Granular construction: designed particles for macro-scale architectural structures. Archit. Des. **87**, 88–93 (2017)
8. Dierichs, K., Schwinn, T., Menges, A.: Robotic pouring of aggregate structures: responsive motion planning strategies for online robot control of granular pouring processes with synthetic macro-scale particles. In: Brell-Çokcan, S., Braumann, J. (eds.) ROB|ARCH 2012: Robotic Fabrication in Architecture, Art and Design, pp. 196–205. Springer, Vienna (2013)
9. Aejmelaeus-Lindström, P., Mirjan, A., Gramazio, F., Kohler, M., Kernizan, S., Sparrman, B., Laucks, J., Tibbits, S.: Granular jamming of loadbearing and reversible structures: rock print and rock wall. Archit. Des. **87**, 82–87 (2017)
10. Aejmelaeus-Lindström, P., Thoma, A., Mirjan, A., Helm, V., Tibbits, S., Gramazio, F., Kohler, M.: Rock print: an architectural installation of granular matter. In: Tibbits, S. (ed.) Active Matter, pp. 287–296. MIT Press, Cambridge (2017)
11. Aejmelaeus-Lindström, P., Willmann, J., Tibbits, S., Gramazio, F., Kohler, M.: Jammed architectural structures: towards large-scale reversible construction. Granul. Matter **18**(2), 1–28 (2016)
12. Fauconneau, M., Wittel, F., Herrmann, H.: Continuous wire reinforcement for jammed granular architecture. Granular Matter **18**(2), 1–27 (2016)
13. Iliev, P., Wittel, F., Herrmann, H.: Discrete element modeling of free-standing wire-reinforced jammed granular columns. Comput. Part. Mech. 1–10 (2018)
14. Ars Electronica Festival 2017: Rock Print: a Manistone, February 2018. https://www.aec.at/ai/en/rock-print/. Accessed 1 Jun 2018
15. Rusenova, G., Wittel, F.K., Aejmelaeus-Lindström, P., Gramazio, F., Kohler, M.: Loadbearing capacity and deformation of jammed architectural structures. In: 3D Printing and Additive Manufacturing (2018, accepted for publication)
16. Giftthaler, M., Sandy, T., Dörfler, K., Brooks, I., Buckingham, M., Rey, G., Kohler, M., Gramazio, F., Buchli, J.: Mobile robotic fabrication at 1:1 scale: the In situ Fabricator. System, experiences and current developments. J. Constr. Robot. **1**, 3–14 (2017)

References

Control and Fabrication

FIBERBOTS: Design and Digital Fabrication of Tubular Structures Using Robot Swarms

Markus Kayser, Levi Cai, Christoph Bader, Sara Falcone(✉), Nassia Inglessis, Barrak Darweesh, João Costa, and Neri Oxman

The Mediated Matter Group, Massachusetts Institute of Technology, Cambridge, MA 02139, USA
{sfalcone,neri}@mit.edu

Abstract. Rapidly increasing demands to enhance speed and safety for on-site and site-specific construction drive the need to develop collaborative and autonomous systems. Such systems generally involve the use of retrofitted gantry and robot arm systems, which have size constraints and are computationally complex to use in collaboration with other machines. This paper describes an alternative multi-robot system built from the ground up to enable collaborative and site-specific construction. The strategy simplifies design workflows while simultaneously maintaining structural, environmental, and robot dynamic constraints. This system of 'swarm fabricators' enables robotic agents to operate in parallel, digitally fabricating independent tubular forms. Each robot controls its position allowing the system to effectively 'grow' a large-scale woven architecture. The robots fabricate by pulling fiber and resin from ground-based storage and winding a composite around their own bodies. Additional sections of a composite tube are appended upon each other, starting from a base. The system relies on an environmentally informed flocking-based strategy to design the structure and inform the robots' trajectories.

Keywords: Swarm robotics · Autonomous construction
Site-specific construction · Composite fabrication · Adaptive fabrication
Fabrication-aware design

1 Introduction

Developments in robotic fabrication tools enable new control over materials, which change our built world. Among the innovations is the development of swarm robotic platforms: systems where multiple robots work together simultaneously to complete a task. However, such systems have not been deployed in real construction settings, and challenges exist in several aspects. Research and development of these systems can generally be separated into four categories: (1) computer-aided design (CAD) environments for designing the structure; (2) computational environments linking the CAD software with the robot software; (3) the physical robotic system; and (4) the materials used to construct the final structure.

Traditional manufacturing processes embody a discrepancy between what can be designed and what can be digitally fabricated and produced [1]. This has led to an

J. Willmann et al. (Eds.): ROBARCH 2018, *Robotic Fabrication in Architecture, Art and Design 2018*, pp. 285–296, 2019.
https://doi.org/10.1007/978-3-319-92294-2_22

emergence of software tools that can decompose high-level design specifications to fabricable instructions, such as in [2, 3]. Bermano et al. [4] classify high-level goals according to appearance, deformation and motion, high-level objectives, domain-specific objectives, and process-specific objectives. They discuss digital representations of these objectives in conjunction with generating fabrication-specific optimizations, such as adjusting tool paths or segmentation of the file to enable parts larger than the build volume of a specific machine. However, these computations scale in complexity with additional robots. In [5], a simulated agent-based swarm algorithm is used to create a structural design, where the agents are limited in possible directions they can traverse. Such a digital limitation encodes the fabrication constraints of the physical robots that produce the structure. The present paper builds on this approach, by developing a hardware system that works more closely with the swarm-based design generation algorithm to enable a larger range of tunable parameters and a more scalable approach.

Several groups have also explored the use of custom-designed, cooperative, multi-robot systems to increase fabrication speed by parallelizing aspects of the construction process. Some strategies are outlined in [6]. Many of these approaches are mobile, which allow structures to be built that are much larger than the robots themselves. Swarm systems can be separated into aerial [7–9] and ground-based platforms [10–14]. Aerial systems can generally traverse larger areas faster, but are more difficult to control, and may require additional infrastructure. Ground-based systems can carry larger pay-loads and therefore carry more power, but collision avoidance and path planning along the surface of the in-progress structure is more constrained.

These approaches can be further separated into assembly-based systems [8, 11–14], where pre-fabricated modules are placed together on-site, and continuous fabrication-based [9, 10] systems, where continuous material is extruded onto the structure as in a FDM 3D printers. These projects explore a variety of trade-offs between structural performance and adaptability of the building materials. Each has added insight to the mobile, multi-robot fabrication domain, but few have simultaneously provided a design framework that allows designers to easily, flexibly, and reliably interact with the system.

The TERMES project [12] is unique in its approach to co-design a high-level specification procedure, fabrication algorithms to coordinate robots at build-time, and a hardware platform that is guaranteed to replicate the desired structure. The system presented here combines the co-design strategy presented in TERMES with the swarm-based design and constraint-inclusion methods of [5] and the use of high-performance and lightweight fibrous composites, as in [9]. The goal is to create an entire workflow, from digital design to physical realization of a structure, that is relatively easy to use for designers and can be built using a parallelized swarm robotic system.

To do this, we developed, built and experimentally evaluated a complete workflow, co-designing hardware and algorithms simultaneously from the ground up. We developed a flocking-based [15] approach towards structural design that enables users to incorporate both design and fabrication constraints at once. Digital designs can then be physically produced by our swarm fabrication platform with minimal correction. Our swarm-based robotic construction system relies on a team of identical robots each fabricating independent, self-supporting tubes. The robots work together to interweave the tubular structures into a woven architecture. Each robot pulls raw material—fiber and resin—from ground-based storage up through the center of the tube it is fabricating and winds a composite segment to extend its tube. After each segment is fabricated the robot moves forward to position itself to wind another segment. At the start of each segment the robots control their orientation to create controlled curvatures in the tubes.

The combined hardware and software innovations allow for designers to create sophisticated and site-specific plans that can be produced on-site, quickly, and efficiently by a robot swarm.

2 Design Software for Swarm Fabrication

Just as the practice of design was altered with the advent of additive manufacturing, it must also be adjusted for digital fabrication with a swarm system. It would be unrealistic for a designer to dictate the motion of each robot within a swarm. With each additional robot, the overall complexity due to robot-robot and robot-structure interactions increases exponentially. For a swarm system to be scalable, additional robots cannot increase interaction complexity. Instead of directly dictating commands to each robot individually, this system allows designer to specify high-level temporal and spatially varying attributes. This system repurposes the designer from an architect who drafts specific elements within the environment into a conductor who starts the motion, controls the tempo, and adds and removes instruments. Figure 1 outlines the design workflow of the system.

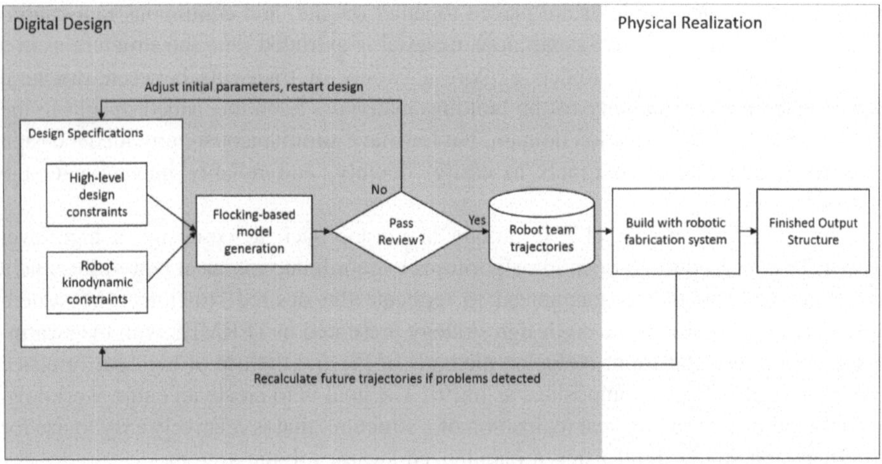

Fig. 1. Proposed end-to-end workflow for the entire system

High-level design constraints can be site-specific, such as trees to avoid, rivers to cross, or preferences such as regions where material density is required or unwanted. Robot kinematic and dynamic constraints are dependent on the physical limitations of the fabrication system. These considerations prevent physically impossible and in-fabricable designs. Typically, generative designs are complex to construct, as they are not tool-specific. Designs are often broken down into smaller pieces that can then be fabricated and assembled into the final form, requiring significant logistic effort and coordination. By coupling the design tool with the fabrication system, this system simplifies the design process while allowing complex, organic, tunable, and site-specific forms.

To integrate generative design protocols with tangible construction, we developed a *flocking-based* design strategy in conjunction with the design of the swarm robotics fabrication system. With this process, the designer specifies a handful of structural constraints such as *adhesion*, *alignment*, *global goals*, *curling bias*, and *global avoidance regions*, and robot constraints such as *local collision regions* and *maximum curvatures*. These properties can be varied regionally or over time to achieve additional high-level design goals. Please refer to Table 1 for a complete list and description of the tunable high-level parameters. With this approach the designer sacrifices detailed control for rapid generation of intricate designs. This method has been shown to produce repeatable designs and the process is intuitive to control and shape them. Currently, the entire structure is predesigned and adjusted in simulations, but it is possible to use this strategy to design and build simultaneously.

Table 1. Design parameters and their effects on path and structure

Algorithmic parameters	Effect on robot paths	Effect on overall structure
Number of robots	Interactions with neighboring robots via other parameters	Determines size, build speed, and potential density
Starting position	Initial density	Controls density at bottom of structure
Adhesion	Instructs robots to approach each other	Increases potential weaving and density
Alignment	Steers robots in similar direction	Decreases weaving
Separation	Steers robots away from each other	Decreases density and weaving
Curling bias	Instructs robots to curl around other tubes	Controls amount/tightness of weaving in a region of space
Vector field	Biases tube alignment with vector field	Controls overall shapes
Virtual walls	A virtual surface to avoid and/or follow	Rigidly controls overall shapes
Max angle	Limits curve radius of a tube	Limits tightness of weave
Collision margin	Volume around robot to prevent collisions	Limits tightness of weave

This implementation models the robots as points $R = \{r_1, \ldots, r_n\}$ with associated properties $A = \{a_1, \ldots, a_n\}$. Where a_i is ever robots, r_i, set of attributes influencing the extent to which the algorithmic parameters in Table 1 take effect. Every robot, r_i, traces out a polygonal trajectory (V_i, E_i), with points $V_i = \{v_1, \ldots, v_n\}$, and edges $E_i = \{e_1, \ldots, e_j\} \in V_i \times V_i$. The movement of each robot is determined by the parameters in Table 1 with collision detection and avoidance. The parameters are denoted by $F_{ij}(\{B\}, a_i)$, where $\{B\}$ is the set given by the n closest elements from R and V to r_i within a given threshold. $F_{ij}(\{B\}, a_i)$ are typical flocking primitives, λ_j denotes a scalar. For example, the curling bias is given by the following, where $q(x, \theta)$ is the quaternion from axis x and angle θ and v_r is the nearest element in $\{B\}$ to one of the E_i.

$$F_{icurl}(\{B\}, a_i)$$
$$= q((v_{r+1} - v_r) - (v_{r-1} - v_r), \theta)(v_r - r_i)q((v_{r+1} - v_r) - (v_{r-1} - v_r), \theta)^{-1}$$

Given a current robot position, r_i, its desired next position is fully determined by a linear combination $d_i = \sum_j \lambda_j F_{ij}(\{B\}, a_i)$. However, d_i needs to be corrected to consider potential collisions with other robots or the already-built structure. For collision detection with the structure, we generate polygonal struts from E and compute intersection with a polygonal cone from r_i to $r_i + d_i$ with increased thickness in the direction of d_i. The intersection contour is used to adjust d_i accordingly. Throughout the process the angle between d_i and $v_n - v_{n-1}$ is restricted by the maximum angle physically achievable by the robots.

While Table 1 lists parameters, Fig. 2 provides visual descriptions of their impacts on the final structure. Panel A shows high alignment of robots with low angle for curling bias and small neighborhood threshold, resulting in a straight structure with localized curling around individual pair of tubes. B in contrast shows minimal alignment of robots with high angle value for curling bias and larger neighborhood threshold resulting in a spread-out structure with curling across multiple tubular elements. C is a variation of A with a single goal, allowing us to steer the structure globally. D - F show high level design elements to guide the structure: D shows steering of the structure along a path, E shows steering of the structure through a vector field, and F shows alignment of the structure to a surface. A hierarchical organization of the tubular structure is shown in G. This is achieved by steering robots in individual clusters and increasing adhesion between centroids of clusters as height increases. H shows collision avoidance that still allows the tubes to curl around the obstacle. Panel I shows a structure taking local deflection into account during construction. Here, deflection under self-load is computed for each new segment a robot builds, showing the potential to take structural considerations into account during the construction process.

3 Swarm Fabricators

The fabrication system was designed and manufactured in-house by the authors in conjunction with the algorithm developed for design of architecture. The system consists of many identical robots that concurrently build a pre-designed, tubular structure from fiberglass composites, shown in Fig. 3. The architectural design is separated into a single tube per robot. Each tube is segmented and fabricated sequentially by a single robot, which appends new segments to the end of the tube. Currently, a central wireless controller communicates with all robots, assigns tasks, monitors their progress, and

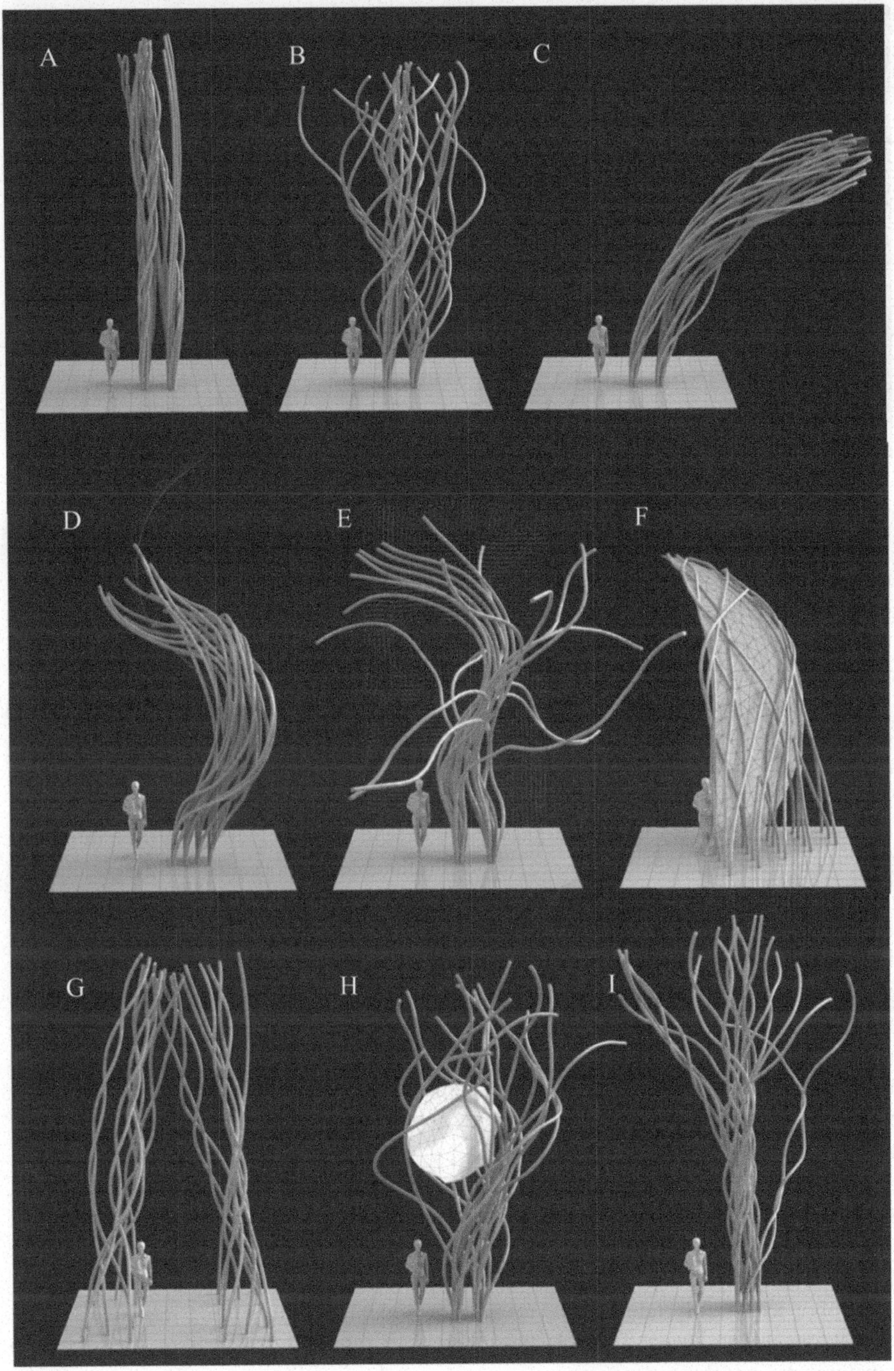

Fig. 2. Variations of generative structures that can be achieved through our system.

Fig. 3. View of a single robotic agent winding a segment of the tube.

tracks their progression. Tasks, or commands, are represented using a custom programmatic interface, similar to G-code. Though the system is autonomous, a user can intervene during the construction process, at any given stage, to pause or change commands to an individual or set of robots, without affecting the other robots.

Each robot works independently and has an individual material power and material feed. The robots avoid collision by following their pre-planned, collision-free trajectories as closely as possible. These criteria allow the system to be capable of adding, removing, or reusing robots mid-fabrication. By completely parallelizing the system, the footprint of the structure can transversely extend without increasing the fabrication time. Every robot has a winding arm, nozzle, inflatable membrane and differential drive system, illustrated in Fig. 4. The arm pulls fiberglass thread and UV-curing resin from material storage in the base. These materials are drawn through the tube, and rotationally deposited on the outside of the inflatable membrane. The wind arm can be finely controlled to tune the thickness, material deposition rate, and patterning of the wetted thread. The inflatable membrane acts as a contractible mandrel that can decrease in diameter, allowing the robot to delaminate from the cured composite. In Fig. 3 the mandrel is inflated as the robot winds a segment of the tube. After the segment is fully wound the mandrel deflates, then the drive system moves the robot up, tilts to a desired orientation, and repeats.

The drive system allows the robot to move up and down the structure and orient itself in the tube to extend the tubes and create curvature, respectively. The robot can localize and follow trajectories by using feedback from an on-board IMU and encoders that are mounted to the drive system. These components are necessary and sufficient for each robot to digitally fabricate its own, independent self-supporting tube.

All constraints from Table 1 are incorporated into the algorithm dictating the structural design. Several constraints arose from physical limitations. For instance, the geometry of the robot restricts the maximum angle it can tilt within the tube to 2.5°. This results in the curvature of the tube being a radius of 0.72 m or greater. The power

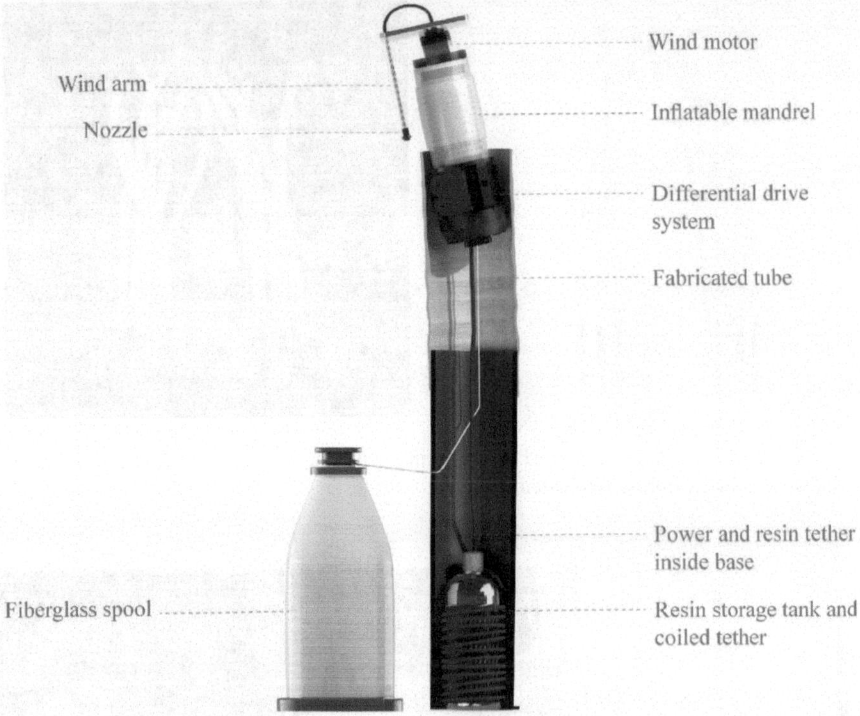

Fig. 4. Overview of a single robot fabricating a tube, the tilt shows how curvature is achieved.

and resin tethers are a fixed length and connected to the base, which limit the height of the tubes. As the winding arm protrudes past the maximum radius of the tube, the tubes need to be separated by at least that distance to avoid collisions. Though the positioning of the bases can be adjusted, they provide a crucial foundation for the architecture and their location defines the starting point of the fabrication.

4 Physical Realization of a Swarm Fabricated Structure

As a case study, we built a 4.5 m-tall pavilion using 16 autonomous robots in parallel over the course of two days. We wanted to create an architectural-scale pavilion that felt like an enclosed space a viewer could enter. Therefore, robots were clustered into four groups of four. Each cluster was then placed in a square configuration 1.75 m apart. An attraction point was added in the design tool 3 m above the center of the clusters. With this configuration robots within clusters interacted on the lower part of the structure and clusters came together to show global motion and create an enclosed space. The thickness of the tubes was designed to decrease with structural height, to minimize material usage and optimize structural performance. Parameters including alignment and curvature were adjusted to generate various iterations of the design and corresponding robot trajectories. The iteration shown in Fig. 5 was selected (Fig. 6).

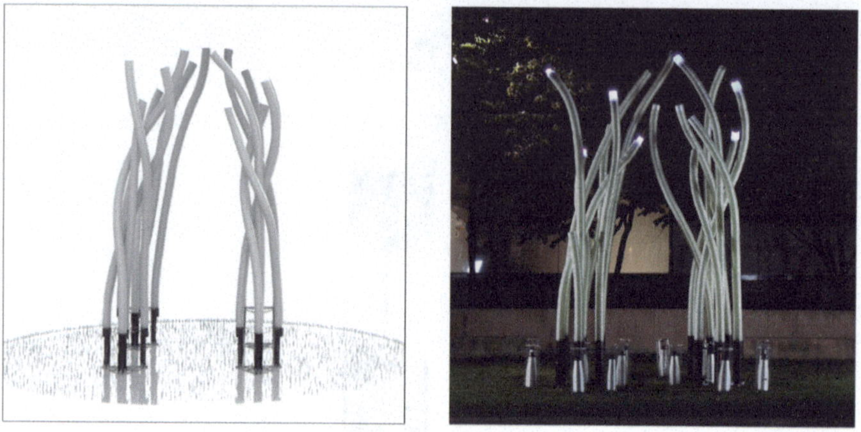

Fig. 5. On the left is the pre-designed structure, incorporating both robot and design constraints. On the right is the resulting fabricated structure.

Fig. 6. Time series demonstrating the time length and scale of the full build

The central control system segments and allocates trajectories to the robots while monitoring overall progress. With the parameters used in this case study, each robot fabricated about 0.8 m of tube per hour. The fabrication speed varies based on the thickness of the tube. This structure was constructed in two days, building for roughly 12.5 h each day. After the robots began construction they operated autonomously, excluding the human intervention needed to monitor and refill the resin reservoirs. In this case study individual robots were stopped several times throughout the build for inspection and repair.

This pavilion stood self-supported in the New England climate for more than 7 months without showing wear. It is important to note that several challenges still exist within our system. Many physical constraints, both robot and environmental, may not be fully captured by the design process, which can lead to robots getting stuck along their trajectories or colliding with unintended obstacles. Ground-plane estimation errors and odometry drift contribute to overall structural errors. Working without absolute external feedback can cause large errors in robot localization, which may lead to collisions and deviations between the structure and the design, as shown in Fig. 5. Error was manually measured and averaged between 0.5–1 m for each tube over the 4 m distance. Additional measurements are needed for a thorough error analysis.

5 Conclusion and Future Work

This work proposed, demonstrated, and experimentally evaluated swarm-aware design and digital fabrication strategies. An architectural scale structure was digitally fabricated using the described design software and swarm fabricators. These tools work hand-in-hand to enable designers to take full advantage of swarm systems without introducing compounding complexity to the design and fabrication processes.

In future work we hope to explore the use of different construction materials, as well as global and local patterning methods to tune the flexibility and durability of both the structure and the individual tubes. Developing a fully decentralized control strategy for the swarm is also of interest. Additional external could mitigate overall localization errors. Extra ports and wires are designed into the current robot specifically to enable this future research. Implementing this control scheme would allow designers to have real-time interactions with the fabrication system. This would decrease project schedules by eliminating the time between design and construction. We believe these innovations push forward a strategy for working with multi-robot systems that will enhance the speed and safety of construction, while enabling the fabrication of forms that were previously not feasible.

Acknowledgments. The authors would like to acknowledge Robert Ricardo I Garriga for fabrication work in constructing the robots and base structure. Thank you to Melinda Szabo for her instrumental work building the software interface and implementation of networking to the robots. This project was made possible in part through generous contributions from GETTYLAB.

References

1. Royo, J.D., Soldevila, L.M., Oxman, N.: Physical feedback in fabrication information modeling (FIM). In: Martens, B., Wurzer, G., Grasl T., Lorenz, W.E., Schaffranek, R. (eds.) Proceedings of the 33rd International Conference on Education and Research in Computer Aided Architectural Design in Europe (eCAADe): Real Time: Extending the Reach of Computation, vol. 2, pp. 299–307. Vienna (2015)
2. Mehta, A.M., DelPreto, J., Shaya, B., Rus, D.: Cogeneration of mechanical, electrical, and software designs for printable robots from structural specifications. In: Proceedings of the IEEE/RSJ International Conference on Intelligent Robots and Systems, pp. 2892–2897. Chicago (2014)

3. Chen, D., Levin, D.I.W., Didyk, P., Sitthi-Amorn, P., Matusik, W.: Spec2Fab: a reducer-tuner model for translating specifications to 3D prints. ACM Trans. Graph. **32**, 1 (2013)
4. Bermano, A.H., Funkhouser, T., Rusinkiewicz, S.: State of the art in methods and representations for fabrication-aware design. Comput. Graph. Forum. **36**, 509–535 (2017)
5. Studio Roland Snooks. http://www.rolandsnooks.com/
6. Ardiny, H., Witwicki, S., Mondada, F.: Construction automation with autonomous mobile robots: a review. In: Proceedings of the 3rd RSI International Conference on Robotics and Mechatronics (ICROM), pp. 418–424. Tehran (2015)
7. Augugliaro, F., Lupashin, S., Hamer, M., Male, C., Hehn, M., Mueller, M.W., Willmann, J. S., Gramazio, F., Kohler, M., D'Andrea, R.: The Flight Assembled Architecture installation: cooperative construction with flying machines. IEEE Control Syst. **34**, 46–64 (2014)
8. Lindsey, Q., Mellinger, D., Kumar, V.: Construction of cubic structures with quadrotor teams. In: Durrant-Whyte, H., Roy, N., Abbeel, P. (eds.) Robotics: Science and Systems VII, pp. 177–184. MIT Press, Cambridge (2012)
9. ICD/ITKE Research Pavilion 2016–17 | achimmenges.net. http://www.achimmenges.net/?p=19995
10. Minibuilders. http://robots.iaac.net/
11. Dogar, M., Knepper, R.A., Spielberg, A., Choi, C., Christensen, H.I., Rus, D.: Multi-scale assembly with robot teams. Int. J. Robot. Res. **34**, 1645–1659 (2015)
12. Petersen, K., Nagpal, R., Werfel, J.: TERMES: an autonomous robotic system for three-dimensional collective construction. In: Durrant-Whyte, H., Roy, N., Abbeel, P. (eds.) Robotics: Science and Systems VII, pp. 257–264, MIT Press, Cambridge (2012)
13. Allwright, M., Bhalla, N., El-faham, H., Antoun, A., Pinciroli, C., Dorigo, M.: SRoCS: leveraging stigmergy on a multi-robot construction platform for unknown environments. In: Dorigo, M., Birattari, M., Garnier, S., Hamann, H., de Oca, M.M., Solnon, C., Stützle, T. (eds.) Swarm Intelligence, pp. 158–169. Springer International Publishing, Switzerland (2014)
14. Jenett, B., Cheung, K.: BILL-E: robotic platform for locomotion and manipulation of lightweight space structures. In: Proceedings of the 25th AIAA/AHS Adaptive Structures Conference, American Institute for Aeronautics and Astronautics (AIAA), pp. 1876–1888. Grapevine, TX (2017)
15. Reynolds, C.W.: Flocks, herds, and schools: a distributed behavioral model. In: Proceedings of the 14th annual conference on Computer graphics and interactive techniques (ACM SIGGRAPH'87), pp. 25–34. Anaheim (1987)

InFormed Ceramics: Multi-axis Clay 3D Printing on Freeform Molds

Minjae Ko, Donghan Shin(✉), Hyunguk Ahn, and Hyungwoo Park

B.A.T., Seoul, Korea
kmj2653@gmail.com,
{donghan.shin,parisene,rhyzome}@b-at.kr

Abstract. This paper addresses a novel technique of 3D clay printing using a robotic arm along with the architectural applicability of the proposed fabrication technique. In the initial phase of research, we selected the clay material to be used for printing in consideration of its viscosity and developed the clay extruder customized for the material. In the late phase, to verify how effective the robotic fabrication technique we propose is, we manufactured the freeform foam mold and tested multi-axis clay 3D printing on it by using three end-effectors consecutively: hotwire cutter, spindle, and clay extruder. Also, applying the proposed technique to a larger-scale ceramic structure consisting of nineteen panels, we checked the feasibility of fabricating actual ceramic building components with this technique. If this robotic clay 3D printing technique is improved more and fulfills the demands of the industry in future, it will enable cost-effective ceramic modular fabrication for a customized design in architecture.

Keywords: Clay 3D printing · Robotic fabrication · Freeform ceramics
Subtractive manufacturing · Additive manufacturing

1 Introduction

1.1 Ceramics in Architecture

In the process of building construction, ceramics have been used as modular units (e.g. bricks, panels, and tiles) which are necessarily mass-produced. Architects have recognized the addressable elemental units of ceramics not so much as the clay itself but as standardized modules, so that the general ceramic components of buildings have been applied in the conventional manner of filling the flat facade elements.

The recent introduction of digital fabrication technology is enabling architects to process and produce ceramics and other existing materials in a new way, entailing such attempts as to escape from the existing standardized modular system. For instance, Andreani and Bechthold (2014) added the robotic brick-forming process to the existing brick production line in their research project, which was to construct parametric masonry walls by mixing general and transformed bricks. Also, Peter (2014) in his "Building Byte" case produced undulating ceramic brick modules by combining a

J. Willmann et al. (Eds.): ROBARCH 2018, *Robotic Fabrication in Architecture, Art and Design 2018*, pp. 297–308, 2019.
https://doi.org/10.1007/978-3-319-92294-2_23

general affordable three-axis 3D printing device and a ceramic extruder. These research cases on digital fabrication with ceramic materials are showing possibilities of affordable production and fabrication of freeform or parametric ceramic components by gradually atomizing the smallest unit of ceramics that the designer can address in an architectural scale.

1.2 Previous Works

Amongst many 3D printing techniques, fused deposition modeling (FDM) is the most popular and affordable technique which is applicable not only to plastic but also to liquid or pasty materials. However, such a 3D printing technique is based on the three-axis gantry system, which has an inherent limitation of one-way deposition along the Z-axis and so is difficult to apply to architectural scales. Some researchers have recently endeavored to overcome such a limitation of the existing system by grafting the technique of industrial robotic multi-axis fabrication to 3D printing. A research project by IAAC and Joris Laarman Studio demonstrated the possibility of 3D printing with free direction of deposition, mainly investigating the rapidly sintered materials and the cooling device (Laarman et al. 2014). Another research conducted at Tongji University was to print the product with acrylonitrile butadiene styrene (ABS) and no support, based on the information of geometry optimized by simulating the structural rigidity of the material (Yuan et al. 2016). Both research cases focused on the process of forming self-supporting structure based on the cooling system and the rapidly sintered material which ensures structural rigidity. Tam (2016) suggested "stress line additive manufacturing" (SLAM) as a technique to overcome the limitations of general FDM in such terms as aesthetic expression, efficiency of material consumption, accuracy, and anisotropic layered structure. It was to aesthetically embody the same form and ensure structural performance by making various patterns of depositions based on the stress lines of the designed surface mold. Also, the project "Support-free Volume Printing by Multi-axis Motion" conducted in collaboration by TU Delft, Tsinghua University, and INRIA (French Institute for Research in Computer Science and Automation) established the Tool-path generation strategy that enables the support-free printing of models with large overhangs and high-genus topology by the control of an industrial robot equipped with the print bed (Dai et al. 2018).

Amongst the precedent projects, the "Thallus" (2017) by Zaha Hadid Architects is one that applied the FDM-based multi-axis 3D printing technique on the styrene mold processed by hotwire cutter, maximizing the DOF (degree of freedom) of deposition processing by fastening the mold onto the rotation table. The Thallus project clearly demonstrates the possibility of aesthetic expression by robotic multi-axis FDM as well as the strength of robotic fabrication for the scale that enables manufacturing.

There are recent papers of research on the multi-axis clay 3D printing, such as "Woven Clay" (Friedman et al. 2014), "Free Form Clay Deposition in Custom Generated Molds" (Dunn et al. 2016), and "Clay Non-Wovens" (Rosenwasser et al. 2017). These projects show the efforts of overcoming the limitations of materiality and escaping general deposition techniques in the methodological terms of depositing clay material. Although the three projects focus on different issues, it is interesting that all of them utilize the curvaceous mold as the bed for the deposition of clay. Friedman used the milled foam for the weaving density of clay coil or its visual effects. Rosenwasser changed the speed of extrusion and used the mold based on sine curves to intentionally generate irregular patterns by the material behavior of clay, which is a natural material, and engender the variation of those patterns at a controllable level. While these two papers focused on the variation of patterns through special techniques of clay deposition, Dunn highlights the efficiency and possibility of the two-step robotic fabrication technique, which includes the robotic milling of a sustainable formwork for the first step and the controlled deposition of liquid clay into a form or mold for the second step.

1.3 Project "InFormed Ceramics"

In this project, we investigated multi-axis clay 3D printing on the freeform mold utilizing an industrial robot and applied it to large-scale fabrication. We intended to carry out free multi-axis robotic 3D printing within the usual material boundary of clay, rather than creating a particular kind of clay or mechanical sintering device in order to secure self-supporting material properties.

This process looks similar to that of Dunn's research, but differs actually in overall aspects such as the clay used, the extruder device, and the material and processing of the mold. For the material, Dunn applied the slip-casting method based on the "slip," the soil-water combination which generally has a high percentage of moisture. In the initial phase of research, we experimented on extruding clays with various viscosities and concluded that utilizing the "slip" which has significantly low viscosity is somewhat ill-suited for the 3D printing material which needs rapid sintering as a prerequisite. We assigned a priority to selecting such a material as to secure productivity and the quality of printing, and then endeavored to develop such an extruder device as to overcome the stress of the highly-viscous "clay body." In the late phase of research, we mainly conducted verification of the composite fabrication process that we proposed to secure productivity and quality. In the process of fabricating the surface mold, the continuous robotic processing based on the consecutive use of the hotwire cutter and the spindle significantly saved the time consumed for the roughing stage.

In this paper, we intend to address the following content in detail: investigation of materials used for this research and fabrication process; development of the extruder; verification of the fabrication process mixed with composite processing and the Toolpath strategy; and the issues arising from post-processing and the countermeasures against them.

2 Material Studies and Robotic Tooling for Clay Extrusion

2.1 Material Selection

In this project, we tested the clay material considering viscosity, form maintenance, applicability, and convenience. Upon determining materiality of the clay, we developed the extruder accordingly. In the early phase, we mixed clay with white earth powder, and water in a certain ratio to make a clay mixture with proper viscosity. This method was that which has been generally used, mixing components of the "slip" except a deflocculant in order to secure viscosity to the extent of enabling deposition in a coiling pattern by adjusting the water content. However, the clay mixture created in this way showed low viscosity generally, so that the height of a layer one could build up at a time was low and the extruded material did not maintain its form but flowed down and caused low quality (Fig. 1).

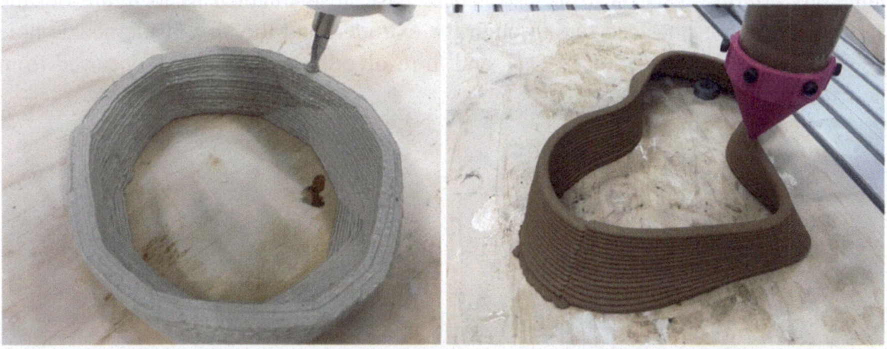

Fig. 1. The clay made by slip-casting and the progressive cavity pump extruder (Left), and the commonly used clay body and the cylinder extruder (Right). It took 11 and 2 min respectively for the Left and the Right to print forms with almost the same height and width.

Although we could have conducted research by significantly reducing the water content for the material mix, this process of making a clay mixture required much time and effort as we had to prepare the materials in an accurate ratio in advance and eliminate the bubbles arising in the mixing process. We also could not produce a large amount of clay mixture at once because long-term storage would change its viscosity. Therefore, we decided to use the commonly available clay body, whose moisture content is around 15%, as the printing material, and developed a new extruder fit for its materiality. For this project, we mainly used Buncheong clay, a kind of earthenware clays. Generally produced with a clay mixer, this clay body has no internal bubble from the beginning but high viscosity, so as to show the better quality of form maintenance after extrusion than clay mixtures do.

2.2 Development of Clay Extruder

We have developed a series of extruder devices according to the printing material (Fig. 2). In the early phase, we used the extruder equipped with a progressive cavity pump to shoot out a tender clay mixture. In the following phase, however, using the 15% water-containing clay body as the printing material to secure printability and convenience, we discovered some problems with the high pressure arising in the extruder. As the cavity-pump-type extruder did not endure the pressure of clay body extrusion, we had to take a new material extrusion technique, the syringe type using a high-torque step motor and a screw jack. For the two-liter extruder, we used the motor whose maximum torque is 200 kgf·cm and the screw jack with the gear ratio of one to three in order to pass low-water-containing clay body through the 5 mm-thick nozzle, while the extruder was made of high-stiffness materials such as aluminum, Polyacetal, and polycarbonate to endure the high pressure arising from extrusion. Using the syringe-type extruder, we were able to secure the quality and stability of clay 3D printing. All works addressed in this paper were carried out with this syringe-type extruder. The developed clay extruder was attached to the robotic arm and underwent numerous printing tests on a flat bed. These basic tests enabled us to figure out the speed of the robot, the feed rate through the extruder, and the size of the nozzle hole required for stable deposition of clay, and to resolve the problems that occurred from the extruder and materials.

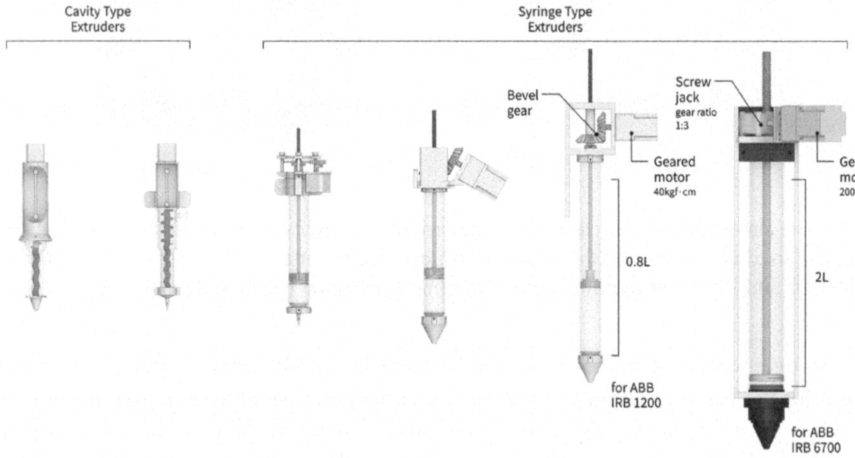

Fig. 2. A series of clay extruders were developed and the last two devices were mainly utilized in this project "InFormed Ceramics".

3 Fabrication Methodology and Application

Prior to the full-scale panel fabrication, we conducted simple experiments of producing prototypes to determine the possibility of a new multi-axis clay 3D printing. Designing a dynamic form of double-curved surface and proceeding with a continuous fabrication process, we attempted to consecutively use three different robotic tools—hotwire cutter, spindle, and clay extruder—for the serial process from mold machining to clay deposition, without changing the mold location (Fig. 3).

The three tools used in this process have different tool center points (TCPs), so that we had to conduct several times of software simulation before physical robotic processing, in order to designate the optimal mold position that can enable all kinds of robot behaviors with each tool mounted and within the scope in which robotic arm can avoid singularity. In this project, we used such Grasshopper-based in-house plug-in "GERTY" and Robot Studio together to conduct accurate simulation of the robot behavior and create the Tool-path and the RAPID code. In this way, the mold location was controlled during the process to maintain the accuracy of multiple robotic

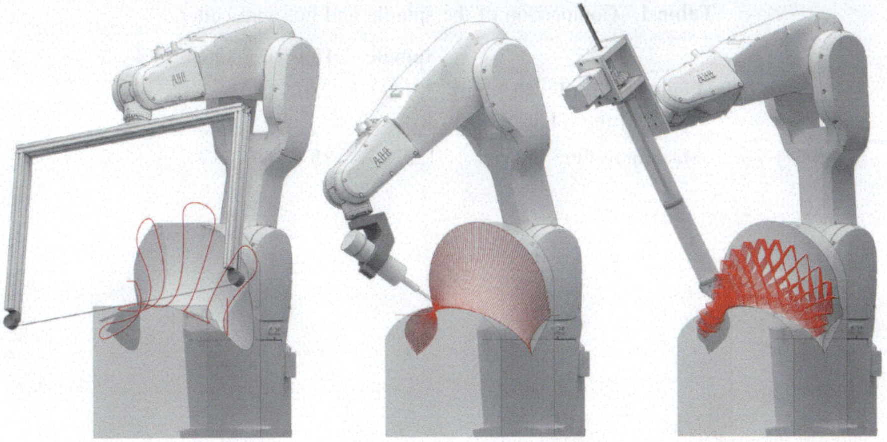

Fig. 3. Utilized three different end-effectors (hotwire cutter, spindle, and clay extruder) and Tool-path strategies

procedures occurring on the mold, without the need of introducing a secondary technique like 3D scanning.

3.1 Freeform Mold

Clay is not quickly dried after extrusion and so makes it difficult to form a self-supporting structure. Even if one develops a mechanic device to dry clay quickly, the quickly dried clay generally entails cracking that weakens the structure. Thus, in order to create hanging shapes by 3D printing with general clay, a supporting structure is inevitably needed that will maintain the form of deposited clay until it is dried enough. In this project, we made a mold as the supporting structure and the formal reference to shape the freeform porous ceramic panels. The surface mold was made of expanded polystyrene (EPS) with the robotic machining technique of hotwire cutting and multi-axis milling. Generally, the machining operation proceeds in two stages: roughing and finishing. Roughing is the first stage of machining whose objective is to quickly remove the bulk of the waste material. This stage is assigned comparatively more time amongst the whole machining process, because the spindle's processing capability at a time is limited to a comparatively small amount due to the diameter and processing depth of its end-mill. In this project, we were able to save much time in mold roughing by using hotwire cutting instead of spindle. For instance, as a consequence of comparing the Tool-path lengths and the processing speeds between the spindle and hotwire cutter in the roughing stage on the same form, it was found that the spindle with a 6 mm diameter end-mill took about 125 min while the hotwire cutter took just 15 min for processing (Table 1, Fig. 4).

Next, finishing is a detailed processing stage that uses the spindle tool: we made the final mold by processing the precision surface as the original form for panel production (Fig. 5).

Table 1. Comparison of the spindle and hot-wire cutter.

	Spindle	Hot-wire cutter
Length of Tool-path (mm)	301111.5	2802.2
Speed of the robot (mm/s)	40	3
Machining time (min)	125	15

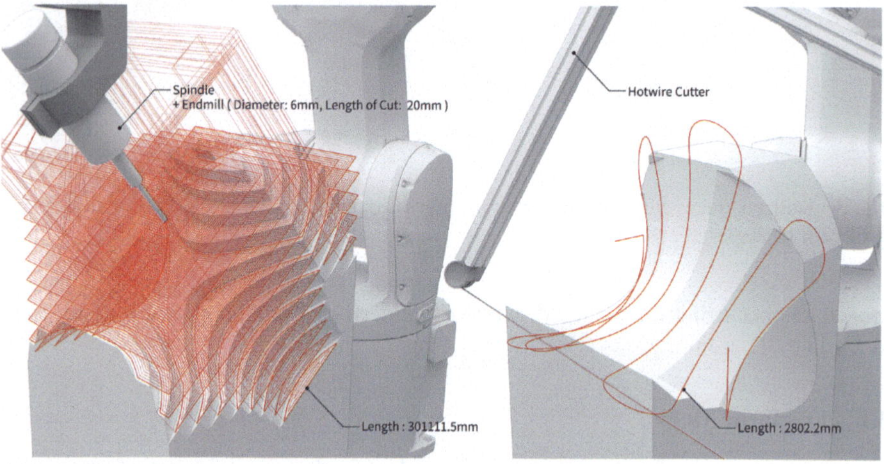

Fig. 4. Tool-path for roughing stage with the spindle (left) and hot-wire cutter (right)

Fig. 5. Roughing with a hot-wire cutter (left) and finishing with a spindle (right)

3.2 Robotic Multi-axis Clay Deposition

Using the developed syringe-type clay extruder, we deposited the porous clay on the freeform mold. The processed panel was designed with a diagrid pattern to secure structural stiffness after drying and kiln firing. The Tool-path is made up of a continuous single curve, and designed to make the extruded clay bodies overlap and conjoin at nodal points. The first clay layer was spaced narrowly from the mold surface with around 1 to 1.5 mm to attach clay well to the mold, and a spray adhesive is used to fix it temporarily. The next clay layers deposited on the first one were set to be 5 mm width and 2 mm height, so that all fifteen layers deposited became about 30 mm thick (Fig. 6).

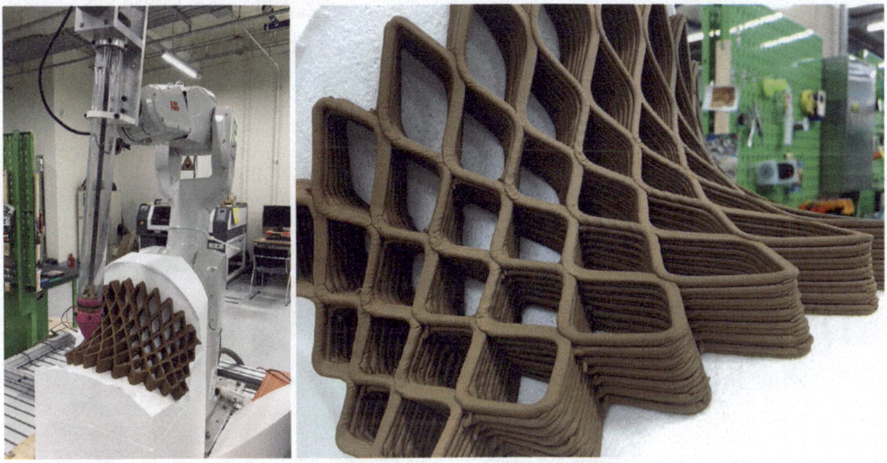

Fig. 6. Multi-axis clay deposition with customized clay extruder

3.3 Application

Through this simple procedure of prototype panel fabrication, we have observed the possibility of manufacturing new ceramic panels in the way of depositing clay on the curved surface, escaping from the standardized modular ones. We intended to apply the new clay 3D printing technique to the larger-scale fabrication project titled "InFormed Ceramics." Using Kangaroo physics engine for Grasshopper, the design team suggested such diagrid paneling as to be filled with the flow of minimal surfaces defined by several sequential boundary curves. The height of the structure right after extrusion is about 1700 mm. In the early phase of paneling design, it was planned to be made as two to three panels with the aid of a large robot (IRB6700). However, the larger the clay structure becomes, the more likely the clay is to have cracks and the more unlikely we are to obtain such a large kiln in reality. We thus produced it separately as nineteen different panel modules (Fig. 7). Two consecutive modules were designed not to break the flow of minimal surfaces, but to interlock with each other. A large robot (IRB6700) was utilized to carry out mold production and clay 3D printing for six to seven modules at a time (Fig. 8). In this way, we made a porous paneling whose depth is around

Fig. 7. Parametric design for the fabrication project

Fig. 8. Clay 3D printing on molds with IRB 6700 and produced nineteen panels

50 mm and consisting of 25 layers deposited, each of which is set to 8 mm width, considering the installation size.

3.4 Post-processing and On-site Installation

The produced panels undergo a slow drying process for about a week, together with the mold in an indoor shady space. The clay contracts by about 10 to 15% as its moisture evaporates in the drying process, so that any unbalanced drying would entail cracking and so be likely to yield the failure of this project. In this project, the deposited clay contracts in such a process but the mold does not, so that partial cracks occurred quite frequently in the previous test prototyping. In the main fabrication process, the panels for a dynamic structure were separated to make more gentle curvatures, thus generating almost no cracking by the difference of contraction rates between the mold and the panel in the drying process. The dried panel modules underwent a firing process at

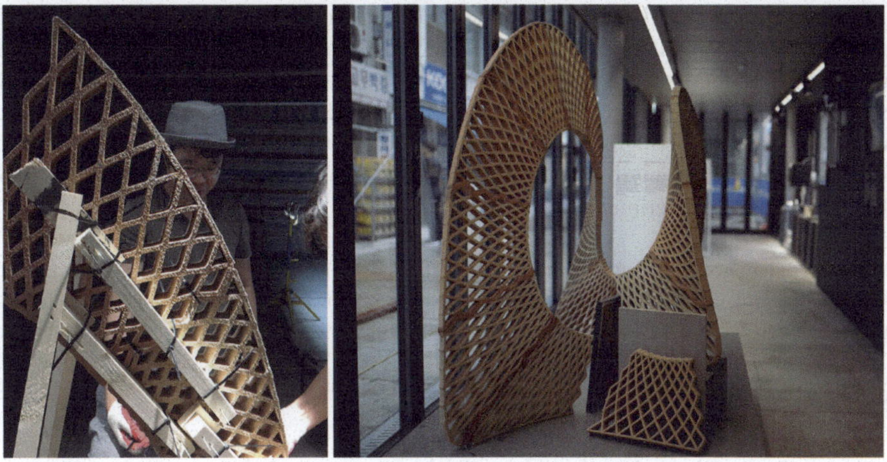

Fig. 9. On-site installation and final result

1250 °C in a kiln for a day, and became ceramic panels, which were installed on site with epoxy bonding (Fig. 9).

4 Conclusion and Future Development

This project successfully demonstrated the large-scale outcome of applying a new robotic 3D printing with clay, but also revealed various problems in the production process. First, the syringe-type extruder used in this project doubles as the clay container, so that all clay bodies to be extruded at a time should be loaded on the robot. The physically large extruder was an element of limiting the robot behavior in the fabrication process, and the overall efficiency of work decreased as one had to stop all the works time and again to replace the polycarbonate pipe of the extruder to fill up with new clay. Second, there occurred deformation of the product in the process of drying and firing after the clay panel formation. The difference of contraction rates between the printing product and the mold, and the high temperature within the kiln yielded subtle deformation of each panel. As a consequence, the end of the last panel amongst the nineteen panels became totally deviated from the original position and unable to be combined, although the combinations between adjacent panels were successful in the last fabrication. We are planning to conduct relevant research and technical improvement continuously to make better end-effectors and prevent such deformation of the product. After this project, we are already endeavoring to secure the degree of freedom of robot behavior and the convenience of material supply by improving the extruder in Bowden's way of separating material supply from the nozzle. Also, we are planning an experiment of producing the mold whose material property can sustain the role of a supporting structure until the drying and firing process.

This research project still has limitations that should be overcome, but successfully demonstrated the possibility that the robotic clay 3D printing technique can apply to the production of new structures and panels in the architecture and construction industry.

If further research satisfies the current needs of the industry in terms of precision and productivity of the final product, this printing technique will be effectively used for the customized production of architectural materials, instead of the existing standardized ceramic modules. Moreover, the new 3D printing technique addressed in this project has a potential of applying not only to ceramics but also to other architectural materials with viscosity, so that it will have positive influence on and be a source of inspiration to the various future experiments of robotic fabrication.

Acknowledgements. The authors would like to thank Ji-Eun Hwang (Professor, Department of Architecture, University of Seoul) for continuous and generous support. Authors' special thanks are extended to the staff of Seoul Biennale of Architecture and Urbanism 2017 for their support.

References

Andreani, S., Bechthold, M.: [R]evolving brick: geometry and performance innovation in ceramic building systems through design robotics. In: Gramazio, F., Kohler, M., Langenberg, S. (eds.) Fabricate: Negotiating Design & Making, pp. 182–191. gta Verlag, Zurich (2014)

Peters, B.: Building bytes: 3D-printed bricks. In: Gramazio, F., Kohler, M., Langenberg, S. (eds.) Fabricate: Negotiating Design & Making, pp. 112–119. gta Verlag, Zurich (2014)

Laarman, J., Jokić, S., Novikov, P., Fraguada, L.E., Markopoulou, A.: Anti-gravity additive manufacturing. In: Gramazio, F., Kohler, M., Langenberg, S. (eds.) Fabricate: Negotiating Design & Making, pp. 192–197. gta Verlag, Zurich (2014)

Yuan, P.F., Zhang, L., Meng, H., Yu, L.: Robotic multi-dimensional printing based on structural performance. In: Reinhardt, D., Saunders, R., Burry, J. (eds.) Robotic Fabrication Architecture, Art and Design 2016, pp. 93–106. Springer International Publishing, Switzerland (2016)

Tam, K.-M.M., Coleman, J.R., Fine, N.W., Mueller, C.T.: Robotics-enabled stress line additive manufacturing. In: Reinhardt, D., Saunders, R., Burry, J. (eds.) Robotic Fabrication Architecture, Art and Design 2016, pp. 350–361. Springer International Publishing, Switzerland (2016)

Dai, C., Wang, C.C.L., Wu, C., Lefebvre, S., Fang, G., Liu, Y.: Support-free volume printing by multi-axis motion. In: ACM Trans. Graph. **37**(4) (2018, in press)

Saunders, S.: TU Delft's Multi-Axis Robotic 3D Printing System Makes it Possible to Print Without Supports. https://3dprint.com/215803/multi-axis-robotic-3d-printer/. Accessed 11 June 2018

Thallus for White in the City. http://www.zaha-hadid.com/2017/04/05/thallus-for-white-in-the-city/. Accessed 11 June 2018

Dunn, K., O'Connor, D.W., Niemelä, M., Ulacco, G.: Free form clay deposition in custom generated mold. In: Reinhardt, D., Saunders, R., Burry, J. (eds.) Robotic Fabrication Architecture, Art and Design 2016, pp. 317–326. Springer International Publishing, Switzerland (2016)

Friedman, J., Kim, H., Mesa, O.: Experiments in additive clay depositions woven clay. In: McGee, W., Ponce de Leon, M. (eds.) Robotic Fabrication Architecture, Art and Design 2014, pp. 261–272. Springer International Publishing, Switzerland (2014)

Rosenwasser, D., Mantell, S., Sabin, J.: Clay non-wovens. Robotic fabrication and digital ceramics. In: Nagakura, T., Tibbits, S., Mueller, C., Ibañez, M. (eds.) Acadia 2017: Disciplines & Disruption, Proceedings of the 37th Annual Conference of the Association for Computer Aided Design in Architecture, pp. 502–511. Cambridge, MA (2017)

Husqvarna
Husqvarna

Altered Behaviour: The Performative Nature of Manufacture
Chainsaw Choreographies + Bandsaw Manoeuvres

Emmanuel Vercruysse[1](✉), Zachary Mollica[1], and Pradeep Devadass[2]

[1] Design + Make, Architectural Association School of Architecture, London, UK
vercruysse@aaschool.ac.uk
[2] Individualized Production, RWTH Aachen University, Aachen, Germany

Abstract. This paper explores the ventures of the Robotic Fabrications AA Visiting School. Cultivating and implementing radical and innovative modes of architectural fabrication, we borrow from neighbouring creative fields such as choreography, performance and martial arts - efficient code being infected with an exciting spatial complexity and positing an artistic engagement within the world of physical production.

1 Introduction

The Robotic Fabrications Visiting School[1] operates as an experimental branch of the Architectural Association's Design + Make programme – cultivating and implementing radical and innovative modes of architectural production. We foster a deep interest in the technical aspects of robotics while ultimately looking to engage it in a wider spatial and cultural discourse. Merging expert knowledge of timber construction with 6-axis robotic fabrication - the creative potential of prototyping complex timber structures has been explored through the progressive manufacture of a Foundry for Hooke Park.

Traditional tools and techniques for making are hijacked, re-invented and applied as innovative processes for architecture. Our two weapons of choice - a bandsaw and chainsaw - gain an augmented level of precision and control when wielded by the Kuka KR150. In converting these tools into end-effectors, their performance is altered. Through rigorous physical testing, we prototype intricate connection details, and apply the extraordinary precision and flexibility of multi-axis robotic machining while ensuring design strategies are firmly rooted within tradition.

[1] Robotic Fabrications is run by director Emmanuel Vercruysse and tutor Zachary Mollica with the robotics team of Pradeep Devadass, Gary Edwards and recently, Michael Arnett. Authors Emmanuel and Zachary are respectively a co-director and studio tutor for the AA's Design + Make programme. At the time of the work presented, author Pradeep Devadass was AA Hooke Park's Robotics Developer as part of a collaboration between the AA, UWE and KTP.

J. Willmann et al. (Eds.): ROBARCH 2018, *Robotic Fabrication in Architecture, Art and Design 2018*, pp. 309–319, 2019.
https://doi.org/10.1007/978-3-319-92294-2_24

If the behaviour of the hijacked tool has transformed, then so has our space of production. As traditional fabrication principles undergo a digital re-interpretation through the implications of robotics – they deviate from the tightly controlled motion paths of industrial applications into the realm of staged events. Borrowing from neighbouring creative fields such as choreography, performance and martial arts, efficient code is infected with an exciting spatial complexity and posits an artistic engagement within the world of physical production. Although the fabrication of complex architectural components remains the core driver, the performative nature of manufacturing has become a key driver in our augmented space of production.

1.1 The Foundry

While timber sourced from the surrounding woodland has served as the primary working material for the architectural constructs developed in Hooke Park, recent investigations into metal casting instigated the development of a Foundry for the campus. The outcome of the first workshop, 'Chainsaw Choreographies,' is an opening move. Four overhanging Douglas fir beams interwoven and supported on three naturally formed tree forks set out a precise circular plan. The design evolves incrementally, through 'Bandsaw Manoeuvres,' 10 distinct components were fabricated which together form a complex ring beam which will receive the structure's eventual shell (Figs. 1 and 2).

Fig. 1. Photogrammetric 3D scan of the Foundry after 'Chainsaw Choreographies'. This precise survey confirmed the accuracy of fabrication and served as a base for further development.

1.2 Material Library

Headquartered in Hooke Park, we are embedded within our primary material library - the forest itself. Beyond its immediate proximity, timber is chosen as the main material for our experiments due to its inherently dynamic properties. The form of each standing tree is a distinct composition of grain that has grown in direct response to its environs.

Fig. 2. 'Bandsaw Manoeuvres' participants inhabit the 10 components of the ring beam.

This form governs how each piece of timber derived from the tree reacts to subsequent operations. Through careful understanding of this variation, it can be accounted for within our methods. "It is important to remember wood is a living thing – one must understand its properties and not oppose them" [1].

In the development of the Foundry, timber has been used in various formats: large rectilinear beams; naturally formed 'Y' shaped forks[2]; and bespoke stocks made through a variety of lamination techniques. The timber used throughout has been 'green' - with a moisture content between 20–70% and therefore an increased predisposition to movement. Restricted by the quantity of energy they could employ to join timbers, traditional craftsmen developed methods which engaged closely with the material's grain and embodied information. While mechanical fasteners have become standard practice for jointing timber assemblies due to their ease and predictable performance, Robotic Fabrications explores the development of complex timber to timber connections inspired by traditional Japanese joinery.

[2] The authors, as faculty for the AA's Design + Make programme, gained significant expertise in working with complex inherent timber forms through the Wood Chip Barn project (2016). 'Chainsaw Choreographies' expanded on the extensive workflows established [2].

2 Altering the Behaviour of Tools

2.1 Act 1: The Chainsaw, Act 2: The Bandsaw

In the laboratory environment of Robotic Fabrications, we foster a critical approach towards design-through-making where making is central to the act of design and learning central to the act of making. The core objective of these 10-day workshops is an exploration – to unravel, discover and assemble an in-depth knowledge of the newly installed end-effector. The pace is furious, producing 10 components in 10 days this year, and through the reinvention and hybridisation of tools and techniques, unique methods of fabrication are exposed (Fig. 3).

Fig. 3. Weapon of choice - bandsaw end-effector used during 'Bandsaw Manoeuvres'

The selection of tool to be employed for each task is significant. Beyond a requirement to achieve a particular duty well, tools are chosen for their temperament, complications and limitations. Borrowed from the world we operate in daily - the wood workshop and surrounding forest - the bandsaw and the chainsaw have become our weapons of choice. Though each of these tools has been applied as a robotic end-effector in previous investigations, they have been employed typically as a means of achieving component shape – with less emphasis given to material properties and the rich history of craftspeople deriving timber joinery directly from these characteristics. In a work titled '7x Stool' [3, 4], Pawlofsky and Weissmahr employed a chainsaw end-effector to cut a pair of stools and a small table from a single piece of round timber. With priority given to the gesture of this act, the anisotropic nature of timber remains neglected in the finished pieces. A bandsaw end-effector has been applied to timber fabrication within two recent projects: Philip Yuan and Hua Chai's 'Robotic Wood Tectonics' [5], and Ryan Luke Johns and Nicholas Foley's 'Bandsawn Bands' [6]. Each demonstrates the ability of this continuous blade to create curved geometries, and while neither engages significantly with timber to timber joinery, the latter's intent to derive forms from wood's own nonstandard grain configuration adds a richness to the work.

Robotic Fabrications use of these two tools is distinguished by two primary motives: to develop complex timber to timber connections rooted in tradition; and to explore performative manufacturing through analogy between robotic trajectories and human motion. Fundamentally setting apart the work presented is its location in Hooke Park. With a diverse team made up of craftspeople, designers, engineers, roboteers, a forester and more specialist knowledge is introduced from each. Working closely with Hooke Park's Workshop Manager Charlie Corry Wright and Forester Christopher Sadd to attach these tools to the robotic arm we begin to dissect and analyse the shifts in behaviour implied by their operator (Fig. 4).

Fig. 4. Weapon of choice - chainsaw end effector used during 'Chainsaw Choreographies'

In the hands of a human operator, the chainsaw becomes a vigilant act of averaging out and balance - the operator's body offering the damping needed to maintain a straight cut through the saw's violent vibrations. Resistance is evaluated throughout, and cut speed adjusted instinctually, with the operator free to move continually in the cut's plane. Directed by the robotic arm, these conventional tools gain an augmented level of precision and control. The behaviour of the tool has changed – and the modes in which the tool may be used fundamentally altered. Some of these deviations reveal themselves instantly and are obvious,[3] others require deeper analysis and are observed through careful examination of film and various surveying mediums.

The chainsaw is installed as an end-effector as-found. Two holes through the saw's bar allow it to be fixed while preventing continuous fine tuning of the chain's tension impacting its calibration. The bandsaw's reinvention is taken a step further. Disassembling an off-the-shelf saw, a bespoke bandsaw was crafted from its remains. Welding together a minimal frame, all non-essential components were left off and its cutting gate widened to maximise cutting access. Wielded by the robotic arm, the bandsaw reverses its traditional relationship to the material it works on – traveling towards it.

Rules emerge governing the forms to be produced by each tool. As for the Japanese carpenter whose complex joints are developed from a series of planar cuts owing to the plane of their pullsaw, each cut in 'Chainsaw Choreographies' is governed by the chainsaw's flat bar – its dimension dictating the size of components and joints.[4] Enacting invisible instructions, the robotic arm has no need to set out its cuts – avoiding a time-consuming process on complex timber. Where the saw is unable to finish a joint unaided due to reach or depth, it is used to provide markings for hard finishing. Working with a thin flexible blade, the bandsaw is able to carefully make singly curved cuts. Carefully sculpted form and complex joint becoming one (Fig. 5).

2.2 Robot Language and Logistics

Restricted by the orientation of existing software packages towards industrial operations, the pace and range of our experimentation demands the development of bespoke toolpath generators. Parameter based offline programming is developed through KUKA|prc's inverse kinematic solver – allowing complex sequences of operations to be programmed by inexperienced participants. For the chainsaw, toolpaths are generated from a minimal input of two curves - the first defining the extreme cutting edge of the saw and the second governing its orientation. For the bandsaw, toolpaths are generated by extracting medial curves from each surface of complex objects modelled in 3D. Retract heights are generated from these guide curves to ensure safe entry and exit from the material, and in areas containing radiuses and curvature, a greater density of planes are generated in order to ensure both smooth and precise cutting.

---

[3] While the robotic arm operates with sub millimeter precision, through close observation, the chainsaw was found to resonate up to 4 mm due to the arm's inflexibility and lack of damping.

[4] The Chainsaw uses a bar of approximately 400 × 60 × 4 mm, while the bandsaw's kerf is much thinner and restricted by the 400 mm cutting gate between its two wheels.

Fig. 5. Analog finishing is applied through both workshops to complete joinery for which the robotic arm has set out the extents (left – kerf to be removed) and to ensure a tight join (right).

As the human operator carefully considers their physical orientation to both tool and workpiece, the mounting of each end-effector to the robotic arm and methods for stock holding are thoroughly thought through in each instance – balancing the need for workpiece stability and maximum access. Constraints of the arm itself are considered in parallel – the chainsaw for instance mounted at an angle of 45° to the arm's 6th axis to avoid wrist singularities.

While the robotic arm allows each saw to approach the workpiece in ways a human operator may not manage, the saw may still struggle through the cut owing to characteristics of their blades.[5] This correspondence between analog and digital workmanship becomes an important component of our investigations – identifying the unique potential offered by mounting each tool as an end-effector.

3 Performative Nature of Manufacturing

The natural habitat of robots is the controlled environment of the factory. Their natural movement strategy the utilitarian tool-path described by the typical requirement to move in prescribed lines or arcs, and optimised for economic efficiency. Investigating the principles underlying Japanese joining we actively pursue the potential of allowing

[5] Chainsaws for example may be mounted with ripping (for cutting along grain) or crosscutting (for cutting across grain) chains – each possessing a different tooth profile. Impractical to switch between each operation, the arms feed rates are instead adjusted to compensate – simulating the human operator. With the bandsaw, a trade off is made between speed of cutting (requiring a large tooth pitch) and minimum turning radius (requiring a finer band/tooth).

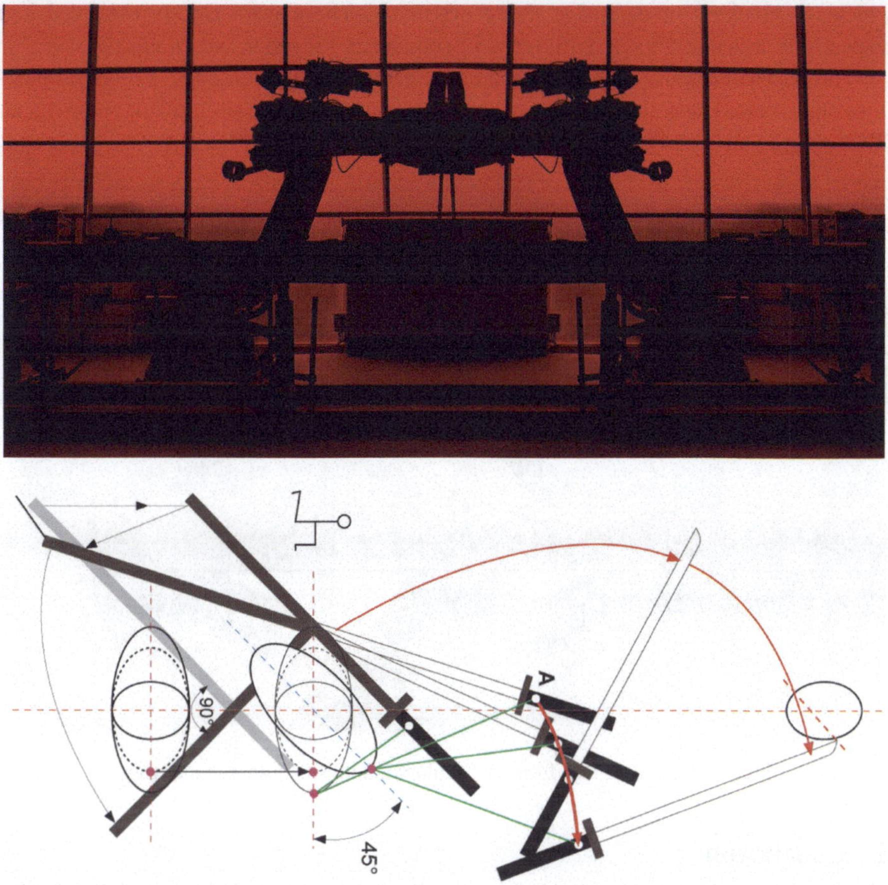

Fig. 6. In Robotic Fabrications, typical lead-in sequences are transformed, taking inspiration from different stances observed in the practice of Iaido (Below [7]).

operations to deviate from the restrained motion paths of industrial applications. Explorations of controlled movement borrowed from the realm of martial arts and performance infect the work with an exciting spatial complexity (Fig. 6).

Operating within the depths of the code, these prescribed movement sequences link code segments that relate to the more functional operations where the chainsaw directly engages with the timber. These highly choreographed movement arrangements serve to connect the primary movement sequences to each other in space and are mainly applied through linking commands.

The conventional home and repose positions became informed by terminology and principle translated from the art of Iaido [8]. This traditional Japanese martial art of the sword strives for smooth, controlled movements of drawing the sword from its scabbard aiming to strike or cut with a dynamic and elegant action. Typical lead-in sequences were transformed by different stances - strategic standing positions from which an attacking or defensive move originates. These were closely observed and respectfully translated towards movement sequences for robot and chainsaw.

These inserted motions produced an amalgamation of technology, timber connections and martial arts, generating an exciting spatial complexity and developing an artistic engagement within the world of physical production (Fig. 7).

Fig. 7. Ornament and form are explored alongside joining strategies.

4 Conclusion

Through this project we are in search of serendipitous moments of fusion between tool, timber construction and performance. These choreographies situate the precision and repeatability inherent in robotics in relation to traditional fabrication techniques and hijacked tools, defining gestural motions not born from efficiency and suggestive but instead of cultural references and occasionally mischief. We are vigorously resisting ideas of full automation - instead we forcefully extend and invent ways to increase the level of interaction with our tools. Sustaining the notion of tools as extensions of our bodily self, the idea that they perform fully independently -as a phantom limb- only advances the disconnect between design and fabrication.

Encouraging a dialog between haptic and scripted operations lets us circumnavigate their individual shortcomings and produces workflows that work complimentary.

Implementing various tools regularly used in conventional timber fabrication, our aim is to use robotic technology not forcefully, for power, repeatability or willful formalism, but delicately, for the strategic augmentation of a naturally complex material. We are concerned with direct applications for industry, but also with the inherent possibilities of beauty and power, theatre and magic [9].

Ultimately, we operate in a territory where robotics in architecture, art, and design can construct a highly original reality and be seen as a powerful instrument for creative speculation.

References

1. Sato, H., Nakahara, Y., Paul Nii, K. (Trans): The Complete Japanese Joinery. Hartley & Marks, Vancouver (1995)
2. Mollica, Z., Self, M.: Geometric strategies for exploiting inherent material form. In: Adriaenssens, S., Gramazio, F., Kohler, M., Menges, A., Pauly, M. (eds.) Advances in Architectural Geometry 2016, pp. 138–153. vdf Hochschulverlag ETH Zurich, Zurich (2016)
3. kkaarrlls homepage. http://www.kkaarrlls.com/index.php?feature=editions,7Xstool. Accessed 12 May 2018
4. Pawlofsky, T.: Rhino2krl. In: Brell-Çokcan, S., Braumann, J. (eds.) ROB|ARCH 2012: Robotic Fabrication in Architecture, Art and Design, pp. 166–171. Springer, Vienna (2013)
5. Yuan, P.F., Chair, H.: Robotic wood tectonics. In: Sheil, B., Menges, A., Glynn, R., Skavara, M. (eds.) Fabricate: Rethinking Design and Construction, pp. 45–49. UCL Press, London (2017)
6. Johns, R.L., Foley, N.: Bandsawn bands: feature-based design and fabrication of nested freeform surfaces in wood. In: McGee, W., Ponce de Leon, M. (eds.) Robotic Fabrication in Architecture, Art and Design 2014, pp. 17–32. Springer International Publishing, Switzerland (2014)
7. "Nuki Tsuke" [digital image] (n.d.). https://shinbukan-paris-iaijo.tumblr.com. Accessed 8 May 2018
8. Salmon, G.: Kendo: A Comprehensive Guide to Japanese Swordsmanship. Tuttle Publishing, Rutland (2013)
9. Davies, K., Vercruysse, E., Svilans, T., Dodd, I.: Instruction and instinct, performing within unordered complexity. In: McGee, W., Ponce de Leon, M. (eds.) Robotic Fabrication in Architecture, Art and Design 2014, pp. 223–232. Springer International Publishing, Switzerland (2014)

Cyber Physical Macro Material as a UAV [re]Configurable Architectural System

Dylan Wood(✉), Maria Yablonina, Miguel Aflalo, Jingcheng Chen, Behrooz Tahanzadeh, and Achim Menges

Institute for Computational Design and Construction, University of Stuttgart, Keplerstr. 11, 70174 Stuttgart, Germany
dylan-wood@icd.uni-stuttgart.de

Abstract. The research presented in this paper describes a new strategy for deploying a cyber-physical macro material combined with aerial construction robots as a reconfigurable architecture. The resulting architectural system is capable of autonomous rearrangement and stagnant operation driven by behavioral design patterns rather than a singular robotic assembly or construction method.

The central part of the paper describes the functional framework for applying the system to a self-supporting roof canopy where the material is comprised of discrete, lightweight, polyhedron shaped, carbon fiber units with onboard sensing, processing, and communication. The material is autonomously reconfigured by calling and coupling with task-specific UAV builders. Programming and hardware for intra- and inter-material, and builder communication and user interaction are described. The system's live decision making, interaction, and continuous reconfiguration based on programmatic input are outlined. The cyber-physical framework is produced and shown in 1:1 functional prototype.

Three potential behavioral scenarios applying the system as a new type of architecture are outlined: adaptive behavior, interactive behavior, and learning behavior. Further, the programming of the existing framework using machine learning and artificial intelligence is proposed as a future development. In conclusion, developments of the system and the impact and disruption of architectural design, planning, and construction processes are proposed.

Keywords: Cyber-physical architecture · Digital materials
Robotic architecture · Reconfigurable architecture · Adaptable architecture
Programmable matter · Robotic materials

1 Introduction

Technological developments in intelligent digital materials and autonomous robotic assembly open new possibilities for cyber-physical construction and behaviorally programmed architecture [1–3]. Building with a digital material that is also capable of on-board sensing, processing, and communication as a system enable buildings to embody new roles but often lack the capacity to physically reconfigure. However, fully distributed systems and forms of classical programmable matter, capable of onboard actuation/movement are today scale limited by mechanics, electronics, costs, and

J. Willmann et al. (Eds.): ROBARCH 2018, *Robotic Fabrication in Architecture, Art and Design 2018*, pp. 320–335, 2019.
https://doi.org/10.1007/978-3-319-92294-2_25

production. In parallel behavioral robotic fabrication processes with sensor feedback have advanced construct components and structures with traditional or digital materials in a sophisticated fashion [1, 4]. Building from these principles a higher level of distributed intelligence and control can be placed within the building material, inverting the relationship between robotic builder and material. In this work the robotic system advances beyond the construction process to become an active part of the architecture, creating a cyber-physical architectural system.

This research combines an intelligent digital material with aerial construction robots to prototype a new breed of reconfigurable architecture (Fig. 1). The system is comprised of lightweight, macro-scale material units with onboard sensors and processing, task-specific UAVs (Unmanned Aerial Vehicles) used as builders, and a communication, control, and programming framework. As a system it can be considered a machine within which the building material and robotic builder are in constant digital communication with each other as well as with the occupants and space which it inhabits. Material is configured and reconfigured in the x-y plane by a UAV and spans between moveable vertical columns. The material, builder and user are strategically physically separated into distinct vertical construction zones (Fig. 2).

Fig. 1. Prototype of a cyber-physical building material for a reconfigurable architectural system. Left - Material arranged as a canopy. Right - a UAV builder relocating material.

Fig. 2. Visualization of the proposed cyber-physical architectural system traversing throughout an open space.

2 Context

Digital materials are specifically designed for reconfiguration [5]. These physical properties combined with developments in smart building blocks as part of the IoT (Internet of Things) and programmable matter present an entirely new perspective on how 'robotics' can be applied to architecture in which the building materials and builders possess a level of unsupervised embedded computation and inherent reconfigurability. Conceptual visions of macro scale, machine-based, spatially reconfigurable architecture has fascinated visionaries looking for a mobile and adaptable architecture [6, 7]. The architectural application of more interact behaviors in these systems is strictly limited by their physical capacities to change configuration resulting in limited to analog movements of typical building components (Ori Systems, 2018) or shape-morphing of single surfaces incapable of typological reconfiguration [8, 9]. Recent advances applying robotic manipulators to reconfigure architecture have proved promising yet allude to the fact that a material-based approach would provide higher resolution and further adaptability [10].

3 State-of-the-Art

Multiple projects have demonstrated autonomous reconfiguration of building materials. Robotic system have been proposed for different types of reconfigurable materials: using strut and voxel-based materials arranged based on structural requirements [11], with construction-specific digital materials arranged by mobile robots in laboratory settings [12], larger scale infrastructures built from carbon fiber digital materials [13], in outer space where manual construction is difficult [14] and assembled by comparably larger gantry and cable robots [15, 16]. From a traditional construction robotic perspective, in situ mobile robotics and aerial construction using sensor feedback and standard building materials such as bricks has been demonstrated [17–19]. Specific to aerial robotics strut assembly and materials with integrated flight mechanisms have been conducted in short duration laboratory tests with non-building materials [20–22] and proposed as architectural concepts in the project 'Omnipresence' (Hyperbody Studio, TU Delft, 2016). Similarly inflatables/balloons have been conceived as lifters and manipulators for large-scale shell structures and as autonomous assemblers in the project 'Aerial Assembly' (Self Assembly Lab, 2015) [23]. In each case, the systems are primarily based on either physical construction sequencing, or conceptualizing distributed robotic reconfiguration behaviors [24]. Architecture can be constructed from robotically and manually assembled discrete elements [25]. The integration of sensors within a reconfigurable component-based architecture has given prototypical buildings the chance to collect and analyze data from users and the environment to adapt fabrication procedures in real time using traditional means of reconfiguration [26].

In parallel, technological advancements in mobile and aerial robotics have pushed construction robots to become increasingly smaller, lighter, accessible, and customizable. Open communication systems have been developed to aid in distributing tasks between multiple robots and low-cost sensing and communication hardware is readily attainable increasing the intelligence that can feasibly be integrated into building materials [27]. Digital fabrication and the development of high performance materials such as carbon fiber reinforced plastics enable the mass production of lightweight and high strength building materials enabling easier transport and customization.

4 Methods

The system is comprised of: intelligent digital material, UAV builders, a communication and programming framework, and human users. It is shown as a functioning physical prototype applied as roof canopy accompanied by proposals for applicable programmable behaviors.

4.1 Material

The main physical component of the system is a custom designed macro scale digital material made of mass producible modular units (Fig. 3). Reconfigurability is the primary design objective; however, the unit must effectively incorporate aspects of architectural performance, transportability, structural capacity, cost, weight, power

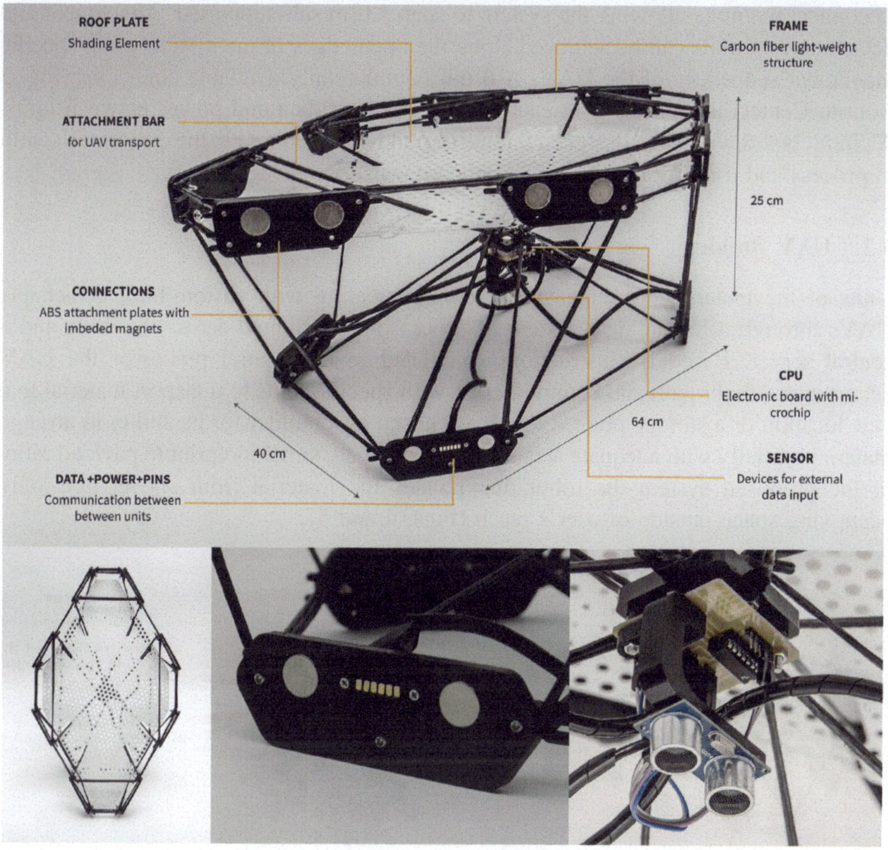

Fig. 3. Cyber-physical macro material unit. Top - integrated functions of the unit. Bottom Left - planar view of the unit. Center - connection and communication plate. Right - onboard CPU and motion sensor stack.

consumption, and aesthetics. In addition, the elements must be able to collect data through sensors and transfer information and power to adjacent units.

Physically the units are comprised of a Carbon Fiber Reinforced Plastic (CFRP) skeleton structure forming a polyhedron space frame with a volume of 0.06 m^3 covering an area of 0.20 m^2 (Fig. 3, Top). The frames are manufactured using a coreless winding process. Connections are possible on the four vertices faces and a polycarbonate sheet is attached beneath the top face to form a closed area below. The polyhedron aggregation strategy is chosen to allow a relatively simple reconfiguration in a single x-y plane and is applicable for roof or canopy structures. However, a similar approach could be applied to a vertical or spatial assembly. The use of CFRP significantly reduces the weight of the unit while enabling a size and structural performance applicable to architecture and reducing the strength required for the connections. As building material the system is lightweight at 698 g per unit with material and hardware costs of 40 euros. In construction terms the material weight is 3.49 kg/m^2 and cost is 200€/m^2. Magnets are used to connect the units allowing the system to span 3.0 m self-supported. Attachment and detachment of units can be achieved along the open edges of the system using specific movement sequences of the UAV. A 6 pin commercially available connector (Fig. 3 Bottom Center) is embedded in each face to transmit data and power between units. A circuit board with a pre-assigned unique digital ID is mounted in the center of the unit to process and transmit data from the onboard sensors.

4.2 UAV Builder

Units of the material are relocated by communicating with custom-built detachable UAVs through a hierarchical system of processors distributed across the units and a central server. Based on instructions calculated in the central processor the UAV determines a flight path and communicates with specific units to transport material to a new location or a storage point. UAV was chosen as a builder for its ability to arrange material spatially with adequate accuracy and relatively small footprint to payload ratio. In the proposed system the robot manipulates the material from above, thus only occupying space outside of user's reach (Figs. 4 and 5).

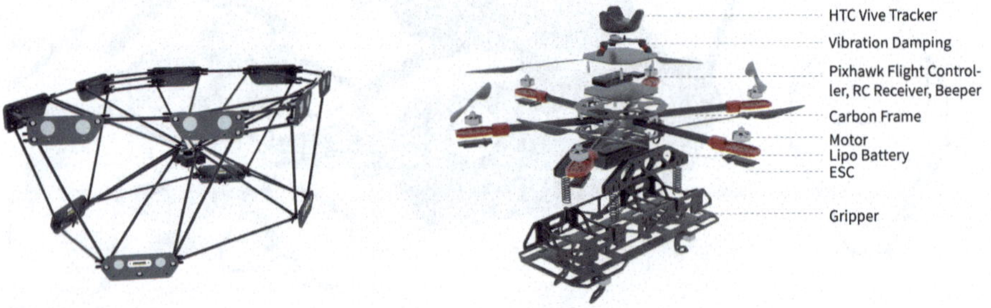

MATERIAL UNIT

Carbon Frames	436 g	Volume	0.06 m^3
Electronics	22 g	Area covered	0.20 m^2
Magnets	80 g	Cost (Materials)	40 Euros
ABS Connections	80 g		
Metal Hardware	80 g		
Total Weight	**698 g**		

UAV Builder

Rotors	6
Diameters	960 mm
Est. Flight Time	10 mins
Weight	2210 g
Payload	1500 g

Fig. 4. Comparison of the cyber-physical macro material unit and UAV builder showing low weight material and higher weight and complexity builder.

Robotic hardware and material were co-designed to optimize for weight and cost. A custom-built hex copter configuration was chosen that provides higher payload and flight stability during the pick and place routine. The builder is equipped with a gripper that uses hooks driven by servo motors to connect to attachment bars on the top of each unit. The gripper, the locomotion body and the localization system are attached to each other via dampening springs that stabilize the movement during takeoff, landing, pickup, attachment and detachment of the units.

A localization and control system were developed. A commercially available external machine vision system for tracking is combined with a secondary on-board IR

Fig. 5. Flight test of UAV builder. Left approaching attachment point of a material unit to overall structure. Material unit coupled to UAV builder during transport.

camera based localization system to allow the builder to localize individual material units locally during the pick and place routines. The secondary system is cost-efficient and lightweight, which allows installing the camera directly on the builder determining relative position of the unit in relationship to the builder. For real-time UAV control a PID control (proportional-integral-derivative control) system was used. This system is often used in industrial applications where a set of unpredictable conditions are to be considered, such as cruise control in vehicles. In this system the PID controller controls the velocity of the builder relying on the localization system for error feedback.

An autonomous navigation software was developed to plan and execute the UAV's local and global actions. Local navigation determines standard flight path planning for each routine within the reconfiguration process: hovering over, approaching, picking, placing and transporting the units. The global navigation system defines overall reconfiguration of the structure, determining the sequence of moving units from one location in the structure to another, considering physical relationships between the units in the assembly. Overall the custom robot, its localization, control and navigation systems allow for autonomous execution of the processes required for assembly and reconfiguration of the material at an architectural scale.

4.3 Communication and Programming

To function, the material and UAV builder communicate through layers of software ranging from low-level communication protocol to higher level algorithms which coordinate each agent based on collected information. Communication between units happens over I2C protocol through two signal wires. These low-level messages are sent to a server computer over a serial communication port and dispatched to other agents including builder over higher-level protocols such as WebSocket and UDP based on client requirements. Specifics of the communication are described:

Unit-Master Unit Communication. A unit's onboard computer (ATTINY84) communicates with the server computer over I2C protocol. I2C (Inter-Integrated Circuit) is a serial messaging protocol for interconnected electrical components in which a master node can send data to each individual slave node by using its specific address. Each unit has a predetermined address and access to two I2C shared line (Data and Clock) through its adjacent units.

Unit-Unit Communication. Apart from the four shared lines between units (+5 v., ground, data and, clock), a fifth line is connecting one pin of each microcontroller to exactly one pin of the adjacent unit's microcontroller. By setting the state of this pin to Low the state of the connected pin on the neighbor unit will be pushed to Low too. This change can trigger a function on the adjacent unit. This method is later used to reconstruct a map of unit's placement.

Master Unit-Server. The Master unit is the one connecting the network of the units to the server. An Arduino board as the driver bridges the Master unit onboard microcontroller to the server computer over Serial. A single-board computer (Raspberry Pi 3) as the server runs more computationally demanding functions such as mapping and coordination of other agents.

Server Software/ROS. A set of software modules running on the server computer are responsible for high-level reasoning and planning and interfacing connected hardware. ROS as the software infrastructure "provides a structured communications layer above the host operating systems of a heterogeneous computer cluster. A node sends a message by publishing it to a given topic and any node that is interested in a certain kind of data will subscribe to the appropriate topic." This project benefits from this architecture in following ways: Multi-hosting nature of ROS allows for more computationally demanding processes such as data mining and reasoning to run remotely while being seamlessly integrated into the master process. Due to the open-source aspect of ROS, available packages, especially in the field of machine vision and machine learning, are used without any further modification. Users can communicate with the ROS server through a set of Web Application over WebSocket protocol. This is provided by Rosbridge package. "Rosbridge provides an application-layer network protocol for robotics, allowing arbitrary, non-ROS client processes (including Web interfaces) to interoperate with ROS processes."

User Interfaces/Server Software. Users can interact with the server through a web application for monitoring and testing.

End-User Interface. Users can monitor the status of the structure as well, manipulate it by defining growth patterns and, access the integrated sensors within the system (Fig. 7).

Developers Interface. I2C interface, developers can use this interface to compose a one-byte message. Also, it can log incoming messages coming from the I2C network. Drone Control Interface, with a real-time 3d view of drone and plots of PID controller status, was used to tune the PID values.

Builder/Server Software. Builder's control system is a separated ROS software consisting of localization algorithms and control node. This cluster interacts with the

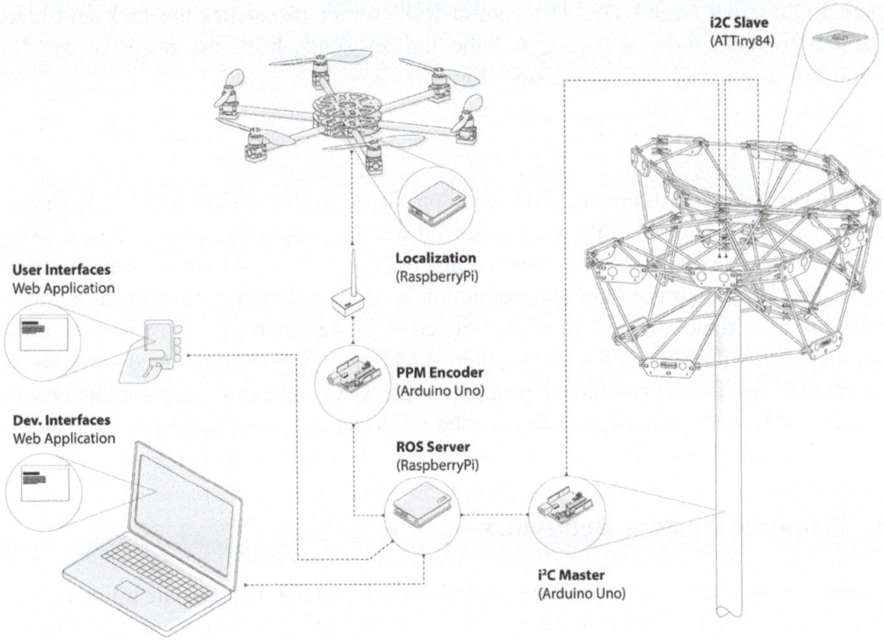

Fig. 6. Diagram of communication within the cyber-physical system

Fig. 7. Field test of cyber-physical architectural system in an outdoor public space.

main ROS server over UDP. The central ROS server dispatches the pick and place position to the builder software and the builder sends back the result of the last command to proceed to the next one (Fig. 6).

4.4 System Tests

The system is demonstrated as a 1:1 scale prototype in the form of a 2.5 m high roof canopy. Twenty material units were produced with full communication and processing functionality (Fig. 7). Field tests were conducted in an open outdoor space in Stuttgart, Germany. Areal configuration and communication were tested by executing a simple behavior of adding material to a leading edge by removing it from another edge, effectively moving the roof in a specified direction following continued presence of occupants. Tests were conducted using a single UAV operating autonomously with manual approval of each flight path for safety. Further tests are planned in large indoor space with public access using two UAV's and an additional 30 units.

5 Proposed System Behaviors

Within the physical and digital framework developed for the system different programming strategies can be proposed. Here we describe three strategies to show the potential of the system to perform as an architectural system.

5.1 Responsive Behavior

The configuration is a direct result of current environmental parameters to achieve a predetermined effect. In this case, the system will be reconfigured slowly so that the light and shadow in the occupied space remains consistent despite changes in the sun position throughout the day (Fig. 8, Left).

5.2 Interactive Behavior

In this mode, the user's presence defines the next configuration of the overall structure. Data that is collected from movement sensors embedded in the material generates a concentration map which becomes input for the pick and place routine. The rule is to move units from less crowded zones toward more populated ones.

5.3 Learning Behaviors

The ease of data collection and reconfiguration of the canopy can provide vast amounts of raw data to study the hidden relationships between the users, environment, space, and the architectural system. Access to enough data and sophisticated statistical analysis, provides the opportunity to map the user flow and even manipulated based on previous interactions. Following this phase, the system would begin to develop new unpredicted behaviors while continuing to refine itself based on new data (Fig. 9). The goal of this mode is to use data to determine the success (fitness value) of the current arrangement. Fitness could be described for example by: Guiding users to a specific

Fig. 8. Potential environmental responsive (left) and interactive behaviors (right) for the architectural system.

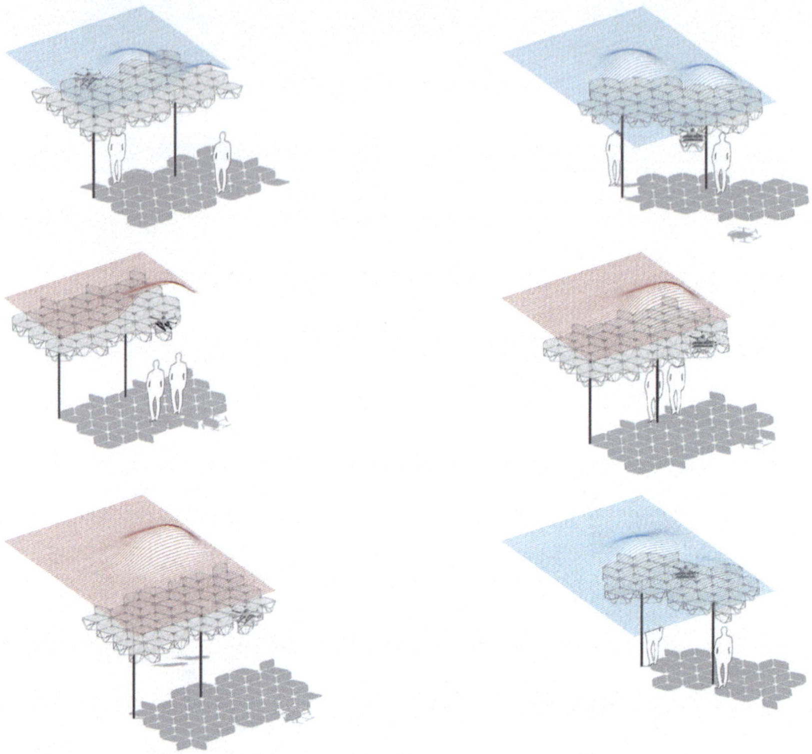

Fig. 9. Proposed programming of 'Learning behaviors'. Data gathering shown in red and dynamic behaviors shown in blue.

position: Users placement can be interpreted as the label of the database records and the rest of parameters will be considered as features. A classification tree could generate the best arrangement based on all parameters.

6 Discussion

The functional prototype of this system provides a new framework for demonstrating the architectural potential of a cyber-physical material system. Further validation and development of the system will require technical improvements as well as a rethinking of the impact of such a system on the role of the designer.

6.1 System Potentials

Architecturally the system presents a combination of digital and physical opportunities. First, autonomous reconfigurability allows the canopy to effectively 'walk' around or dynamically move through its environment, fragment into smaller sub-structures, or disassemble entirely into individual units. This can be conducted continuously or in intervals at variable speeds without additional equipment or labor. Second, unlike a typical static building, the canopy can be programmed with different reconfiguration behaviors. With the range of configurations vastly exceeding that of morphing surfaces and the ability to interact with occupants and environment through sensors, it represents a promising application of robotics compared to automated robotic fabrication of static building materials.

The quantity of material and builders is scalable within the framework to achieve larger and/or faster systems with similar behaviors. The strategic separation of the material and actuator/movement creates a system where only the functionality required for continuous operation is embedded in the architecture itself, while the energy expensive movements are performed by external manipulators that become the mobile muscles of the overall system. Specific to the time scales of building movements these factors efficiently enable both quick adaptation and retention of a static configuration for long durations.

6.2 Further Development

The system is limited in two areas, its ability to reconfigure in only a single horizontal plane and its structural stability which relies on magnetic connections. Similar units could be designed for arrangement in x-y-z planes which would reduce the need for separate columns and increase the flexibility of the system. In this case, physical interaction between the UAV and occupants would need to be considered. For connections a combination of magnetic and structural joining details could be implemented to improve span and structural continuity while maintaining configurability and increasing the complexity of the attachment sequence. Electromagnets could be used to rely less on the movement of the UAV to detach units at the cost of increased weight and power.

The current separation of the builder as the external locomotion provider and the material is one of the fundamental features of this project which radically decreases the

Fig. 10. Visualization of the architectural system in a public space.

complexity of the material and makes the system more robust and scalable. While the opposite system (embedded locomotion) can result in quicker changes, it increases inefficiency as A. Due to the inherent complexity of an internal location system, robustness will decrease. B. Internal locomotion implies additional load to the overall structure which would require more robust materials throughout the system. While functional, the resolution of the system could be further adapted and made adjustable by the number of units carried by the builder (Fig. 10).

7 Conclusion and Outlook

Smart digital materials complemented with an integrated distributed mobile robotic assembly method can be deployed as a cyber-physical architecture in situations benefiting from high adaptability and temporary conditions. In this context the project demonstrates a functioning framework for implementation in which the robotic system emerges as an active and influential architectural system open to future programming and development. The physical adaptability of the system and programmability opens it to spatial and architectural application of advances in artificial intelligence where a building can physically learn and adapt. The Introduction of such systems in an architectural context requires significant rethinking of architectural design methods towards a behavior-based architecture, focusing on continuous post-occupancy design. This architecture forms a highly disruptive notion considering the current linear and lethargic design, construction and occupancy sequence.

References

1. Menges, A.: The new cyber-physical making in architecture: computational construction. Archit. Des. **85**(5), 28–33 (2015)
2. Oosterhuis, K.: Hyperbody: First Decade of Interactive Architecture. Ram Publications, Garland (2012)
3. Tibbits, S. (ed.): Autonomous Assembly: Designing for a New Era of Collective Construction. Architectural Design, vol. 87, no. 4 (2017)
4. Menges, A. (ed.): Material Computation: Higher Integration in Morphogenetic Design, vol. 82, no. 2. Wiley, London (2015)
5. Popescu, G.A., Gershenfeld, N.: Digital materials, pp. 16–19, Massachusetts Institute of Technology (2009)
6. Friedman, Y.: Manifesto de l'Architecture Mobile. Congrès International d'Architecture Moderne nr. 10, Dubrovnik (1956)
7. Negroponte, N.: The Architecture Machine: Toward a More Human Environment, 1st edn. MIT Press, Cambridge (1973)
8. McGee, W., Velikov, K., Thun, G., Tish, D.: Infundibuliforms: kinetic systems, additive manufacturing for cable nets and tensile surface control. In: Sheil, B., Menges, A., Glynn, R., Skavara, M. (eds.) Fabricate: Rethinking Design and Construction, pp. 4–6. UCL Press, London (2017)
9. Ahlquist, S.: Sensory material architectures: concepts and methodologies for spatial tectonics and tactile responsivity in knitted textile hybrid structures. Int. J. Archit. Comput. **14**(1), 63–82 (2016)

10. Frearson, A.: Robots will be used to construct BIG and Heatherwick's Google HQ. Dezeen 55 (2015)
11. Melenbrink, N., Kassabian, P., Menges, A., Werfel, J.: Towards force-aware robot collectives for on-site construction. In: Nagakura, T., Tibbits, S., Mueller, C., Ibañez, M. (eds.) Acadia 2017: Disciplines & Disruption, Proceedings of the 37th Annual Conference of the Association for Computer Aided Design in Architecture, pp. 382–391. Cambridge, MA. (2017)
12. Petersen, K., Nagpal, R., Werfel, J.: TERMES: an autonomous robotic system for three-dimensional collective construction. In: Durrant-Whyte, H., Roy, N., Abbeel, P. (eds.) Robotics: Science and Systems VII, pp. 257–264, MIT Press, Cambridge (2012)
13. Jenett, B., Cellucci, D., Gregg, C., Cheung, K.: Meso-scale digital materials: modular, reconfigurable, lattice-based structures.In: Proceedings of the 11th International Manufacturing Science and Engineering Conference (ASME 2016), Volume 2: Materials; Biomanufacturing; Properties, Applications and Systems, pp. 1–11. Blacksburg, VA (2016)
14. Jenett, B., Cheung, K.C.: BILL-E: robotic platform for locomotion and manipulation of lightweight space structures. In: Proceedings of the 25th AIAA/AHS Adaptive Structures Conference, American Institute for Aeronautics and Astronautics (AIAA), pp. 1876–1888. Grapevine, TX (2017)
15. Carney, M., Jenett, B.: Relative robots: scaling automated assembly of discrete cellular lattices. In: Proceedings of the 11th International Manufacturing Science and Engineering Conference (ASME 2016), Volume 2: Materials; Biomanufacturing; Properties, Applications and Systems, Blacksburg, VA (2016)
16. Dierichs, K., Menges, A.: Granular construction: designed particles for macro-scale architectural structures. Archit. Des. **87**, 88–93 (2017)
17. Augugliaro, F., Lupashin, S., Hamer, M., Male, C., Hehn, M., Mueller, M.W., Willmann, J., Gramazio, F., Kohler, M., D'Andrea, R.: The flight assembled architecture installation: cooperative construction with flying machines. IEEE Control Syst. **34**(4), 46–64 (2014)
18. Giftthaler, M., Sandy, T., Dörfler, K., Brooks, I., Buckingham, M., Rey, G., Kohler, M., Gramazio, F., Buchli, J.: Mobile Robotic Fabrication at 1:1 scale: the In situ Fabricator (2017)
19. Latteur, P., Goessens, S., Breton, J.-S., Leplat, J., Ma, Z., Mueller, C.: Drone-based additive manufacturing of architectural structures. In: Proceedings of the International Association for Shell and Spatial Structures Symposium (IASS): Future Visions, Amsterdam (2015)
20. Lindsey, Q., Mellinger, D., Kumar, V.: Construction of cubic structures with quadrotor teams. In: Durrant-Whyte, H., Roy, N., Abbeel, P. (eds.) Robotics: Science and Systems VII, pp. 177–184, MIT Press, Cambridge (2012)
21. Saldaña, D., Gabrich, B., Salda, D.: ModQuad: the flying modular structure that self-assembles in midair, In: Proceedings of the IEEE International Conference on Robotics and Automation, ICR '2018, Brisbane (2018)
22. Oung, R., D'Andrea, R.: The distributed flight array. Mechatronics **21**(6), 908–917 (2011)
23. Otto, F.: IL 13 Multihalle Mannheim, Universität Stuttgart (1978)
24. Yiannoudes, S.: An application of swarm robotics in architectural design. In: Intelligent Environments 2009, vol. 2, pp. 362–370, January 2009
25. Retsin, G.: Discrete and digital - a discrete paradigm for design and production. In: TxA 2016 (2016)
26. Prado, M., Dörstelmann, M., Menges, A., Solly, J., Knippers, J.: Elytra filament pavilion robotic filament winding for structural composite. In: Sheil, B., Menges, A., Glynn, R., Skavara, M. (eds.) Fabricate: Rethinking Design and Construction, pp. 224–233. UCL Press, London (2017)
27. Rubenstein, M., Cornejo, A., Nagpal, R.: Programmable self-assembly in a thousand-robot swarm. Science **345**(6198), 795–800 (2014)

DryLin T by igus
6

Adaptive Robotic Carving

Training Methods for the Integration of Material Performances in Timber Manufacturing

Giulio Brugnaro(✉) and Sean Hanna

The Bartlett School of Architecture, UCL,
22 Gordon St., London WC1H 0QB, UK
g.brugnaro@ucl.ac.uk

Abstract. The paper presents the developments of a series of methods to train a fabrication system for the integration of material performances in timber manufacturing processes, combining robotic fabrication together with different sensing strategies and machine learning techniques, and their further application within a prototypical design to manufacturing workflow. The training cycle, spanning from the recording of skilled human experts to autonomous robotic explorations, aims to encapsulate different layers of instrumental knowledge into a design interface, giving designers the opportunity to engage with material and tool affordances as process driver. The training methods are evaluated in a series of experiments and design iterations, proving their potential in the development of customized design to manufacturing workflows and integration of material performances, with a specific focus on timber.

Keywords: Material behaviors · Machine learning · Instrumental knowledge Subtractive manufacturing

1 Introduction

The paper presents the developments of a series of methods to train a fabrication system for the integration of material performances in timber manufacturing processes, combining robotic fabrication together with different sensing strategies and machine learning techniques, and their further application within a prototypical design to manufacturing workflow.

Such methods question the linear progression from the design intention to its materialization within current production practices which determines a lack of feedback between the different stages of the process. This forces designers to consider materials as passive receivers of a previously generated ideal form stored in a digital model (DeLanda 2004) and reinforces the separation between the act of designing and making (Carpo 2011). Consequently, design practices can only engage with a limited range of standard manufacturing methods and homogeneous materials. Furthermore, the homogenization of natural material results in heavy industrial processing and material waste.

J. Willmann et al. (Eds.): ROBARCH 2018, *Robotic Fabrication in Architecture, Art and Design 2018*, pp. 336–348, 2019.
https://doi.org/10.1007/978-3-319-92294-2_26

The goal of the research is to develop a computational framework which allows designers to engage with the properties of heterogeneous materials and the affordances of non-standard tools as process drivers, extending the design moment toward the fabrication stage to explore novel design opportunities. One of the most pressing problems in working such a process is the assessment of inherent variation, represented in this paper by differing wood grain and chisel cuts. There is a range of possible cuts that are feasible, *i.e.* that actually remove material while not damaging either tool or wood, and within this there is a range that may be considered optimal, *e.g.* they remove the maximum amount of material in a given time. Human experts navigate and anticipate this range intuitively, after training and experience. It is a far greater problem for a machine.

This paper investigates how this range and optimum can be quantified and mapped for use in determining appropriate cuts by a robot. The central question is: how accurately are the effects of a given cut predicted based on inputs easily available to the robot? We also investigate whether the domain training inputs makes a significant difference, by comparing data generated by a methodical, parametric, exploration of grain angles and cuts by the computer against data drawn from an expert human user. Is it better for the machine to learn from a skilled teacher, or from its own experience?

2 Context

The methods are based on a series of training and datasets curation procedures, where the instrumental and material knowledge, acquired from both skilled human experts and robotic carving sessions, is captured, transferred, augmented and finally integrated into an interface that makes this knowledge available to the designer.

In this regard, one of the questions is whether is possible to encapsulate, at the least partially, this instrumental knowledge in the technological means for fabrication available to us, making it "*easily accessible, communicable, repeatable, hackable, and transformable*", in the same way, for instance, as 3D modelling software encapsulates knowledge of calculus-based mathematics (Witt 2010). However, in the accumulation of human experience through making and the interaction with materials, there is a tacit dimension that is difficult to capture, formalize and share (Polanyi 1967).

In the history of automation, there have been previous attempts towards this direction, such as the "*Record/Playback*" system to generate machine operations, developed in the 1940s by General Electric. Within this system, a machinist was able to operate a modified machine tool to produce an artefact and get the totality of his motions recorded on a magnetic tape which could be automatically reproduced later by the machine. In those recorded motion was captured not only the gears' mechanical displacement but also the machinist's intelligence, skills and tacit knowledge (Noble 1984; Callicot 2003). More recently, similar methods, focusing on the value of human's action analysis to inform robotic fabrication tasks, have been successfully applied for robotic reproduction of traditional stone surfacing techniques (Steinhagen et al. 2016) and to reconstruct ancient technical gestures associated with the use of tools and the development of the related cognitive functions (Pfleging et al. 2015).

The encapsulation of instrumental knowledge based on sensor measurements of machine operations, rather than human demonstration, finds precedence in a series of industrial manufacturing applications where machine learning techniques, such as Artificial Neural Networks (ANN), have been utilized in a series of machining operation for the optimization of fabrication parameters (e.g. surface roughness, tool wear) and cost reduction (Al-Zubaidi et al. 2011). Recently, CITA presented a similar approach to increase the accuracy of robotic incremental sheet forming through the acquisition of scan data and supervised learning (Zwierzycki et al. 2017).

The research proposition integrates both approaches for instrumental knowledge encapsulation and combine them in a two-stages robotic training process for the development of custom design to manufacturing workflows, with a specific focus on timber.

3 Robotic Training Methods

This paper tests the accuracy of predictions derived from machine learning on prior example chisel cuts, which are intended in practice to fit into a larger overall workflow (Fig. 1). This consists of a series of training methods which combines the recording of skilled human experts performing subtractive operations with a set of traditional carving tools, such as chisels and gouges, on wooden boards together with autonomous robotic carving sessions.

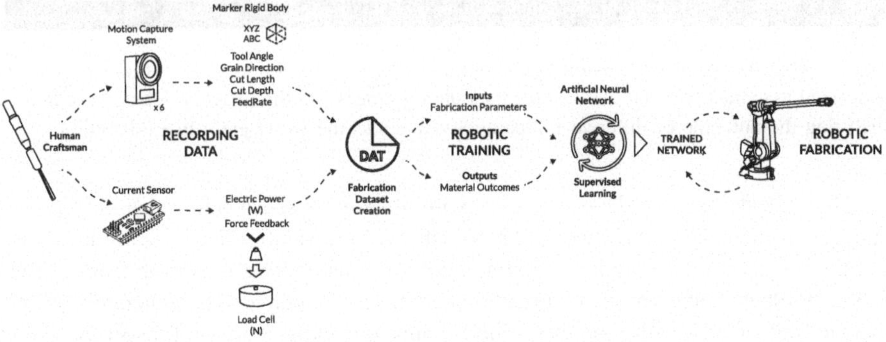

Fig. 1. The robotic training methods are structured around three main stages: (1) Recording fabrication parameters and material outcomes into datasets (2) Encapsulating instrumental knowledge through the training of an Artificial Neural Network (3) Using the trained network to inform a robotic fabrication task.

3.1 Training Tools

For each recording session, the combination of different sensing strategies, such as motion capture cameras and force-feedback sensor, allows collecting fabrication data simultaneously with the performing of the carving operation and compiled it into an ongoing dataset.

Motion-capture cameras are used to track the position and orientation (with a precision of ~0.2 mm) of 3D printed custom markers applied on the carving tools and workpiece. This allows to stream and reconstruct in real-time in the digital design environment (Rhinoceros 3D/Grasshopper) the carving operations and toolpath sequence that generated them.

Following the session, a photogrammetric reconstruction of the training boards is performed to store a precise 3D model of the material outcomes and extract relevant features for the training (Fig. 2).

Fig. 2. After the recording session, a photogrammetric reconstruction of each training board is performed to extract relevant features for the training process. In the image, a selection of boards analyzing the outcome of different toolpath carved across the wood grain main direction.

The collected sensor dataset is used within a supervised machine learning procedure, an Artificial Neural Network (ANN) with backpropagation-based learning, whose learning objective is to predict the simulation of a subtractive operation from a user-defined toolpath and a series of fabrication parameters, or conversely, generate a robotic toolpath out of a carved geometry. The training procedure utilizes Tensorflow (GPU version) as machine learning framework combined with the integration of Keras and Scikit-Learn libraries to respectively generate the ANN architecture and measure the performances of the system.

The input parameters are the tool/workpiece angle, tool/grain direction angle, force feedback, feed rate, target cut depth, target cut length. The recorded material outputs are: depth, length and width of the cut. Both the recorded and robotic toolpath are composed of a sequence of target frames, each storing the local parameters and fabrication outcome information and recorded as a single entry of the total dataset.

During the training, the performance of the network is evaluated following a K-fold cross validation (with k = 5) procedure, where the dataset is split in k subsets, called folds, and the algorithm is trained on k-1 folds, with each time one of the folds left out to be used to test the system with "unseen" data. The k performance scores obtained at the end of the validation are summarized in the measure of the Mean Absolute Error (MAE) of the prediction and its Standard Deviation (SD).

Starting the robotic training process with the human demonstration allows collecting quickly and efficiently information on how to operate a tool with a specific material, mediated by human experience acquired throughout the years and provide a strong foundation to inform robotic fabrication tasks with a similar set of non-standard fabrication tools and wood species.

The trained network, based on human experts, is used to provide guidance for autonomous robotic training sessions, efficiently narrowing down the search space through the definition of domain boundaries in the selected features rather than arbitrary defining an operational range or relying on a reinforcement learning procedure that would have to learn from a series of potentially dangerous "mistakes" during the fabrication training.

The definition of the search space through human demonstration presents to designers a curatorial approach toward the design of the fabrication process: rather than envisioning a universal machine able to operate any tool on any material, the idea is to tune the system to a very specific set of fabrication affordances and design intentions.

Therefore, the initial dataset is extended through the robotic production of a series of cuts where the parameters investigated are finely interpolated across the training board and the obtained data used for the training of the network informing the actual robotic fabrication task.

3.2 Design to Fabrication Workflows

The aim of encapsulating instrumental knowledge is its integration into a design interface which makes it accessible to designers and present them the opportunity of using materials behavior into their design workflow as process drivers. Once the network has been trained and the correlations between fabrication parameters and carved geometries are established, it's possible to translate back and forth between the two sets and customize the network topology to a specific design application (Fig. 3).

The three main modes of applications explored in the experiments are:

- *From robotic toolpath to simulation of the carved geometry.* While conventional digital Boolean operations result insufficient in calculating the outcome of subtractive operations with non-standard tools on heterogeneous materials, the trained network provides a more accurate simulation based on actual material properties and tool affordances. Designers can directly test how individual fabrication parameters affect the operation result and evaluate how these could be tuned to match their design intention. The prediction could be applied to multiple cuts at the same time, each with different parameters, and used to generate the overall appearance of the cutting pattern.
- *Individual Parameters Optimization.* Utilizing the same set of training inputs and outputs is possible to create labels (or Boolean flags) to predict a series of event thresholds based on sets of fabrication parameters, such as the successful removal of material or the correct extraction of the tool from the workpiece. Moreover,

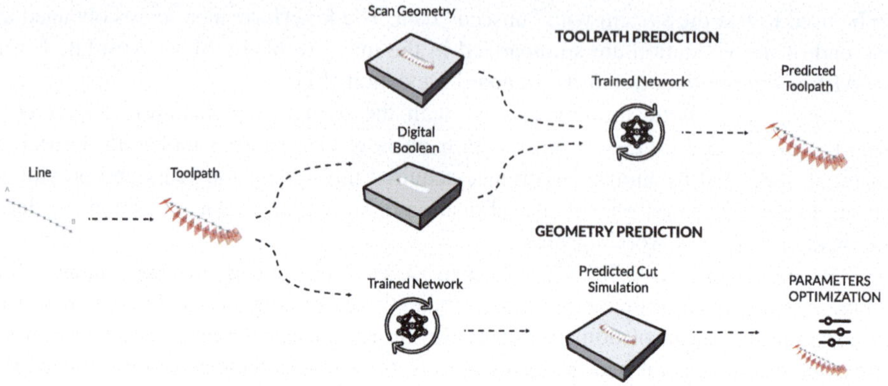

Fig. 3. The instrumental knowledge encapsulated in the trained network is integrated into a design to fabrication workflow based on the opportunity of translating back and forth between robotic toolpaths and carved geometries.

additional labels could be created by the designer to describe formal preferences (e.g. surface roughness, edges definition), curating the training dataset along a specific design direction. The event threshold is predicted using a ANN for binary classification and the prediction accuracy is evaluated with the same cross validation method previously described.

- *From carved geometry to robotic toolpath.* Extracting fabrication data out of the scanned model of a previously carved workpiece to reconstruct the robotic toolpath that has generated it. Alternatively, the same method could be applied starting from a digital geometry obtained through a subtractive Boolean operation as a way of matching a formal design intention in the fabrication stage.

4 Experiments and Results Discussion

As part of their development, the training methods have been iteratively tested to inform the robotic fabrication of a series of carved panels using different wood species and carving tools (Fig. 4).

One of the key things evaluated has been how the integration of human instrumental knowledge compares to the robotic training process. To perform this evaluation, two different training cycles have been set up, one starting with the recording of skilled human expert manually performing carving operations on a series of boards, the other directly with the industrial robotic arm (ABB IRB1600) generating a collection of cuts within an arbitrarily defined range of fabrication parameters.

In the first experiment, extending the findings of previously published work by the authors (Brugnaro and Hanna 2017), the craftsman created an initial collection of 50 carving operation on a series of European Lime wood boards using a traditional carving gouge (Stubai 9/20). The dataset has been generated concurrently with the carving itself following the methods previously described. The human expert, thanks to his

Fig. 4. In the robotic fabrication stage, the trained network is used to inform the robotic toolpath to carve the previously simulated geometries.

experience, has been able to avoid operations where the tool was not actually cutting the material or, on the opposite, digging too much into it and consequently forcing to stop the operation. Moreover, once individuated a preferred carved result, the cutting goal has been adjusted through human judgement to achieve a similar qualitative result in different lengths, depths and wood grain directions.

In the second stage of the experiment, the initial human-based dataset has been used to narrow-down the parameters domain and systematically explore it with the industrial robot through a collection of 215 cuts.

While the final prediction goal of the training cycle is the simulation of a carved geometry according to a set of given fabrication parameters, the first stage, based on human expert data, is aimed toward the generation of robotic toolpaths for additional subtractive operations that would extend the initially acquired dataset.

In the specific, the dataset of the first stage has been used to: (1) Analyze how the domain boundaries of the key parameters of tool/surface angle variation and "depth" profile of the toolpath changes according to different cut lengths, widths and wood grain directions. (2) Train a network and use it to predict such fabrication parameters to generate the robotic toolpath for the 215 operations within the dataset defined boundaries. The prediction error for the toolpath generation stage was the following: Depth: MAE = 0.56 mm, SD = 0.12 mm; Tool Angle: MAE = 2.12°, SD = 0.35°.

In the second experiment, counting the same number of cuts and using the same type of wood and carving gouge to compare it to the previous, the range of the fabrication parameters to explore has been defined without the guidance of a human expert interacting with the materials and tools but directly with the robotic operations stage. The focus of the training session has been on the variation of the material outcomes in respect of the angle between tool, workpiece surface and grain direction applied to different length of the cuts.

It's important to note that in the second experiment, the user applied his undirect intuition and understanding of the task in the programming of the robotic actuation, while for the first experiment direct real-world fabrication data have been used to inform it.

The output of the experiments are two networks trained with datasets counting the same number of cuts but different prediction boundaries. As a consequence, they perform differently in the prediction of the material outputs (length, depth and width relative to each individual target frame) necessary to simulate the fabrication outcome of the chosen operation.

In the specific, the two-stages (human + robot) network has been trained toward a specific design intention and performs better within its narrowed-down fabrication parameters domain (Depth: MAE = 0.35 mm, SD = 0.04 mm; Width: MAE = 0.83 mm, SD = 0.13; Length: MAE = 0.28 mm, SD = 0.05 mm).

The single stage experiment (robot only), while in general performing worse than the former (Depth: MAE = 0.53 mm, SD = 0.12 mm; Width: MAE = 1.02 mm, SD = 0.32; Length: MAE = 0.78 mm, SD = 0.32) and resulting less efficient for some operations described below, is able to cover a more extended and generic range of parameters prediction and could result more useful in those cases when the design task is not strictly defined from the training stage.

The dataset generated in the second experiments presents some set of parameters which does not generate any material removal on the board or damage the tool not allowing to complete the operation. In this case, an important step has been introducing two different labels for the definition of event thresholds: (1) "Tool Damaging" and (2) "Material Removal". Once these labels have been assigned, it's been possible to train the network for a binary classification process of the two event states (Fig. 5). The accuracy of the prediction for the two labels were respectively of 92.45% (Standard Deviation: 4.96%) and 84.58% (Standard Deviation: 3.20%). Combining the binary classification prediction with the trained network of the second experiment has been possible to efficiently tune the individual parameters and filter out dangerous or inefficient robotic operations (Fig. 6). The prediction of the range of successful cuts allows to explore its boundaries with confidence and optimize individual parameters to achieve a specific material outcome or increase the efficiency of the process, maximizing, for instance, the removal volume (Fig. 7).

Finally, the dataset generated during the robotic training have been used to reverse the network topology of the previous experiments and predict the tool/workpiece angle parameter and input cutting length necessary to reconstruct a toolpath used to generate a carved geometry. The prediction error was respectively 1.69° (SD = 0.42°) and 3.37 mm (SD = 1.23 mm) and the overall process performs slightly worse in the reconstruction of the toolpath than the network trained in the opposite direction for geometry simulation, probably due to hardware noise and imprecisions in the physical measurements, confirming what already pointed out by Zwierzycki et al. (2017) in their research.

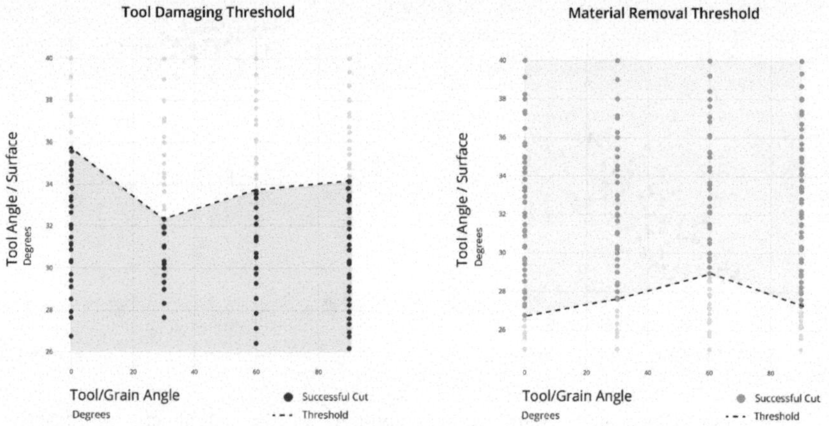

Fig. 5. "Tool Damaging" and "Material Removal" event thresholds are predicted with an ANN for binary classification and allow to optimize individual fabrication parameters to avoid potentially dangerous or inefficient operations.

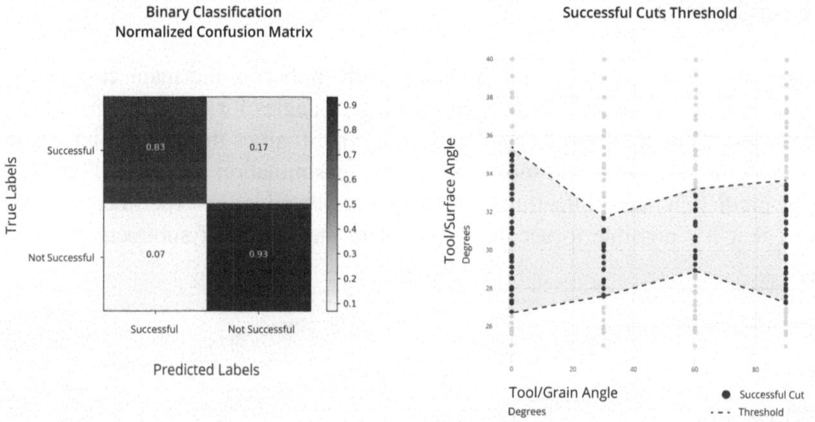

Fig. 6. The event threshold for "successful" cuts is predicted combining the previous event labels, defining the range of optimal cuts. A normalized confusion matrix shows the prediction rate for the "Successful" and "Not Successful" labels, with an accuracy of 83% and 93% respectively.

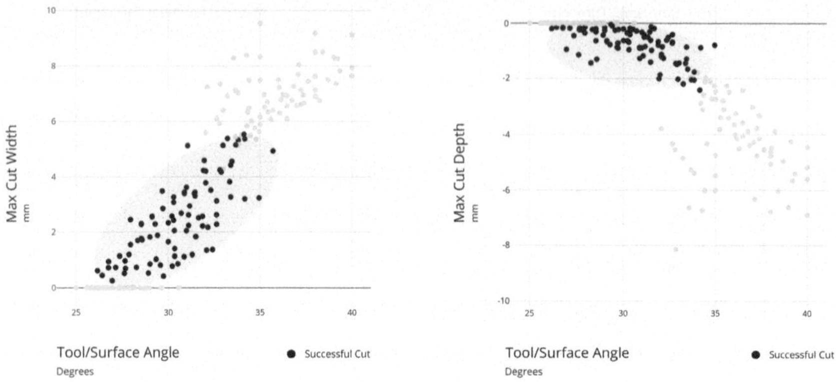

Fig. 7. The tool/surface angle value at the beginning of the cut affects significantly the successful result of the operation. Such parameter could be optimized toward a specific material outcome, such as maximizing the depth and width of the cut to increase the efficiency of the subtractive process.

5 Conclusions

Focusing on robotic subtractive fabrication with timber as the main case study, the paper presents the potential of machine-learning strategies for design to manufacturing applications, as a way to explore novel design opportunities through the integration of actual material properties and tool affordances for simulation purposes (Figs. 8, 9).

The prediction rates of the trained networks presented in the experiments discussion suggest that it's possible to accurately simulate the results of subtractive operations

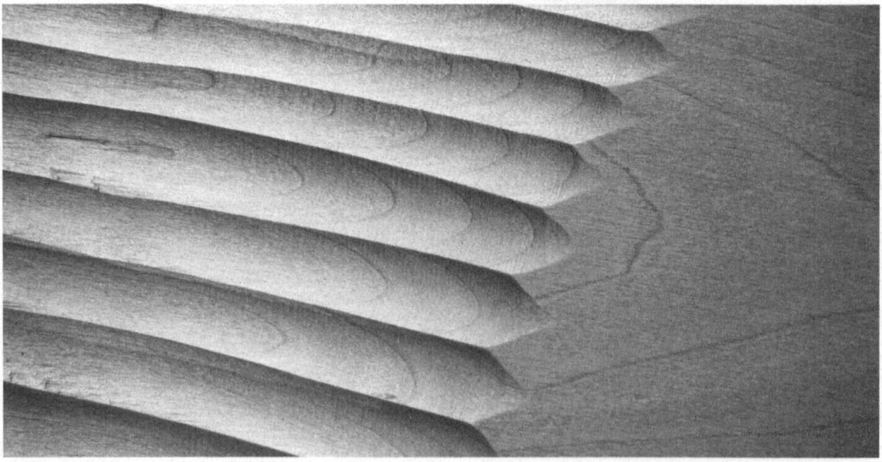

Fig. 8. The training process allows to tune the fabrication system to a specific set of tools and type of wood to explore design opportunities that would become evident only in the fabrication stage.

Fig. 9. Combining together prediction of cuts with different fabrication parameters, it's possible to simulate the final carved geometry in several pattern configurations.

based on a given set of fabrication parameters and use the encapsulation of such instrumental knowledge to translate back and forth between robotic toolpath data and geometry prediction.

The experiments focused on two different workflows for the generation of fabrication datasets and curation of the training process, examining whether is better for the robotic system to learn from a skilled human expert or autonomous training sessions. The lower prediction error of the network based on the combination of human and robotic training should be attributed to the steering action of the human expert, operating within a narrowed down parameters range, which excluded inefficient or dangerous cuts, and leading the robotic training sessions to gather data only within an optimal fabrication range. The more systematic and wider range of the second experiment doesn't allow to have the same resolution in the more relevant areas of "successful" cuts, resulting in lower prediction performances.

To conclude, the manipulation of knowledge across distinctly operating domains such as human making and industrial robotic manufacturing presented the opportunity to develop an approach for human-machine interaction which questions current industrial method of knowledge transfer and will be further explored in future steps of the research.

Acknowledgments. The project is part of ongoing Ph.D. research conducted by Giulio Brugnaro, supervised by Prof. Bob Sheil and Dr. Sean Hanna, at the Bartlett School of Architecture, University College of London, within the framework of the "InnoChain Training Network," supported by the European Union's Horizon 2020 research and innovation programme under the Marie Sklodowska-Curie grant agreement No. 642877.

References

Al-Zubaidi, S., Ghani, J.A., Haron, C.H.C.: Application of ANN in milling process: a review. Model. Simul. Eng. **2011**, 9 (2011)

Brugnaro, G., Hanna, S.: Adaptive robotic training methods for subtractive manufacturing. In: ACADIA 2017: Disciplines and Disruption, Proceedings of the 37th Annual Conference of the Association for Computer Aided Design in Architecture (ACADIA), Cambridge (MA), pp. 164–169 (2017)

Callicott, N.: The tacit component and the numerical model: representation in computer-aided manufacture and architecture. J. Archit. **8**(2), 191–202 (2003)

Carpo, M.: The Alphabet and the Algorithm. MIT Press, Cambridge (2011)

DeLanda, M.: Material Complexity. In: Leach, N., Turnbull, D., Williams, C. (eds.) Digital Tectonics, pp. 14–21. Routledge, Chichester (2004)

Noble, D.F.: Forces of Production. Knopf, New York (1984)

Pfleging, J., Stucheli, M., Iovita, R., Buchli, J.: Dynamic monitoring reveals motor task characteristics in prehistoric technical gestures. PLoS ONE **10**(8), 1–20 (2015)

Polanyi, M.: The Tacit Dimension. Anchor Books, Garden City (1967)

Steinhagen, G., Braumann, J., Brüninghaus, J., Neuhaus, M., Brell-Çokcan, S., Kuhlenkötter, B.: Path planning for robotic artistic stone surface production. In: Reinhardt, D., Saunders, R., Burry, J. (eds.) Robotic Fabrication Architecture, Art and Design 2016, pp. 122–135. Springer International Publishing Switzerland (2016)

Witt, A.J.: A machine epistemology in architecture. Encapsulated knowledge and the instrumentation of design. Candide J. Archit. Knowl. **3**(3), 37–88 (2010)

Zwierzycki, M., Nicholas, P., Thomsen, M.R.: Localised and learnt applications of machine learning for robotic incremental sheet forming. In: De Rycke, K., Gengnagel, C., Baverel, O., Burry, J., Mueller, C., Nguyen, M.M., Rahm, P., Thomsen, M.R. (eds.) Humanizing Digital Reality, Design Modelling Symposium Paris 2017, pp. 373–382. Springer, Singapore (2017). https://doi.org/10.1007/978-981-10-6611-5_32

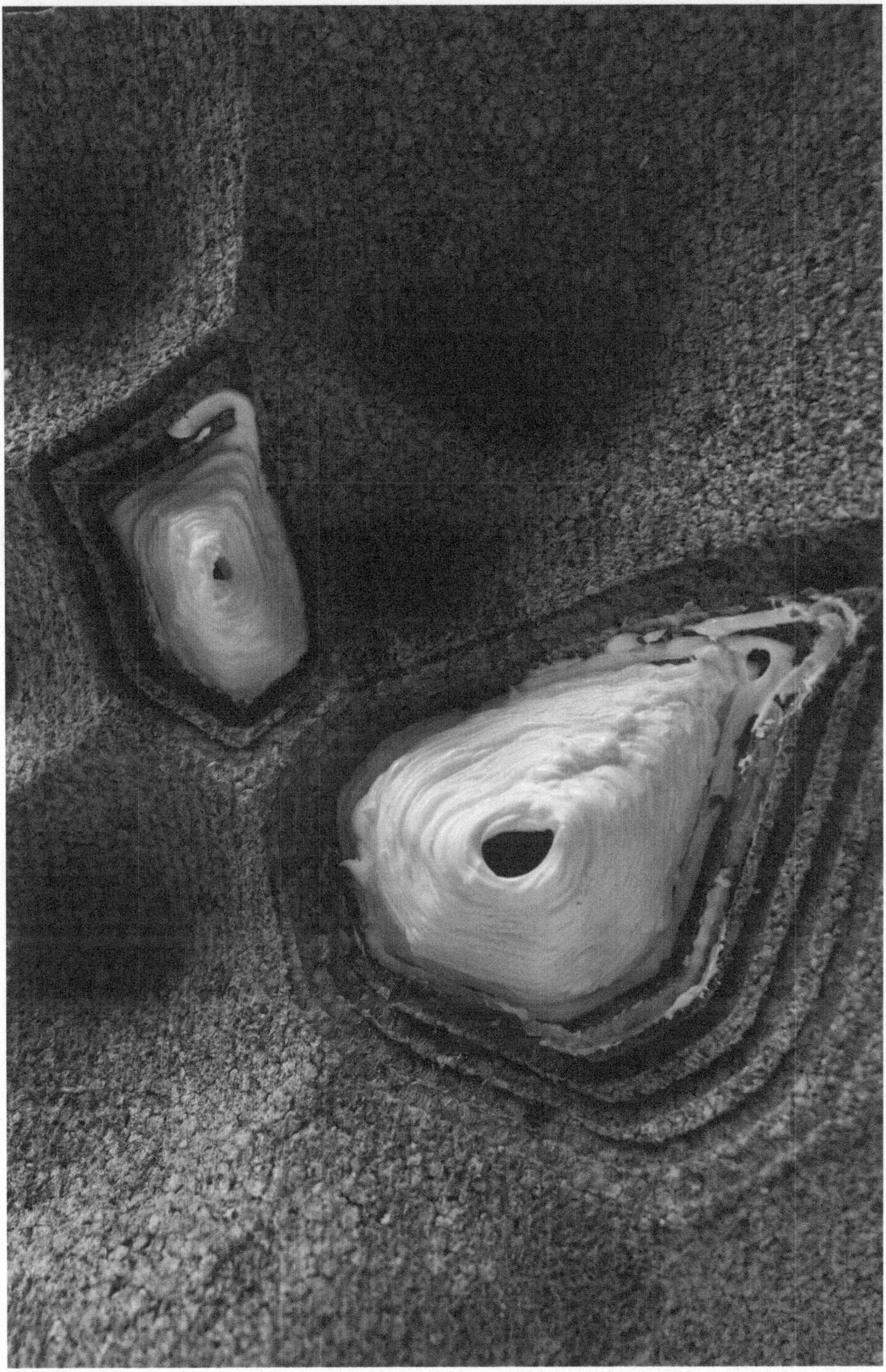

Multimode Robotic Materialization

Design to Robotic Fabrication Method of Integrating Subtractively Produced Hard Components and Additively Deposited Soft Silicone

Sina Mostafavi[1,2(✉)], Benjamin N. Kemper[1], and Daniel L. Fischer[1]

[1] Delft University of Technology, Delft, Netherlands
s.mostafavi@tudelft.nl
[2] Dessau Institute of Architecture, Dessau, Germany

Abstract. This paper explores and discusses the implementation of multimode, subtractive-additive, robotic production methods. The research addresses the opportunities and challenges in the integration of subtractive and additive robotic manufacturing routines for architectural scale applications. It specifically presents a customized computational design to robotic fabrication workflow of a material system composed of subtractively produced expanded polystyrene and additively manufactured silicone. The integration of these two sets of production methods allows for a symbiosis of hard and soft materials. Through initial case studies and a pilot subtractive-additive prototype, this paper discusses how multimode robotic production methods, with the goal of purposefully expanding the potential of isolated fabrication techniques, results in multiscale and efficient design materialization. Furthermore, this paper demonstrates the design, computation, materialization, and prototyping process of a chair project with multiple materials in more detail.

Keywords: Multimode robotic production · Subtractive manufacturing Additive manufacturing · Multi-materiality · Hybridity · Silicone 3D printing

1 Introduction

The flexibility and programmability of industrial robots allow for the implementation of multiple manufacturing strategies in one or several production setups. This integration has fundamentally changed fabrication procedures and is growing in industries such as automotive and electronics (Gemma and Litzenberger 2016; Michalos et al. 2010). In the building sector and the architectural practice, research and education, the applications of different fabrication techniques has been extended in recent decades (Oosterhuis and Bier 2013; Gramazio and Kohler 2014). Robotic subtractive manufacturing routines extend the capabilities of two or three-axis computerized numerical control (CNC) tooling, allowing for more degrees of freedom in volumetric cutting, milling, and production of complex geometries (McGee et al. 2012). Advancement and the proliferation of additive manufacturing enable the realization of efficient building systems potentially with higher three-dimensional resolution (Khoshnevis 2004;

J. Willmann et al. (Eds.): ROBARCH 2018, *Robotic Fabrication in Architecture, Art and Design 2018*, pp. 349–362, 2019.
https://doi.org/10.1007/978-3-319-92294-2_27

Dini et al. 2006; Lim et al. 2012). This research explores the integration of subtractive and additive processes as two major sets of production techniques. Moreover, a developed robotic additive manufacturing method for printing silicone combined with robotic hot wire cutting and milling is presented.

Incorporating different methods of production facilitates the utilization of multiple materials. This integration introduces material hybridity that allows for going beyond available building material palettes. In material science, hybrids are defined as engineered materials, mostly at micro scale, that extend the different material properties of two or more separate substances and fill the empty holes of property-space (Ashby 2011). Correspondingly, in architecture and product design, integration of multiple materials at various scales is fundamental to achieve a functional hybrid outcome (Mostafavi and Anton 2018). In this context, this research discusses the development and utilization of presented multimode robotic fabrication methods.

Each manufacturing process delivers a certain range of tooling freedom, resolution, tolerance and workable production space. Embedding the fabrication constraints and simulation of multiple processes establish different types of feedback to a digital design to physical production workflow. A set of selected projects as background studies and a prototype design elaborate and conclude on the correlations between digital design, material properties, and production spaces with a focus on multimode, subtractive-additive, and robotic production methods.

2 Multimode Robotic Production Methods

Being able to design and customize different types of end-effectors to be integrated into a robotic production setup introduces gradients of varying material handling and processing approaches for building applications. This flexibility makes it difficult to provide a universal taxonomy of different robotic fabrication methods. However, surveying the background and state-of-the-art subtractive, additive and formative methods are the three main categories. With a focus on subtractive and additive approaches, this research provides a set of background examples of multimode robotic production and a concluding design-to-prototyping process of a chair. In each of these experiments, the goal is to emphasize how the process of design materialization is influenced by the established feedback loops of robotic fabrication, and how both subtractive and additive methods combined are approached or customized differently for more effective production. Key topics discussed are roughing, material hybridity, volumetric design, and development of an additive silicone printing system to be integrated with a robotic subtractive manufacturing procedure.

2.1 Subtractive-Subtractive

Roughing or rapid material removal in subtractive manufacturing is required to achieve a more efficient and faster production process. In cases that the geometric logic of the designed surface or volume follows the topological characteristics of the tool, i.e., developable for wire cutting, a finishing pass is not required. However, whenever it is not possible or intended to embed this constraint, an alternative combination and application of subtractive processes can be considered. To test this, in the design and mock-up prototyping process of a case study (Fig. 1), robotic hot wire cutting and milling are integrated in such a way that part of the designed surface is developable and part of it is approximated through milling. Milling is applied to fine-tune and produce a smooth curved shell. In this example, robotic milling shapes the specific edge areas to match the neighboring components. This multimode robotic production results in a more efficient process of subtractive production by rapid volumetric material removal, followed by high-level detailing processed through milling where needed. The robotic strategy enables a high curvature geometry to be produced with an optimum manufacturing effort.

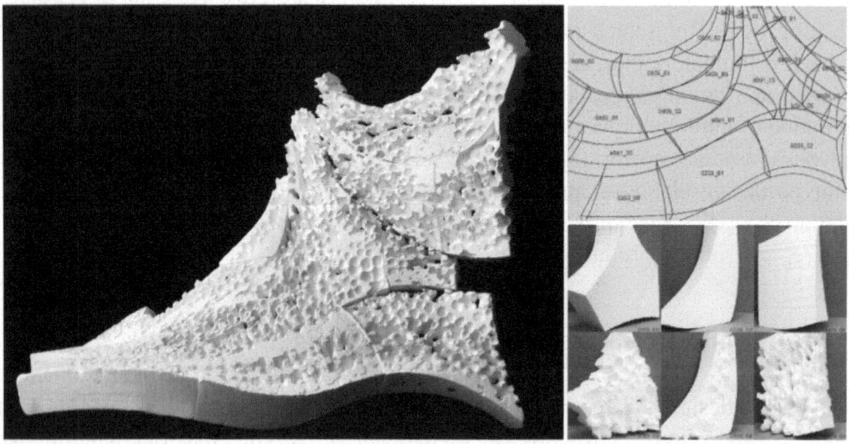

Fig. 1. Roughing and multimode, hot wire cutting-milling robotic manufacturing, the assembled prototype with surface areas untouched and areas finished with milling (left); components with a saddle point in the middle (top-right); produced components in sequence (bottom, right)

Hybridity, or multi-materiality, is obtainable through multimode methods of robotic production. An application of a soft and hard hybrid system is studied in a project that combines two different approaches to mill two materials: EPS foam and rigid cork (Fig. 2). In this project, through the incorporation of stretched cork on EPS, the goal is to enhance individual physical properties. Even if both chosen materials share similar attributes, such as rigidity, granulation, and density, the result of robotic production changes the physical behaviors in favor of the expected design performances. Carving the planar rigid cork board from multiple sides, results in a double curvature element with flexibility that is structurally supported by the hard polystyrene components. This symbiosis of soft and hard materials is further explored and discussed in the prototype chair presented that employs both subtractive and additive robotic fabrication.

A **volumetric** approach to design and production is implementable through subtractive methods of manufacturing such as hot wire cutting. However, this frequently results in treating the exterior side of the surface and stacking of non-porous solid

Fig. 2. Hybridity achieved through robotic milling of foam combined by robotic carving out of cork with varying angles-depth (top left); cork gains intended flexibility (bottom left & right)

components. Combined with milling, this leads to roughing followed by higher resolution finishing. As a prototypical example in a multimode production project (Fig. 3) the approach to the sequence and combination of two subtractive methods is reexamined. Internal cavities are introduced by removing material from inside the foam block with hot wire cutting. These cavities are then connected to holes on the volume on both sides of the components that have developable conical shapes to be removed by milling. Consequently, there is no roughing and material crushing in milling.

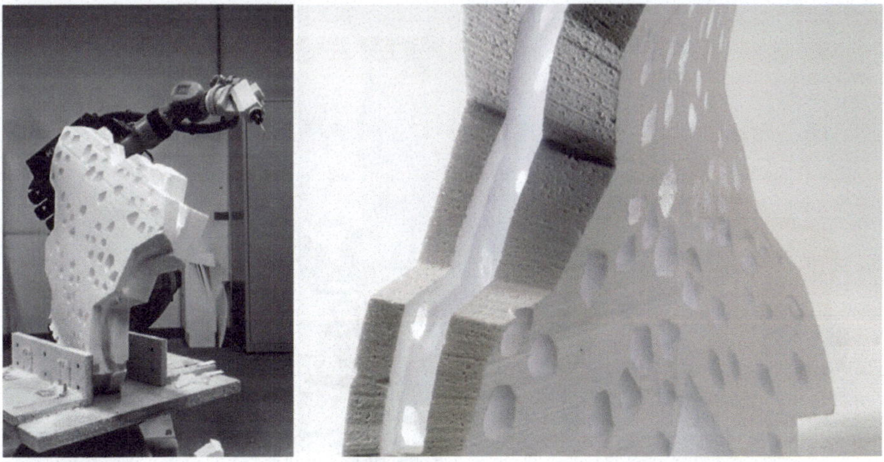

Fig. 3. Creating internal cavities using multimode hot wire cutting robotic milling (right); component produced with seven axis production setup with multiple end-effectors (left)

2.2 Subtractive-Additive

The physical state of the material in most of the additive production processes such as Fused Deposition Methods, Selective Laser Sintering, Stereolithography, or even casting, changes from one state to another. The phase change makes it difficult to simultaneously or progressively combine two processes of subtractive and additive manufacturing. In the field of architectural robotics, there are some research and design experiments that investigate and apply both types of methods. Relevant to this paper are: “wiggle wall” in which fast printing of foam is followed by robotic milling (Pigram and McGee 2011); and “Compound Fabrication” in which a subtractive routine follows an additive method for fine-tuning (Keating and Oxman 2013). Other cases are “Woven Clay” in which a previously milled foam mold is temporarily used as an undulating printing bed (Friedman et al. 2014) and robotic ceramic printing on subtractively produced components with a hotwire cutter (Mostafavi and Bier 2016). The following presented research investigates robotic printing of silicone as an adhesive material that can be printed on different surfaces and does not need heat to solidify. Moreover, the elasticity of various shapes and thicknesses is studied and evaluated for different potential design applications. With these objectives, an extruder for printing silicone is designed (Fig. 4). Exploiting the movement capacity of a six-axis arm, the extruder with two changeable material containers, i.e., translucent and opaque silicone is located on top of axis three. Therefore the specific design of the extruder allows for a short connection to the nozzle, directly on the tip of axis six. Between the calibrated tip of the tool and the flange, a ball bearing is integrated that allows for rotation of the slender funnel. This is why the connecting pipes from the cartridges to the nozzle faces upwards during the movement. On the one hand, this short connection enables higher ranges of three-dimensional movement of the nozzle on complex surfaces and on the other hand, a lower pressure is required to push or stop the extrusion.

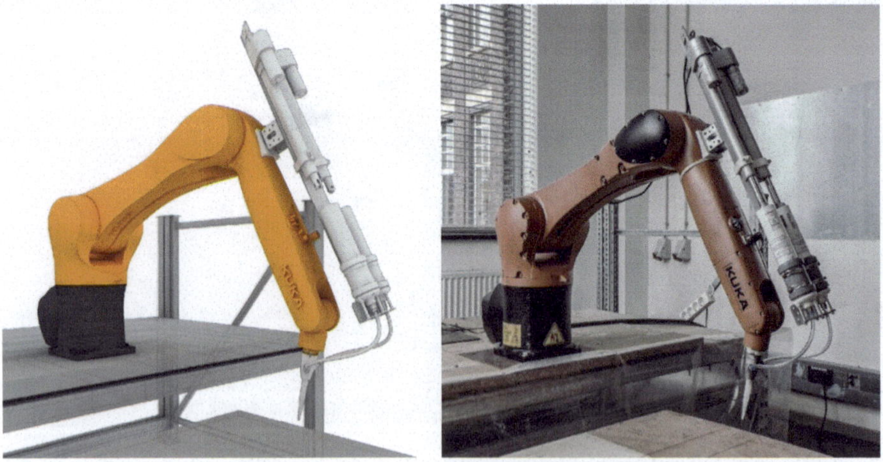

Fig. 4. Parametric simulation with in KUKA|prc with integrated digital model of the extruder (left); Robotic additive manufacturing setup for silicone printing with KR10-1100 (right)

The research evolves along a series of experiments on silicone behavior to understand the additive production of a semi-flexible material, as well as the properties and performance of the prototypes (Fig. 5). This results in two main categories of cellular and linear silicone robotic toolpaths and ranges in between. As the goal is to incorporate two production methods, sets of additive experiments are tested on free-form shapes. The outcomes are first, the understanding of the constraints and correlation between material capacities and second, the movement range of the arm and printing angles with no support structure. Categories of experiments in summary are:

Fig. 5. Robotic silicone printing experiments, linear continuous printing (top); cellular printing on a free from fabric (middle); prototype testing height, cantilevering and size ranges (bottom).

1. Linear printing on double curved fabric with continuous toolpath using an external fix extruder placed next to the manipulator (Fig. 5, top); This resulted in finding the values for the correlation of the robot speed, the power supply voltage for the external extruder, and an indication for the layer distance and offset. Further, a clarification of the silicone properties, such as pot life, hardening duration, viscosity and the first estimation of 3D printing capacities is achieved.
2. Flat cellular printing on double curved fabric results in extruder modification and further material properties modifications (Fig. 5, middle); The main target is to solve the problem of the start and the stop of the material flow and achieve a constant material flow. This concluded in reducing the distance between the

external extruder and the printing fabric, which is resulted in a shorter tubing system. The solution is a custom build extruding system mounted on top of the robot to provide the shortest distance of tubes as possible.
3. Medium size cellular printing on flat fabric with five types of toolpaths with the mounted extruder on the manipulator; printing of angled and cantilevered cells with internal reinforcement; diameter and height ranges of the printed cells are: 25 mm < D < 80 mm & 11 mm < H < 45 mm (Fig. 5, bottom); This iteration also is a feedback for estimation of the maximum angle for cantilever printing, heights and wall thicknesses.
4. Large size cellular printing on flat fabric with the mounted extruder to test, wall thickness and cell size; diameter and height ranges of the printed cells are: 35 mm < D < 105 mm & 35 mm < H < 125 mm (Fig. 5, bottom); This experiment resulted in sufficient printing quality of medium to large cell shapes and a verification of the previous tested specific printing values.

Through these experiments the overall printing quality is improved. The nozzle has a diameter of 3 mm, which results in a print layer height of 2.4 mm and a range of one to five layers for wall thickness. The maximum printing angle can exceed 45 degrees. The printing angle is in correlation to the following factors: wall thickness, stickiness of the silicone type, overall topology, and the mass of material to be printed on top. Therefore, an exact value is always specific for a certain shape. As seen in iteration three, the material and printing technique has the potential to print cantilevered parts. Silicone reacts with air after extruding, within 15 min the shape begins to harden. Furthermore, fully hardened silicone can be welded together with fresh silicone. These attributes allow for taller prints with maximum cantilevering angles.

3 Design to Fabrication Workflow and Prototyping

The workflow is extended, explored, and demonstrated in the design-to-fabrication process of a prototype chair (Fig. 6). This process considers the key topics discussed with regards to subtractive production in combination with the opportunities of silicone printing. The integrated workflow establishes interconnected feedback loops between digital modeling, design computation, material properties, and a multimode robotic production method. The project proposes a hybrid system composed of high-density polystyrene as a hard and silicone as a soft material. The macro scale geometry of the chair is designed in foam with developable surfaces. In micro scale, a distribution pattern in relation to the human body is applied while considering robotic milling, which is linked to the additive manufacturing process of the soft material.

3.1 Digital to Physical Materialization Procedure

The macro and micro scales form-finding to design materialization methodology of the prototype chair includes three main feedbacks from the multimode robotic fabrication (Fig. 6). As these three processes are considered and simulated in one seven-axis setup with a rotary table, it is essential to iteratively evaluate the constructability of the design

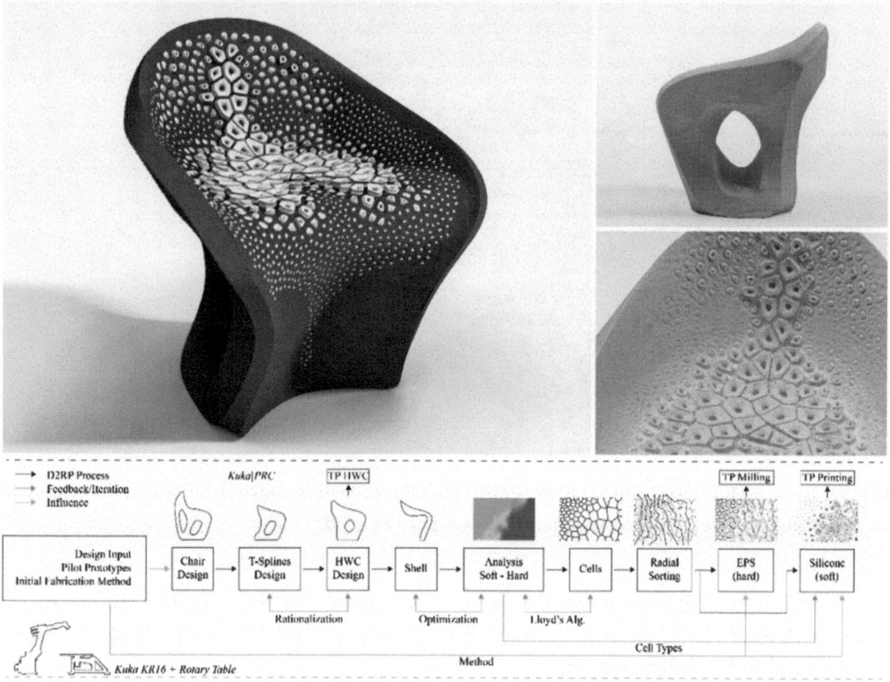

Fig. 6. Prototype chair visualization (left); printed scale one-to-five model in PLA (right); design-fabrication flowchart with rationalization and optimization feedback loops (bottom)

by examining the overlaps between the optimum production-space of each method. The distribution of the silicone cells with differentiated sizes and typologies is implemented according to contact with the human body as it pertains to the seating and leaning areas on the front side of the chair. In similar scales, multi-materiality is explored and tested in chair design projects. Amongst them are the Gemini chaise (Oxman et al. 2014) with a focus on acoustical performance and the multicolored multi-material ZHA chair (Bhooshan et al. 2017) with an emphasis on structural efficiency gained through multi-material printing in a layer-by-layer printing fashion with high resolution. In the prototype chair presented in this paper, flexible material with a feasible resolution for silicone printing is considered to be robotically deposited directly on the subtractively produced volume with three-dimensional surface tectonics, produced with robotic milling.

3.2 Robotic Production Simulation and Toolpath Strategy

The macro-scale modeling is mainly done with Autodesk T-Splines in Rhinoceros 3D. The output of this modeling process is one digital model which is then rationalized to four continuously developable surfaces that approximate the design (Fig. 7). The result of this approximation is then translated to an initial parametric model which is linked to the robotic production simulation that allows for minor parametric customization of the

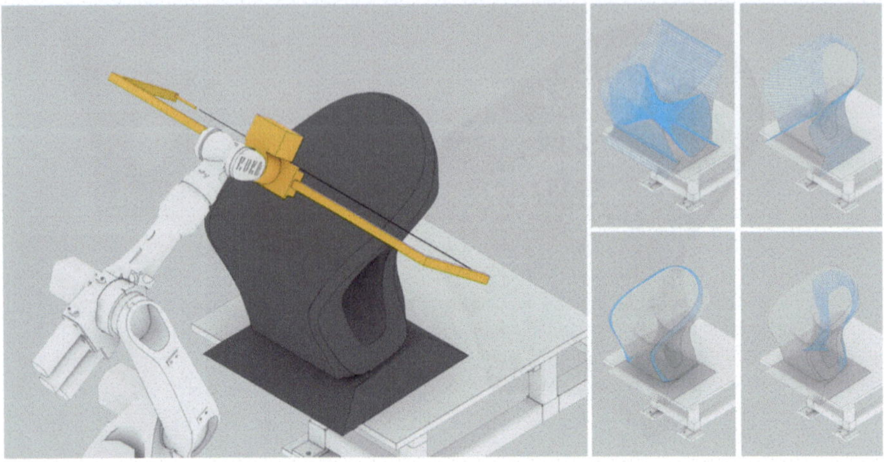

Fig. 7. Robotic hot wire cutting of overall form, with only four cuts out of which one side will be milled for more elaborated required details where needed.

design in macro scale. To make the volume weight of the chair lighter, similar to the volumetric approach described in Sect. 2.1, an internal hole with an adjustable three-dimensional twist is introduced.

The next mode of production is robotic milling on only the front face of the prototype chair, which has a concave curvature (Fig. 8). While the macro design shape is produced by hot wire cutting, milling is used to further define the seating area. Roughing is necessary and only applied in this area to speed up the process. The robotic milling toolpath follows the cellular logic of the front surface that varies in size

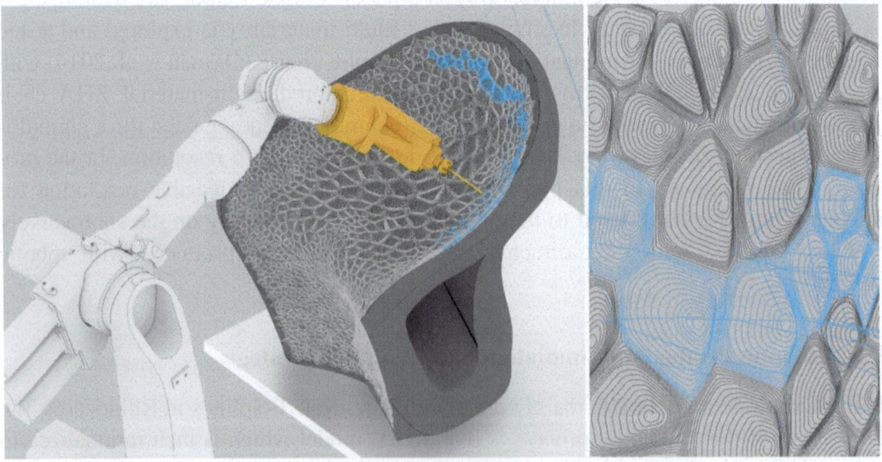

Fig. 8. Robotic milling on the concave part of the prototype chair (left); incremental radial material removal strategy to fabricate the cellular pattern (right)

and depth according to the distribution of the soft silicone cells. To stay as perpendicular as possible to the surface during the milling process, the toolpath is parametrically generated based on the original cellular logic of the geometry. Each cell has a local entering and exiting safe point above the surface to be followed by radial incremental material removal. In this process, instead of a conventional line-by-line removal, the milling follows the cellular and thus a radial logic.

Two subtractive processes are followed by an additive method. Silicone cells with varied sizes, depth and typologies are distributed on top of the three dimensional concave front surface of the chair (Fig. 9). The printability of the cells is decided based on a series of experiments on the fabric presented in 2.2. The toolpath generation follows a similar cellular logic applied in robotic milling in the previous step. In this process, the continuity of the printing path is essential. Continuity in this stage of production means that after finishing the printing of one cell, the toolpath always continues to print a neighboring cell and avoids hovering above the surface until all cells are produced.

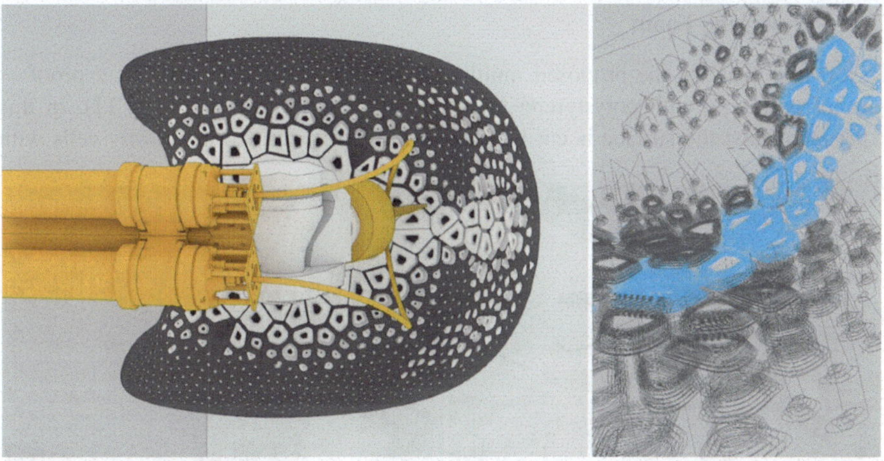

Fig. 9. Robotic 3D printing of silicon on subtractively produced front concave surface of the prototype chair (left); continuous printing toolpath (right)

The production time of high resolution milling and printing is reduced by an optimized robotic milling toolpath. Due to the differentiation in the amount of neighboring cells and the gradient in size, a one-directional sorting technique is not applicable. The nature of cell distribution on the prototype chair demands for a tailored sorting approach that results in a continuous cell sorting with short travel time. Therefore, the outer edge of the chair shell is considered as reference for a radial sorting from outside to inside (Fig. 10).

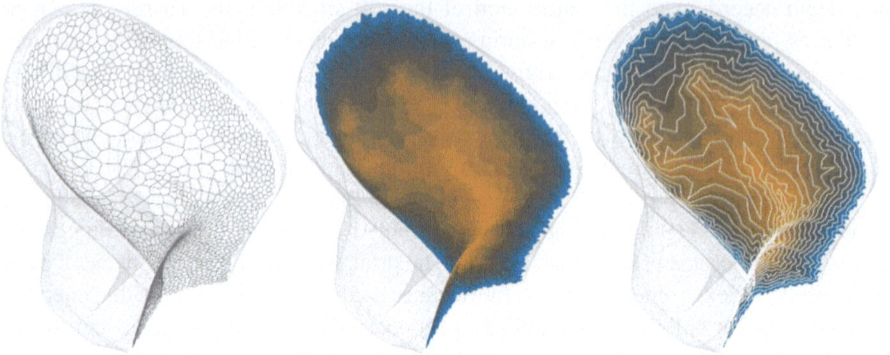

Fig. 10. Resulting cell distribution after human body analysis (left); toolpath optimization following a radial logic (middle); continuous toolpath travels through all cells without hovering above the surface (right)

3.3 Prototyping Test of Multi-materiality: Soft Silicone and Hard Polystyrene Foam

To test and evaluate the proposed multimode robotic production method, a proof of concept prototype with polystyrene foam and silicone is produced (Fig. 11). In this experiment, first the surface is cut by hot wire, then it is milled, and lastly cells with

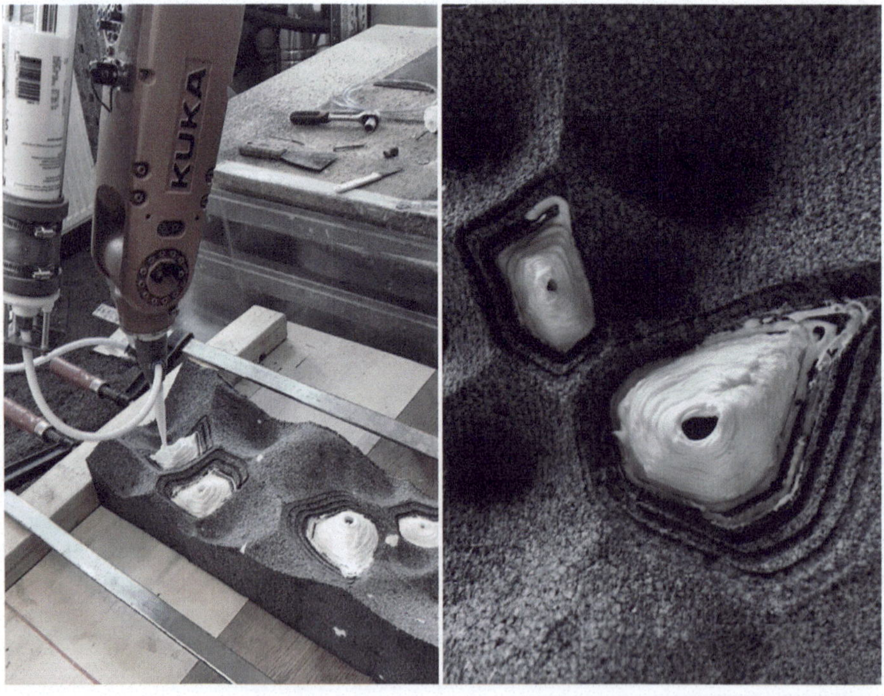

Fig. 11. Additive material deposition on a subtractively produced component (left); zoom of the printed soft white silicone cells on hard black polystyrene (right)

similar geometry that vary in sizes and depth are printed on top of the undulating surface. The dried outcome of the printed cells demonstrates the desired elastic behavior while it firmly stays glued to the foam. While the foam has structural properties, the elastic cells ensure comfortable seating. Through the combination of these two soft and hard materials, different functions and material properties are merged into a unified hybrid material system.

4 Conclusion and Future Works

The methods and experiments presented in this paper discuss multimode subtractive-additive robotic fabrication. The research elaborates how multimode methods facilities materialization of multi-materiality in production. The resulting material hybridity extends the opportunities of working with a combination of materials with varying properties such as elasticity and rigidity. This is demonstrated in the chair design to production project, that combines two materials to achieve better functionality. From a design methodology point of view, this requires an integrated workflow that establishes feedback loops between digital modeling, design computation, material, and production space. This iterative approach results in an understanding of the material behaviors and production space.

As the next stage of this research, the goals are to advance the extruding system and include 3D scanning and sensing, as well as tool changing for more efficient building processes into the overall workflow. The extruding system used demands one continuous toolpath. An improved version with full control of the material flow would increase the efficiency and freedom of material positioning in different sequences. Further studies are implanted by scanning of a milled or an existing surface, with the robotic arm, which can be integrated into multimode robotic production processes. More specifically the presented method of silicone printing demonstrates new design materialization opportunities to be implemented into design and architectural applications.

Acknowledgments. We acknowledge the essential contributions of all current and former members of Robotic Building team and Henriette Bier, Dessau Institute of Architecture as well as Hyperbody group at TU Delft. We would also like to thank MSc students at BK City of TU Delft and M. Arch students in DIA at HS Anhalt, who have contributed to the design, research and production of some of the background works discussed in this paper. Further, we thank Evelyn Fischer and Sean Eddings for their kind proofreading of the text. We would also acknowledge the support of ABB Benelux and KUKA Benelux in some of the presented projects.

References

Ashby, M.: Hybrid materials to expand the boundaries of material-property space. J. Am. Ceram. Soc. **94**(s1), s3–14 (2011)

Bhooshan, V., Fuchs, M., Bhooshan, S.: 3D-printing, topology optimization and statistical learning. In: Turrin, M., Peters, B., O'Brien, W., Stouffs, R., Dogan, T. (eds.) 2017 Proceedings of the Symposium on Simulation for Architecture and Urban Design, SimAUD 2017, Society for Modeling & Simulation International (SCS), pp. 107–114. Toronto (2017)

Dini, E., Nannini, R., Chiarugi, M.: Method and device for building automatically conglomerate structures (2006). Accessed 16 Mar 2006

Friedman, J., Kim, H., Mesa, O.: Experiments in additive clay depositions. In: McGee, W., Ponce de Leon, M. (eds.) Robotic Fabrication in Architecture, Art and Design 2014, pp. 261–272. Springer International Publishing, Switzerland (2014)

Gemma, J., Litzenberger, G.: International Federation of Robotics (IFR) Secretariat World Robotics Report 2016: European Union Occupies Top Position in the Global Automation Race Frankfurt am Main (2016)

Gramazio, F., Kohler, M. (eds.): Made by robots: challenging architecture at the large scale. Archit. Des. **84**(3), 136 (2014)

Keating, S., Oxman, N.: Compound fabrication: a multi-functional robotic platform for digital design and fabrication. Rob. Comput. Integr. Manuf. **29**(6), 439–448 (2013)

Khoshnevis, B.: Automated construction by contour crafting—related robotics and information technologies. Autom. Constr. **13**(1), 5–19 (2004)

Lim, S., Buswell, R.A., Le, T.T., Austin, S.A., Gibb, A.G.F., Thorpe, T.: Developments in construction-scale additive manufacturing processes. Autom. Constr. **21**(January), 262–268 (2012)

Pigram, D., McGee, W.: Formation embedded design. In: Taron, J.M. (ed.): Acadia 2011: Integration through Computation, Proceedings of the 31st Annual Conference of the Association for Computer Aided Design in Architecture, pp. 122–131. Calgary/Banff (2011)

McGee, W., Feringa, J., Søndergaard, A.: Processes for an architecture of volume. In: Brell-Çokcan, S., Braumann, J. (eds.) ROB|ARCH 2012: Robotic Fabrication in Architecture, Art and Design, pp. 62–71. Springer, Vienna (2013).

Michalos, G., Makris, S., Papakostas, N., Mourtzis, D., Chryssolouris, G.: Automotive assembly technologies review: challenges and outlook for a flexible and adaptive approach. CIRP J. Manuf. Sci. Technol. **2**(2), 81–91 (2010)

Mostafavi, S., Anton, A.: Materially informed robotic fabrication, architectural robotics and multiscalar material architecture. In: Daas, M., Wit, A.J. (eds.) Towards a Robotic Architecture, pp. 88–99. ORO Editions, Novato, CA (2018)

Mostafavi, S., Bier, H.: Materially informed design to robotic production: a robotic 3D printing system for informed material deposition. In: Robotic Fabrication in Architecture, Art and Design 2016, 338–349. Springer International Publishing, Switzerland (2016)

Oosterhuis, K., Bier, H. (eds.): IA #5: Robotics in Architecture. Jap Sam Books, Heijningen

Oxman, N., Dikovsky, D., Belocon, B., Carter, W.C.: Gemini: engaging experiential and feature scales through multimaterial digital design and hybrid additive–subtractive fabrication. 3D Printing Add. Manuf. **1**(3), 108–114 (2014)

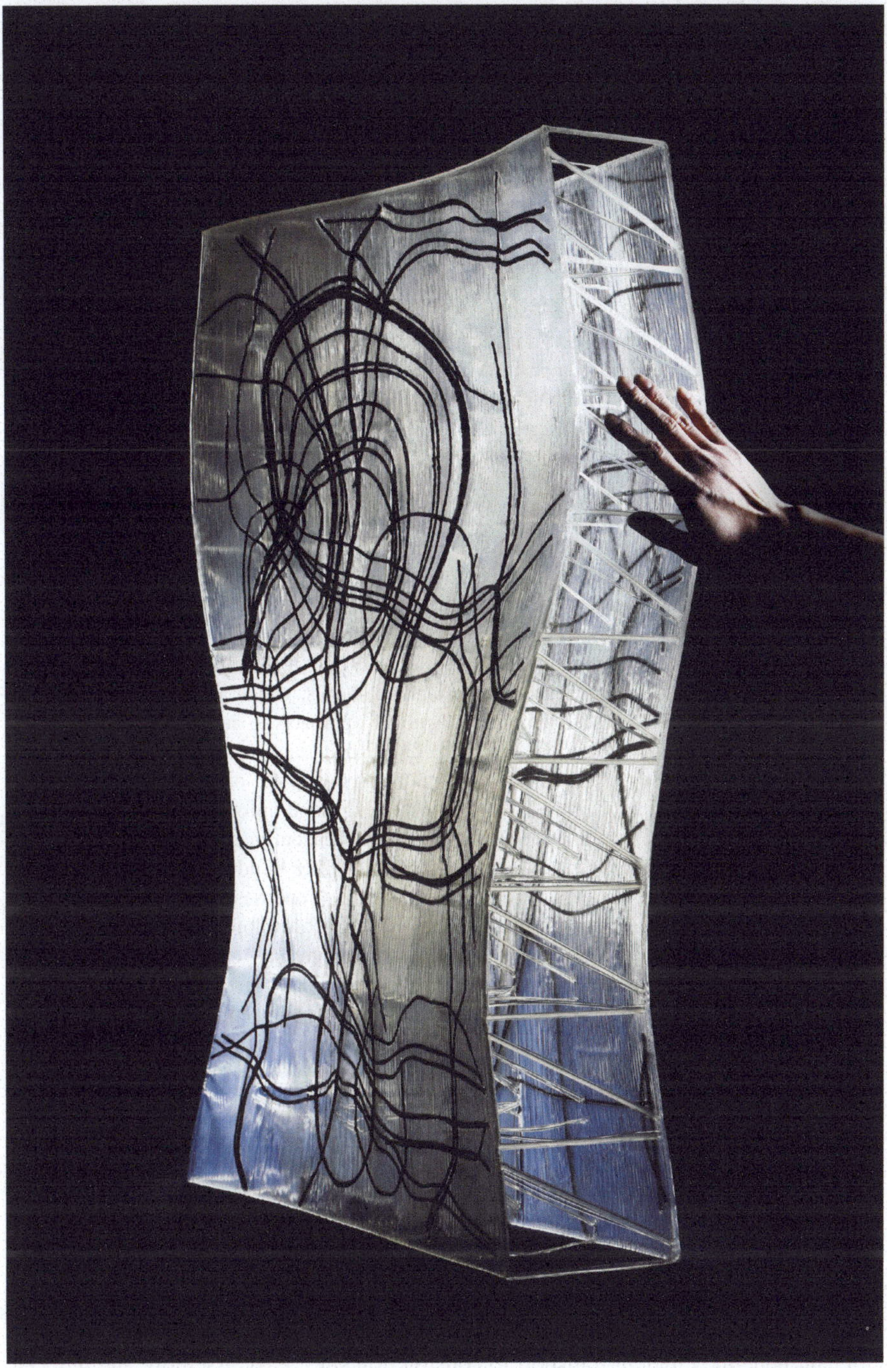

Digital Composites: Robotic 3D Printing of Continuous Carbon Fiber-Reinforced Plastics for Functionally-Graded Building Components

Hyunchul Kwon[1(✉)], Martin Eichenhofer[2], Thodoris Kyttas[1], and Benjamin Dillenburger[1]

[1] Chair of Digital Building Technologies, ETH Zurich, Zurich, Switzerland
{kwon, kyttast, dillenburger}@arch.ethz.ch
[2] Laboratory of Composite Materials and Adaptive Structures, ETH Zurich, Zurich, Switzerland
martieic@ethz.ch

Abstract. Carbon fiber-reinforced thermoplastic (CFRP) can impact the built environment significantly. In addition to its extraordinary weight-to-strength ratio, CFRP provides numerous benefits in architecture, including its plasticity and aesthetics. In theory, CFRP can take any shape, but the fabrication constraints of mold making currently limit the design. The mold is the most resource-intensive element of CFRP fabrication, and it is particularly challenging and costly for the bespoke non-standard shapes, which are often found in architecture. This research presents a novel approach to utilizing robotic three-dimensional printing (3DP) to make CFRP materials accessible in architecture by combining layered 3DP of neat polymers with add-on 3DP of CFRP materials. This paper begins by providing background information and an overview of state-of-the-art CFRP fabrication. It then outlines the challenges related to the fabrication process and design of 3DP CFRP building components. Taking into account the design parameters, material properties and fabrication constraints, the paper describes the fabrication process development, computational design tool development and demonstrative physical experiment. The presented research is the first part of an interdisciplinary research involving architecture and mechanical engineering in order to investigate the potential for large-scale mold-less fabrication of CFRP components.

Keywords: Carbon fiber-reinforced plastics
Computerized materials and structures · Robotic 3D printing

1 Introduction

1.1 Background

In architecture, using lightweight materials for prefabricated building components has significant benefits: efficient transport to construction sites, easy assembly and reduced embodied energy. Meanwhile, the smaller dead-loads are structurally more efficient,

J. Willmann et al. (Eds.): ROBARCH 2018, *Robotic Fabrication in Architecture, Art and Design 2018*, pp. 363–376, 2019.
https://doi.org/10.1007/978-3-319-92294-2_28

and the lightweight building components allow longer structural spans and gravity-defying cantilevering structures with minimal material [1].

The most efficient lightweight materials are composites, two or more materials combined to form a new, useful material. Composites have significantly different properties compared to their base materials, usually inheriting the best qualities of each [2]. Fiber-reinforced plastic (FRP) is a composite that boasts decades of successful use in demanding applications in the aerospace and automotive industries due to its excellent material properties: formability, lightness and strength [3]. Particularly, carbon fibers have a very high weight-to-strength ratio in comparison to other fibers, such as glass and aramid. Carbon fibers also are 50% lighter and three times stronger than steel [4].

Lynn eloquently describes the current trend of metals being gradually replaced by carbon fiber-reinforced plastic (CFRP) to produce more efficient lightweight structures:

> "The contemporary racing sailboat is an example of a building-scale object built with the plastic sensibility of multiple geometries. [...] Increasingly, functional, aesthetic, and structural components are integrated into fewer and fewer surfaces. There are very few pieces of metal which have a low strength-to-weight ratio in comparison to fibers on a racing sailboat. Nor is metal used much anymore in high-performance automobiles; Formula 1 race cars are now built almost entirely out of epoxy resin and carbon-fiber-reinforced plastics." [5]

This trend affects architecture as well, as can be seen in the Steve Jobs Theater unveiled in 2017. Designed by the British firm Foster + Partners, it showcases how the strength and lightweight of CFRP roof components diversify the spatial vocabulary of architecture. Supported only by a glass structure, the roof could not have been realized without CFRP [6].

In spite of its beneficial properties, CFRP remains rarely used in architecture, primarily because of its costly fabrication method involving a significant amount of manual labor [7]. Superior properties of CFRP can enable freeform lightweight architecture, yet remain weighed down by the need to create a mold. Mold fabrication becomes a cost driver, especially when component design requires unique, one-of-a-kind shapes [8]. Molds usually are made through subtractive fabrication processes with substantial amounts of waste materials and limited geometric freedom. This issue represents one of the largest challenges in CFRP manufacturing for architecture [9]. Therefore, it is necessary to cost-effectively fabricate various shapes of CFRP components for lightweight architecture by simplifying the mold-making process in order to enable the production of freeform building components.

This research proposes the use of additive manufacturing (AM) technologies as a cost-effective solution for CFRP fabrication. It aims to create lightweight-custom-freeform building components at a reduced cost and with less labor when manufacturing molds. The proposed method completely eliminates the need for expensive negative molds (mold-less fabrication), necessitating minimal, temporary and reusable scaffolding/support.

1.2 State-of-the-Art

This section presents a brief overview of recent developments in robotic fabrication and three-dimensional printing (3DP) for CFRP.

Robotic Fabrication of CFRP. There has been a great deal of recent progress in existing robotic fabrication technologies for CFRP materials, such as Automated Tape Placement and Laying (ATP and ATL). Simultaneously, new opportunities for large-scale CFRP components have arisen [10]. In the field of architecture, the Institute for Computational Design and Construction (ICD) at the University of Stuttgart has showcased the potential of fabricating large-scale components by aligning multiple composite materials (CFRP and glass FRP) [11].

Significant innovations in the integrated fabrication process have enabled the localized and successive deposition of materials, but ATP and ATL of CFRP materials still rely on customized steel frames. As a consequence, they can create only a limited variety of differentiated elements.

3DP of CFRP. AM, or 3DP, involves the process of materializing physical artifacts from digital data by adding material using a computerized numerical control (CNC). A 3D model of an object is sliced, usually with parallel horizontal planes. Each digital slice contains the path for the tool-head, which gradually stacks the material in layers. The family of AM processes includes a variety of different technologies [12]. Collectively, they promise the following:

- Almost no geometric limitations and no extra cost for complexity,
- No additional cost for customization without subtracted waste, and
- A wide variety of materials that can be printed (to work within a single process).

Of the different AM technologies, the research interest lies in fused deposition modelling (FDM) that can directly build up both polymer and fibrous materials.

FDM is a widely-available 3DP method in which molten plastic material is extruded and hardens instantly after deposition. FDM is able to utilize a wide variety of plastics, and its hardware setup is simple. Among the different plastic 3DP technologies, FDM is unique in its capability to produce in large-scale (>405 m^3) with reasonable speed (20–100L/h) and sufficient resolution (smallest feature 0.1 mm).

Nevertheless, FDM 3D printed artifacts are rarely used alone in architecture due to their limited tensile strength [13]. FDM has a fundamental limitation: the layer adhesion along the z-axis is weak [14]. The low-bending strength in this direction results from the weak bond between the layers, and architectural components require equivalent strength in all directions. As a possible solution to overcome the low-bonding strength, recent research conducted at the MIT Digital Structures has shown how multi-directional 3DP enables the addition of multiple materials along mutually perpendicular stress-lines [15].

Due to increasing demand for stronger 3D printed plastic elements, new approaches to using CFRP for FDM are emerging. The versatile FDM extrusion process allows for modifications in order to strengthen the printed parts.

In 2015, Skidmore, Owings & Merrill unveiled its design for the Oak Ridge National Laboratory. The project, Additive Manufacturing and Integrated Energy

(AMIE), showcases the capabilities of AM for building facades [16]. AMIE was printed with short CFRP, thereby enabling rapid production of sturdy and lightweight parts [17].

However, the full potential of the structural capacities of CFRP can be achieved only with continuous fibers. In 2014, MarkForged unveiled the world's first continuous CFRP FDM 3D printer, Mark One, for commercial use. However, the machine can only fabricate parts up to 0.007 m^3, which limits its practicality for architecture [18].

In 2016, the Composite Materials and Adaptive Structures laboratory (CMASLab) at ETH Zurich developed a continuous CFRP FDM method for the extrusion of high performance CFRP materials. In contrast to the MarkForged 3D printer, the method of CMASLab uses an industrial robotic arm to increase fabrication volume. This CFRP 3DP technology shows great potential due to the high performance materials used (fiber volume content > 50%) and high build rates (>1 kg/h) [19].

2 Free CFRP Fabrication

The large number of seemingly disjointed methods discussed in the previous section is relevant when it comes to the aim of this research: enabling lightweight, strong and freeform architecture. Of the methods discussed, AM is the most promising in achieving unconstrained fabrication freedom of CFRP.

With the advantages of 3DP in mind (customized formability and integrative multi-material fabrication process), this research aims to expand the field of CFRP application in architecture through a custom CFRP 3DP system enabling locally-differentiated material compositions. This system allows the accurate placement of the CFRP in a structurally-informed manner. Thus, the desired shape can be manufactured with minimal costs in terms of mold fabrication and labor. Furthermore, a fabrication process based on 3DP allows material gradations, i.e., different densities and varying carbon-fiber orientations. By placing the carbon fibers only where needed, the amount of required material can be reduced, allowing the efficient use of resources in architecture.

This unconstrained fabrication process for functionally-integrated CFRP building components opens up new possibilities for non-developable lightweight freeform buildings. From an architectural point of view, the proposed approach to CFRP fabrication can be applied in several cases, wherever a geometric- or material-based differentiation is needed. For example, it would be well-suited for applications requiring freeform-double-curved surface elements. In architecture, these shapes often are praised for their aesthetics and for their unique structural properties. Architects like Zaha Hadid and Frank Gehry integrated such shapes into their iconic styles [20]. The proposed system offers the possibility of creating freeform geometries with ease, allowing for the construction of optimized, thin panels. The opportunity to cost-effectively produce one-of-a-kind, customized, lightweight elements can significantly benefit a wide spectrum of building elements, from self-supporting external building skins (e.g., 3D façade elements, including balconies and balustrades) to partition walls. Moreover, even load-bearing elements – such as lightweight roofs for cantilevering and long-span structures – could be fabricated using this method. Indeed, the hybridization of all these elements could be possible.

3 Questions

In order to achieve the vision, the following questions are posed:

- Scalability. How can continuous CFRP 3DP be scaled up for mold-less fabrication of lightweight building components?
- Structural integrity. How can one overcome the low-bonding strength of FDM?
- Precision and material deformations. What are the preventive and reactive measures that can ensure reliable, high-quality printing?
- Graded components. To what extent is it possible to differentiate components based on local material composition and load-tailored design (e.g., stress-lines) to achieve a higher degree of structural efficiency?

4 Approach

The approach synthesizes knowledge from the cross-disciplinary fields of digital architecture and lightweight engineering: the fabrication process development of the prototypical 3DP system and fabrication-aware computational design tool development for this system.

4.1 Fabrication Process

This study adapts the CFRP extrusion system from CMASLab [19] with the help of the industrial partner 9T Labs in combination with a polymer extrusion system.

Fabrication Strategy. The presented fabrication strategy consists of the following consecutive manufacturing steps: (1) the 3DP of successive layers of polymer to create a base structure, (2) the repositioning (90° rotation) of the initially-printed base structure using temporary 3D printed support rests and (3) the local reinforcement of the substrate with add-on CFRP 3DP. This reinforcement improves the z-direction bonding strength with strands of continuous CFRP material (see Fig. 1). The proposed system has the advantage of printing on existing 3D prints in layers as well as in optimal structural directions of highly carbon fiber filled polymers. This method – which is similar to established CFRP manufacturing methods based on quasi-isotropic lamination [4] – has noteworthy implications for reinforcing the bonding strength of 3D printed components, achieving homogenous structural behavior.

In general, robotic fabrication has excellent accuracy; however, because of the nature of the process, distinct cases of deviation arise. During the 3DP of planar layers (1) where the build material solidifies and cools down quickly – local anomalies appear due to uneven shrinkage during thermal contraction. Immediately following the vertical repositioning, (2) these aberrations can additively result in a deformed substrate for the following step of add-on CFRP 3DP (3). Consequently, precision represents one of the largest challenges, because significant deviations in some parts of the base structure can compromise the adhesion of the add-on CFRP (3). The proposed method integrates a path adaptation strategy between the fabrication steps through a distance-measuring

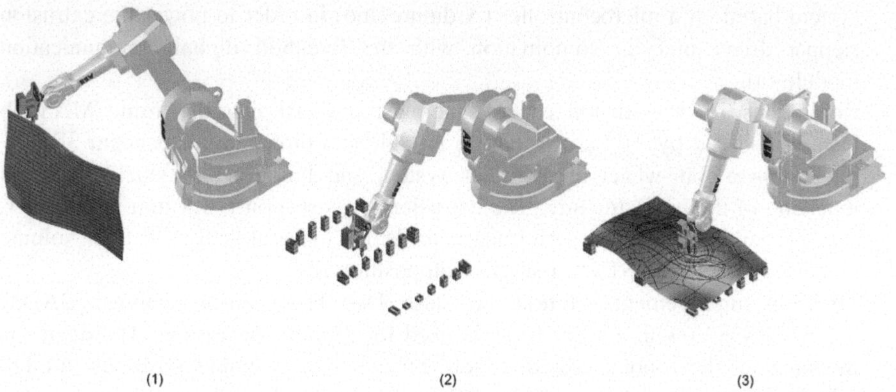

Fig. 1. Prototypical 3DP system: (1) 3DP of plastic base structure; (2) 3DP of plastic temporary supports and vertical repositioning; (3) continuous CFRP add-on 3DP.

mechanism in order to meet precision requirements. This auxiliary feedback process aims to revise the paths of the add-on CFRP 3DP.

Hardware Development. To develop the system, a medium-scale 6-axis industrial robot arm setup is used (ABB IRB 1600), integrating the thermoplastic and CFRP extrusion systems. The setup includes the following subcomponents (see Fig. 2):

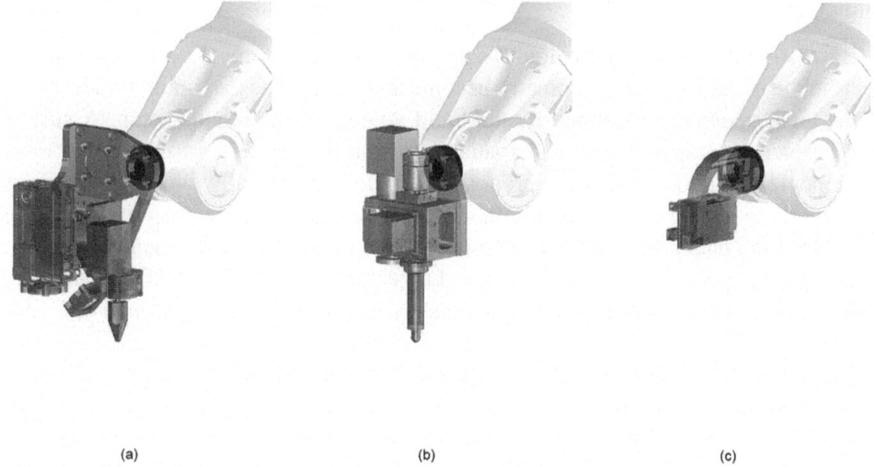

Fig. 2. Custom tool-heads: (a) plastic extrusion; (b) CFRP extrusion; (c) laser measuring.

- Plastic extrusion system. A BondTech QR 3.0 Universal extruder is mounted to the robot arm with a customized bracket. In order to make plastic FDM feasible for architectural applications, a nozzle is customized (2.5 mm ø, >40 mm/s) for high-speed extrusion. Custom software is developed to control the plastic extrusion system based on a microcontroller (Arduino Uno) in order to power the extrusion stepper driver unit, in combination with the Beckhoff digital communication module (a).

- Continuous CFRP extrusion system. The existing CFRP extruder from CMASLab [19] is modified by 9T Labs to ensure reliable and direction-independent 3DP by integrating a four-wheel drive puller system and hot gas heat shield for local reheating of the base structure. The extrusion cross section is 1.4 mm in diameter, and the build for the high performance feedstock material is at 52% fiber volume content with a PA12-STS40 material composition (b).
- Precision improvement solution. A Class Two laser sensor (Baumer OADM 13U6475/S35A) with a point beam is used for distance measuring. The device is mounted on the robotic arm to ensure precision in sampling positions through robotic actuation. Using an Arduino Uno for selective measuring based on the robotic arm signal input, the sensing occurs within a range of 50-100 mm, resulting in a measured resolution of 0.05 mm in analog signal output (c).

4.2 Fabrication-Aware Computational Design Tool

In parallel with the fabrication process development, the research develops a parametric design tool for building elements. This computational tool uses double-surface geometries as input for generating fabrication code for the base structure, temporary support, laser measuring and CFRP reinforcement. Tool development focuses on the following:

- Base structure. Limited to double-curvature, non-uniform rational basis splines from the edges of surface geometry become input data. The maximum dimension of the input geometry is defined by the build volume of the robotic setup. A custom slicer generates layer-based and spatial extrusion path code.
- Temporary support. The study proposes FDM to economically fabricate temporary supports to reposition the initially fabricated base structure. The material amount is reduced dramatically compared to conventional molds for CFRP manufacturing.
- Design strategy for structurally-informed trajectories for CFRP. CFRP is distributed in efficient layouts based on selected iso-stress-curves determined through structural analysis.
- Integrating fabrication constraints in the design tool. These fabrication constraints primarily include the imprecision of the add-on 3DP. Distance measurements are synchronized with 20 ms internal circuit values. To normalize the 600,000 probed measurements, recursive value filtering is applied according to the standard deviation of the values. To compensate for the deviations of the base structure, the tool interpolates the normalized values as z-values in the corresponding coordinates of original points.

5 Results

Initially, the first case study is conducted to prove the concept of add-on 3DP out of short CFRP on a small-scale (26.5 cm × 26.5 cm × 3 cm). One of the main objectives of the study is to investigate the material design in terms of the relationship between the CFRP alignments and surface geometry enabling effective reinforcement (see Fig. 3).

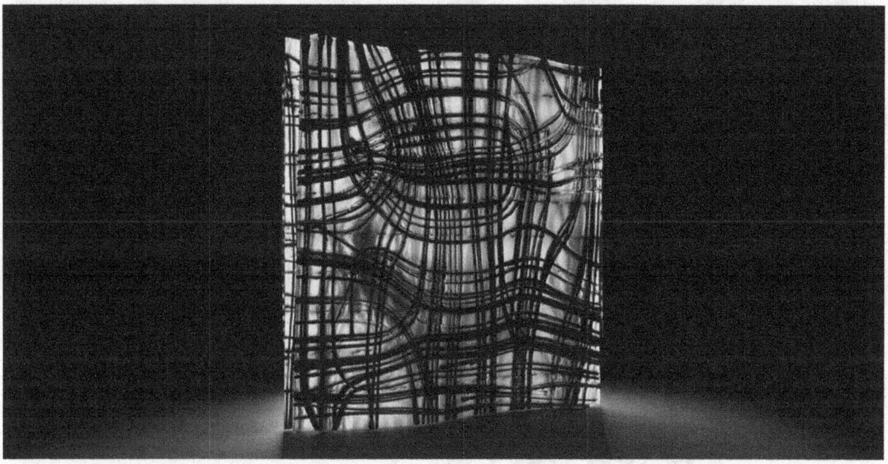

Fig. 3. Add-on 3D printed CFRP sandwich panel in 1:5 scale.

To investigate CFRP applications in architecture, a self-supporting double-curvature sandwich-panel facade element is prototyped at a medium-scale (0.9 m 0.6 m × 0.25 m) following the identified processes (see Fig. 4):

- Freeform plastic sandwich panel. Layer-based 3D printed surfaces and spatially extruded isotropic lattice structures are fabricated in a single process (A).
- Temporary support. According to the shape of the panel, the temporary supports are generated and fabricated automatically (B).
- Measuring device. In combination with the measurement strategy, the path is revised within a resolution of 0.05 mm fostering CFRP adhesion (C).
- CFRP printing. By setting the panel as a discretized part of the whole building skin, load and support cases are identified by the surrounding ones. Accordingly, the stress-lines are generated and compiled to the tool-path. The CFRP extruder follows the selected trajectories keeping the head perpendicular to the tangential vector of the base surface to locally deposit single strands of CFRP (D).

Empirical observation demonstrates the value of the custom fabrication strategy and the computational tool, which reinforces 3D printed plastic components by locally applying material strands of CFRP material in optimal directions. Together, they increase the inherently low z-direction bonding strength while achieving a higher structural efficiency for the global structure (see Fig. 5).

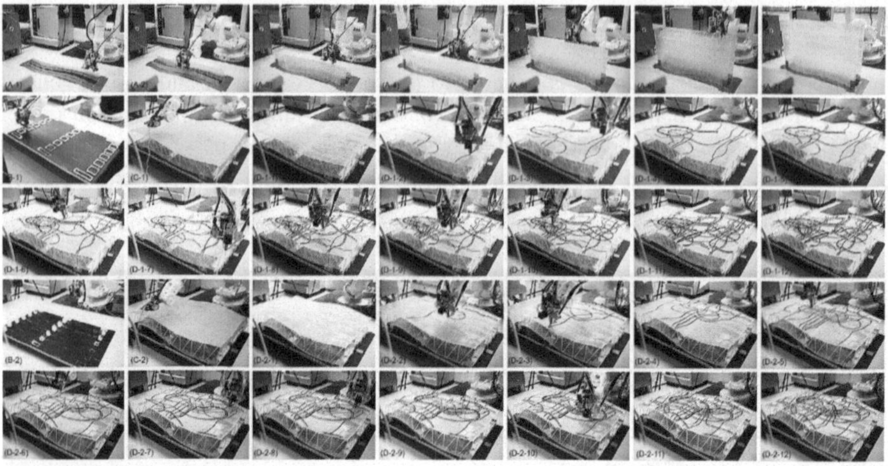

Fig. 4. Façade-element fabrication processes: (A) base panel; (B) temporary supports; (C) measuring; (D) CFRP add-on 3DP.

Fig. 5. Details of add-on 3D printed CFRP trajectories along stress-lines.

While the method presented above increases the build volume by 94.8% compared to the one of the Mark One 3D printer, the overall amount of material used is reduced significantly. This material reduction results from the fact that the temporary scaffolding needs 97.2% less material than a classical mold (from $\approx 0.0501\ m^3$ to $\approx 0.0014\ m^3$).

6 Discussion and Outlook

During the CFRP add-on 3DP process, two distinct improper behaviors are observed: delamination at sharp corners of the paths and decreased precision on junctions due to multiple overlaps. Because these behaviors affect global structure performance, further research will investigate the minimum curvature of CFRP that can be 3D printed as well as strategies to avoid a large number of junctions. These fabrication constraints will be integrated into the computational design tool. In parallel, an optimization strategy for the fiber layout will be developed to further improve structural performance.

The low z-direction bonding strength of the FDM technology appears to be overcome by the local add-on 3DP of CFRP material. Future research will quantify the structural benefits of CFRP. More specifically, three-point bend tests and/or compression tests will quantify and evaluate comparatively at two distinct levels:

- Quantification of the improvement in cross-layer bonding strength along the z-direction. In order to quantify the performance benefits of the add-on CFRP 3DP, two of each five identical samples will be structurally tested: the layer-based plastic 3D printed specimens and the ones with CFRP add-on reinforcement.
- Comparative evaluation in reference to alternative approaches and materials. In order to establish the applicability of the proposed fabrication method in architecture, it is compared to: (1) conventionally fabricated CFRP specimens using the same amount of carbon fibre regarding its flexural strength, and (2) steel, concrete and wood specimens, regarding their weight-to-strength ratio.

For the purpose of this publication, the three different tool-heads is swapped manually to demonstrate their feasibility. Nevertheless, the manual swapping introduced imprecisions and delays in the process, making it necessary to automate these steps through an integrative multi-material tool-head. The tool-head will be developed and customised for efficient multi-material fabrication. This will allow the change of multiple extrusion systems by rotating the sixth axis of the robotic arm, thus enabling the fabrication of multiple materials in an uninterrupted, single 3DP process.

Regarding scalability, the focus so far has been on parts that fit the size of the medium-scale robotic setup. This paper demonstrates a medium-scale building component, and larger scale components will be tested to broaden the applicability to architecture. The flexibility of the proposed fabrication system allows it to be adapted to various types of robotic arms. Therefore, the next step is to implement it in a large-scale robotic setup, the Robotic Fabrication Lab at ETH Zurich, with a fabrication volume of up to 4,590 m^3. This will include a parallel design research considering the development of an integrative assembly method and a load transferring system for the assembly of the discretized components. For the large-scale demonstrator, the following steps are scheduled:

- Prefabrication of discretized 3D printed lightweight plastic sandwich units that easily can be assembled on-site for a building-scale structure.
- Investigation of an integrated joint strategy to connect the units by adopting existing methods, such as male-female locking [21]. Here, the joint features will be integrated into the 3DP process.

- Add 3DP of meter-long continuous CFRP onto the assembled structure for mitigating force transfer. This add-on 3DP will provide the stiffness and structural integrity of the global structure.

3DP not only enables material gradation but also facilitates the fabrication of geometrically complex building components. Therefore, the design research also will explore new aesthetics through ornamentation, innovative shapes, new types of architectural components and functional integration of transparency out of various plastic materials.

For large-scale 3DP, a measuring process is necessary. However, the measuring process takes almost as long as the CFRP fabrication process itself ($\approx$5 h 47 min as opposed to $\approx$5 h 54 min). Therefore, further research should consider the development of a reliable real-time measuring strategy. Furthermore, the amount of shrinkage increases proportionally with the volume of the print, making investigation of various extrudable plastic materials valuable in identifying low-shrinkage options. More specifically, this research will further investigate these aspects:

- Handling deformation. This paper represents a promising first step in combining high-resolution sensing with precise adaptation to an unknown substrate. Nonetheless, there are limitations to the fragmented measuring and revision, particularly in terms of the overall production deceleration and limited adaptability in micro deformation. In this sense, the research will present an advancement of the revision strategy in real-time communication between robotics and the revision software. Together, they can couple highly detailed path revision and filtering with advanced adaptation.
- Prevent deformation. Further research has to answer the question of which plastic materials are best suited for large-scale FDM. Such investigations need to consider the shrinkage and crystallization of the polymer to prevent warpage as well as the printed bonding capabilities between the paired materials. CFRP has a unique advantage: it minimizes warpage through its increased compressive strength after instant solidification, a feature that is particularly promising for extremely large 3DP applications. Spatially-extruded lattice structures also can be printed out of CFRP to prevent shrinking and promote stiffness, which can potentially be graded based on structural analysis and relative design system development.

7 Conclusion

This paper has presented a method for producing freeform and non-standard CFRP building elements with minimal molds. The novelty of this approach is twofold. On the one hand, it provides a new 3DP process for continuous CFRP, and on the other, it highlights a graded material distribution that is precisely designed based on computational structural analysis. This method promises a more sustainable construction process, with no waste material, easier on-site assembly with lightweight components, and greater design freedom for CFRP elements.

Last but not least, integrated functionality will be investigated and integrated in this research. Buildings cannot exist without a set of performative and functional layers, which increases the need for more on-site labor. Enabling the fabrication process to materialize multiple layers of functionality within one setup can allow more complex architectural components while reducing complexity on the construction site.

Acknowledgements. The authors would like to thank a number of partners and collaborators whose dedication helped us fulfill the research described in this paper:

- Andrei Jipa, Pietro Odaglia, Dr Mania Aghaei Meibodi and the rest of the Digital Building Technologies personnel at ETH Zurich
- Professor Dr Paolo Ermanni (CMASLab, ETH Zurich)
- David Jenny (Gramazio Kohler Research, ETH Zurich)
- Michael Lyrenmann and Philippe Fleischmann (technical support, ETH Zurich)
- Ma Xijie (student assistant, ETH Zurich)
- 9T Labs, Extrudr (industry partners) and HAL Robotics (software support)

This research is supported by the NCCR Digital Fabrication, funded by the Swiss National Science Foundation (NCCR Digital Fabrication Agreement #51NF40-141853).

References

1. Whalley, A.: The Eden Project glass houses world environments. In: Barnes, M., Dickson, M. (eds.) Widespan Roof Structures, pp. 75–84. Thomas Telford, London (2000)
2. Jones, R.M.: Mechanics of Composite Materials. CRC Press, Boca Raton (1998)
3. Chung, D.D., Chung, D.: Carbon Fiber Composites. Butterworth-Heinemann, Oxford (2012)
4. Mallick, P.K. (ed.): Composites Engineering Handbook. CRC Press, Boca Raton (1997)
5. Lynn, G.: Good-bye tectonics, Hello composites. In: Bell, M.J., Buckley, C. (eds.) Permanent Change: Plastics in Architecture and Engineering, pp. 86–111. Princeton Architectural Press, New York (2014)
6. Foster + Partners Homepage. https://www.fosterandpartners.com/projects/steve-jobs-theater/. Accessed 09 Mar 2018
7. Prudon, T.H.: From boat to bust: the Monsanto House revisited. In: Bell, M.J., Buckley, C. (eds.) Permanent Change: Plastics in Architecture and Engineering, pp. 60–67. Princeton Architectural Press, New York (2014)
8. Åkermo, M., Åström, B.T.: Modelling component cost in compression moulding of thermoplastic composite and sandwich components. Compos. Part A Appl. Sci. Manuf. **31** (4), 319–333 (2000)
9. Pronk, A., van Rooy, I., Schinkel, P.: Double-curved surfaces using a membrane mould. In: Domingo, A., Lazaro, C. (eds.) Proceedings of the International Association for Shell and Spatial Structures Symposium (IASS): IASS 50th Anniversary: Evolution and Trends in Design, Analysis and Construction of Shell and Spatial Structures, pp. 618–628. Valencia (2009)
10. Chawla, K.K.: Composite Materials: Science and Engineering. Springer, Berlin (2012)
11. Doerstelmann, M., Knippers, J., Koslowski, V., Menges, A., Prado, M., Schieber, G., Vasey, L.: ICD/ITKE research pavilion 2014–15: fibre placement on a pneumatic body based on a water spider web. Archit. Des. **85**(5), 60–65 (2015)
12. Ryder, G., Ion, B., Green, G., Harrison, D., Wood, B.: Rapid design and manufacture tools in architecture. Autom. Constr. **11**(3), 279–290 (2002)

13. Jipa, A., Bernhard, M., Ruffray, N., Wangler, T., Flatt, R., Dillenburger, B.: skelETHon formwork. In: Proceedings of the XXI Conference of the Iberoamerican Society of Digital Graphics (SIGraDi): Resilience Design, pp. 22–24. Concepción, Chile (2017)
14. Peters, B.: Solar bytes pavilion. In: Reinhardt, D., Saunders, R., Burry, J. (eds.) Robotic Fabrication Architecture, Art and Design 2016, pp. 327–337. Springer International Publishing, Switzerland (2016)
15. Tam, K.M.M., Coleman, J.R., Fine, N.W., Mueller, C.T.: Robotics-enabled stress line additive manufacturing. In: Reinhardt, D., Saunders, R., Burry, J. (eds.) Robotic Fabrication Architecture, Art and Design 2016, pp. 350–361. Springer International Publishing, Switzerland (2016)
16. Holshouser, C., Newell, C., Palas, S., Love, L.J., Kunc, V., Lind, R.F., Peter, W.H.: Out of bounds additive manufacturing. Adv. Mater. Process. **171**(3), 15–17 (2013)
17. Biswas, K., Lind, R., Post, B., Jackson, R., Love, L., Green Jr, J., Guerguis, A.M.: Big area additive manufacturing applied to buildings. In: Thermal Performance of the Exterior Envelopes of Whole Buildings XIII International Conference, pp. 583–590. Clearwater, FL (2016)
18. Mark, G.T., Gozdz, A.S.: U.S. Patent No. 9,579,851. U.S. Patent and Trademark Office, Washington, DC (2017)
19. Eichenhofer, M., Wong, J.C., Ermanni, P.: Continuous lattice fabrication of ultra-lightweight composite structures. Addit. Manuf. **18**, 48–57 (2017)
20. Pottmann, H.: Architectural geometry as design knowledge. Archit. Des. **80**(4), 72–77 (2010)
21. Miller, F.: U.S. Patent No. 5,094,051. Washington, DC: U.S. Patent and Trademark Office (1992)

Robotic Extrusion of Architectural Structures with Nonstandard Topology

Yijiang Huang(✉), Josephine Carstensen, Lavender Tessmer, and Caitlin Mueller

Massachusetts Institute of Technology,
77 Massachusetts Avenue, Cambridge, MA 02139, USA
yijiangh@mit.edu

Abstract. This paper presents a fast and flexible method for robotic extrusion (or spatial 3D printing) of designs made of linear elements that are connected in nonstandard, irregular, and complex topologies. Nonstandard topology has considerable potential in design, both for visual effect and material efficiency, but usually presents serious challenges for robotic assembly since repeating motions cannot be used. Powered by a new automatic motion planning framework called *Choreo*, this paper's robotic extrusion process avoids human intervention for steps that are typically arduous and tedious in architectural robotics projects. Specifically, the assembly sequence, end-effector pose, joint configuration, and transition trajectory are all generated automatically using state-of-the-art, open-source planning algorithms developed in the broader robotics community. Three case studies with topologies produced by structural optimization and generative design techniques are presented to demonstrate the potential of this approach.

Keywords: Robotic extrusion · Motion planning · Topology optimization

1 Introduction

Architectural robotics has proven a promising technique for assembling nonstandard configurations of building components at the scale of the built environment, complementing the earlier revolution in generative digital design. However, despite the advantages of dexterity and precision, the time investment in solving the construction sequence and associated robotic motion grows increasingly with the topological complexity of the target design. This gap between parametric design and robotic fabrication congests the overall digital design/production process and often confines designers to geometries with standard topology. In order to close this gap and enable more possibilities for discrete architectural robotic assembly, a more systematic and explicit computational exploration of constraints and robotic motion planning is needed.

This paper presents a new way to apply automated robotic assembly sequence and motion planning to robotic extrusion of geometries with nonstandard topologies. The case studies presented serve to demonstrate the computational planning system's power

J. Willmann et al. (Eds.): ROBARCH 2018, *Robotic Fabrication in Architecture, Art and Design 2018*, pp. 377–389, 2019.
https://doi.org/10.1007/978-3-319-92294-2_29

to generate feasible robotic instructions and how its integration into existing digital design workflows can resurrect topology as a fundamental design variable on designers' palette for robotic assembly.

1.1 Complexity and Topology

Assemblies of discrete elements, such as trusses, space frames, masonry vaults, and stacked blocks, have been explored repeatedly in the architectural robotics domain. All discrete structures of this type can be represented by three design characteristics: size, shape (or geometry), and topology. This terminology originates from structural optimization of trusses in the 1960s–1980s (Spillers and MacBain 2009), but can be used generally to describe any discrete structure.

Within this framework, size refers to a cross sectional property of an individual element in the structure (e.g. cross sectional area of a linear truss element, or width of a brick in a stacked wall). Shape refers to the locations of internal and external points, lines, curves, and surfaces in a design. Finally, topology refers to the connectivity relationships between elements in the structure, and is the most fundamental; there is no easy way to morph a design from one topology to another (unlike with size and shape). Topology therefore offers both design opportunities and fabrication challenges: the largest impacts in visual effect and efficiency are possible (some examples are given in Fig. 1), but complex topologies can be nontrivial to assemble.

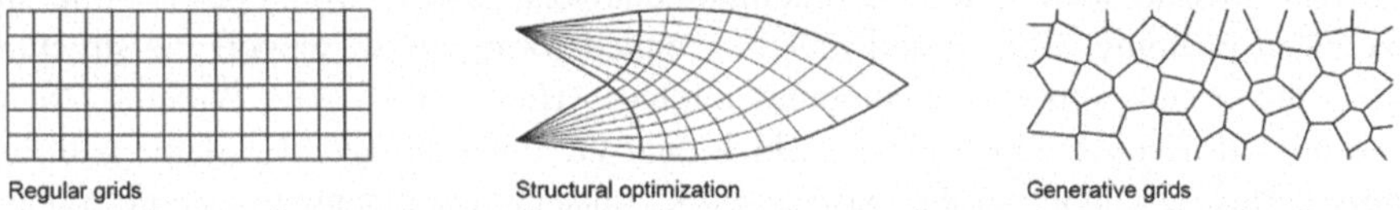

Fig. 1. Examples of different topologies resulting from different design techniques

Research in architectural robotics has included explorations in materializing design-driven complexity at all three of these levels (Gramazio et al. 2014). Most of this existing work involves generating a tool center point pose on the geometry to be assembled, and using the industrial robot's built-in interpolation method to compute transition trajectories with a certain safety factor. Recently, Søndergaard et al. propose an incremental search algorithm to find a construction sequence for a large-scale topology optimized space frame, while guaranteeing the existence of node-specific, collision-free assembly motion (Søndergaard et al. 2016). However, robotic configuration's feasibility is not considered during construction sequence searching, and robot's trajectories are resolved by inserting custom unwinding positions. While this geometry and machine-specific approach is feasible for designs with simple and sparse topologies, the construction sequence and robotic motion planning is much more nuanced for designs with denser material distribution and non-standard topologies. The computational complexity of "custom iterative path planning" in a densely populated environment has proved to be a major technical challenge that confines designers in the design domain of regular topologies (Eversmann et al. 2017).

In recent years, there has been some success in tackling this problem using automated motion planning by using a single-query robotic motion planning algorithm (Parascho et al. 2017) or an online control strategy (Giftthaler et al. 2017) to compute transition trajectory between pre-programmed assembly primitives. However, the construction sequence in the existing work was still assigned manually, taking advantage of either the sparse or the repetitive topology of the target geometry.

1.2 Topology in Robotic Extrusion

This paper focuses on one particular method for robotic assembly of discrete structures for the sake of specificity. Robotic extrusion (sometimes called spatial 3D printing) involves extruding a thermoplastic along linear paths, typically to form a mesh or grid structure, using robotic motion. This fabrication technique has been presented as an alternative to layer-based additive manufacturing, with advantages both in terms of mechanical properties and speed of construction (Gramazio et al. 2014; Yuan et al. 2014). The flexibility of industrial robotics has mostly been deployed to facilitate complexity in shape (as opposed to size or topology); morphed grids with standard topologies have been shown to be useful both for formal variation (Branch Technology 2018; Soler et al. 2017) and structural efficiency (Tam et al. 2018).

There has been some research in robotic extrusion for nonstandard topology, but all has required a time-consuming, non-automated robotic motion planning process. In the architectural robotics domain, (Tam et al. 2016) present robotic extrusion along lines of principal stress to achieve desired structural behaviors, but this work does not offer an automated planning solution and relies on milled formwork to support the structure during the printing process. In computer graphics, (Huang et al. 2016; Wu et al. 2016) have printed irregular topologies in which only the outer surface of a shape is materialized. However, none of the existing work considers the planning of entire robotic trajectories. Nor does any existing process explicitly demonstrate its ability to efficiently handle the construction sequencing and motion planning problem for designs with intricate volumetric topology patterns that are of interest in architecture.

1.3 Research Aim

In response to this need for easier robotic programming for complex, dense topologies in the architectural domain, this paper introduces a new automatic motion planning system called *Choreo*, which removes the need for human intervention in the tedious and nontrivial tasks of assembly sequence definition, collision detection, and trajectory planning.

2 Automatic Motion Planning System

This section lays out the general framework for robotic extrusion of nonstandard topologies. As shown in Fig. 2, there are four broad steps used in this workflow. First, a designer generates an overall concept using discretized linear elements; the design is defined in terms of topology and shape (or geometry). The second and third steps are carried out by Choreo without the need for human intervention: a feasible assembly sequence is automatically generated, and then the robot's path and instructions are automatically planned. In this third path planning step, the robot's trajectory is generated for both joint configurations and associated end-effector poses for each element's extrusion, and for transition between adjacent extrusions. The planning output is tagged with metadata so that users can easily weave hardware IO commands and micro path modifications using any programming platform, including Grasshopper. Taking advantage of existing robot brand-specific post processors (such as KUKA|prc 2018), an executable robot instruction file can be harvested and uploaded to a robot controller for execution. These four steps are explained in more detail in the following subsections.

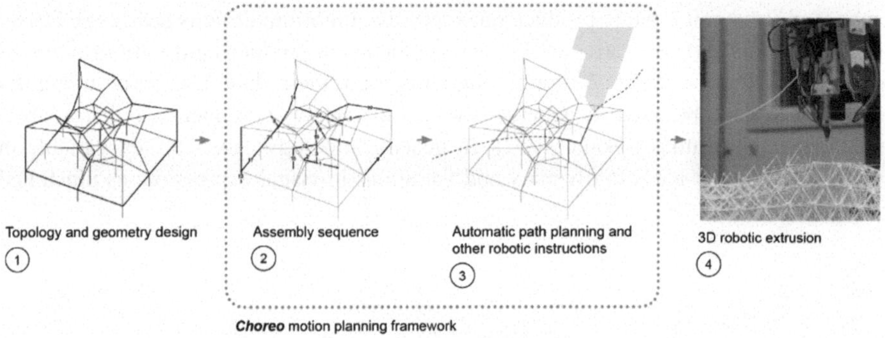

Fig. 2. Overview of the robotic extrusion workflow including the automated planning system

2.1 Step 1: Ground Structure Topology Optimization

One option for generating the initial design to be robotically extruded is topology optimization, which is especially attractive when structural efficiency is important. Topology optimization is a design approach that finds the best material distribution within a discretized design domain according to structural criteria (Bendsøe and Kikuchi 1988). The majority of approaches that use truss elements for the discretization are based on the so called *ground structure* method where nodes are distributed throughout the design domain and potential bars are defined between them (Dorn et al. 1964). Using a mathematical programme, the bars in the domain are sized to obtain the least weight design with a user-specified stiffness. By letting the smallest allowable bar area approach zero, the structural topology is obtained.

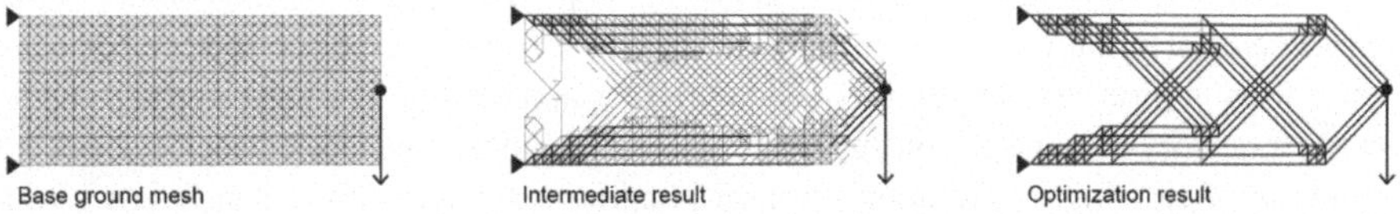

Fig. 3. Topology optimization process

Figure 3 shows an illustration of this process applied to a simple 2D cantilever problem. In the figure, a bar element's line width denotes cross section that is between 0 and 1 times of the desired constant cross section A_0. Elements with a thick line width have area A_0 and elements with areas approaching 0 are removed. This same topology optimization method is applied to a more complex 3D case in Sect. 3.1.

2.2 Step 1: Other Topology-Generating Methods

There are many other design-driven methods for generating complex topologies algorithmically of interest in architecture. For example, Stiny and collaborators have shown how to generate designs using shape grammars, including frames inspired by Chinese ice ray lattices (Stiny 1977). More recently, grammars have been used with embedded structural logic to produce unexpected equilibrium designs (Lee et al. 2016). Islamic patterns and generative tools to create them can produce culturally meaningful topologies (Khouri 2017). Voronoi diagrams, and their dual Delaunay triangular meshes, can each be used to generate meshes that seem biomimetic or that address other aesthetic agendas (Okabe 1992). In general, topology can be a key variable in creative design that leads to diversity and variations in visual expression, as illustrated in Fig. 4.

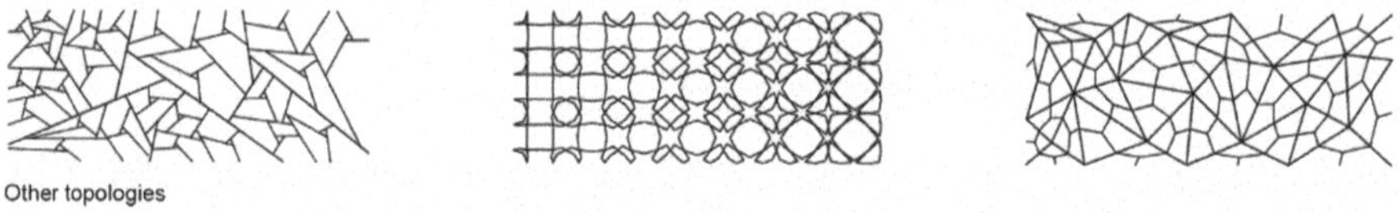

Fig. 4. Visually diverse topologies produced through generative design.

2.3 Steps 2 and 3: Assembly Sequence and Motion Planning with Choreo

The complexity of topology introduces significant challenges for finding a collision-free extrusion sequence and robot trajectory. The robotic extrusion planning problem requires finding a chronological construction sequence of motions to extrude each element, as well as moving through free space to connect adjacent extrusion processes.

To solve this combined task and motion planning problem, a three-layer computational hierarchy is proposed and implemented in Choreo to gradually narrow down the computational complexity along the search tree. First, a constraint-based sequence planner is introduced to search the construction sequence, while guaranteeing the intermediate construction's stability and stiffness, and the existence of collision-free robot kinematics solution at each extrusion step. Then, a sampling-based semi-constrained Cartesian planner is used to compute the robot's joint configuration during each extrusion process. Finally, an off-the-shelf motion planner is called to compute the robot's trajectory to navigate through free space to connect adjacent extrusions. Taking advantage of existing single-query motion planning packages, users can interact with multiple state-of-the-art motion planners and choose the one to balance their needs between the optimality, smoothness, and speed. Figure 5 highlights the three different

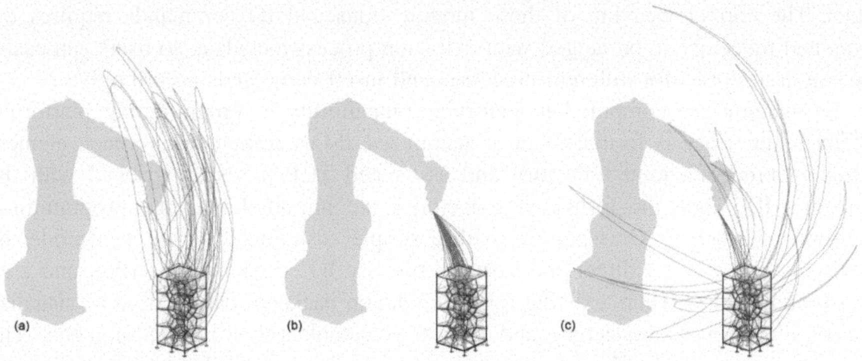

Fig. 5. Transition trajectory computed by state-of-the-art motion planners: (a) STOMP (Kalakrishnan et al. 2011) (b) CHOMP (Note that a resetting home pose is inserted in the transition trajectory if initial direct transition planning fails. Thus, the results shown here indicate that the CHOMP planner is not as good in finding a feasible trajectory as the other two planners presented here.) (Ratliff et al. 2009) (c) RRT* (Karaman et al. 2011)

planning results for a 3D Voronoi (Sec. 3.2), using three different state-of-the-art open-source motion planners developed in the robotics community. An in-progress paper (Huang et al. 2018) gives a more detailed description of the assembly planning algorithms used in Choreo.

Choreo is implemented in C++ on Robot Operating System (ROS) (Quigley et al. 2009), integrating functionalities from ROS-Industrial (ROS-Industrial 2018) and Moveit! framework (Sucan et al. 2018). Choreo's system architecture is designed to be modularized and adaptable. This modularized system feature offers users and researchers the flexibility to plug in and experiment their customized sequence or motion planner without changing the entire system's codebase. Finally, Choreo can be configured to support 6 or 7-axis industrial robots of any brand with any user-defined end-effector[1].

2.4 Step 4: Post-processing and Robotic Extrusion

The generated robotic trajectory from Choreo is geometrical and without timestamp information. In order to generate instructions for the robot to interact with the physical world, the user needs to weave IO commands to synthesize the robot's motion and its end effector's behavior. In addition to this, the variation of extruder design and extrusion material properties requires the incorporation of ad hoc fabrication logic to achieve the desired visual results (Hack and Lauer 2014) or increase the product's structural performance (Tam et al. 2018). These fabrication logics derived from physical extrusion experiments usually involve local modification of an end effector's pose, such as pressing or extruding following small circular movements at structural

[1] The mechanical parts of the extrusion system used in this work's case studies are developed by Archi-Solution Workshop (http://www.asworkshop.cn/).

joints. The correct weaving of these modifications and IO commands requires the computed trajectory to be tagged with extrusion process metadata, so users can easily separate trajectories for different processes and insert commands accordingly.

To increase the computed trajectory's compatibility to programming platforms, Choreo's trajectory is formatted in a customized JSON format, where each element extrusion process's joint trajectory and associated TCP poses are packed with the element's ID. Then the formatted trajectories are imported into any programming environment, such as Grasshopper, with a simple customized parser, to decode the JSON file and allow a direct and visually friendly IO commands insertion and path micro-modification. Then, existing robot simulation packages can be used to visualize and simulate robot's trajectory and export executable robot instruction code. The fabrication parameter calibration process can go back and forth between Grasshopper and physical tests, keeping the overall robot trajectory unchanged.

3 Case Studies

This section presents three robotic extrusion case studies of different topologies. Computation time on assembly planning and fabrication results are presented in Table 1, and overall shape and topology properties are given in Fig. 6. These case studies demonstrate Choreo's power in automatically generating executable robotic extrusion trajectory in a reasonable amount of time.

Table 1. Computation statistics of the case studies. All computational experiments were performed on a Linux virtual machine with 4 processors and 16 GB setup on desktop PC with a quad-core Intel Xeon CPU. [+]Extrusion planning time is specified by users.

Model	Node count	Element count	Sequence planning time [s]	Extrusion planning time [s][+]	Transition planning time [s]	Fabrication time [hr]	Size [mm]
Topopt vault (Sec 3.1)	114	205	1346	1200	843 (STOMP) 1211 (RRT*) 1511 (CHOMP, 9 fails)	3	200 × 200 × 200
Voronoi (Sec 3.2)	148	292	2299	1200	846 (STOMP) 1286 (RRT*) 945 (CHOMP)	3.2	150 × 150 × 320
Mars habitat (Sec 3.3)	86	214	1498	1200	800 (STOMP) 918 (RRT*) 1054 (CHOMP, 4 fails)	3	180 × 180 × 155

3.1 Topology Optimized Vault

Using the ground structure topology optimization method described in Sect. 2.1, a 3D-trussed vault was generated. With this approach, it was possible to remove 91% of the material initially included in the ground structure.

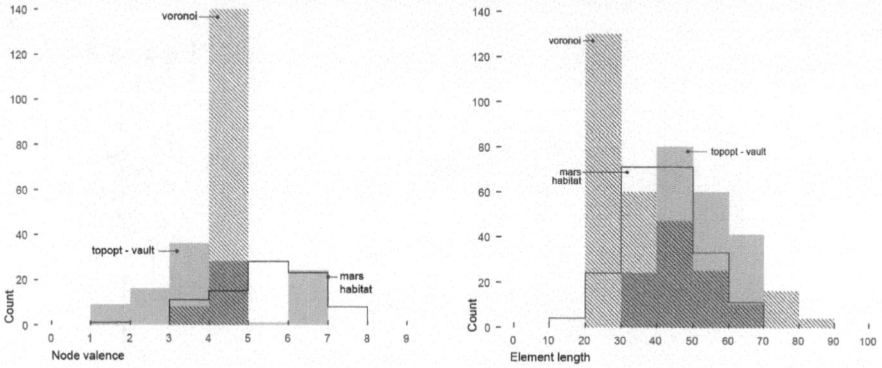

Fig. 6. Node valence and element length histograms of the case studies.

The average element length is long, and element length variation is low because the design is generated from a regular base mesh. However, the geometric configuration generated from these elements is not trivial. 12% of the nodes have valence of six, tending to create narrow pathways for robot in transition planning. The trajectory highlighted in Fig. 7 shows the corresponding tool center point traveling trajectory from the transition planning result, indicating that the robot's configuration changes significantly between many pairs of adjacent extrusions, requiring the planner to output a long and unintuitive trajectory to stay within joint limitations and stay clear from collisions.

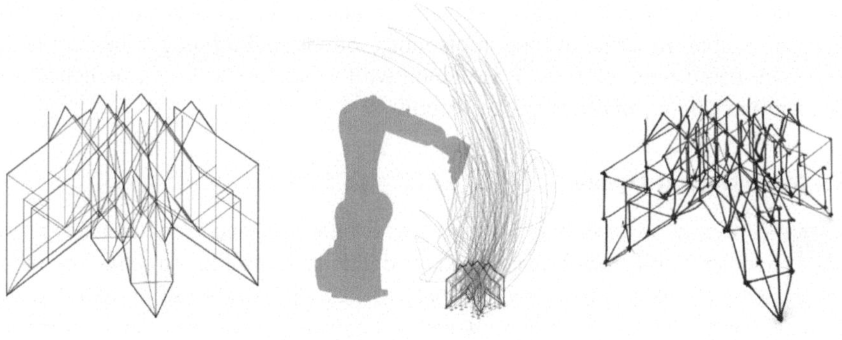

Fig. 7. Topology optimized vault, robotic trajectories with STOMP, and final extruded result.

3.2 3D Voronoi

The 3D Voronoi design was generated by randomly sampling points within a rectangular solid, and then using the 3D Voronoi component in Grasshopper together with Kangaroo 2. A sphere collision algorithm was used to force the elements lengths to have a distribution with lower variance. Figure 8 shows the design and fabrication of this case.

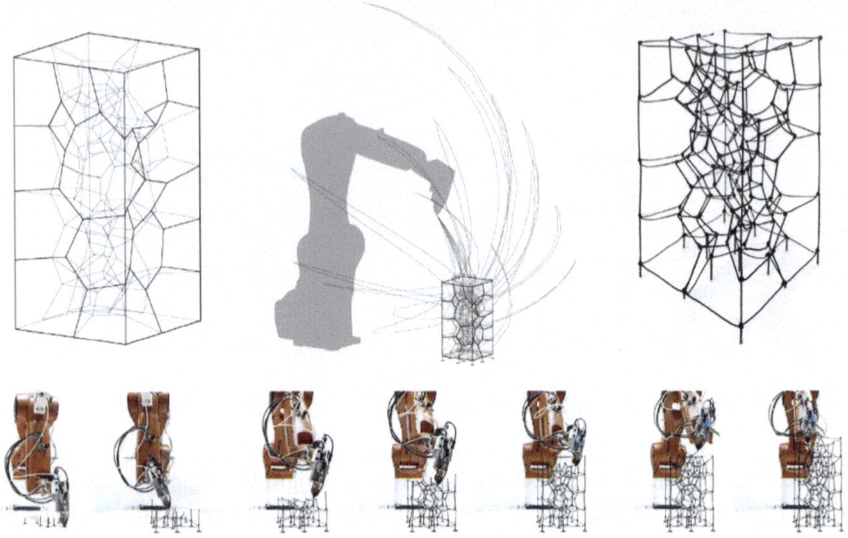

Fig. 8. 3D Voronoi design, robotic trajectories with RRT*, and final extruded result.

Because of the Voronoi-generating algorithm, there is low variation in node valence, with most nodes having four elements, a relatively low number, connecting. In this design, elements are well supported during each construction step, and there are few very long elements. However, the long elements at the boundary have smaller node valences, and the resulting material warping and sagging can sometimes prevent the robot from locating and connecting to these elements even though the computed trajectory is feasible. In terms of motion planning, the internal topology does not have a trivial layer-based pattern. Thus, it is unintuitive for humans to find a sequence manually, and the Choreo platform proves useful.

3.3 Mars Habitat Design

The third case study is a model of a pressurized habitat designed for a human colony on Mars. An outer dome membrane, discretized into a mesh-like structure, is helped structurally by an internal tree structure that acts like a tension spoke system to anchor the membrane to the ground. Figure 9 shows the design and fabrication of this case.

The construction sequence alternates between outer and inner structure to gradually close the membrane at the top. Nodes on the stem of the internal tree structure have highest node valence. The outer layer needs to be built before the internal tree elements, but introducing more surrounding collision objects leaves narrow pathways for the robot to enter. This forces the planner to find long trajectories to allow specific joints to have sufficient rotation in open space to approach the desired joint configuration.

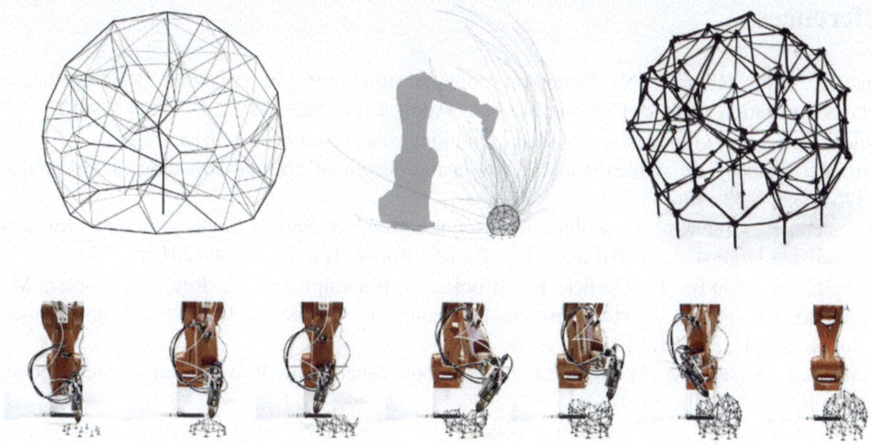

Fig. 9. Mars habitat design, robotic trajectories with RRT*, and final extruded result.

4 Conclusions

This paper has demonstrated a new path planning framework, Choreo, and used it to show how opening up topology as a design variable in robotic extrusion offers opportunity for more efficient structures and more creative flexibility. Three case studies were presented, each of which has nonstandard topology and more than 200 elements. Because of the flexibility and speed of Choreo, the trajectories in all of the case studies were computed in a little more than an hour, with three additional hours needed for the actual robotic execution. This timescale suggests an exciting future possibility: fabrication logic related to robotic constructability could be integrated as a driver in iterative conceptual design, pushing the role of technical assessment from checking a nearly finalized design to an early-stage decision-making aid.

Although the approach presented in this paper was applied to the specific method of robotic extrusion at a relatively small scale, the Choreo framework is very flexible. Because the underlying algorithms state-of-the-art, they are fast enough to generate robotic sequences to be used in production. Choreo could also be applied to other aggregations of linear elements beyond extrusion with similar benefits, or more broadly to assembly problems in general (e.g. masonry structures with nonstandard topologies). Because Choreo is independent of robot brand or even numbers or types of axes, it can work with many different robotic set ups, including those with additional external linear axes or turntables. The broad future vision for this work is a better way for designers to interact with robots, shifting the machine programming experience back to high-level tasks in the architectural language of shape and topology.

References

Bendsøe, M.P., Kikuchi, N.: Generating optimal topologies in structural design using a homogenization method. Comput. Methods Appl. Mech. Eng. **71**(2), 197–224 (1988)

Branch Technology. https://www.branch.technology/. Accessed 8 Mar 2018

Dorn, W., Gomory, R., Greenberg, H.: Automatic design of optimal structures. J Mecanique **1964**(3), 25–52 (1964)

Eversmann, P., Gramazio, F., Kohler, M.: Robotic prefabrication of timber structures: towards automated large-scale spatial assembly. Constr. Robot. **1**(1–4), 49–60 (2017)

Giftthaler, M., Sandy, T., Dörfler, K., Brooks, I., Buckingham, M., Rey, G., Kohler, M., Gramazio, F., Buchli, J.: Mobile robotic fabrication at 1:1 scale: the In situ Fabricator. Constr. Robot. **1**(1–4), 3–14 (2017)

Gramazio, F., Kohler, M., Willmann, J.: The Robotic Touch: How Robots Change Architecture. Park Books, Zurich (2014)

Hack, N., Lauer, W.V.: Mesh-mould: robotically fabricated spatial meshes as reinforced concrete formwork. Archit. Des. **84**(3), 44–53 (2014)

Huang, Y., Garrett, C., Mueller, C.: Automated motion planning for robotic assembly of discrete architectural structures. Constr. Robotics (2018, in preparation)

Huang, Y., Zhang, J., Hu, X., Song, G., Liu, Z., Yu, L., Liu, L.: FrameFab: robotic fabrication of frame shapes. ACM Trans. Graph. (TOG) **35**(6), 224 (2016)

Karaman, S., Frazzoli, E.: Sampling-based algorithms for optimal motion planning. Int. J. Robot. Res. **30**(7), 846–894 (2011)

KUKA|prc. http://www.robotsinarchitecture.org/kuka-prc. Accessed 4 May 2018

Khouri, N.K.: Structural grid shell design with Islamic pattern topologies, Master's thesis, Massachusetts Institute of Technology (2017)

Kalakrishnan, M., Chitta, S., Theodorou, E., Pastor, P., Schaal, S.: STOMP: stochastic trajectory optimization for motion planning. In: Proceedings of the IEEE International Conference on Robotics and Automation, ICRA 2011, pp. 4569–4574. Shanghai (2011)

Lee, J., Mueller, C., Fivet, C.: Automatic generation of diverse equilibrium structures through shape grammars and graphic statics. Int. J. Space Struct. **31**(2–4), 147–164 (2016)

Okabe, A.: Spatial Tessellations. Wiley, New York (1992)

Parascho, S., Gandia, A., Mirjan, A., Gramazio, F., Kohler, M.: Cooperative fabrication of spatial metal structures. In: Sheil, B., Menges, A., Glynn, R., Skavara, M. (eds.) Fabricate: Rethinking Design and Construction, pp. 24–29. UCL Press, London (2017)

Quigley, M., Conley, K., Gerkey, B., Faust, J., Foote, T., Leibs, J., Wheeler, R., Ng, A.Y.: ROS: an open-source robot operating system. In: ICRA Workshop on Open Source Software, vol. 3, no. 3.2, p. 5 (2009)

Ratliff, N., Zucker, M., Bagnell, J.A., Srinivasa, S.: CHOMP: gradient optimization techniques for efficient motion planning. In: Proceedings of the IEEE International Conference on Robotics and Automation, ICRA 2009, pp. 489–494. Kobe (2009)

ROS Industrial website. https://rosindustrial.org/. Accessed 8 Mar 2018

Spillers, W.R., MacBain, K.M.: Structural Optimization. Springer, New York (2009)

Stiny, G.: Ice-ray: a note on the generation of Chinese lattice designs. Environ. Plan. B Plan. Des. **4**(1), 89–98 (1977)

Sucan, I.A., Chitta, S.: Moveit! http://moveit.ros.org. Accessed 8 Mar 2018

Søndergaard, A., Amir, O., Eversmann, P., Piškorec L., Stan, F., Gramazio, F., Kohler, M.: Topology optimization and robotic fabrication of advanced timber space-frame structures. In: Reinhardt, D., Saunders, R., Burry, J. (eds.) Robotic Fabrication in Architecture, Art and Design 2016, pp. 190–203. Springer International Publishing, Switzerland (2016)

Soler, V., Retsin, G., Jimenez Garcia, M.: A generalized approach to non-layered fused filament fabrication. In: Nagakura, T., Tibbits, S., Mueller, C., Ibañez, M. (eds.) Acadia 2017: Disciplines & Disruption, Proceedings of the 37th Annual Conference of the Association for Computer Aided Design in Architecture, pp. 562–571. Cambridge, MA. (2017)

Tam, K.M.M., Coleman, J.R., Fine, N.W., Mueller, C.: Robotics-enabled stress line additive manufacturing. In: Reinhardt, D., Saunders, R., Burry, J. (eds.) Robotic Fabrication in Architecture, Art and Design 2016, pp. 350–361. Springer International Publishing, Switzerland (2016)

Tam, K.M.M., Marshall, D., Gu, M., Kim, J., Huang, Y., Lavallee, J., Mueller, C.: Fabrication aware structural optimisation of lattice additive-manufactured with robot-arm. Int. J. Rapid Manuf. (2018, in press)

Wu, R., Peng, H., Guimbretière, F., Marschner, S.: Printing arbitrary meshes with a 5DOF wireframe printer. ACM Trans. Graph. (TOG) **35**(4), 101 (2016)

Yuan, P.F., Meng, H., Yu, L., Zhang, L.: Robotic multi-dimensional printing based on structural performance. In: Reinhardt, D., Saunders, R., Burry, J. (eds.) Robotic Fabrication in Architecture, Art and Design 2016, pp. 92–105. Springer International Publishing, Switzerland (2016)

On-Site Robotics for Sustainable Construction

Alexandre Dubor[1(✉)], Jean-Baptiste Izard[2], Edouard Cabay[1], Aldo Sollazzo[1,2,3], Areti Markopoulou[1], and Mariola Rodriguez[4]

[1] Institute for Advanced Architecture of Catalonia (IAAC), 08005 Barcelona, Spain
alex@iaac.net
[2] Tecnalia France, 950 rue Saint-Priest, 34090 Montpellier, France
[3] Noumena, Gran Via de les Corts Catalanes 690, 08010 Barcelona, Spain
[4] Tecnalia, Mikeletegi Pasealekua 7, 20009 Donostia - San Sebastián, Gipuzkoa, Spain

Abstract. Although additive and robotic manufacturing, is considered a technology with lots of potentials in the construction industry, its deployment has not yet reached wide applications for on site construction of sustainable architectural structures. This paper focuses on the deployment of a 3D printing technology that combines robotics with natural materials for the construction of environmentally performing small scale buildings. Cable robots are explored for 3D printing with adobe, while drones are explored for real-time monitoring technologies of the construction process. A full scale prototype of the technology has been deployed during 15 days at an international construction fair, demonstrating its potential by producing live a 20 m^2 pavilion. The paper describes and discusses the experience, feasibility and limitations of the technology operating on site and in a direct collaboration with human operators and craftsmen. The prototype demonstration presented in the paper has led to the conclusion that there is a significant potential of using the technology for large scale sustainable architectural constructions on-site.

Keywords: Construction site automation · Additive manufacturing
Cable robot · Earth architecture · Monitoring by drone
Sustainable construction

1 Introduction

1.1 Background

In a context of demographic explosion, resource scarcity and global warming, the construction sector needs to evolve towards faster, cleaner, more efficient and more customisable building systems.

Past research in robotic fabrication for architecture has shown the potential to challenge the way we design and build, bringing new opportunities for the construction sector to improve productivity and expand design possibility towards a potentially more performative building [1]. In this field, additive manufacturing for on-site construction processes such as contour crafting [2] have been explored in research and

J. Willmann et al. (Eds.): ROBARCH 2018, *Robotic Fabrication in Architecture, Art and Design 2018*, pp. 390–401, 2019.
https://doi.org/10.1007/978-3-319-92294-2_30

developed in multiple context, often focusing on cementitious materials [3], while few example also focus on alternative construction material such as soil based material or adobe [2, 4] that can potentially limit the embodied energy of construction by being found as local material.

Yet difficulties are encountered in terms of equipment installation, printing quality at large scale and design integration [3], needing more research in this field.

1.2 Motivations

Combining technological advances in robotics (cable robot and drones), natural materials and generative design, the aim of the project is to explore the potential of additive manufacturing on the construction site, allowing the production of high-performance buildings with sustainable materials.

The following paper discusses a prototype demonstration constructed in 2017 by the authors (and an extending team of collaborators) cumulating multiple years of investigation into adobe 3D printing and cable robotics. The project, within an indoor exhibition space, achieved to erect robotically a 15 m^2 pavilion in just 15 days, including the installation and disassembling of the necessary machinery on-site.

This paper focus on the robotic and technical aspect of the project, while new possibilities of design and materiality will be the subject of a separate paper. Here, the authors discuss in detail success and issues of the robotic setup, as well as the opportunities for improvement and further development of the technology.

2 Cable Robotics for On-Site 3D Printing

Numerous works are going on the definition of 3D printing material and processes for construction, but today one of the main limiting factor for deployment to the construction site is that there are no suitable large-scale equipment existing for this task [3]. The main problem with the current design is either the size and weight, in the case of gantry robots, or the complexity when a robotic arm is installed on a mobile base. On the other hand cable robotics have been introduced for applications involving manipulation of large payloads [5, 6] and/or large workspaces [7, 8] in a manner similar to a crane, with the addition of the control on the 6 degrees of freedom of the load without sway. Most of live demonstrations have been performed for industrial applications [9]. Applications in the field of construction have been foreseen as well [10, 11]. More recently, and more specifically to the problem at hand, cable robots have been showcased as a potential solution for 3D printing [12] operating on the base of lightweight components laying on the ground and without the need of exteroceptive sensors.

The present section describes the feasibility of cable robotics for on-site construction by detailing the actions (logistics, calibration) that were necessary for relocating a large scale cable robot to the demonstration space at a construction fair. Details on printer operation show how craftsmanship of operators and digital design are combined to complete a complex print combined with the installation of modular parts.

2.1 Transporting and Reconfiguring a Cable Robot

In their essence, cable robots are constituted of winches controlling the length of a set of cables, running through sheaves along a specific geometric pattern in space and fixed to an end-effector platform which controls the position in all 6 degrees of freedom. All these components are lightweight, easily relocated as they can pack in small volumes, and can take advantage of existing structure on or around the work site for installation. In addition, operation of a cable robot requires having an accurate measure of where cables are fixed to the platform (the fixing points) and the position of the pulleys attached to the structure (the drawing points), which can be measured on site using usual construction tools like a total station. These properties make cable robots easily relocalizable from one site of operation to another.

The cable robot COGIRO (shared property of TECNALIA and LIRMM) used for the research project presented here is constructed in a modular way. Its 8 identical winches, capable of pulling up to 5000 N, are placed by sets of two on a pallet-sized steel part laying on the floor at each corner of the workspace. These parts also serve as anchor points for the structure. At each top corner, two swiveling sheave assemblies are positioned for guiding the cables from the winches to the moving platform (Fig. 1). Finally, once the cables are disconnected, the platform is easy to move around on a pallet.

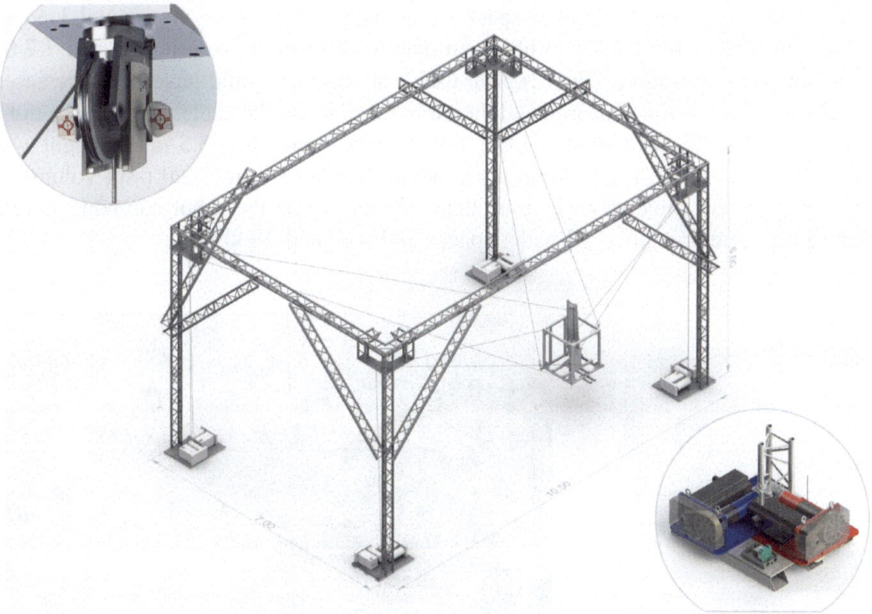

Fig. 1. Cable robot overview, with detail of the winches and drawing points.

Robot's winches, control cabinets, drawing points and platform have therefore been disconnected to the robot's structure and sent to the demo space. In the meanwhile, another similar structure is assembled at the site, making it ready for the winches and drawing points assembled to it and connected to the control cabinet. At that point, the position of the drawing points and the platform at home position have been performed at the demo space by using a Leica Builder 309 total station using reflective targets fixed to the sheave parts and to the platform. The routines of the robot are finally tested with the platform, and the robot is ready to operate.

The whole operation took 2 working days and 2 robotics operators on the demo site, in addition to the transport time. An additional day has been required to set up the demonstration itself.

2.2 Support Structure Design

In the case of this robot, the support structure is closed, meaning that the only forces transmitted to the ground are the dynamic forces on the cable robot platform, which are usually low. It is therefore not necessary to bolt the structure to the floor. An aluminium structure, easy to be assembled and easy to be found in the market), was set up in function of the requirements at the demo space. The aluminum truss parts materialize the edges of a parallelepiped, its top angles being reinforced by angled side bars fixed using shell clamps.

In practice, however, it proved difficult to install 3 shell clamps per joining on the angled bars: one of the joining among 16 was fixed with two clamps only, resulting in a lower first mode of vibration than expected.

Over the available horizontal space of 10.5 by 7.0 m, carrying a total load of 120 kg, an area of 6.6 by 4.4 m has been determined usable for printing, and 7.7 by 4.8 m for other operations (Fig. 2). The usable workspace could have been increased by adding weight to the platform, and therefore tensions on the cable, but this was not a requirement for this experiment. Over these areas, more than 3.5 m of height was accessible for the print head within the structure height of 6 m. Total print volume of the demonstration cable robot is more than 100 m^3, while the robot components (excluding the structure) fitted 10 pallet spaces and weighed 2800 kg.

Fig. 2. L: workspace for the 3D printing cable robot (red: for printing, orange: access only). R: general view of the structure during assembly.

Most of the weight in the current configuration comes from the pallet parts themselves that have not been optimized for their mechanical purpose as they act as ballast, allowing the robot to be simply laid on the floor without bolting it. Oversized motors for the task at hand, in particular in terms of speed, is another factor. A cable robot specifically built for 3D printing, with the same footprint, the same winch force rating (leading to the same workspace), with a lower speed rating and winches and structure anchored directly to the floor has been estimated at 840 kg (structure weight excluded). Besides, it would have been possible to increase the reachable workspace by increasing the weight of the platform with limited impact on the reachable height.

2.3 Material and Extrusion Technology

Using a previously developed pneumatic extrusion system [16], a specially developed earth mix (or adobe) is extruded into thin surfaces of 15 mm width by 3.5 mm height at a speed of 100 mm/s. The reduced capacity of the extrusion system (approx 20 kg) induce an important work of refilling every 20 min. A specific recharge system using hermetic premix cartridges have been developed to reduce downtime of the machine, allowing to recharge the machine in 5 min and keep the system printing 80% of the time.

With such prototype system, the project achieved a limited production rate of 40 kg/h in optimal conditions, still capable to extrude 1000 m per day. Conscient of this limitation, continuous extrusion system compatible with the span of a cable robot is currently being developed to rise up this printing rate by potentially 400% (160 kg/h), but have not been used in this project.

Another speed limitation came from the slow hardening process of adobe, related to the evaporation rate of the water contained in the material, a rate highly dependant on the geometry, site temperature and humidity as well as dependant of the ventilation flows acting on the piece. The drying process require therefore a specific attention as to analyse the hardening condition, guarantee stability and avoid cracks.

Lab experiments and repetitive testing permit the team to settle a maximum 3D printing height limit of 60 cm/day, with a more conservative limit for practice of 20 cm/day. Being able to monitor the drying process is crucial to push the speed of production, as required in this experiment.

3 Drone Scanning

To monitor the printing process, custom designed autonomous robots have been developed by NOUMENA to navigate freely within the complex space of cable robotics (Fig. 3). While frequent 3D scanning of the wall's fabrication allowed the team to detect misalignments occured between the G-code instructions and the deposited material, a special focus was given on the capacity to analyse the hardening of the material over time. To achieve this task, each drone is equipped with a specific sets of cameras capturing pictures to digitally reconstruct the built wall and collect thermal data to register the hardening process for the printed clay.

Fig. 3. Drone with protective frame in polypropylene and carbon fiber

Those inputs are necessary not only to monitor the threshold among digital and physical models but also to control and inform the fabrication process, observing through thermal data the hardening process of every single layer of clay extruded, hence understanding how to regulate increasing payload due to continuous printing. Correlation between thermal value and hardening state can be easily read as water evaporation naturally cools down the material, therefore helping to identify "dry" parts (room temperature) from "wet" parts (cold temperature).

3.1 Monitoring Solutions

Each drone is equipped with a multispectral and a thermal camera. The first one is a Parrot Sequoia, a camera which can collect more than 5 different bands as separated channels: R, G, B, Infrared, stored as TIFF and RGB pictures in jpg format. Our thermal camera is also from a commercial product call Flir Vue Pro camera, capable of registering operating temperature ranges between −20 °C to +50 °C, with a spectral band between 7.5–13.5 μm. These information becomes crucial to evaluate the essication process of the 3D printed wall, calculating how many extra layer of material each specific section can support.

To rebuild 3D models with spatial information, all images taken by the camera are processed using Agisoft Photoscan, a commercial software that performs photogrammetric processing of digital image. Output is a digital model from which it is possible to extract thermal and geometrical data. scope of this application was to compare its spatial digital reconstruction with the programmed path produced by GCODE. This process is performed in the design environment of Rhino Grasshopper developed by McNeel. Using Volvox a Grasshopper add-on, we can import the point cloud information and compare these data with the original G-code generated from the 3 dimensional geometry.

3.2 Results

Resulting scan for the monitoring (Fig. 4) allowed the design and trajectory specialist to monitor possible deviations (max. ± 10 mm in XY plane) and constantly adjust the 3D model of the missing part as well as adapt the 3D printing speed in relation with the load bearing capacity of the part already printed. While data have been gathered every hours with a drone flight, the photogrammetry calculation to obtain a 3D model with thermal value have taken many more hours and allowed the team to properly assess the material state only once a day.

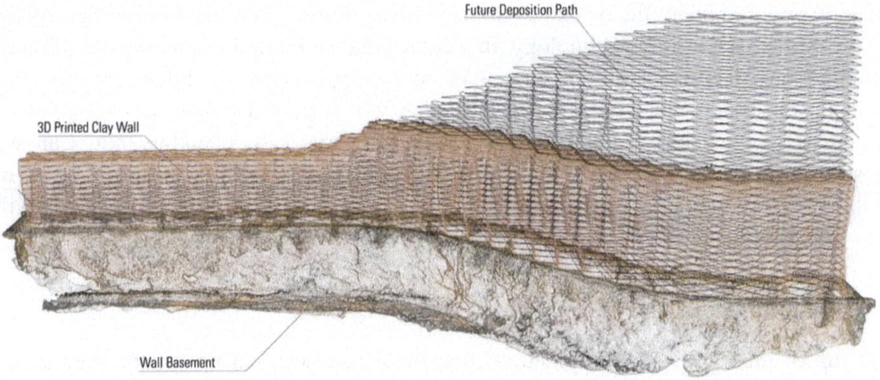

Fig. 4. Resulting scan from the monitoring with drone, compared with planned production file.

3.3 Next Challenges

More powerful hardware for calculation could allow for a shorter lead time between drone flight and model analysis. Aerial robots path planning could also be automated and would requires a robust autonomous system capable to read and react to the cable robot inputs position as well as a precise map of obstacles and printed artefacts. Further development based on ROS [Robotic Operating Systems] could be integrated converting as connected modules the drones, the cable robot system and other possible sensors to monitor the physical performances of the 3D printed artefact in real time.

4 Human-Machine Interface

While automation brings new potentials, human presence on site is still required. Human intervention in this experiment have been crucial for design and planning decisions, material and process analysis, as well as to take care of tasks not yet automated such as refill and maintenance of the extruder, and finishing task on the printed artefact itself. To better integrate the coordination between machines and humans, a set of Human-Machine Interface (HMI) have been developed in the context of this experiment.

4.1 CNC Control

For the purpose of the demonstration, a trajectory specialist was in charge of generating the G-code for the current cartridge. When extrusion stopped, for refilling the extruder with material, a new G-code file was generated on the base of the current index. This new trajectories were constantly updated according to the design evolution and synchronized with the collected data by the drone. The file was fed to the cable robot by the robot specialist after reloading and repositioning of the platform by the process specialist.

While part of the G-code generation and loading have been automatized, a crucial decision was taken constantly at this stage: whether continuing the printing process as plan or stop and adapt the design to the new data coming from the monitoring drones. In the current state of the technology, this crucial decision requires an understanding of structure (e.g. Is the material dry enough or not to support extra load?), materiality (e.g. Will the material bind with existing layer and will it give a proper surface quality), design (e.g. How can I adapt the design to new criteria?), planning (e.g. Can we postpone the printing process?) and socio-political factor, (e.g. Can we take the risk of failing or being late?); a complex decision process that can't be automated yet and will most probably still require human intervention in the future.

4.2 On-Site Quality Control

On top of that, a specific automatized maneuver has been set up for the purpose of halting the print, reloading a material cartridge and resuming the print where it stopped. A human-machine software interface features all the functionalities for running the cable robot in different modes: point-to-point movements, jog mode and CNC operation along trajectories defined using G-code. The cable robot was operated by a robot specialist.

At the same time, a print process specialist was in charge of optimizing the print parameters (feed, synchronization of extrusion and movement, position corrections) to achieve the best print quality. In addition to a set of cameras showing the extruder tip, and, when required, direct visual assessment of the situation close to the platform, his tool was a wireless remote control with different functionalities implemented for the demonstration (Fig. 5): halting and resuming CNC operation (a), halting and resuming extrusion (b), feed speed setting (coarse/fine) (c), Z position correction for start of print (d), Start reload maneuver (e).

4.3 Cable Robot as a Precise Crane

The operation of the cable robot during the demonstration took 9 days of printing to achieve the 6.25 by 3.75 by 2.4 m pavilion. On the next day the extruder was removed to let the cable robot pick wooden roof elements to be placed on top of the freshly printed wall, to complete the pavilion (Fig. 6). On that operation the cable robot was operated in manual jog mode, similarly to a crane. This capability shows the potential of cable robots for performing assembly of unique elements (windows, lintels, functional wall panels such as solar panels) in sequence with a 3D printing process.

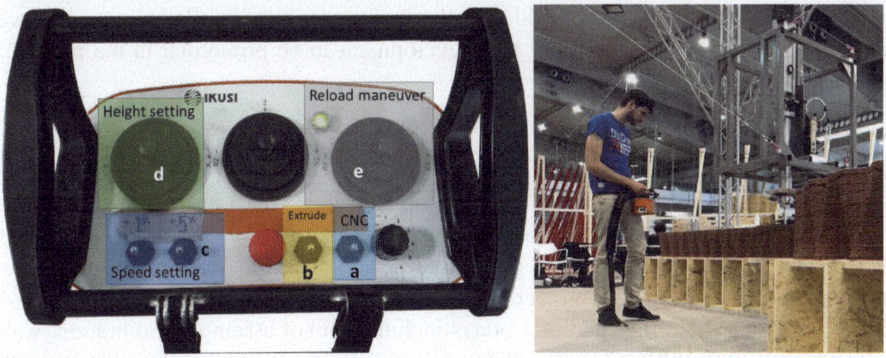

Fig. 5. L: detail on the remote control and the functionalities. R: operation with the remote control.

Fig. 6. Use of the COGIRO robot as a precise teleoperated crane to position prefabricated roof elements.

4.4 Conclusions

In this project, collaboration between human and machine where primordial, combining craftsmanship and digital tools to ensure a good workflow and quality in the finished construction. Further development need to be addressed to improve security and collaboration.

In the current state of art of machine security, such a machine should be operating in a space that would not be accessible to humans, unless drastic security measures were taken. Recent updates in the norms (ISO TS 15066:2016 on collaborative robots) open the way for free access to the workspace, conditioned to the presence of workspace scanning sensors. Considering the importance that parallel operations have on a construction site, we consider that development to be primordial in the future.

5 Conclusion

Construction using large scale 3D printing on-site with 100% natural material have been demonstrated feasible, combining technological advances in robotics (cable robot and drones), natural materials and generative design. The full scale prototype of the technology deployed in this project have shown the possibility to 3D print onsite a 20 m^2 pavilion within the span of 15 days including robot assembly, calibration, wall printing, roof assembly and dismantling. Adaptation of hardware, software, material, and design have proven to make possible such system prototype demonstration in a relevant environment (TRL 6). Direct collaboration with human operators and craftsmen with the digital and robotic system have shown to be crucial to the good execution of the project.

Yet, the experience described in this paper also exhibit the limitations of the technology operating on site, needing further research and development to improve the human-machine collaboration in terms of security, improve material, increase the extruder throughput and the overall printing productivity as well as automate further the monitoring by drone. Furthermore, the technology have yet to be tested in outdoor conditions where weather conditions will affect the behaviour of this natural material during printing and drying. In this context, the direct connection among material performance, environmental conditions, design and fabrication would led the real-time monitoring and adaptation to enhance the feasibility of the technology.

The prototype demonstration presented in the paper has led to the conclusion that there is a significant potential of using the technology for big scale sustainable architectural constructions on site, opening new opportunity for the design of performative building.

Acknowledgments. "OnSite Robotics" is a collaborative project between Institute for Advanced Architecture of Catalonia (IAAC) and TECNALIA. IAAC team (Concept, Coordination, Design, Extrusion, Material & Sensors): Areti Markopoulou, Aldo Sollazzo, Alexandre Dubor, Edouard Cabay, Raimund Krenmueller, Ji Won Jun, Tanuj Thomas, Kunal Chadha, Sofoklis Giannakopoulos. TECNALIA Robotic Team (Cable Robot): Mariola Rodríguez, Jean-Baptiste Izard, Pierre-Elie Herve, Valérie Auffray, David Culla, Jose Gorrotxategi. TECNALIA Construction Division team: Mikel Barrado, Idurre Fernandez, Juan José Gaitero, Elena Morales, Iñigo Calderón, Amaia Aramburu. NOUMENA Team (Drone development, Data collection): Starsky Lara, Chirag Rangholia, Daniele Ingrassia (Fab Lab Kamp Lintfort), Marco Sanalitro, Eugenio Bettucchi, Andrea Melis, Adrien Rigobello. Other partners: LIRMM (Robot research center), Luciano Carrizza (Technical adviser), Joaquim Melchor - Art Cont (Material expert), Lenze (Automation), Wam (Solid process), NanoSystem (Material nanotechnology), Nicolas Weyrich (Video animation).

References

1. Brell-Çokcan, S., Braumann, J. (eds.) ROB|ARCH 2012: Robotic Fabrication in Architecture, Art and Design. Springer, Vienna (2013)
2. Kwon, H.: Experimentation and analysis of Contour Crafting (cc) process using uncured ceramic materials. Ph.D.thesis, University of Southern California (2002)
3. Wu, P., Wang, J., Wang, X.: A critical review of the use of 3D printing in the construction industry. Autom. Constr. **68**, 21–31 (2016)
4. WASProject. http://www.wasproject.it/w/en/about-us/. Accessed 1 Mar 2018
5. Albus, J., Bostelman, R., Dagalakis, N.: The NIST robocrane. J. Robot. Syst. **10**(5), 702–724 (1993)
6. Culla, D., Gorrotxategi, J., Rodríguez, M., Izard, J.B., Hervé, P.E.: Full production plant automation in industry using cable robotics with high load capacities and position accuracy. In: Ollero, A., Sanfeliu, A., Montano, L., Lau, N., Cardeira, C. (eds.) ROBOT 2017: Third Iberian Robotics Conference, vol. 2, pp. 3–14. Springer International Publishing (2018)
7. Nan, R., Li, D., Jin, C., Wang, Q., Zhu, L., Zhu, W., Qian, L., et al.: The five-hundred-meter aperture spherical radio telescope (FAST) project. Int. J. Modern Phys. D **20**(06), 989–1024 (2011)
8. Skycam. http://skycam.tv/. Accessed 2 May 2018
9. Tecnalia: Cable Driven Parallel Robotics for industrial applications (2015). https://youtube/px8vwNerkuo. Accessed 2 May 2018
10. Bosscher, P., Williams II, R.L., Bryson, L.S., Castro-Lacouture, D.: Cable-suspended robotic contour crafting system. Autom. Constr. **17**(1), 45–55 (2007)
11. Izard, J.B., Gouttefarde, M., Baradat, C., Culla, D., Sallé, D.: Integration of a parallel cable-driven robot on an existing building façade. In: Bruckmann, T., Pott, A. (eds.) Cable-Driven Parallel Robots, pp. 149–164. Springer, Berlin, Heidelberg (2013)
12. Tay, Y.W.D., Panda, B., Paul, S.C., Noor Mohamed, N.A., Tan, M.J., Leong, K.F.: 3D printing trends in building and construction industry: a review. Virtual Phys. Prototyp. J. **12**(3), 261–276 (2017)
13. Izard, J.-B., Dubor, A., Hervé, P.-E., Cabay, E., Culla, D., Rodriguez, M., Barrado, M.: Large-scale 3D printing with cable-driven parallel robots. Constr. Robot. **1**(1–4), 69–76 (2017)
14. Minke, G.: Building with Earth, Design and Technology of a Sustainable Architecture. Birkhäuser, Berlin (2006)
15. Fashami, N., Jokic, S., Lara, S.N.: FabClay. https://fabbots.wordpress.com/2012/12/09/fabclay-2/. Accessed 1 Mar 2018
16. Gianakopoulous, S.: Pylos. http://pylos.iaac.net/. Accessed 1 Mar 2018
17. Dubor, A., Cabay, E.: Energy efficient design for 3D printed earth architecture. In: De Rycke, K., Gengnagel, C., Baverel, O., Burry, J., Mueller, C., Nguyen, M.M., Rahm, P., Thomsen, M.R. (eds.) Humanizing Digital Reality, Design Modelling Symposium Paris 2017, pp. 383–394. Springer, Singapore (2017)
18. Lagüela, S., González-Jorge, H., Armesto, J., Arias, P.: Calibration and verification of thermographic cameras for geometric measurements. Infrared Phys. Technol. **54**(2), 92–99 (2011)
19. Olson, E.: Robust and efficient robotic mapping. Ph.D. thesis, Massachusetts Institute of Technology (2008)

Application and Practice

Tailored Structures, Robotic Sewing of Wooden Shells

Martin E. Alvarez[1(✉)], Erik E. Martínez-Parachini[1], Ehsan Baharlou[1], Oliver David Krieg[1], Tobias Schwinn[1], Lauren Vasey[1], Chai Hua[2], Achim Menges[1], and Philip F. Yuan[2]

[1] University of Stuttgart, Keplerstrasse 11, 70174 Stuttgart, Germany
martin.alvarez@icd.uni-stuttgart.de,
martinez.erik.e@gmail.com

[2] Tongji University, 1239 Siping Road, Shanghai, China

Abstract. This paper investigates the use of robotics with sensing mechanisms in combination with industrial sewing techniques to explore new strategies for the fabrication of thin wooden shells.

The investigation is characterized by a parallel theoretical and prototype-based methodology, the latter serving as a vehicle to further the technical development, which could ultimately enable novel architectural qualities.

The development unfolds in four interdependent avenues: (1) The transfer of textile patterning techniques used in garment production to inform the design of flexible 3 mm beech plywood segments; (2) The capacity of wood to be elastically bent and connected into geometrically stable structures; (3) the use of sewing as a new construction joint for thin material; and (4) The integration of sewing into an automated and adaptive robotic fabrication workflow enabled by sensing and scanning with the capacity to join complex three-dimensional curved structures at an architectural scale.

Keywords: Wood · Timber construction · Robotic sewing · Adaptive robotics Sensing · Textile techniques

1 Introduction

Acting as the closest protective layer to the human body, textiles share essential functions with architecture. As such, garment design, textile techniques and their fabrication technologies are relevant to architectural design.

Many aspects of production processes in garment production remain manual operations. This problem, characterized by the lack of information from physical conditions in digital tooling processes, has pushed the development of materially aware sewing techniques, and has catalyzed progress in automating parts of garment production (The Economist 2015).

M. E. Alvarez and E. E. Martínez-Parachini—First Authors.
E. Baharlou, O. D. Krieg, T. Schwinn, L. Vasey, C. Hua, A. Menges and P. F. Yuan—Second Authors.

J. Willmann et al. (Eds.): ROBARCH 2018, *Robotic Fabrication in Architecture, Art and Design 2018*, pp. 405–420, 2019.
https://doi.org/10.1007/978-3-319-92294-2_31

Recent developments in architectural design showcase the relevance of incorporating garment design and sewing techniques into architectural design (Bechert et al. 2016). The concurrent complexity of these new material behavior and construction techniques renders conventional simulation and pre-calculation approaches obsolete, making new cyber-physical fabrication strategies necessary (Menges 2015).

Concurrently, standard joinery techniques in timber construction rely on substantial material overlap at connection points to meet regulation standards, therefore research in new joinery techniques is increasingly relevant.

The paper will therefore discuss innovation within four fields of research in the context of two recently finished architectural demonstrators: (1) The transfer of textile patterning techniques in order to inform the shaping of thin plywood segments; (2) The capacity of wood to be elastically bent and connected into geometrically stable structures; (3) Sewing as a distributed and easily applied joinery technique for three-dimensional structures made from thin plywood; (4) Adaptive robotics enabled by sensory equipment (Fig. 1).

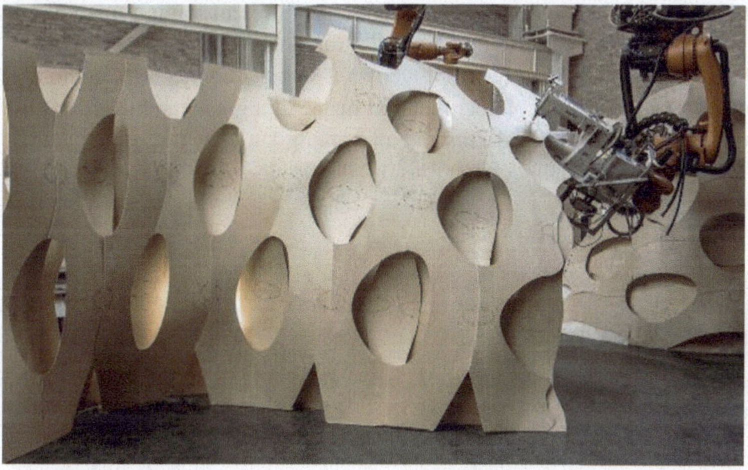

Fig. 1. Robotic sewing fabrication process.

2 Background

2.1 Garment Design and Patterning in Architecture

Patterning is the principle of transferring flat cut patterns into three-dimensional structures. In architecture, it has remained within the design realm of tensile membranes and pneumatic structures. However, fashion designer Tomoko Nakamichi's use of rigid textiles suggests a crossover from fashion to architecture (Nakamichi 2010). As suggested by her work, this project explores the innate potential of patterning with materials stiffer than fabric and their potential to create stable three-dimensional structures from bent developable surfaces.

2.2 Patterning Techniques in Wood and Architecture

The elasticity and commercial availability of thin plywood make it an ideal material to investigate the use of patterning design strategies from textile materials for extremely thin wooden shell structures. Its higher stiffness however, does not allow a direct transfer of textile patterning strategies, and must consider higher stiffness and developability. Ray and Charles Eames's unrolled Chair and Splint prototypes, for example, show the procedure of designing flat patterns to be cut and formed to fit the human body (Eames and Herman Miller Furniture Company 1946). More recently, the ICD/ITKE Research Pavilion 2010 used plywood strips in a similar manner (Fleischmann et al. 2012).

2.3 Sewing Wood

Traditional fasteners provide reliable solutions for structural timber connections. However, the slenderness of thin plywood significantly changes the relationship between the size of the connection and the material thickness. Because of their impact on material performance and structural continuity, different joining methods should be considered for extremely thin wooden shell structures.

In this regard, sewing can be employed as a suitable connection method. The very small punctures required to connect material makes it a great candidate for joining thin plywood (Bechert et al. 2016). Instead of relying on few large connection points, sewing allows many small connections to distribute loads more uniformly over a predefined area (Fig. 2).

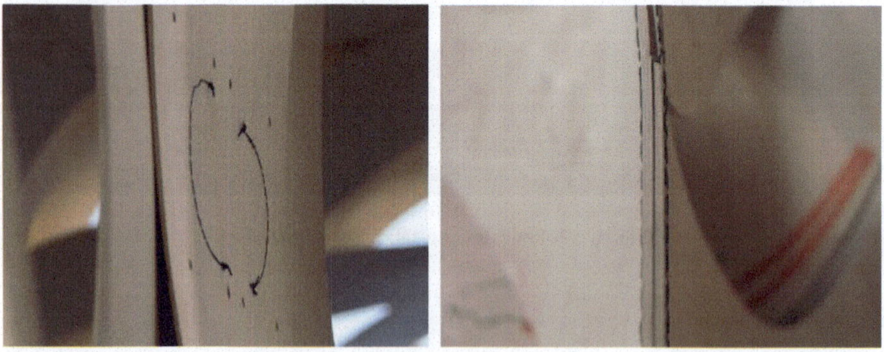

Fig. 2. Details of sewn connections.

2.4 Adaptive Robotics

The complexity of manipulating fabric in garment production has made the process difficult to automate. However, SofWear, a company based in Atlanta, USA, utilizes high speed cameras to register locations of threads, thus enabling precise manipulation and sewing of garments (The Economist 2015).

In-situ robotics for architecture and construction have had similar challenges and developments. In an industrial context, the implementation of sensor based technology has allowed KUKA, an industrial robot manufacturer, to develop the KMR Quantec system to work autonomously on large scale industrial applications, such as polishing an entire aircraft (KUKA Roboter GmbH 2018). In architecture and construction, researchers at ETH Zurich continue to develop autonomous mobile robots for construction sites (Dörfler et al. 2016).

The similar development within these two industries is used as an analogue in the presented fabrication system, utilizing sensing processes like those employed by the garment production industry to enable adaptive robotic construction processes for materially complex assemblies.

2.5 State-of-the-Art

Buckminster Fuller's PlyDome in 1957 was one of the first projects which showcased the potential of elastically bent thin plywood to generate structurally stable architectural enclosures (Fuller and Marks 1973) (Fig. 3b). Continued research on this topic has been developed by Marcus Hudert at EPFL, through his woven timber structures (Fig. 3a) (Hudert 2012). Hudert's work showcases the qualities textile design and patterning techniques can bring to architectural design.

Fig. 3. (**a**, left) Marcus Hudert, Timber Fabric, (**b**, center) Buckminster Fuller, 1957 plydome and (**c**, right) ICD/ITKE Research Pavilion 2015–16 all employ the elastic properties of wood for constructing three-dimensional structures.

The ICD/ITKE Research Pavilion 2015–16 (Fig. 3c) also showcases this potential, however, additional development in robotic sewing processes allowed for an automated fabrication of complex three-dimensional shapes, and for using sewing as fabric connections in thin materials (Bechert et al. 2016).

3 Methodology

The following section describes the tools and methods used to develop the four research paths outlined in the first section of this article.

3.1 Patterning Design

The tectonic and assembly systems presented in this paper were developed with the use of strategies from textile patterning techniques and their possible translation into stiffer materials such as plywood.

The use of patterning techniques designed for textiles cannot be directly translated into plywood since the materials exhibit different stiffness properties. The exploration of these patterning techniques in paper models served as a stepping stone to move from fabric to wood (Fig. 4).

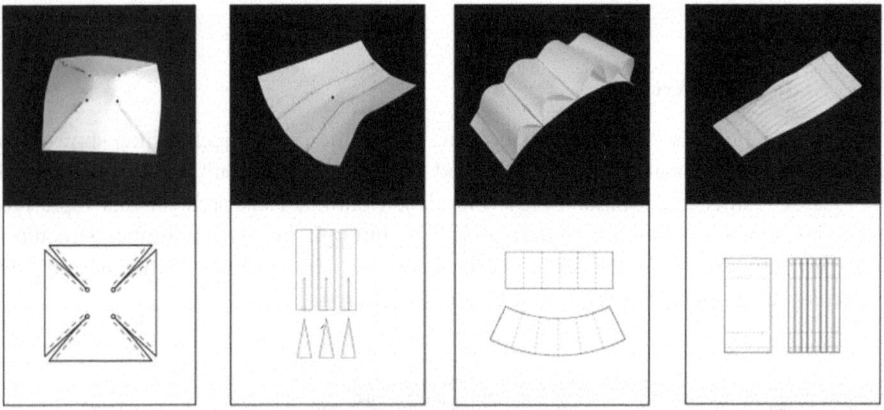

(a) (b) (c) (d)

Fig. 4. **a, b, c, d** Paper models above and corresponding cutting patterns below: **(a)** Darting. **(b)** Godet. **(c)** Flounces and **(d)** Pleating.

The following patterning strategies were investigated: Darting (Fig. 4a), which consists of removing V-shaped darts and joining the resulting edges. When used in paper and plywood, it creates synclastic curvature. The reverse method, Godet (Fig. 4b), consisting of adding material to a similar V-shaped cut, results in anticlastic curvature. Flounces (Fig. 4c), are used to manipulate curvature of two or more layers. This method consists of connecting two linear segments of different lengths at pre-determined locations to create controlled curvature. Finally, Pleating (Fig. 4d), or the act of folding to create depth and structure, was applicable as an edge stiffening method (Wolf 1996).

While the implementation of textile patterning techniques provides viable strategies for the strengthening and assembly of thin plywood from flat sheets, the methods were further developed into tectonic systems, which considered the development of a specific robotic fabrication setup.

3.2 Wood Forming: Species Bending Radii and Bending Tests

Beech is one of the most flexible species of wood with a minimal steam bending radius of 370 mm at 25.4 mm thickness (Stevens 2007). An empirical testing method was established to extrapolate a mathematical relationship between the width and length of a piece of plywood and its maximum bending capacity before failure. The gathered information yielded the maximum bending capacity of a piece of 3mm plywood in relation to its aspect ratio, which was used to ensure no segment of the full-scale demonstrators described in Sect. 3.10 was beyond this limit.

3.3 Microscopic Behavior of Fibers Through Puncture

The effect of puncturing a piece of 2 mm beech veneer with a 1.4 mm sewing needle was observed at a microscopic level (Fig. 5). The objective of this test was to understand the behavior of the wood fibers before and after being impacted by the sewing needle. Microscopic images show that, while some of the fibers in the middle of the impact area are broken, most fibers move aside to make space for the needle. The continuity of fibers around the puncture hole is very beneficial for load transfer as it minimizes potential breaking points.

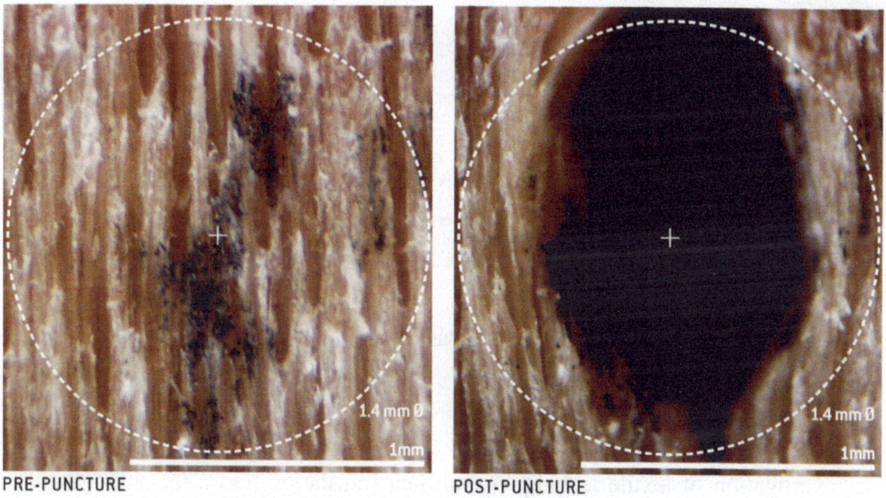

Fig. 5. Microscopic images of beech veneer, before (left) and after (right) puncturing with a sewing needle.

3.4 Tension Load Test

To establish the performance of sewing as a joinery method in construction, it was tested against a traditional bolted connection. A tensile load test was carried out with lapped 3 mm beech plywood samples measuring 100 × 150 mm. Bolted testing samples were prepared with M6 and M8 standard steel bolts with rows of 2 and 3 bolts at different distances. Sewn samples were prepared with 1, 2, and 3 rows of stitching (Fig. 6). All testing samples oriented the plywood's main grain direction parallel to the tension forces applied. All samples were loaded until material failure.

The results of the 2xM8@32 mm and the 3 Sewing Line samples (Highlighted in Fig. 6) both reached approximately 5000 N and deformed about 9 mm before failure, indicating that the same load performance can be achieved with both connection methods. Bolted connections did remain stiffer during deformation while sewn ones stretched the string until failure. Each stitch can carry loads ranging from 140 to 160 N on average when considering the total maximum load before failure and the number of stitches in each connection, suggesting the potential for calibrated connections.

Testing provided background to show that sewing can replace bolted connections in thin materials, with the advantage that its small punctures allow for joining very closely to the edges of the material, and that distributed connections prevent continuous fractures (Shown on the left in Fig. 6).

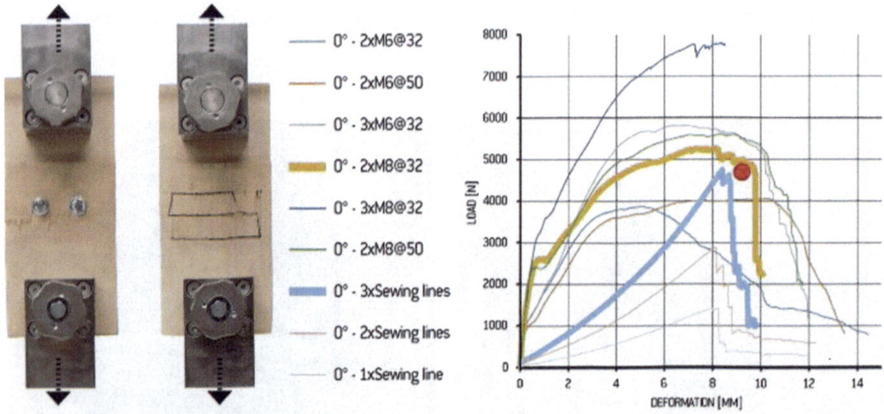

Fig. 6. Tensile load test of Sewn vs Bolted connections in 3 mm beech plywood.

3.5 Digital Design Techniques

The investigation of textile techniques and their translation into a tectonic system in paper (as described in Sect. 3.1) necessitated the development of a computational modeling tool. It was used to generate the undulating segments as developable surfaces and to check for fabrication limitations.

The tool was developed using the parametric plugin Grasshopper, in the Rhinoceros 5 modeling software. The modelling allowed to relate a NURBS surface as a design guide to a number of segments populating the surface accordingly. During this integrative design process, constraints such as developability, maximum bending (Sect. 3.2) and stock material fitting were checked together with the physical fabrication constraints derived from the geometry of the robotic effector in relation to the geometry of the design itself. See Sect. 4.4 for explanation of each demonstrator's design constraints.

3.6 Fabrication Process Overview

The elastically bent enclosure designed with the tool described in Sect. 3.5 was stabilized by bending thin plywood segments against each other.

The workflow is summarized in the following steps. (1) Marking plywood segments with markers where sewing connections are necessary, (2) Milling pattern outlines of each segment from flat sheets of plywood, (3) Bending and temporarily clamping segments at indicated points into a three-dimensional structure, (4) Scanning constructed portion, (5) Maneuvering and entering the material, (6) Sewing along markers, (7) Retract and exit the material, and (8) Repeat procedure from step 4.

The fabrication setup was developed based on the possibilities afforded by integrating robotic manufacturing with a traditional sewing machine setup. Instead of moving pieces through a sewing machine as it is done traditionally, mounting the machine as an end-effector on a 6-axis robotic arm reverses the relationship of the machine to the work piece, allowing it to be used on larger segments or potentially in-situ.

3.7 Connection Design and Sensing

Precise shaping and positioning of each connection used in the bent plywood pieces is essential since it ultimately defines the shape of each neighboring segment, and consequently determines the three-dimensional arrangement of the complete structure. Therefore, prior to milling out each of the segments, a marker is CNC drawn onto the wood sheets, indicating the position and geometry of each sewn connection. After cutting each segment, the markers are used to drive the autonomous robotic sewing process and sew the necessary connections during assembly (Fig. 9, step 5).

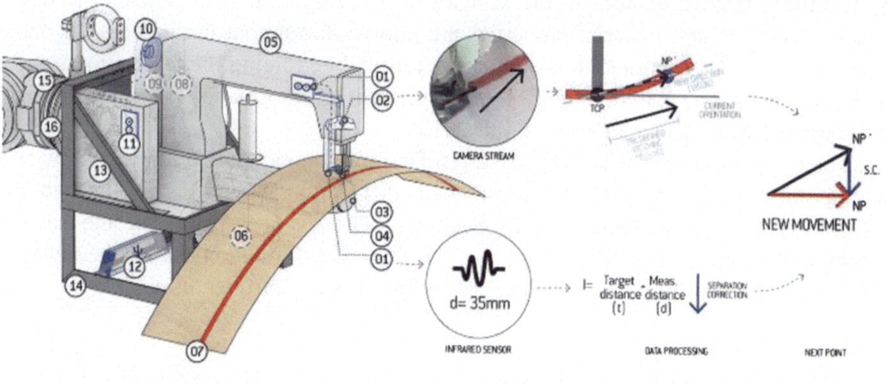

Fig. 7. Left: sewing effector: (1) VL6180 Infrared distance sensor & Arduino Control, (2) Havit hv-n5086 usb webcam (3) Bobbin (4) Sewing needle (5) Hightex CB4500 Sewing Machine (6) Sewing Machine motor (7) Sewing path, (8) Pneumatic control (9) Compressed air intake (10) Sewing machine encoder, (11) Sewing machine encoder Arduino control (12) Microsoft Kinect 1, (13) Sewing machine motor stepper control (14) Effector steel frame (15) Schunk Pneumatic Plate (16) Kuka KR-125 industrial robot arm. Right: sewing effector sensors.

While sewing, an RBG webcam is used to detect the markers and map them to the Tool Coordinate System from which a directional vector is extracted. Simultaneously, an infrared time-of-flight distance sensor is used to detect the distance from the arm of the sewing machine to the plywood to avoid colliding with the structure.

The sensor's value is translated into a vector perpendicular to the sewing arm plane. This vector is then added to the direction vector from the camera to produce a combined movement vector for the next movement step. The vector's amplitude is 10 mm, an optimal stitch length. This procedure eliminates the need for any pre-programmed code.

Camera calibration and the client computer′s ability to parse live stream and photographic data became difficult process in some fabrication scenarios. The use of consistent indoor lighting was helpful to show the contrast between the wooden surface and the marker for easier detection (Fig. 7).

3.8 3D Scanning

A Microsoft Kinect 1.0 device with the Kinect Fusion scanning software package is used to obtain a colored mesh of the structure before sewing a new piece onto it. By using 3 markers with known positions in the workspace, the scanned geometry is oriented digitally through a manual procedure, in relation to the position of the robot, to prevent collisions and to generate a robotic tool path for guiding the sewing machine to the first point to be sewn. While the Kinect scan was also meant to track the deformation of the piece through construction, the slow process of scanning and re-orienting the geometry made it impossible to apply at each step of the fabrication process.

3.9 Robotic System Integration

In repetitive factory manufacturing scenarios, pre-programming is suitable as workspace conditions and material properties are known throughout the process. In materially inconsistent scenarios, this approach is unsuitable, forcing the fabrication system to rely on an on-line network architecture, dependent on information about the physical environment from sensors (Vasey et al. 2015).

The project used the ICD/ITKE 2015/16 Pavilion robotic fabrication workflow as a base point for development (Schwinn et al. 2016), but extended this through the integration of sensors, described in Sect. 3.7. The cyclical nature of the sewing machine from stitch to stitch, determines the flow of information within the network (Fig. 8). A client computer is connected to the robot controller through a software

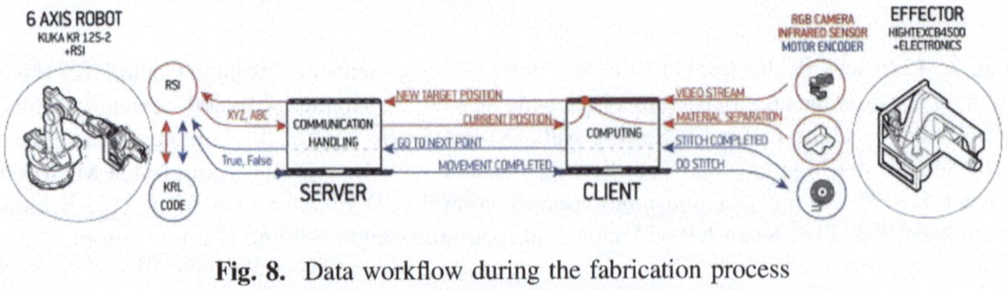

Fig. 8. Data workflow during the fabrication process

interface that allows real-time (12 ms) communication through UDP packages. The Robot controller sends the robot's position and receives information that the Client computer forwards. The client computer receives a Boolean value confirming a cycle is completed, the camera stream, and the distance to the material value, from the sewing effector. The client computer integrates those three values to generate a new target position (Sect. 3.7).

3.10 System Integration

Following the material and fabrication methods described, the entire approach was developed and evaluated through the construction of two full scale architectural demonstrators (Figs. 11 and 13) described in Sect. 4.4. The system's demonstrators were modelled using the tool described in Sect. 3.5, the digital geometry of the patterns was unrolled in preparation for milling with sewing markers and lines for sewn connections. Assembly was carried out following the steps described in Sect. 3.6.

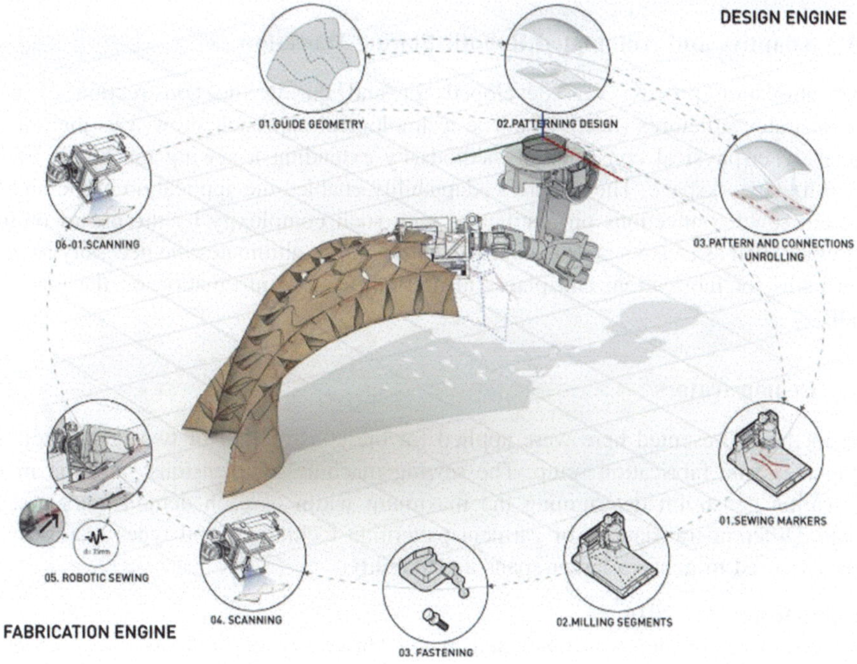

Fig. 9. Design and fabrication workflow overview.

4 Results and Discussion

4.1 Patterning and Forming

The translation of garment patterning strategies into plywood opened a field of tectonic systems where structures are strengthened through elastic bending revealing their potential to create stable free-standing structures.

4.2 Sewing

To the authors' knowledge this is the first time sewn connections have been employed at an architectural scale and used as the main joinery strategy throughout the structure. While the ICD/ITKE Research Pavilion 2015–16 showcased the potential of using fibrous connections in a larger structure, the sewing was secondary to a laced connection. The research presented here shows the load bearing capacity of sewn connections without using glue, and their potential for the joinery of extremely thin wooden shell structures.

4.3 Adaptive and Automated Robotic Sewing Workflow

The fabrication process was developed for and during the construction of two demonstrator structures (Sect. 4.4). The technological approach showcases the potential of cyber-physical construction methods by extending a sewing machine's traditional fixed workspace. The system′s adaptability enabled the application of the highly efficient sewn connections on a structure with such complexity by the use of online strategies such as sensors and scanners. Additionally, it eliminates the necessity to pre-plan paths for fabrication, and places all the fabrication information on the material itself.

4.4 Demonstrators

The methods presented here were applied for the construction of two demonstrators using the same fabrication setup. The sewing machine's dimensions were the main fabrication constraint determining the maximum width of each demonstrator's segments. Different translations of garment patterning techniques and types of plywood were explored to achieve larger spans and stability.

Demonstrator A - 2016

This first demonstrator was fabricated at the University of Stuttgart using 3.5 mm Beech three-layered plywood. The setup included a KR-120 6 axis robot and the work piece was moved around manually in order to keep the robot within reach. The tectonic system alternated single and double-layer areas to gain structural depth and used undulating segments. These maximized the system's coverage while allowing the sewing machine to enter at each valley and sew a connection to the previous piece in tune with an industrial robot's maximum reach (Fig. 10).

Fig. 10. Tectonic system in relation to fabrication constraints. Demonstrator A - 2016.

Fig. 11. Interior (left) and exterior (right) of demonstrator A.

Demonstrator A had a construction weight of 60 kg, a total covered Area of 6.4 m^2, a structural weight of 2.4 kg/m^2, a maximum span of 3.7 m, a material thickness of 3 to 6 mm and a total stock material volume of 0.083 m^3.

Demonstrator B - 2017

The second demonstrator was fabricated at the facilities of Tongji University in Shanghai, China, in collaboration with the University of Stuttgart.

Due to the increase in size, this second translation aimed at taking the undulating logic into a fully interconnected double layered structure in order to accomplish higher stiffness (Fig. 12). In addition to the parameters mentioned in Sect. 3.5, the shape of the patterns was designed to filter out direct light into the structure.

The demonstrator's parts were fabricated with a 9-axis robotic setup that had a KR-120 robot arm mounted on a 3-axis gantry accessing an area of 11 × 6 × 3.6 m. Such large dimensions afforded the possibility of pre-fabricating pieces up to 9 × 5 × 3 m. Nevertheless, because of shipping constraints the demonstrator was split in 6 smaller pieces. Demonstrator B employed a standard 3.5 mm plywood with a core of Lauan wood and thinner finishing faces of Birch veneer. The Pavilion had a construction

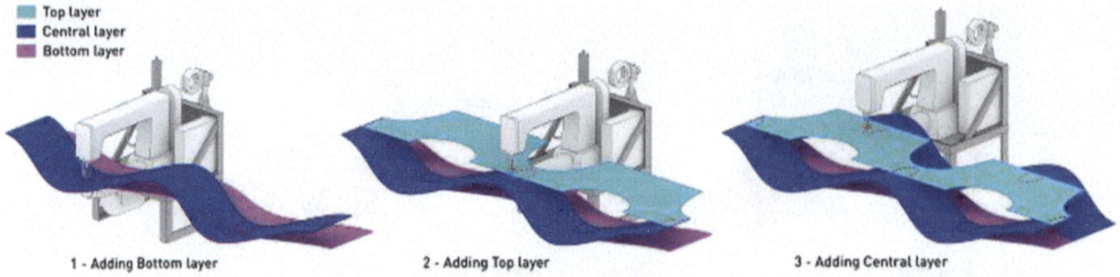

Fig. 12. Tectonic system in relation to fabrication constraints. Demonstrator B. Red represents the current row of connections

Fig. 13. Interior of Demonstrator B - 2017 (left), architectural Demonstrator B - 2017 (right).

weight of 178.3 kg, a total covered Area of 17.85 m^2, a structural weight of 1.74 kg/m^2, a maximum span of 5 m, a material thickness of 3 to 6 mm and a total stock material volume of 0.509 m^3.

5 Conclusion

The innovative aspect of these projects is the combined and integrative approach of their development and fabrication, which furthered several areas of research.

The project enhanced a robotic sewing workflow into an autonomous and adaptive robotic sewing procedure using sensing and vision coordinated with a robot sensing interface. The cyber-physical construction methods showcased here challenge traditional pre-planned fabrication involving writing code, controlled fabrication environments and digital material simulation. The simplicity of embedding the structure's segments with connection information through physical markers, automated parts of production and accounted for construction unpredictability, material imperfections, deflection and work-piece repositioning.

The design of an integrative and material-aware tectonic system was developed, considering bending capacity, sewing machine workspace, and architectural performances. The translation of textile patterning strategies into stiff materials was explored for the use of thin plywood, showcasing their architectural potentials. Finally, sewing was developed as a viable load bearing construction joint for thin shells and was validated through structural testing. While Demonstrator B was significantly larger than Demonstrator A, the researchers believe B was at the limit of this system's size, structurally and shipping wise. The thinness of the material, and the strength it acquired in the elastically bent assembly reached a limit.

Further research could be carried out by testing material sagging and weathering over time, including sewing loosening and its resulting effect on the structure. Overall structural capacity of an entire assembly, including the internal elastic bending behavior of the structure could be analyzed. Additionally, individual joints could be engineered to local loads, based on more in-depth testing as shown in Sect. 3.4. Further development into the robotic fabrication system could provide a robust enough fabrication method to be utilized in-situ with mobile robots.

Acknowledgements. Demonstrator A was developed and constructed within the ITECH MSc programme by the first authors. Demonstrator B – 2017 was supported by the Sino-German Center for Research Promotion project (GZ 1162) where the where the Institute for Computational Design and Construction (University of Stuttgart) and the Digital Design Research Center (Tongji University) act as co-principal investigators. Additionally, it was exhibited as part of the inaugural exhibition 'Mind the digital' at the Design Society Museum, Shenzhen, China.

References

Dörfler, K., Sandy, T., Giftthaler, M., Gramazio, F., Kohler, M., Buchli, J.: Mobile Robotic Brickwork. Robot. Fabr. Archit. Art Des. **2016**, 204–217 (2016). https://doi.org/10.1007/978-3-319-26378-6_15

Bechert, S., Knippers, J., Krieg, O., Menges, A., Schwinn, T., Sonntag, D.: Textile fabrication techniques for timber shells: elastic bending of custom-laminated veneer for segmented shell construction systems. In: Adriaenssens, S., Gramazio, F., Kohler, M., Menges, A., Pauly, M. (eds.) Advances in Architectural Geometry 2016, pp. 154–169. vdf Hochschulverlag AG ETH Zurich, Zurich (2016). ISBN 978-3-7281-3778-4

Fleischmann, M., Knippers, J., Lienhard, J., Menges, A., Schleicher, S.: Material behaviour: embedding physical properties in computational design processes. In: Architectural Design, vol. 82 no. 2, pp. 44–51. Wiley Academy, London (2012). ISBN 978 0470973301

Fuller, R.B., Marks, R.W.: Dymaxion World of Buckminster Fuller. Anchor Books, New York (1973)

Eames, C., Herman Miller Furniture Company: Patent US2548470 - Laminated Splint. US (1946)

Hudert, M.: Structural timber fabric: applying textile principles in building scale. Ph.D. thesis, École Polytechnique Fédérale de Lausanne, Lausanne (2012)

KUKA Roboter GmbH: KMR QUANTEC. Mobile robotics for the precision machining of XXL components. https://www.kuka.com/-/media/kuka-downloads/imported/9cb8e311bfd744b4b0eab25ca883f6d3/kmr-quantec-en.pdf. Accessed 28 July 2018

The Economist: Made to Measure, a robotic sewing machine could throw garment workers in low-cost countries out of a job. The Economist (2015). http://media.economist.com/news/technology-quarterly/21651925-robotic-sewing-machine-could-throw-garment-workers-low-cost-countries-out

Nakamichi, T.: Pattern Magic, 1st edn. Laurence King Publishing, London (2010)

Schwinn, T., Krieg, O., Menges, A.: Robotic sewing: a textile approach towards the computational design and fabrication of lightweight timber shells, in posthuman frontiers: data, designers, and cognitive machines. In: Proceedings of the 36th Conference of the Association for Computer Aided Design in Architecture (ACADIA), Ann Arbor, pp. 224–233 (2016)

Stevens, W.C.: Wood Bending Handbook, p. 96. Fox Chapel Publishing Company, East Petersburg (2007)

Vasey, L., Baharlou, E., Dörstelmann, M., Koslowski, V., Prado, M., Schieber, G., Menges, A., Knippers, J.: Behavioral design and adaptive robotic fabrication of a fiber composite compression shell with pneumatic formwork. In: Combs, L., Perry, C. (eds.) Computational Ecologies: Design in the Anthropocene, Proceedings of the 35th Annual Conference of the Association for Computer Aided Design in Architecture (ACADIA), University of Cincinnati, Cincinnati, OH, pp. 297–309 (2015). ISBN 978-0-69253-726-8

Wolff, C.: The Art of Manipulating Fabric, 2nd edn. Chilton Book Co., Baltimore (1996)

Menges, A.: The new cyber-physical making in architecture: computational construction. Architectural Des. **85**(5), 28–33 (2015). http://doi.wiley.com/10.1002/ad.1950

Dynamic Robotic Slip-Form Casting and Eco-Friendly Building Façade Design

Lei Yu(✉), Dan Luo, and Weiguo Xu

Architecture School of Tsinghua University, Haidan District, Beijing, China
asworkshop@vip.163.com

Abstract. Robot arm technology has been increasingly considered a future solution for the building industry, which is keen to develop eco-friendly and labor-efficient processes. The automobile industry provides a powerful reference for such a revolution. However, in contrast to the automobile industry, which mostly uses standard industrial materials, such as metal, the building industry still relies on a variety of traditional materials, including concrete cement, brick and clay. Although precast building components have been highly promoted, they present innate constraints, including limitations associated with transportation, storage, and installation capacity, which greatly impact the overall budget. In this paper, research on a dynamic slip-form concrete extrusion method based on a robot arm technique will be presented within the scope of laboratory experiments, on-site fabrication and a design-oriented installation that carefully considers energy efficiency and lighting optimization.

Keywords: Slip-form · Dynamic concrete cast
Automatic robotic manufactory pipeline · Eco-friendly computational design

1 Introduction

The traditional technique for casting concrete is formwork in which a mold composed of timber, plywood or metal sheet is installed to hold a well-mixed liquid concrete solution for a certain amount of curing time. After removing the mold, the solid concrete is released and used as structural or ornamental elements, such as columns, beams, slabs, and shading louvers. However, slip-form represents another casting method that has been widely used in the construction of roads, towers, and heavy off-shore platforms. Since the 20th century, slip-form has been developed as a cost-, labor- and time-efficient construction technique.

The core of our project lies in the flexible formwork, which is based on Smart Dynamic Casting (SDC). Several different SDC solutions have been based on recent developments by researchers at ETH Zurich and designed to achieve sectional transition during the extrusion process, and they include the earlier prototype of a four-actuator driven formwork (Lloret et al. 2), a V-shaped formwork, and dynamic formworks composed of rigid boxes with two metal plates (Lloret et al. 1). A different technical solution to achieving such sectional flexibility has been designed with a specific goal and resulted in sets of extruded dynamic columns with recognizable features.

J. Willmann et al. (Eds.): ROBARCH 2018, *Robotic Fabrication in Architecture, Art and Design 2018*, pp. 421–433, 2019.
https://doi.org/10.1007/978-3-319-92294-2_32

Based on laboratory research work conducted at ETH Zurich, we have taken one step further towards applying dynamic slip-form casting to a real project and created a flex-wall installation of 115 pieces, which has the potential to serve as an angle-dependent smart shading device for a building façade. The fabrication process is technically developed and refined for rapid manufacturing, thus serving a practical requirement based on the specific project and design. The research developments mainly encompass the following aspects:

1. To reduce the overall casting time, a larger slip-form mold is used to carry the entire amount of mixed concrete;
2. A central steel core is fixed at the center of the mold to ensure strength, and it resembles the reinforcement system of conventional load-bearing concrete;
3. The fiber-reinforced concrete is strategically stirred with a mixture of additives to avoid collapsing during the extrusion process, and a considerably reduced sitting time is observed;
4. A continuous production pipeline is established to preserve a steady workflow that maintains the robot arm working nonstop by shifting between different sessions, including pouring concrete, vibrating the concrete, locating on a railing for concrete curing, slip-forming, and concrete sitting for hydration.

2 Background

Since Contour Crafting first attempted to utilize a robotic arm for concrete fabrication (Khoshnevis 2004), robotic fabrication of concrete has been progressing at a rapid speed. However, most research studies are focused on additive manufacturing methods based on 3D printing, which includes the layering of materials (Huang and Khoshnevis 2004). Recent examples of robotic concrete fabrication, such as the XtreeE (Gosselin et al. 2016) and 3DCP (Bos et al. 2016), also follow this path. However, our research focuses on a separate path based on a dynamic formwork method.

In 1910, MacDonald published a paper entitled “Moving Forms for Reinforced Concrete Storage Bins,” and it described the use of molds for moving forms composed of jacks and concrete, which formed a continuous structure without joints or seams. This paper described in detail the concept and procedure for creating slip-form concrete structures. On May 24, 1917, a patent was issued to James MacDonald of Chicago (James 1910) “for a device to move and elevate a concrete form in a vertical plane.”

Slip-forming enables continuous, noninterrupted, cast-in-place, “flawless” (i.e., no joints) concrete structures that present superior performance characteristics to piece-wise construction using discrete form elements. The building of silos, a core building within a high-rise tower that presents a homogeneous cross-section from the bottom to the top, has adopted slip-forming casting as a very productive solution. This process is different from conventional static concrete formwork because it moves semi continuously with respect to the concrete surface being formed and form ties are not used (Risser 1995). Such technology could potentially reduce costs by up to 30–40%.

In the process of slip-forming, the concrete is poured at a predetermined rate on top of a continuously moving form and then emerges in a quick sitting state with strength.

Therefore, the concrete component is shaped by the section profile of the slip-form. The speed of slip-forming must meet the following two criteria.

- When the consolidated (by vibration) concrete emerges from the moving form, it should be strong enough to retain its shape, carry its own weight and carry the weight of the new concrete above it as well. Any drastic deformation will cause an unpredictable collapse.
- The freshly set concrete should be flexible enough to permit the slip-form to move upward without too much friction in case the form and concrete stick together during the process, which would lead to failure of the whole piece.

The National Centre of Competence in Research (NCCR) was founded in 2014 and has been engaged exploring the potential uses of robotic technology in traditional building materials and methodologies. Since 2013, projects pioneered by ETH Zurich have introduced multi-axis robot arms that combine slip-forming techniques. The digital fabrication group launched a cross-disciplinary research study in concrete casting with robotic fabrication. The group, led by Matthias Kohler and Fabio Gramazio at ETH Zurich used a 6-axis robot arm to obtain dynamically casted structural concrete columns (Lloret et al. 2014). This project could be perceived as an alternative to additive manufacturing, which is usually termed 3D printing technology (Fig. 1). 3D printed mesh-molds, which are plastically extruded mesh frameworks, are also used to indirectly generate double-curved concrete components (Hack et al. 2016). In recent developments, a feedback system has been added to continuously monitor whether the mixed material is adequate in strength for slip casting (Lloret et al. 2017).

Fig. 1. Dynamic concrete slip-forming by multi-axis robot arm ETH Zurich

3 Modification of Dynamic Robotic Casting

Regarding the slip-form technique, the key challenge often lies in reducing the sitting time to match the speed of slipping. Conventionally molded concrete requires at least 2–3 days of sitting time before the formwork is removed, another a week to reach a full solid state, and a month or so to reach complete structural strength under the desired moisturized condition. Chemical additives are incorporated into the new formula to reduce the sitting time so that each column can be fabricated within 30–50 min, which is associated with the requirement of the extruding speed and to ensure that the concrete can be self-supported after lifting the slip-form mold. Although the additives accelerate the hardening speed, the overall casting time is still not short enough for production. Approximately 2 or 3 h are required to cast a two-meter-tall column according to the ETH Zurich lab report.

Another core of our research is the structural quality of the robotic cast concrete component. Although fiber-reinforced concrete has better physical properties than traditional concrete, bonding with rebar should be performed to ensure the structural performance under pressure and tensile force.

Another crucial issue that must be optimized is the human labor factor. To fully exploit the multi-axis industrial robot arm, the fabrication process should be automated to an extent that minimizes the involvement of labor. The automobile industry is a good example in which human involvement is minimized. In the case of this project, the robot arm not only works on the slip-forming task but also the delivery of the mixed concrete, which needs to be performed between every session and is extremely labor intensive due to the weight of the concrete. Thus, a highly efficient continuous production activity similar to an automatic assembly line is formed, and the robot arm plays a key role.

With above considerations, we set up a new type of the robotic slip-form system based on the precedence from ETH Zurich:

1. To reduce the overall casting time, a larger mold is chosen to carry entire amount of mixed concrete
2. A steel tube is fixed in the center of the mold to ensure the strength and durability.
3. The fiber reinforced concrete is strategically stirred with proportional additives to avoid collapsing during the extruding process.
4. A continuous production pipeline is established to preserve a nonstop workflow.

To validate the proposal and techniques, a real project was commissioned, designed, fabricated and installed within 2 months. The task was to create a landscape screen wall (2 m tall and 60 m long) composed of 115 concrete columns cast and formed by a robot arm. This linear queue of deformed up-rising columns with half-meter spacing defines an obscure borderline between a main pedestrian entrance and a parking area. A KUKA KR150 R3100 Prime robotic arm was used and equipped with a custom-made air-pressure-control clamp, and it was capable of carrying and casting over 15 tons of concrete.

3.1 Fabrication Pipeline

A well-designated workflow was set up to meet the scheduled deadline, and it consists of two sections

1. Concrete mix and infill. In this section, most of the work is performed by human labor.
2. Dynamic robotic slip-form. In this section, most of the work is performed by an industrial robot arm, which exhibits the power and flexibility of digital fabrication.

These two sections form a looping cycle that repeats as necessary (Figs. 2 and 3).

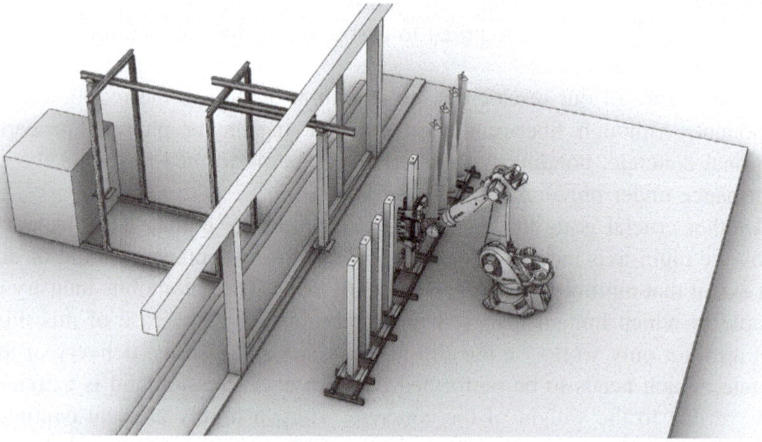

Fig. 2. Robotic slip-form pipeline

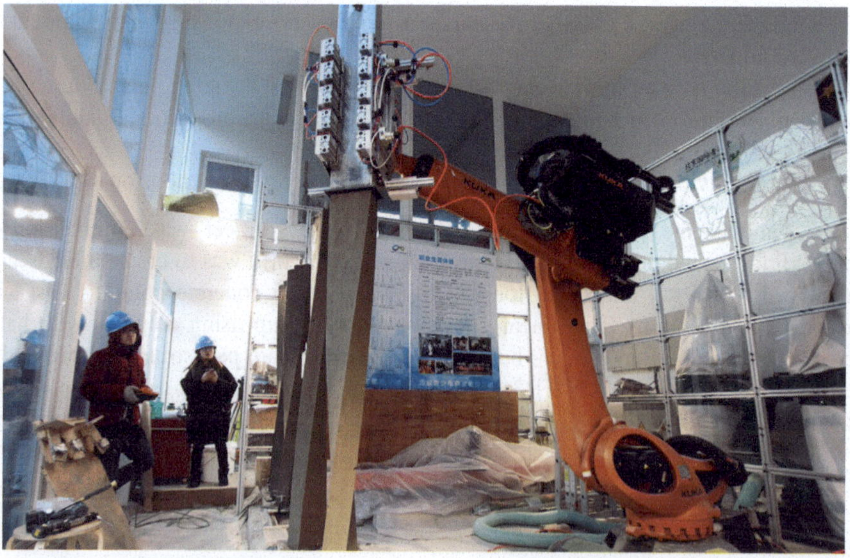

Fig. 3. Dynamic concrete casting process

Since each concrete column (200 mm × 100 mm × 2200 mm) plus the stainless-steel mold weigh over 100 kg, a ceiling rail was installed between the concrete pouring workstation and the robot slip-form workstation. After the mold was fully filled with concrete and evenly vibrated to eliminate air bubbles, it was delivered to the robot working range via the rail.

Another 6 m long ground rail was installed in front of the robot arm. Six carts were slid on top of the high-quality lineal rail with a tolerance of 10 μ. The robot end-effector was a heavy-duty clamp driven by compressed air above 0.8 MPa. The clamp was designed to open and close in the X&Y directions to firmly catch and release the 3 mm thick stainless-steel mold. The air compressor was switched by robot I/O via two electronic compressor valves.

As long as the fully loaded mold was in the correct location, the programme for fabricating a specific column would be launched to perform a series of operations.

1. The robot clamp will catch and carry the load to a cart on the rail, and the position of the column in the queue is determined by the sitting time. Each column is on record starting at the first infill of concrete. Every key point of the timeline is strictly monitored.
2. When one of the molds is ready for slip-forming, the robot will carry it to a designated place, usually on the left side from the center axis, where a simulation is run to ensure that it is free of singularities.
3. After the mold is released from the base plate, the dynamic slip-form process begins. The slip-forming time of each column is limited to 10 min each.
4. After the slip-form is completed, the concrete column, which is still soft and wet, is pushed to the other side of the rail to continue curing.
5. The empty mold is sent back on the ceiling rail and cleaned by a water jet for the next session.
6. By the next day, the concrete column will be hard enough for the final curing phase, which is at least one-week long.

3.2 Material and Formula Composition

During slip-forming, an accurate time-control process for hydration is critical. If the concrete mixture is too soft after slipping from the mold, it will be deformed by gravity, instantly lose its retaining ability, and severely collapse as a consequence. However, if the concrete mixture is too hard, then the twist force from the robot arm will crack the concrete columns, resulting in an undesirable surface quality. Compared with that of the conventional concrete framework, the time window for dynamic slip-forming is very narrow. The experimentally developed basic mixture formula is as follows: one batch of material contains 32.3 kg Portland cement, 30.5 kg fine sand, 11.5 kg of water, 80 g Nylon fibers at 30 mm, 80 g Nylon fibers at 20 mm, 40 g Nylon fibers at 8 mm, 105 g Superplasticizer #5, and 82 g Accelerator. The mixture is rapidly stirred for 5 min, well vibrated and allowed to sit for 50 min before it is ready for extrusion.

In addition to the basic formula, adjustments will have to be made for different batches of material orders. The water amount, for example, should represent the most critical parameter related to the concrete sitting time. However, the moisture inside the sand should also be accounted for in the total. As a result, to guarantee desirable slip conditions, hand-testing must be performed first to ensure that the proper curing time is in an acceptable range.

Temperature is another important consideration. During construction in Beijing in November, the outdoor temperature dropped from 10 °C down to −5 °C, which led to a tremendous decrease in the early hydration speed, thus increasing the sitting time from 30 min to 50 min. If this timing-sensitive fabrication is an on-site activity, then maintaining the fabrication environment represents a critical factor. The on-site environment parameters for fabrication are as follows: humidity 16.1%, air temperature 8.1 °C, water temperature 9.8 °C, and water PH 6.4.

A substantial amount of air bubbles usually become mixed together with the concrete in the stirring workstation, and bubble-free additives are not completely able to remove all air chambers, which causes unexpected porous surfaces. Therefore, physical vibration is applied during the concrete-infill phase. A tube vibrator was employed and lifted simultaneously while the concrete was gradually piped. A lubricant was brushed on the inner surface of the slip-mode as another solution to increase the finish quality.

4 Design Process Driven by Data Flow

4.1 Computational and Interactive Design

In this design, 115 slip-formed columns are situated at every half meter on a 60 m long lineal flat steel base (Fig. 4). Each column has a certain vertical twist angle and is algorithmically related to its neighbors. This layout formats column rows into an undulating panorama, although they stand in a straight line.

Fig. 4. Flex-wall installation

4.2 Data Flow from Design to Fabrication

A parametric 3D model created in Rhinoceros was used to connect 115 sets of data packages for the columns (in Cartesian coordinates) with the design parameters. This reaction is automatically transferred to the robot arm simultaneously from the

Grasshopper environment. KUKA|prc (Braumann and Brell-Çokcan, 2011), a plugin in Grasshopper, is the application used to generate the.src file, which is the file format of the KUKA robotic system. This digital design-fabrication solution attains at least three advantages compared with the conventional method.

1. The path from design to fabrication is highly interrelated via the data flow. If the form is adjusted by changing the parameter for design evaluation, then the fabrication data are automatically updated. For example, if the distance between two columns is tweaked with different numbers, the entire set of fabrication data will be updated accordingly.
2. The data sets, including the 3D model and robotic G-code, will be documented after construction in case any column needs to be replaced for maintenance. Simply by sending the data package to the robot, a brand-new column could be produced.
3. Data flow runs seamlessly between the schematic design and the fabrication machine, thus representing a brand-new process in which conventional human involvement is a supplement.

5 Scenario of Eco-Friendly Building Design and Fabrication

The design and form of a building façade has a significant impact on the heating, cooling, and lighting loads as well as the thermal and visual comfort in the perimeter zones of a commercial building (Chan et al. 2012). With the effective distribution of solar energy via passive façade elements, energy consumption can be significantly reduced (Tzempelikos and Athienitis 2007). Additionally, the lighting quality can be improved to address the need for different programs and spatial comforts (Atmaca et al. 2007).

Beyond serving as a mere installation for a soft division of urban space, the design of such units and the associated techniques have a greater objective of potentially become a passive smart shading device to adjust solar gains and interior reactions based on the nuanced relationship between the building and the sun.

5.1 Smart Shading Component

Motivated by the dynamic concrete slip-form technique, the curtain wall shading system represents a new element that excels in both aesthetic and functional aspects. This digital fabrication technique could change the role of concrete, a traditional building material, and allow it to perform on a new level in the near future. Compared with GRC (Glass Fiber-Reinforced Concrete) technology, which still uses static formwork to produce a surface curvature, robotic slip-forming provides an effective solution to the challenge of free-forming concrete construction. Overall, this new technique has considerable potential for use.

Based on the outlined smart fabricating solution, the twist-formed concrete components could serve as an eco-friendly concrete façade structure for use in more-refined lighting planning for interior spaces and an expressive aesthetic element developed via the data-flow gradient and robotic fabrication pipeline.

Although the concrete façade element has more self-weight than a typical curtain wall element, if strategically applied, it will only exceed the area weight of a typical glass curtain wall by 30%. Thus, the extra cost of reinforcement to the entire structure could be compensated by the reduced operational costs as an energy efficient building component. The benefit of SDC as a decorative façade element is enhanced by considering energy efficiency in a green building (Fig. 5). In on-going research, we are investigating methods of lowering the impact of self-weight by redefining joints and connections.

Fig. 5. Building elevation scenario with concrete components

5.2 Eco-Friendly Concrete Screen

As a further development of the previously described "Flex-wall", we propose a mixed-use building scheme that implements this robotic technique together with eco-friendly design strategies. Initially, four basic requests are registered.

1. The concrete component should be able to hide the facilities and equipment attached to the exterior of the building, such as AC workstations, water shafts, etc.
2. This concrete installation should provide a comfortable shading solution based on the statistics of the local climate and environment.
3. The shading solution should also meet the goal of improving the interior lighting conditions.

The solar distribution of the designated building is affected by the surroundings, which may result in an inappropriate pattern of energy gain (Fig. 6), with certain regions receiving too much radiation (colored in red in the figure) and others requiring an enhanced light supply (colored in blue).

5.3 Lighting Planning

Data for Beijing are used as an example, and the GH programme sets up a corresponding relationship between the solar analysis and building skin. If the building skin

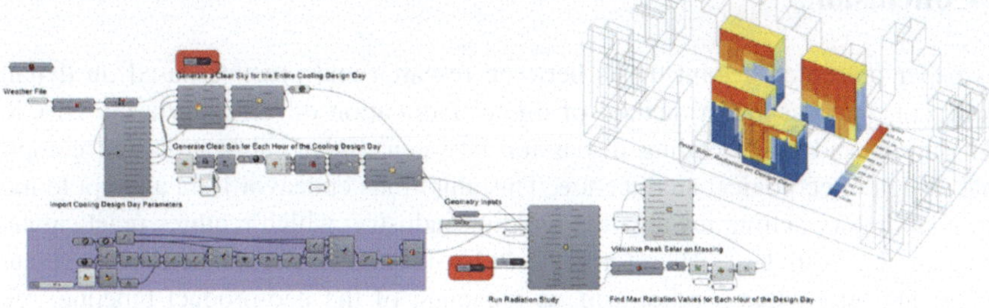

Fig. 6. Solar analysis from Ladybug in Grasshopper

is unfolded for a total evaluation, then the unique factors of different orientations and drastic change around the edges can be easily identified. Therefore, the expectation for the first iteration is mainly to reduce the over-heated zones. By changing the rotation angle, the shading texture will present a gradual density shift. This gradual evolution will provide diverse field-views from different directions.

A comparison of the lighting conditions at the ceiling and the right wall with the simulated lighting quality before and after the concrete installation shows equivalent results and indicates areas where direct light cannot reach (Fig. 7). These results show that the concrete components block direct light from the east and gain light from the south-east and mirror it up to the ceiling and the sidewall via a designated rotation position and appropriate angles. The reflected amount could also be controlled to accommodate the surface quality, e.g., from rough to polished, white paint to gray paint, etc.

Fig. 7. Rendering tests before and after irradiance is included

6 Conclusions

This paper presented a joint thesis between research and practice based on dynamic robotic concrete casting, a method of digital fabrication developed via the NCCR at ETH Zurich, and an extending discussion between digital fabrication and computational design of ecological architecture. This ambitious endeavor is an attempt to move from a laboratory setting towards the building industry, which requires practical cases to breach the boundary. Therefore, the "Flex-wall" is selected both as a research exercise and proof that a robot arm, as a captain of the neo-product pipeline, offers considerable flexibility and efficiency for linking a chain of different disciplines, and it will eventually be driven by data flow from the early design phase all the way to fabrication.

Acknowledgments. We would like to thank the "Flex-wall" production team, who completed this intensive task in only two months, including Wang Yanxin, Yang Daoqian, Xu Boyang, Tong Yue, Yang Nan and the other construction workers who worked in very cold climates; Vintage Park Ltd, who sponsored this project and provided the space; and the National Natural Science Foundation of China (NSFC), who provided the financial support (Project No. 51538006).

References

Lloret Fritschi, E., Reiter, L., Wangler, T., Gramazio, F., Kohler, M., Flatt, R.J.: Smart dynamic casting: slipforming with flexible formwork - inline measurement and control. In: Second Concrete Innovation Conference (2nd CIC), Paper no. 27, Tromsø, Norway (2017)

Lloret, E., Shahab, A.R., Mettler, L., Flatt, R.J., Gramazio, F., Kohler, M., Langenberg, S.: Complex concrete structures: Merging existing casting techniques with digital fabrication. In: Gu, N., Watanabe, S., Erhan, H., Haeusler, M.H., Huang, W., Sosa, R. (eds.) CAADRIA 2014: Rethinking Comprehensive Design: Speculative Counterculture, Proceedings of the 19th International Conference of the Association for Computer-Aided Architectural Design Research in Asia, pp. 613–622. Kyoto (2014)

Khoshnevis, B.: Automatic construction by contour crafting-related robotics and information technologies. Autom. Constr. **13**(1), 5–19 (2004)

Huang, D., Khoshnevis, B.: Concrete wall fabrication by contour crafting. In: Proceedings of the 21st International Symposium on Automation and Robotics in Construction, ISARC 2004, Jeju, South Korea (2004)

Gosselin, C., Romain, D., Roux, Ph, Gaudillière, N., Dirrenberger, J., Morel, Ph: Large-scale 3D printing of ultrahigh performance concrete – a. new processing route for architects and builders. Mater. Des. **100**, 102–109 (2016)

Bos, F., Wolfs, R., Ahmed, Z., Salet, T.: Additive manufacturing of concrete in construction: potentials and challenges of 3D concrete printing. Virtual Phys. Prototyping **11**(3), 209–225 (2016)

MacDonald, J.: Moving forms for reinforced concrete storage bins. In: 7th Annual Convention by National Association of Cement Users, New York, p. 554 (1910)

Risser, B.: Advances in vertical slip-form construction. Aberdeen's Concrete Construction **40** (10), 40–45 (1995)

Braumann, J., Brell-Çokcan, S.: Parametric robot control: integrated CAD/CAM for architecture design. In: Taron, J.M. (ed.): Acadia 2011: Integration through Computation, Proceedings of the 31st Annual Conference of the Association for Computer Aided Design in Architecture, pp. 242–251. Calgary/Banff (2011)

Hack, N., Lauer, W.V., Gramazio, F., Kohler, M.: Mesh mould: differentiation for enhanced performance. In: Gu, N., Watanabe, S., Erhan, H., Haeusler, M.H., Huang, W., Sosa, R. (eds.) CAADRIA 2014: Rethinking Comprehensive Design: Speculative Counterculture, Proceedings of the 19th International Conference of the Association for Computer-Aided Architectural Design Research in Asia, pp. 134–148. Kyoto (2014)

Lloret, E., Gramazio, F., Kohler, M.: Complex concrete constructions. In: Stouffs, R., Janssen, P., Roudavski, S., Tunçer, B. (eds.) CAADRIA 2013: Open Systems, Proceedings of the 18th International Conference of the Association for Computer-Aided Architectural Design Research in Asia, pp. 613–622. Singapore (2013)

Chan, Y., Tzempelikos, A.: A simulation and experimental study of the impact of passive and active façade systems on the energy performance of building perimeter zones. ASHRAE Trans. **118**, 149–156 (2012)

Tzempelikos, A., Athienitis, K.: The impact of shading design and control on building cooling and lighting demand. Sol. Energy **81**, 369–382 (2007)

Atmaca, I., Kaynakli, O., Yigit, A.: Effects of radiant temperature on thermal comfort. Build. Environ. **42**, 3210–3220 (2007)

Ceramic Constellation | Robotically Printed Brick Specials

Christian J. Lange(✉), Donn Holohan, and Holger Kehne

Faculty of Architecture, The University of Hong Kong,
Pok Fu Lam, Hong Kong SAR
cjlange@hku.hk

Abstract. The fired clay brick, a mainstay of Chinese construction, has for many years languished. Rough cast and hidden behind layers of stucco, it plays a supporting role to reinforced concrete and steel - rarely however in a literal sense. This has not always been the case - highly decorative, geometrically complex, perforated or materially modified in response to specific structural or environmental conditions, the brick has played an instrumental and expressive role in the built history of China. This paper describes a means by which a traditional craft and its inherent material intelligence may be transformed into a flexible and performative system of construction through the integration of parametric design software and 3D robotic clay printing technology, ultimately offering a culturally rooted yet modern material solution for a modernizing China.

Keywords: Robotic manufacturing · Parametric design · Digital fabrication 3D clay printing · Brick specials

1 Background

1.1 Challenges and Opportunities

In China, brick remains one of the most widely used building materials, with between 700 and 800 billion bricks produced each year (Murmu and Patel 2018). The infrastructure and expertise that supports this industry is substantial. However, while highly productive, the manner in which bricks are manufactured today has not changed significantly in centuries (Shakir and Mohammed 2013). The most common means of brick production utilizes a fixed die extrusion process, a method that limits formal variation. Nonstandard bricks, when specified, are generally moulded - a process which necessitates the construction of a formwork. This is a task which requires skilled labour and which, as a consequence, is expensive and generates long lead times.

This project seeks to challenge this paradigm through a reimagining of the manufacturing platform. Industrial Robots offer a solution, since they have the power to alter standardized means of production in the discipline of architecture and construction (Gramazio and Kohler 2008).

Through an integrated, as opposed to modular, approach, the project outlined seeks to demonstrate that the gains in productivity synonymous with the brick assembly line can be offset through the intelligent design of each element – which can be made to

J. Willmann et al. (Eds.): ROBARCH 2018, *Robotic Fabrication in Architecture, Art and Design 2018*, pp. 434–446, 2019.
https://doi.org/10.1007/978-3-319-92294-2_33

respond to specific performative criteria in addition to a set of socio-economic, labour, and site conditions. This optimization offers potential reductions in material usage and weight, and increased usability and fitness for purpose.

1.2 The Brick Special

The current state of the industry can trace it roots to "Brick Boredom", a result of the continual pursuit of technical perfection which was first recognized around the turn of the 20th century (Deplazes 1997) which, despite a strongly influential arts and craft movement, became ever more extreme towards the 1950s. Polemical modernist ideologies, with their disdain for ornamentation, coupled with emerging reinforced concrete and steel technology, had the effect of further stripping the brick of its innate adaptability or expressive potential.

Latterly, the perception of the brick as unsophisticated and monotonous has pervaded the collective consciousness. Historically however, we encounter the reverse - traditional, handmade bricks vary in tone, texture and dimension, a consequence of their imperfect fashioning. However, rather than being perceived as flaws, these individualities were accepted and even considered desirable. The Iron Pagoda, Kaifeng City Henan (1049) exemplifies the historical plasticity of the brick - its exterior is elaborated using over 50 unique modules in various configurations (Fig. 1).

Fig. 1. Iron Pagoda Kaifeng City, Henan province (built in 1049) (left and center), final brick pavilion (right).

Technically, a key driver of the modern tendency toward brick standardization has been the difficulty in predicting and managing the geometric distortion and resultant construction tolerances - a result of the firing process. Open source parametric software, coupled with robotically controlled 3D printing, has the potential to reconcile the necessity for dimensional accuracy without compromising either formal elasticity or material expression – allowing precise and variable control of all aspects of production – toolpath, extrusion rate, nozzle diameter, print speed etc.

The project outlined in this paper explores this in practicality through the fabrication of a full-scale structure which, through its construction, examines the relationship between the individual components to each other and to the overall form.

As such, it can be seen to build on much of the recent research in ceramic deposition techniques that seek to upscale the process and to deliver viable real-world construction systems. E.g. Research work conducted by think tank "Emerging Objects" (Rael and San Fratello 2017), which investigates the potentials offered by systems of unique elements but is limited in that those elements are not relational (can be assembled in a wide number of configurations) and require an additional physical, organizational structure.

However, rather than focusing on novel materials, this research seeks to embrace an existing industrial base and material supply chain and to develop a system that can respond to its inherent variability.

2 Objectives

The aim of the presented research is to develop a synthesis between the historical promise of specificity and expression offered by the brick special with the advantages and potentials presented by robotic fabrication and generative parametric design. Robotic technologies provide the ideal platform for developing fabrication processes in an experimental, iterative framework, without reinventing the machines of production (McGee and Ponce de Leon 2014). The potential for morphological differentiation enabled by robotic fabrication, in combination with the robotic reinterpretation of the clay die-extrusion process, forms the premise for the exploration of a 3D printing workflow and the construction of a full-scale pavilion.

Specifically, the following objectives are defined as part of this research. (1) To develop a tolerancing system which predicts the geometric distortion of individual discrete elements and allows for the accurate physical realization of a defined form - thereby defining a parametric fabrication system which relates to the material and process limitations of the fired brick. (2) To enable rapid and automatic toolpath generation, integrated within the same digital environment. (3) To develop a brick construction system in which each element is unique and responds to a given set of site conditions. (4) To develop an indexing and organizational procedure that enables rapid identification of elements and the simple assembly of a structural system containing a large number of fractionally differentiated units.

3 Methodology

3.1 Generating Specificity

To test the objectives outlined above, the team developed a specific design, manufacturing and assembly methodology. The project took a generative parametric design approach, while the manufacturing process involved 3D robotic clay printing in conjunction with CNC milling. The final outcome of the project is a large-scale pavilion constructed entirely from brick specials.

To ensure the specificity of each brick, the project team used the Grasshopper plug-in for Rhino to create a parametric model – exploring complex geometries generated in response to a set of specific environmental constraints.

The team developed a custom-made definition that was used to generate the overall form, the design and distribution of the brick geometry, and its response to environmental forces. Conceptually, the pavilion references the typical plans of generic tower configurations common in China, and utilizes the Euclidean geometry of a rhombus as a generator profile (Fig. 2a).

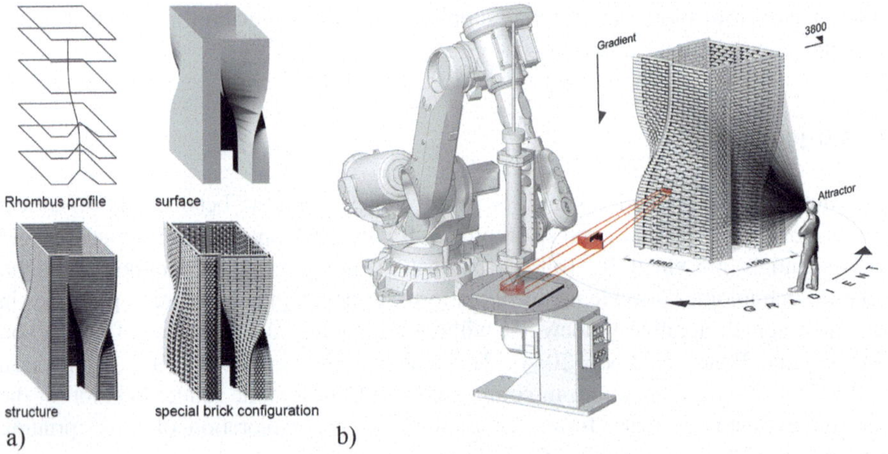

Fig. 2. Concept diagram (left), Attractor influence for continuous differentiation (right)

To achieve a design that continuously differentiates in space; the project team generated a complex surface that functions as the base device to control the location and size of the brick components. This poly-surface was generated through a specific lofting technique that resulted in a spiraling form - creating an underlying three dimensional point grid with incremental variations and no equivalencies.

This point grid controlled the distribution and scale of the bricks, which were organized in a traditional running bond. Each of the eight surfaces of the pavilion had the same alternating system of four full bricks on layer A, and three full bricks plus two half-bricks for the corners on layer B.

Each brick special is influenced by and relates to a specific viewpoint, defined by site conditions and approach angles. This was achieved in the model space through the application of a point attractor system that influenced the dimension and form of each brick (Fig. 2b). This system also allowed for variable openings to be developed within the brick skin, creating a visual correspondence between the interior and exterior faces. The resultant configuration, in its continual transformation, articulates a range of conditions - from enclosed to open, planar to double curved, with varying porosity across the brick façade.

The brick itself has two different faces - the outer varies across the structure, from flat to peaked geometry and fluctuates in width from 60 to 395 mm and in depth from 20–85 mm, while the inner face is standardized, with the variation occurring through continuously differentiated spacing (Fig. 3). This dual aspect defines the transparency of the system, while also exploring ideas of difference and repetition.

Fig. 3. Exterior of pavilion (left), Interior assembly (right)

3.2 Digital Fabrication Strategy

Informed by the preceding stages, and in concert with a material study, a digital fabrication strategy was developed.

In order to fabricate each brick, the project team used a Deltabots linear ram extruder with a capacity of 5500 ml (Fig. 4), attached to a standard ABB 6700 industrial robot with 200 kg payload and 3000 mm reach. The robot was arranged in conjunction with a one axis rotational table that served as a printing platform, allowing for a 450 mm by 450 mm horizontal printing space. For the ABB Rapid code generation, the team utilized the HALPlug-in for Grasshopper, which allowed for a smooth and effective translation of line geometry to target planes for the robotic protocols. Each brick was automatically translated into 234 target planes, representing the essential control points of the geometry.

From the outset, the intention of the project team was to create a self-supporting brick structure. However, due to planning, structural and legislative issues surrounding the place of exhibition, and the certification required, it was necessary to construct a lightweight timber framework, in order to ensure compliance. The timber structure was designed to emulate the performance of brick coursing, and acts in compression, significantly reducing the overall weight of the pavilion.

Fig. 4. DIW process with linear ram extruder attached to ABB 6700

As a result of the timber framework, the bricks could be designed with an additional degree of flexibility – an advantage in a project with a build time of just 21 days.

With the amount of bricks to be printed and fired numbering 1870, it was necessary to minimize the amount of printing time for each component. Outlining the geometry of each brick (Fig. 5) was the most minimal approach to generating a printing path and was also a good fit for the clay extruder, which uses a Direct Ink Writing (DIW) printing method.

As the printer is engineered in such a way that the extrusion process cannot be interrupted instantaneously, it was necessary to generate a strategy for the printing path that resulted in maximum continuation if undesired material extrusions in the final outcome were to be avoided. In order to overcome this issue, and to deliver the most accurate results, the project team utilized a custom-made Grasshopper definition that automatically generated an efficient single-line print path.

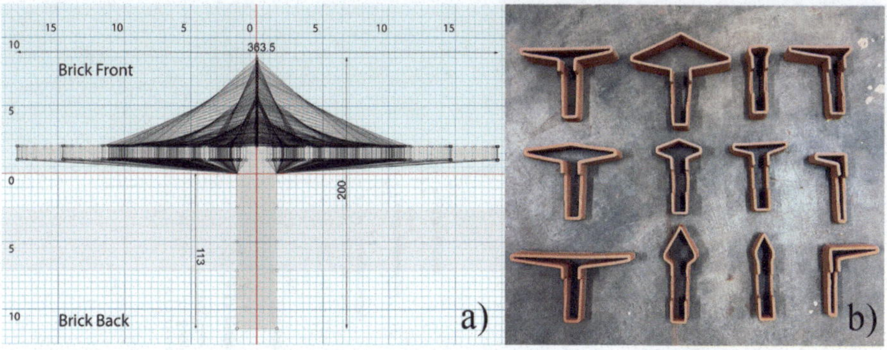

Fig. 5. Superimposed outline of all bricks (left), Selection of continuous types (right)

3.3 Feedbacks and Inputs

As part of the development of a digital fabrication strategy, and as a consequence of the differentiation of individual bricks, a series of feedback loops were necessary to ensure that each met its specific performance criteria and could be printed using the method outlined.

As an extrusion material, the project utilized a standard low-fire terracotta clay (P1331, Potterycrafts ltd.) with a shrinkage factor of approximately 11% at 1100 °C. This shrinkage, and the DIW process, makes clay printing a more complex material process than typical 3D printing techniques such as FDM or SLA processes. Several factors had to be taken into account in order to achieve a successful result.

In order to gain an overall understanding of the limit and range of formal deformation after firing, print time and the structural strength of each brick, and taking into account the large number of individual components, the team utilized a sampling and testing methodology that examined tool path length, area, projection and front face angle (Fig. 8c) in respect to the their maximum and minimum values.

The key factors were the layer height between the horizontal printing path, the extrusion rate of the printer, as well the speed of the robot. Although the set-up used a 5 mm nozzle, the wall thickness and structural strength of the print was dependent on these three parameters. In order to achieve a 5 mm wall thickness in the final fired print that was also structurally sound; the project team used the following framework;

The layer height of each print was set at 3 mm. This delivered good binding and created a uniform cross section (Fig. 6). Increased layer heights resulted in low adhesion and overall lower resolution of the part, while smaller layer heights, though resulting in a higher part resolution, were susceptible to inconsistent wall thicknesses.

TCP velocity and motor velocity parameters define the extrusion and feed rates, and are in a dependent relationship. Following a series of tests, the optimum settings were achieved - motor velocity of the printer was set at 0.15 mm/s (−1500), while the TCP velocity ran at 60 mm/s. Higher speeds resulted in lower print accuracy, while lower speeds resulted in much thicker walls. With optimum settings, print times varied between 1:20 min and 4:30 min for each brick, toolpaths varying in length from 49 cm to 165 cm.

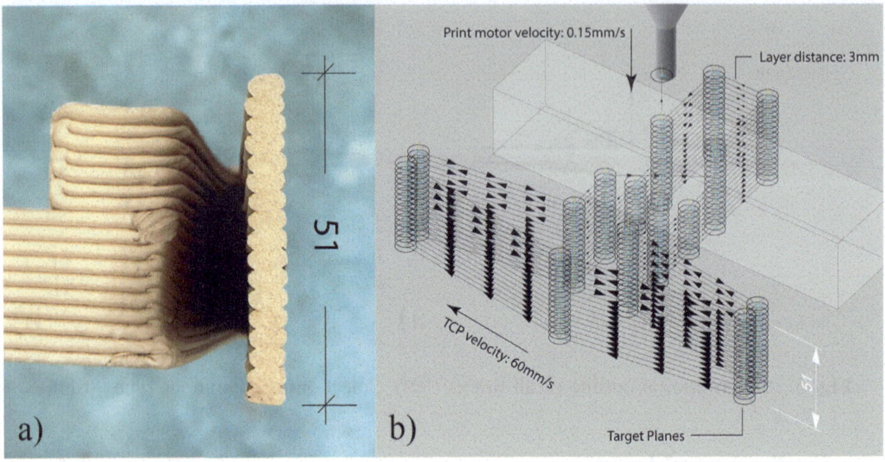

Fig. 6. Section through fired brick showing consistency (left), Printing path of brick with target planes (right).

4 Pavilion Fabrication and Assembly

Approximately 700 kg of raw terracotta clay was used in the construction of the pavilion. The median weight of each brick measures 10% of an equivalent mass produced unit (223 mm × 176 mm × 73 mm). The printing paths for the bricks were manually nested for the size of the print table. There were a total of 294 print files, ranging from four to twelve bricks per file.

All bricks were fired at 1025 °C (Cone 05), using an electric kiln. Because of the sampling methodology, unforeseen issues were encountered relating to production at scale. This was a consequence of the number of bricks fired within the same kiln. Although the kiln itself is highly accurate, when full, and due to the variations in spacing (a product of the irregular forms), variations in air flow and temperature within the kiln result. The greatest problems were encountered with bricks fired close to the openings in the kiln. This would sometimes lead to highly distorted bricks, which would need to be re-fabricated.

The number of unique bricks made it necessary to develop a comprehensive indexing strategy. As the running bond system does not follow a standard grid that would allow for a typical matrix approach, the project utilized a concept that was based on calendar dates. Each brick was manually indented with a date stamp that correlated to a specific position in the pavilion.

The timber framework which supports the pavilion consists of a series of stacked horizontal elements, which interlock at each corner with a lap joint – modulated to approximate the form of the structure. Each horizontal timber layer substitutes a brick course. The timber framework simultaneously resulted in a simplified assembly by dispensing with the need for formwork and mortars, reduced the overall weight, provided a distributed loadbearing and removable connection to the ground and allowed the structure to achieve compliance with building regulations.

Fig. 7. Connection of special bricks with timber section.

Each timber element is a standard square 40 × 40 mm, with a notch depth of 2.5 mm. The resulting framework provided a 35 mm gap in which the bricks were fixed under compression (Fig. 7). In total, there are 108 layers with four wooden elements per layer. The intelligent assembly of the framework was facilitated by individual variation within the jointing system - in which the global geometry was embedded. Each element had only one possible location within the structure. These components were cut using a standard three-axis CNC milling machine.

Over a period of six days, all components were assembled into three larger units, each weighing 350 kg. These were then transported to the construction site and assembled with a crane.

5 Conclusion

The scale and antiquated nature of the brick industrial complex offers significant opportunity for growth and redevelopment. The outlined project attempts to reconcile an imprecise material process with a robotic workflow. 3D printing technology, though increasingly popular amongst researchers in the industry, still has a long way to go to become a viable manufacturing process. While the critical discourse of modernism uses the notion of the standard as a measurement for its success, the trajectory of robotic fabrication will need to define the standards necessary in order to achieve consistent, high-quality outputs in relation to the continuous variation of the building block. These standards are particularly important in relation to clay, as the unpredictable nature of the material has the potential to undermine the precision of computational design. The method described in this paper has been proven to achieve relatively consistent outcomes when it comes to brick quality, printing method, tolerances and overall finish of the brick. The method, in this sense, describes a novel test case which serves to establish the necessary standards, computational workflows and logistical sequencing required to achieve the desired outcome. The sheer scale of the prototype, with its 1870 unique bricks, has allowed it to function as a large scale testing ground - providing a basis for the continued improvement of the methodology.

However, while successful in terms of the stated objectives and in terms of the architectural exploration, as well as the design process and the overall robotic manufacturing, there are aspects to each which, when addressed, would likely have a positive impact on the continuing development of this research. These can be broadly categorized under three main headings; Performance and Logistics, Material Feedback and Computation, and Framework for Certification.

5.1 Performance and Logistics

The performative aspect of this outcome extended only to abstract force. The variable manufacturing platform described, however, offers greater potentials. Further iterations should respond to both environmental and structural constraints.

Logistics became a key point in the project and have proved to be one of the most challenging aspects of the real world implementation of a structure comprised of continuously differentiated brick specials. Further studies will need to address indexing and supply chain issues, in addition to assembly methodologies.

5.2 Material Feedback and Computation

Until now, the set-up to produce these special bricks has been relatively improvised. Ultimately the goal is to utilize the material system for a real scale solution. Hence, it would be advisable to develop a better feedback system, which allows for consistent improvement and optimization of the brick.

The following procedural issues arose as a result of the relationship between material properties and the robotic manufacturing methodology;

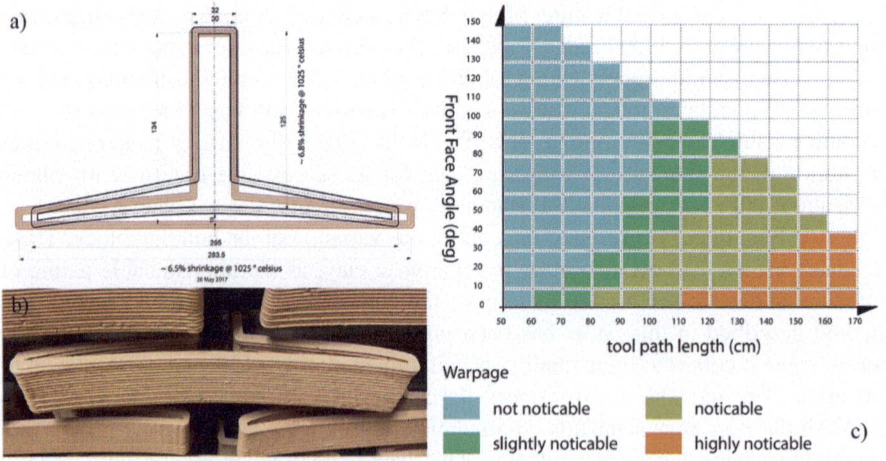

Fig. 8. (a) Superimposition of printed geometry vs. (b) Warped brick after firing. (c) Relationship between front face angle and tool path and warped deformation.

Following printing, each brick was left to dry at room temperature (approximately 22 °C) for at least 24–36 h before it could be fired in a kiln. Although most bricks dried and shrank evenly, some components were adversely affected during this process. These bricks had bent and warped in the XY-plane, which could be described as a banana effect (Fig. 8). The longer the brick, the more noticeable was the distortion. It was also found that a variation in printing speeds within a single brick resulted in an uneven drying process, which in turn led to a deformation of the tile in the z-direction. The best results were achieved when a constant printing speed was applied throughout the entire print run, while smaller bricks had better results than larger bricks.

In addition, it was noted that variations in temperature and humidity during the curing cycle resulted in some components drying faster than others, which affected both workflow and deformation. In order to improve the method it is necessary to further understand the relationship and dependencies between the geometry, extrusion thickness, and drying environment on one side and the deformation/ warp effect on the other. This would be best achieved through a series of analytical and comparative tests.

Although fundamentally a digital process, there were many manual aspects to the production of the bricks. Errors in the manual loading of the clay into the extruder resulted in unavoidable inconsistencies within the print, such as the occurrence of air cavities within the load. This resulted in the discharge of compressed air which would inevitably affect the overall accuracy and quality of the print. This could be avoided with the purchase of a pugmill that de-airs the material and produces a compacted load. However, it can be argued that the resultant imperfections within this material system also have a certain attraction. They serve to add a similar procedural trace as we know it from handmade products.

5.3 Framework for Certification

Ultimately, the successful implementation of any future projects within the public realm will require the development of a testing procedure both for individual bricks and the resultant forms. Due to the fact that each element is unique, it is necessary that this takes place in a simulated digital environment, prior to the fabrication of any part. This will be required in order to obtain approval from the various official government institutions.

Acknowledgments. The research has been significantly supported by a donation from Sino Group. The design team for the Ceramic Constellation Pavilion consisted of Christian J. Lange, Donn Holohan and Holger Kehne. The authors would like to thank the following students who assisted in various aspects throughout the research: Tony Lau, Anthony Hu, Teego Ma Jun Yin, Ernest Hung Chi Lok, Chau Chi Wang, Ren Depei, Mono Tung, He Qiye, Henry Ho Yu Hong.

References

Deplazes, A.: Constructing Architecture: Material Processes Structures: a Handbook, p. 28. Birkhäuser, Basel (1997)

Gramazio, F., Kohler, M.: Digital Materiality in Architecture. Lars Müller, Baden (2008)

McGee, W., Ponce de Leon, M. (eds.): Robotic Fabrication in Architecture, Art and Design 2014, pp. vii–x. Springer International Publishing Switzerland (2014)

Murmu, A., Patel, A.: Towards sustainable bricks production: an overview. Constr. Build. Mater. **165**, 112–125 (2018)

Rael, R., San Fratello, V.: Clay bodies: crafting the future with 3D printing. Archit. Des. **87**(6), 92–97 (2017)

Schröpfer, T., Carpenter, J., Kennedy, S., Margolis, L., Mori, T., Tehrani, N., Yeadon, P.: Material Design (n.d.)

Shakir, A., Mohammed, A.: Manufacturing of bricks in the past, in the present and in the future: a state of the art review. Int. J. Adv. Appl. Sci. **2**(3), 145–156 (2013)

Robotic Fabrication of Bespoke Timber Frame Modules

Andreas Thoma[1(✉)], Arash Adel[1], Matthias Helmreich[1], Thomas Wehrle[2], Fabio Gramazio[1], and Matthias Kohler[1]

[1] ETH Zurich, Zurich, Switzerland
thoma@arch.ethz.ch
[2] ERNE AG Holzbau, Stein, Switzerland

Abstract. This paper presents methods and techniques to robotically prefabricate timber frame modules. The key challenge of this research lies in enabling the digitally informed and fabricated spatial assembly of timber beams into prefabricated timber frame modules. The project combines the fabrication and the spatial assembly of timber beams into one fully integrated robotic fabrication process. A cooperative robotic construction procedure that minimises the need for scaffolding and allows for the informed assembly of spatial structures with non-planar geometries was developed. This required the examination of suitable timber joining methods, assembly sequencing, as well developing appropriate and novel strategies to register and handle material deviations and construction tolerances. The physical implementation of the research in multiple experiments and finally, a full-scale building project validates the approach.

Keywords: Cooperative robotic assembly · Computational design
Timber bar structures · Prefabrication · Building scale · Additive construction
In place milling · Timber frame modules

1 Introduction

The implementation of digital fabrication technology in timber construction dates back to the 1980s. Inventions of specialised digital joinery machines and CNC milling machines (Schindler 2009) enabled the digital production of complex timber parts and are well established in the construction industry. However, the actual assembly of these parts predominantly is still done manually (Willmann et al. 2016), requiring extensive logistics. Timber modules, typically being assembled flat on the ground in various steps and comprising of standardised planar geometry are mostly assembled by hand however, there are a handful of machines which can automatically assemble simple standardised timber frames: exceptions such as window or door frames are still added manually.

The implementation of industrial robots into the fabrication loop, performing a variety of programmed tasks, allows not only the production of individual parts but also the digital assembly of them. The Sequential Roof project (Apolinarska et al. 2016), developed by Gramazio Kohler Research at the ETH Zurich and robotically

J. Willmann et al. (Eds.): ROBARCH 2018, *Robotic Fabrication in Architecture, Art and Design 2018*, pp. 447–458, 2019.
https://doi.org/10.1007/978-3-319-92294-2_34

constructed by ERNE AG Holzbau, marked the first step into the digital fabrication of a non-standard timber assembly at a building-scale.

By shifting the fabrication scenario from The Sequential Roof's layer-based assembly method to a method of spatial assembly, this research expands the possible timber architecture repertoires and applies it at a building-scale to the typology of the timber module: a prefabricated volumetric unit of timber framing which is subsequently assembled on-site. The spatial assembly method builds upon previous work demonstrating the construction of stable spatial structures without the need for any scaffolding by means of a cooperative robotic sequencing (Mirjan 2016; Parascho et al. 2017) by shifting from an assembly logic that constantly requires the robotic support of parts during assembly to one where parts only require support at specific steps of the assembly process.

The DFAB HOUSE (NCCR Digital Fabrication 2018), a three-story building project initiated by the Swiss National Centre of Competence in Research (NCCR) Digital Fabrication, allows for the transfer of this research to its building-scale application. The timber modules have to reach outside the research setting and are subjected to multiple real building challenges such as fire code, engineering code, acoustics, transport logistics, interfaces with neighbouring building components and finally the test of time while being used and exposed to the elements.

2 Techniques

2.1 Timber Frame Construction System

In this section, we present the constructive system for the research. The constructive system is a modification of the well-established timber frame construction that includes vertical timber beams in combination with structural plates (i.e. balloon frame or platform frame construction). This constructive system is informed by the robotic fabrication constraints and the required assembly sequence of the structure. The structure consists of generic timber beams with rectangular profiles that have only one simple cut (at varying angles) at each end. The connection between two timber beams consists of either one or two pairs of screws allowing for tension, compression and shear. The connection can be reinforced where needed by introducing a steel rod. The cutting planes for the beams at each end are generated algorithmically, and processed automatically by the robot in cooperation with the CNC saw. Concurrently, the algorithm generates the milling and drilling vectors for the screws or the tension rod. This approach ensures the possibility of assembling complex structures from generic timber beams without the need for complicated CNC joinery.

This constructive system has several advantages. First, structural plates are not necessary since the arrangement of the timber beams can provide lateral stiffness through triangulation. Therefore, this constructive system provides the opportunity to introduce suitable surfaces based on architectural requirements such as glass, translucent membranes, and openings without compromising the structural stability. Second, since the structure is assembled spatially, it is not necessary to have two corner beams as is the case with prefabricated plate components[1]. Therefore, the spatially assembled corner only requires one beam which is also free to rotate in plan, thus allowing for non-orthogonal configurations without the need for CNC manipulation along the length of the beam.

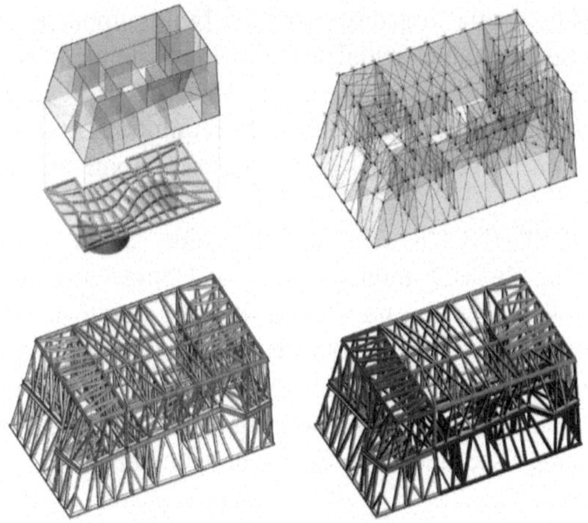

Fig. 1. Illustration of the computational design process/feedback. Diagrams: a. Architectural Inputs: the loadbearing ribs of the lower floor, programmatic organisation, and the exterior envelope b. Structure topology generation (Network graph) c. Generation of Beam class instances d. Modules.

2.2 Computational Design

A bespoke computational design and robotic fabrication workflow is developed for this research project (see Fig. 1). The computational design starts with a set of inputs including support condition (in the case of DFAB HOUSE, the loadbearing ribs of the lower floor), programmatic organisation, and the exterior envelope (see Fig. 2a). This workflow is organised around two main classes: "Building" and "Beam". The Building class is a non-manifold graph[2], which includes methods and attributes to generate and store the topological information of the building structure. The Beam class includes methods and attributes to generate and store all necessary geometrical, structural, and fabrication information for an individual beam.

The timber frame structure is generated in two sequential steps. Based on the inputs previously mentioned, an instance of the "Building" class is initialized. Successively, the methods of this class generate and store topological information of the timber

[1] These components each require an edge beam which, when fully assembled result in a double corner beam.

[2] The Building class is built on top of the "Network" class from "COMPAS". For more information about the Network class please see: http://block.arch.ethz.ch/docs/compas/core/pages/core/compas.datastructures.network.html.

structure in the attributes of the graph: vertices and edges. The topology of each beam in the Building class is represented as an attribute of the class "bar" with an identifier of the two end vertices. In the Building class, each bar has all the necessary information in relation to other beams and the overall structure such as neighbours, parents, and assembly order. After the generation of the building structure topology a "Beam" object for each bar is instantiated. Each generated timber beam (an instance of the Beam class) has the necessary information for fabrication such as the gripping plane, end cut planes, milling and drilling vectors and diameters, connections typology, cross-section etc. Implementation.

2.3 Experimental Setup

One requirement of the fabrication setup (see Fig. 2) is the ability to produce prefabricated timber modules at a building-scale. The setup in general and the engineered tools are designed with a dynamic planning process in mind e.g. changing material dimensions and assembly strategies throughout the course of the project due to changes in the planning process and findings made along the way. Certain timber beam constellations require cooperative robotic manipulation, meaning, two robots work together on the same task during assembly. The same robots also prefabricate (cut, mill and drill, the timber) beams.

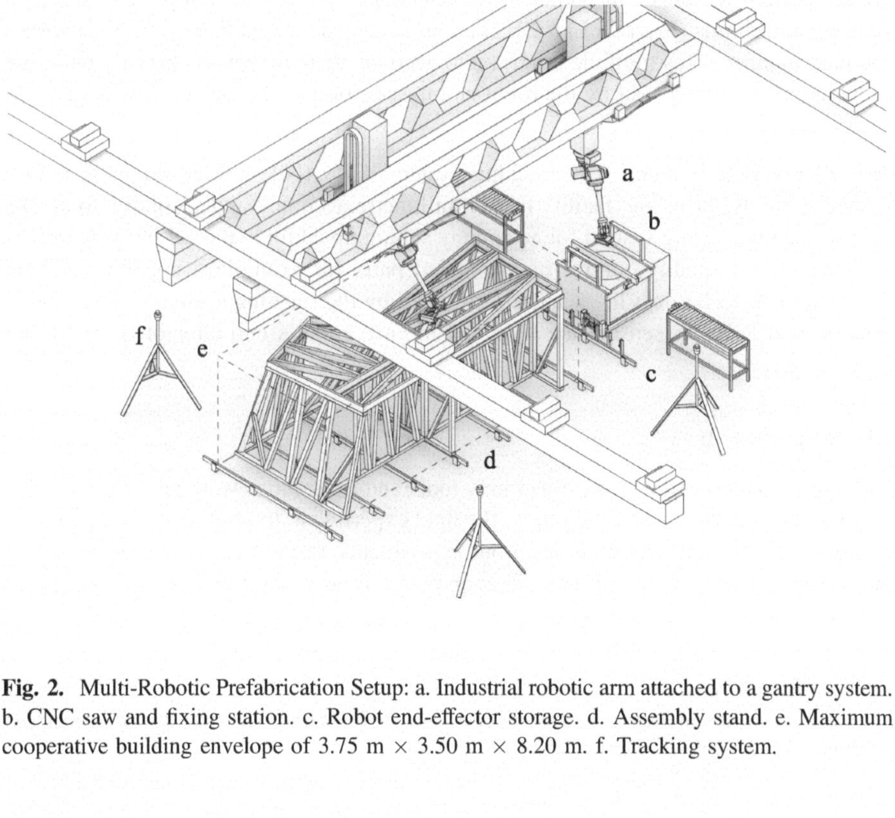

Fig. 2. Multi-Robotic Prefabrication Setup: a. Industrial robotic arm attached to a gantry system. b. CNC saw and fixing station. c. Robot end-effector storage. d. Assembly stand. e. Maximum cooperative building envelope of 3.75 m × 3.50 m × 8.20 m. f. Tracking system.

Multi-Robotic Prefabrication Setup. All of the tests and experiments are fabricated using ETH Zurich's Robotic Fabrication Laboratory (RFL). The setup has a total work space of 45 m × 17 m × 6 m. The robotic setup consists of two six-axis industrial robotic arms, each attached to a base with three axis of movement, x, y and z.

A three-axis CNC saw[3] is implemented for cutting the timber beams and serves simultaneously as a fixing station for the further processing of beams. The robots are equipped with automatic tool changers allowing them to switch easily between multiple tools during a fabrication sequence without the need of human interference. A custom gripper comprising of mostly off-the-shelf components was designed to pneumatically grip timber beams ranging from 60 mm to 220 mm in width and depth, thus enabling the robots to grip all of the timber beams on either their short or long side. This flexibility in gripping allows beams to cut in one orientation and placed in another. In other words beams can firstly be oriented so that cuts shallower than the saw's minimum pitch can be made, secondly the beam can be gripped from a different orientation to avoid collisions with neighbouring beams or cooperating robots during placement. The screw connection detailing required the design of a double-spindle tool. One spindle being equipped with a mill bit while the other houses a long drill for drilling holes at shallow angles.

Path Planner. The robotic fabrication and assembly of bespoke spatial structures leads to the requirement of bespoke robot path planning in order to manoeuvre a part from a start position to its final position in the structure without colliding with the already built structure or the other robot (Gandia et al. 2018). Sampling-based path planning algorithms (Kavraki Lab 2012) are used to generate collision-free trajectories. The path planner takes multiple criteria into account such as: what is already built, the timber currently being held by the robot and the position of the other robot as to avoid collisions.

RFL Correction System. The tracking system iGPS (Nikon Metrology) is implemented to firstly, bring the facility to sub-millimetre accuracy and secondly, to enable an absolute referencing system for accurately setting up additional machinery as well as allowing for the quality control of assembled parts for example using the handheld probe. In order to harness the accuracy provided by the tracking system in any robotic process, two iGPS i5 sensors are mounted to each robot's end-effector on the robot flange side.

2.4 Experiments

Throughout the course of the project new tools and techniques were tested on a series of empirical experiments leading up to the final experiment, the full-scale fabrication of the DFAB HOUSE timber modules. Four experiments, respectively Experiment 1, 2, 3 and 4 were undertaken to validate the cooperative robotic assembly of a corner, beam

[3] The saw has a 650 mm blade and three axis of movement: 0°–360° of rotation, 90°–25° of pitch and −10 mm–230 mm of linear movement.

processing and robotic pose correction, in place milling and finally the integration and realisation of a complete module.

Fig. 3. Cooperative assembly sequence

Experiment 1: Cooperative Robotic Assembly of a Corner. The first experiment aims to verify the assembly sequencing and fabrication strategy. By cooperatively assembling the timber beams with multiple robots, a building-scale timber frame module can be assembled in space without the need for additional scaffolding.

As the fabrication setup was still being developed, the beams were ordered pre-cut by a joinery machine. Despite the robots being accurately measured in and parts being fabricated to an accuracy of plus-minus 1 mm, the assembly proved to have gaps of up to 5 mm, especially in cooperative scenarios where both robot coordinate systems would need to align. Assembly is split into two distinct scenarios, singular, requiring one robot or cooperative, requiring two (see Fig. 3). First the base is assembled using a single robot on which corners are cooperatively assembled. Each is corner is constructed of three beams where the first robot brings the initial beam and supports it while the cooperating robot places the second and subsequent third beam. Once all three beams are connected, the robots can release the corner.

The assembly sequencing was adjusted as it became apparent that it was necessary to leave a "corridor" parallel to the gantry's y-axis free during assembly to allow both robots to avoid collisions with the structure when moving the robots from one side of the assembly stand to the other. The fixed base-to-wrist path planning approach proves to work in most cases but can lead to collisions and situation specific adjustments. There were collisions between the robot and the already assembled structure as well as self-collisions between the robot and the beam it was carrying due to the beams length of up to 3.3 m. The fixed base-to-wrist ratio also requires situation specific tweaking for different beam types e.g. the lower beams require the robot's base to be located above the wrist whereas the ceiling joists require the base to be approximately at the same height, as the z-axis of the robot's base was close to its limit. This shows that the implementation of bespoke path planning would be of assistance during the assembly.

Fig. 4. Beam Cutting and Detailing Sequence: 1. The robot positions the beam and saw executes the first cut. 2. The robot positions the beam and the saw executes the second cut. 3. The beam is repositioned and the robot proceeds to mill and drill the required detailing.

Experiment 2: Beam Processing and Robotic Pose Correction. This experiment aims to implement and test the robotic fabrication of timber beams based on the computational design and reduce tolerances encountered in Experiment 1 via the introduction of the RFL correction system to the robot's end-effector.

In contrast to Experiment 1 the timber beams in this experiment are cut by the robots based on the embedded fabrication information supplied by the computational design model's "Beam" class. The fabrication process (see Fig. 4) consists of multiple tasks carried out by either one or two robots depending on the assembly sequence of the beam. Firstly, a stock timber beam is placed by hand on the saw's conveyor belt. From there on the saw's grippers centre the beam and the robot proceeds to pick up the beam. In collaboration with the saw, the robot places the beam for both end cuts. After cutting has been completed, the beam is placed on the saw and supported again by the saw's grippers allowing the robot to automatically change tool from the gripper to the double-spindle tool and proceed with the milling and drilling of the required connection detailing. The beam is then picked up again by the robot and manoeuvred to its final place in the assembled structure with the aid of the path planning algorithms. Finally, double-threaded SFS structural screws are screwed manually into the predrilled connection detailing to fix the beam. Depending on the beam's assembly, the robot can now either release the beam and continue fabricating the next beam or continue supporting the beam while the other robot prepares subsequent beams. The RFL correction system is also implemented for the first time in conjunction with the robot's end-effector to reduce tolerances during fabrication and assembly. This requires the robot to pause and measure its position for approximately 2 s at every position requiring accuracy[4].

[4] A typical fabrication cycle requires accuracy at the following positions: initial pick-up, placement for the first cut, placement for the second cut, placement of the beam on the fixing station, milling approach, drilling approach, subsequent pick-up and final placement.

Experiment 2 shows if each tracking system sensor has visual contact to 4 or more transmitters, an end-effector accuracy of below 1 mm can be achieved. It also needs to be noted that despite the end-effector being positioned under 1 mm accurately in space and the beam cuts being accurate to 1 mm of one another while in the fixing station, some beam connections were up to 5 mm away from their target geometry. This deviation is due to timber tolerance in beams longer than 2.5 m and the robot gripping the centre of the beam, the point farthest from the connection. The insertion of joists into the perimeter of rim joists proved to be difficult due to tight material tolerances.

Fig. 5. In Place Milling: The robot mills the required façade detailing into the already assembled structure.

Experiment 3: In Place Milling. The aim of Experiment 3 is applying the findings of the previous experiments to a full-scale mock-up module from the DFAB HOUSE. The experiment also aims to integrate all of the necessary facade detailing that is required for final production.

The façade's geometry required certain lower and upper chords to be milled to a ruled surface. This requirement made way for the development of a novel method of fabrication "In Place Milling" (see Fig. 5). In place milling has several advantages over prefabricated milling. If the required ruled surface were to span multiple beams and were to be prefabricated, then a high level of accuracy would be required to ensure that the parts line up again smoothly once assembled. On the other hand in place milling allows for the fabrication of a continuous smooth surface across multiple beams without the need for high accuracy during placement of the beams. It allows for the decoupling of assembly and detailing tolerances. To tackle the tolerances occurring when inserting the joists into the perimeter of rim joists a feedback loop was integrated. Using the handheld probe, strategic points are measured on the rim joists and are fed back into the fabrication loop. This allows for the exact positioning of the rim joists to be updated thus updating the joists and their respective fabrication data.

In place milling greatly increased the accessibility for fabrication as obstructions such as grippers for fixing the part were not required. The ability to mill in place also delivers an absolute level of accuracy as it buffers out additive tolerances. The milled surface is within 1 mm accuracy of the target geometry. The feedback loop proved to be necessary when dealing with material tolerances and tight-fit insertion paths.

Fig. 6. Photo of the RFL during the production of DFAB HOUSE timber modules.

Experiment 4: Integration and Realisation of a Complete Module. The DFAB HOUSE timber modules aim to integrate all of the findings from the previous experiments and reach outside the research setting by taking real building challenges into account. The DFAB HOUSE includes 487 beams with unique geometries. The timber frame structure is divided into six modules (volumetric) and three flat panels.

The transfer of the research to a real world application led to multiple adaptations of the required detailing and subsequent fabrication. The precise positioning of the double threaded screws in the timber frame butt-connections lead to revision of their insertion approach. The screws are no longer inserted from above but rather from below. Based on the engineer's calculations, some of the connections would be exposed to tension forces above 14 kN[5] leading to the introduction of a rod and steel-plate connection.

The fabrication of the DFAB HOUSE timber modules (see Fig. 6) proves that the developed tools and techniques can successfully fabricate timber frame modules consisting of up to 99 beams and measuring 8.1 m × 3.6 m × 2.8 m (length × width × height) as well as flat elements such as the roof and the back walls. The robots are able to fabricate and manoeuvre beams weighing up to 55 kg and measuring

[5] 14 kN was calculated to be the highest tension force for two pairs of double threaded SFS screws.

8.1 m × 0.16 m × 0.08 m (length × width × height). This being said, beams of this size incur large material tolerances and require manual intervention when fixing.

3 Conclusion

The experiments throughout this project have shown that the robotic fabrication of spatial timber frame modules allows for the fabrication and assembly of geometrically complex and bespoke timber structures. The flexibility of the setup allowed for a seamless integration of a wide range of fabrication processes such as sawing, milling, drilling, in place milling and in place drilling. This in turn permitted the introduction of new details at late stages of the project without delaying fabrication, showcasing the advantage of the continuous digital chain. The setup allowed for the fabrication of timber modules at full-scale, thus expanding the range of prefabricated timber architecture.

The developed assembly logic worked in principle but still required adjustments as this proved to be more complex than anticipated. The parts require high tracking system transmitter visibility when being placed, cannot be occluded by previously assembled parts and need to be sequenced so that the robots can reach them. The positioning of the robot and the robotic fabrication of the parts themselves was highly accurate. However, the assembly of the parts proved to be challenging due to material tolerances, the robot's mechanical stiffness and the force-inducing nature of the screw connection. These tolerances were taken care of through human intervention by supporting the part by hand or with clamps and controlling accuracy visually or with the handheld probe.

Strategies to tackle these tolerances such as follows could be developed in the future, making the robust, swift and accurate assembly of spatial timber structures possible. A focus could be set on local accuracy of joining rather than global accuracy. In other words, the detail where two parts come together would be where accuracy is needed rather than the overall position of the part. This could be achieved by the robot gripping close to the connection rather than the centre of the part. The accuracy of the detail would be increased and more stiffness would be provided during assembly as less force would be applied to the robot. Force and/or proximity sensors could be implemented to allow the robot to feel or see its surroundings and place the part accordingly. In place detailing rather than prefabricated detailing could improve connection accuracy and make way for robot-driven methods of detailing such as nailing or doweling.

Acknowledgements. This research project is a result of an interdisciplinary collaboration between a team from ETH Zurich and the industrial partner ERNE AG Holzbau. The authors thank the other involved team members including Augusto Gandia, Gonzalo Casas, Matteo Pacher, Moritz Späh, Dr. Thomas Kohlhammer, Dr. Volker Helm, Dr. Ammar Mirjan and Aleksandra Anna Apolinarska from Gramazio Kohler Research, Chair of Architecture and Digital Fabrication at ETH Zurich. They also thank Michael Lyrenmann, Philippe Fleischmann and Lukas Stadelmann from the Robotic Fabrication Laboratory, ETH Zurich. This research project is supported by the Institute of Technology in Architecture at ETH Zurich, NCCR Digital Fabrication, ERNE AG Holzbau and a Commission for Technology and Innovation CTI grant.

References

Schindler, C.: Ein architektonisches Periodisierungsmodell anhand fertigungstechnischer Kriterien, dargestellt am Beispiel des Holzbaus, p. 194, Ph.D. thesis, ETH Zurich (2009)

Willmann, J., Knauss, M., Bonwetsch, T., Apolinarska, A., Gramazio, F., Kohler, M.: Robotic timber construction: expanding additive fabrication to new dimensions. Autom. Constr. **61**, 16–23 (2016)

Apolinarska, A., Bärtschi, R., Furrer, R., Gramazio, F., Kohler, M.: Mastering the "Sequential Roof": computational methods for integrating design, structural analysis, and robotic fabrication. In: Adriaenssens, S., Gramazio, F., Kohler, M., Menges, A., Pauly, M. (eds.) Advances in Architectural Geometry 2016, pp. 240–258. vdf Hochschulverlag ETH Zurich, Zurich (2016)

Mirjan, A.: Aerial construction: robotic fabrication of tensile structures with flying machines, pp. 157–165. Ph.D. thesis, ETH Zurich (2016)

Parascho, S., Gandia, A., Mirjan, A., Gramazio, F., Kohler, M.: Cooperative fabrication of spatial metal structures. In: Sheil, B., Menges, A., Glynn, R., Skavara, M. (eds.) Fabricate: Rethinking Design and Construction, pp. 14–16. UCL Press, London (2017)

DFAB HOUSE Homepage. dfabhouse.ch. Accessed 11 Feb 2018

Gandia, A., Parascho, S., Rust, R., Casas, G., Gramazio, F., Kohler, M., Automatic path planning for robotically assembled spatial structures. In: Robotic Fabrication Architecture, Art and Design 2018, submitted

Şucan, I.A., Moll, M., Kavraki, L.E.: The open motion planning library. IEEE Robot. Autom. Mag. **19**(4), 72–82 (2012)

Nikon Metrology. https://www.nikonmetrology.com/en-gb/product/igps. Accessed 12 Feb 2018

Large-Scale Additive Manufacturing of Ultra-High-Performance Concrete of Integrated Formwork for Truss-Shaped Pillars

Nadja Gaudillière[1,2(✉)], Romain Duballet[1,3], Charles Bouyssou[1], Alban Mallet[1], Philippe Roux[1], Mahriz Zakeri[1], and Justin Dirrenberger[1,4]

[1] XtreeE, 18–20 rue du Jura, CP 40502, 94623 Rungis, France
nadja.gaudilliere@gmail.com
[2] Laboratoire GSA, Ecole Nationale Supérieure d'Architecture Paris-Malaquais, 75006 Paris, France
[3] Laboratoire Navier, UMR 8205, Ecole des Ponts, IFSTTAR, CNRS, UPE, 77455 Champs-sur-Marne, France
[4] Laboratoire PIMM, Arts et Métiers-ParisTech, Cnam, CNRS, UMR 8006, 75013 Paris, France

Abstract. In the present paper a new additive manufacturing processing route is introduced to produce ultra-high-performance concrete complex architectonic elements, by printing integrated formwork. Interdisciplinary work involving material science, computation, robotics, architecture and design resulted in the development of an innovative way of 3D printing cementitious materials. The 3D printing process involved is based on a FDM-like technique, in the sense that a material is deposited layer by layer through an extrusion printhead mounted on a 6-axis robotic arm. An architectural application is used as a case-study to demonstrate the potentialities of the technology. Along with the detailed description of the design and construction process, a description of the responsibilities and their distribution amongst the stakeholders involved in the project is given. The steps taken to include the 3D printed element in an authorized regulatory context are presented as well. The structural elements produced constitute some of the largest 3D printed concrete parts available until now. Multi-functionality was enabled for structural elements by taking advantage of the complex geometry which can be achieved using our technology for large-scale additive manufacturing. The proposed process succeeds in solving several of the current issues problems that can be found in the production of 3D printed architectural features for an AEC industrial context and therefore suggests an immediately viable route for industry assimilation.

Keywords: 3D printing · Concrete · Cementitious materials
Large-scale additive manufacturing · Architecture · Design
Truss-shaped pillars

J. Willmann et al. (Eds.): ROBARCH 2018, *Robotic Fabrication in Architecture, Art and Design 2018*, pp. 459–472, 2019.
https://doi.org/10.1007/978-3-319-92294-2_35

1 Introduction

Until recently, additive manufacturing (AM) techniques were confined to high value adding sectors such as the aeronautical and biomedical industries, mainly due to the steep cost of primary materials used for such processes. In the last decade, the development of large-scale AM in such domains as design, construction and architecture, using various materials such as polymers, metal, and cementitious materials, has surged.

Historically, the first attempt at cement-based AM was made by Pegna [12] using an intermediate process between the classical powder bed and inkjet head 3D printing (3DP) [14] and fused deposition modelling (FDM) [5], in order to glue sand layers together with a Portland cement paste. Many groups have been involved with the development of large-scale AM for construction applications, all of which have been using processing routes derived from FDM or 3DP, although varying depending on the chosen material and targeted application.

Three pioneering technologies can be mentioned regarding cement-based AM. Contour Crafting technology [11] is based on a printhead mounted on a crane, extruding simultaneously two layers of material intended to be used as formwork. The main drawbacks of Contour Crafting technology are its limitation to vertical extrusion and a complex implementation of the system for industrial production. Loughborough University concrete printing technology [2], using a similar system with a printing nozzle and a crane, uses a material with higher performances and to a small extrudate, allowing a good geometrical control. However, the printing process is relatively slow and the use of an overhead crane both discourages the printing of complex geometries and complicates the industrial development. Finally, D-Shape technology [4] consists in a large-scale sandbed locally solidified by deposition of a binding agent, layer by layer. Although designed for off-site production of prefabricated elements, D-Shape technology currently aims to demonstrate the feasibility of the process on-site. More detailed information regarding existing cement-based AM techniques can be found in the literature [3, 7].

In the past few years, several other technologies have been developed for cement-based AM techniques and applications in construction, both by university research groups and new industrial players. Landmarks of AM construction applications have been produced during this same period, such as housing printed by the Chinese company Winsun and other small-scale building printed by companies such as Apis Cor (Russia) or CyBe (Holland). Major achievements also include two bridges, one by Acciona in Madrid and the other by TU Eindhoven in the Netherlands.

Beyond the mention in the Contour Crafting patent of the use of AM as integrated formwork for construction, only few constructive experiments based on this technique have been documented. AM of formwork can be seen in the Batiprint printing process, though the material used is foam, making the formwork disposable. The case study presented in this paper is the first using method UHPC concrete to produce integrated formwork for an actual building project using the XtreeE 3D concrete printing system.

The aim of the present work is to shed a new light on the perspective of 3D concrete printing in the construction industry, specifically using the integrated formwork technique, by describing this project. With many new industrial players but not many large-scale realisations yet, cement-based AM manufacturing appears to be at a critical point of development. The present paper seeks to discuss the industrial assimilation of these new construction techniques and the issues currently at stake in this matter. First, an introduction on designing structures for large-scale additive manufacturing is given. In Sect. 3, the concrete formwork 3D printing technique is presented, followed by the case study demonstrating the potential advantages of such a technique for the construction industry. Finally, conclusions are drawn regarding the results obtained in this work.

2 Design for Additive Manufacturing

Generating and modelling shapes for additive manufacturing follows specific rules, coming from both processing constraints, and functional objectives. According to [10], the concept of freeform previously used in the literature is not adequate nor sufficient for describing concrete 3D printing. For a given printing process and automation complexity, one can attain specific types of topologies within a given time-frame and performance criterion for the structure. Although out of the scope of the present paper, design conditions for large-scale additive manufacturing depend on many other parameters than just the properties of extruded cementitious materials; parameters such as the printing spatial resolution, overall size of parts to be printed, the environment, the presence of assembly steps, etc. A classification of such relationships between geometrical complexity, processing, and design is proposed in [7].

The processing constraints depend mostly on the fresh material properties in its viscous state, as well as early-age behaviour, in interaction with the building strategy and the stiffness of the structure being built. On the other hand, functional requirements will depend on the properties of the hardened material as well as the structural geometry for effective stiffness, and other functional properties such as thermal and sound insulation. See [10] for a geometrically induced thermal insulation case study. Both types of constraints must be considered at the design stage. Material properties of 3D printed concrete used by XtreeE, including in the case-study presented in this paper, are highlighted in [10] as well. More information on 3D printed concrete in general can be found in the very recent review published by [13].

Printing path generation is a critical step to be considered in early design phase. It is indeed the prime problem, embracing together machine, material and structural requirements. 3D-to-2D slicing, which is by far the most common method adopted in the context of 3D printing, yields planar layers of equal thickness built on top of each other. This approach is not optimal from a design and structural viewpoint as it will induce shifts when two consecutive layers have different sizes and limit the attainable geometries. In the context of 3D printing. Other approaches to tool-path generation have been developed, such as the tangential continuity method introduced in [10] to optimise the structure being built by creating layers of varying thickness. These layers exhibit a maximized surface area of contact between each other, hence stabilizing the overall structure. However, in the present case-study, as no other reliable software allowing the use of other methods existed yet, the first method was used.

3 Concrete Formwork 3D Printing

The process of large-scale concrete 3D printing developed by XtreeE has been mentioned previously in [10], the overall process is summarized in Fig. 1.

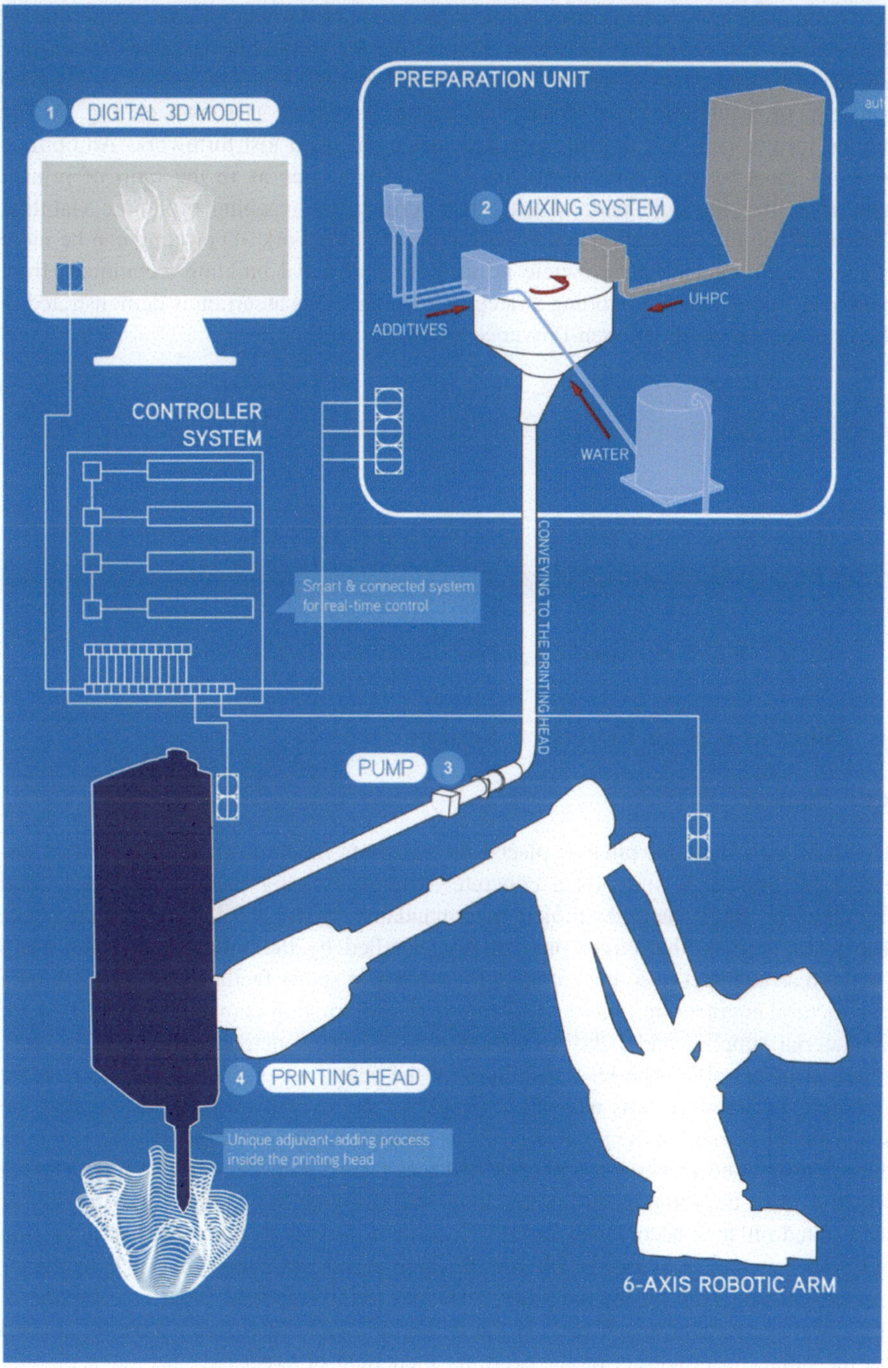

Fig. 1. Workflow of the large-scale concrete 3D printing process.

Based on this process, a construction strategy can be derived for concrete formwork 3D printing. The principle consists in 3D printing the formwork necessary for casting another structural material such as ultra-high-performance concrete for fibre-reinforced concrete, or insulation material such as foamed concrete, as shown on Fig. 2. The printed formwork is left in-place and becomes a so-called lost formwork. An optimal trade-off must be considered from the early design steps as to the ratio of printed material within the built part, which can be critical for reaching economic viability. Depending on the application considered, concrete formwork 3D printing can be more efficient than either all-3D concrete printing, or traditional building techniques, from both an economic and/or building strategy viewpoint. This assertion is demonstrated in the next section on the Aix-en-Provence post case study.

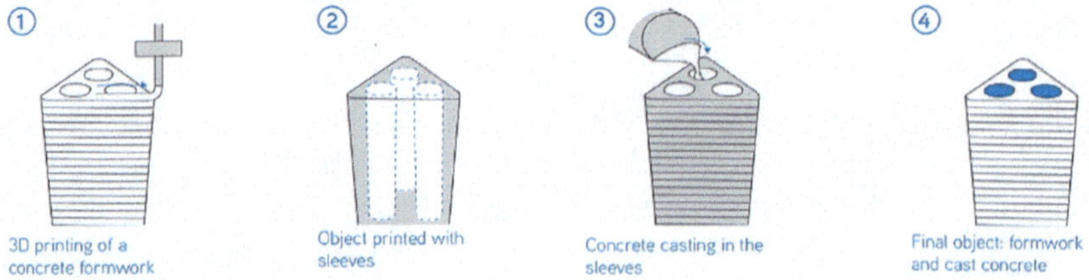

Fig. 2. Schematic view of the concrete formwork 3D printing.

4 Pillar in Aix-en-Provence, France

4.1 Context

This 4 m-high freeform pillar is placed in the sports facilities of a school in Aix-en-Provence, France. It supports a concrete awning covering part of the playground (Fig. 3). The sports facilities project was mandated by the Aix-Marseille Metropolis. The pillar, part of this larger project, was handled by the following people: Marc Dalibard as architect (also the architect for the whole sports facilities building), Artelia as structural engineering office, AD Concept as construction company, LafargeHolcim as material supplier and Fehr Architectural as UHPC concrete caster. For the construction of the pillar, the responsibilities of the actors were divided as follow: Marc Dalibard, as architect, was also the manager of the overall project, and defined the shape and placement of the pillar. Artelia, as structural engineer, supported XtreeE both during design and construction phases, and was tasked with defining the load case on which to base the topological optimization and verifying the strength and stability of the printed pillar in accordance with the load case. LafargeHolcim supplied a specific 3D print concrete, developed with XtreeE in an earlier collaboration and Fehr Architectural casted UHPC inside the pillar, a task requiring a specific licence. Each one of the key players supported XtreeE in the definition of the fabrication strategy adopted for the pillar, by providing input regarding their field of expertise.

Fig. 3. Pillar in Aix-en-Provence, photo by Lisa Ricciotti.

XtreeE identified a fabrication strategy for the post and adapted the printing system developed earlier (presented in Sect. 3) according to the fabrication strategy and its requirements. During the design stages, XtreeE co-defined the load case with structural engineer Artelia and designed an exact shape for the pillar through structural design. In the fabrication stages, XtreeE coded the manufacturing files using HAL Robotics for the printing system and performed the manufacturing, before co-supervising the placing of the pillar on site with architect Marc Dalibard.

Although relying on the skills of well established players of the construction market and its various subgroups, it is interesting to note that the Aix-en-Provence pillar studied in this paper presents a new workflow between the stakeholders of the project. This new workflow is established due to the appearance of a new player, XtreeE. XtreeE plays in this case the roles of technical expert in 3D printing, giving advice for every step of the conception and manufacturing process, of designer of the final shape and of manufacturer for the 3D print elements. While the new industrial protagonists in the field of AM manufacturing for construction applications might not, in the future, play such polyvalent role in projects, the Aix-en-Provence pillar gives an example of the array of possible interventions, and therefore of the many changes that could be implemented, not only with AM, but also with various rising digital conception and manufacturing tools.

4.2 Design

In the initial project designed by architect Marc Dalibard, a complex truss-shaped pillar supporting the roof was already planned, as shown on Fig. 4. But though the idea of a complex truss-shaped pillar was featured, no viable design for the pillar existed at this stage of the project. XtreeE came in at this point and took over the design of the pillar, based on the sketches provided by the architect.

Fig. 4. Initial sketch for the pillar.

The design of the pillar is based as much on the formal intention highlighted in these sketches as on the constraints fixed by the building regulations in effect at the time and by the 3D printing manufacturing method.

As no building regulation existed regarding 3D print items integrated in buildings at the time of construction, and to stick to the projected schedule, the choice was made to use the lost formwork manufacturing method, as introduced in Sect. 3. Instead of having to validate the pillar and its manufacturing method by using an ATEx (Experimental technical appreciation), a long and expensive procedure for experimental construction in France, the lost formwork made it possible to rely on existing building regulations on UHPC concrete. Furthermore, having recourse to 3D printing technology for the fabrication of architectonic elements offers the possibility to build non-standard geometries, with fine and curved elements. However, the complexity of those shapes proscribes implementing steel reinforcements inside. This impossibility, along with the fact that the shape of the pillar as designed by the architect had for main characteristic the slenderness of its parts, also called for the use of UHPC inside the formwork.

The critical design constraint for layer by layer extrusion techniques is the maximum inclination if the geometry. In the case of the Aix-en-Provence pillar, this issue was avoided by printing supports at the same time than the geometry to enable any angle for the parts of the truss. This has been commented in [7].

To define the precise shape of the pillar, we relied on an optimization method, to ensure an optimal use of matter in the truss by refining the limbs and taking into account the fabrication constraints as well as the wishes of the architects regarding the shape. The entire circular volume containing the pillar is used as research space for the truss to develop, and the applied load case included the weight of the concrete awning supported by the pillar as well as site-specific constraints (wind, etc.). The resulting final shape is shown on Fig. 5. A more thorough examination of the possibilities offered by topology optimization in the context of 3D concrete printing is available in [8].

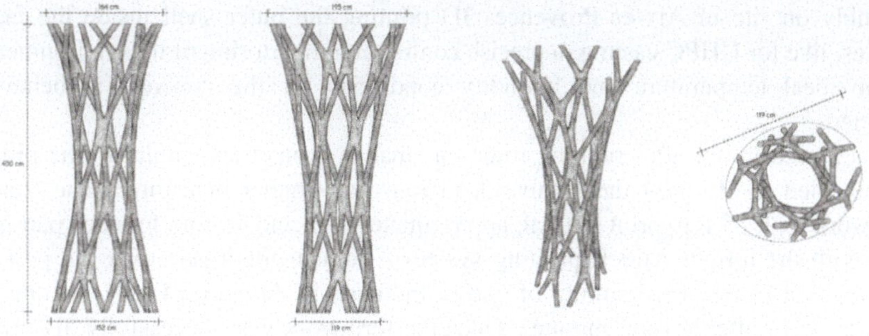

Fig. 5. Final design of the pillar after topology optimization.

Given the selected approach of lost formwork printing, the pillar is made of an outer shell that is 3D printed and later filled with UHPC. The pillar is divided in three smaller parts (cf. Fig. 6), both for transportation constraints and to reduce hydrostatic pressure when self-compacting concrete is casted inside. Each part is to be filled with concrete and then assembled together to form the whole element. During casting, metallic connectors are inserted in the concrete at each end of the parts to ensure reinforcement continuity of the construction joints. Finally, chemical glue is used to seal the connection.

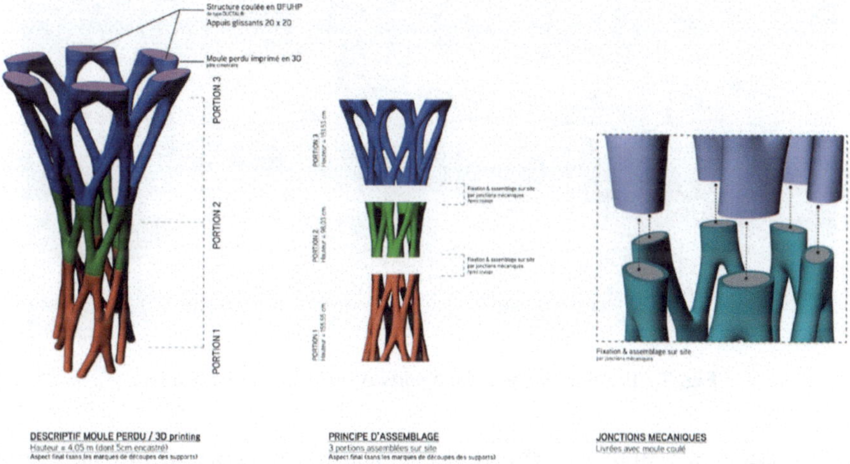

Fig. 6. Splitting and assembling principles for the 3D printed pillar system.

4.3 Manufacturing

Manufacturing of the pillar includes several stages: 3D printing the outer shell, at XtreeE headquarters in the south of Paris, France, casting the UHPC and integrating the connectors, at the Fehr Architectural production facility in Germany, and final assembly on site in Aix-en-Provence. 3D printing the outer shell inside the facility enables, like for UHPC casting, a precise control and monitoring of the environment, to ensure ideal temperature and humidity conditions for the concrete to behave as expected.

As a precaution, after running trials on smaller geometries similar to the pillar, it was decided to 3D print the formwork in four parts rather than three. The concrete formwork took 15 h to print in total, approximately 3 h and 45 min for each part of the post, with the former XtreeE printing system – adjustments made over the past year now make it faster. The printing of one of the parts is shown on Fig. 7. Setting time starts to occur after several minutes. Once the formwork was successfully 3D printed, an assembly trial, shown on Fig. 8, was conducted at our facility to ensure the results were as precise as expected before shipping the parts.

Fig. 7. Printing of one of the parts of the pillar's formwork.

The casting of UHPC in each part of the pillar was then operated by the team of Fehr Architectural. To resist the hydrostatic pressure resulting from the casting, supports printed with the pillar were left in place until the UHPC set, as shown on Fig. 9. The supports were then cut, as shown on Fig. 10, and the parts were shipped on site to Aix-en-Provence. The definitive assembly of the parts was performed there, before installing the pillar in place, with sliding supports on top. Finally, on request of the architect, the pillar was given a smooth finish by coating it to cover the line pattern specific to 3D printed objects. The difference of surface roughness is shown on Fig. 11.

Fig. 8. Assembly trial after printing of the four parts.

Fig. 9. UHPC casting inside the envelope, with supports left in place.

Fig. 10. Cast and lost formwork assembly from which the printing supports were cut.

Fig. 11. Before (left) and after (right, photo by Lisa Ricciotti) surface smoothing through manual coating.

5 Conclusions and Outlook

The various advantages of large-scale additive manufacturing of ultra-high-performance concrete, as well as the concrete formwork 3D printing technique, have been reviewed based on the analysis of the case study of a complex pillar part of an industrial construction projects in France, performed using the XtreeE 3D printing technology. A final significant element of comparison can be given by confronting the total production price of the Aix-en-Provence pillar to the total production price of a traditionally built complex pillar: a 62,5% total price gain is obtained, based on our information for the price of the Aix-en-Provence pillar, and quotes obtained for a traditional manufacturing. One of the main reason for this difference, on top of the gain identified on time, materials, and workforce, is the absence of a specific material and shaping for the mould, hence eliminated by using the lost formwork method. The multiple aspects of potential socio-economic gain for relying on additive manufacturing are three-fold: (1) materials saving by using the right amount of matter where needed, given that a topology optimization computational framework is available; (2) time saving by reducing the number of steps in the construction process, as well as being BIM-compatible for construction-planning strategies; (3) workforce saving by limiting on-site manual building steps, therefore enhancing safety on the construction site and implementing a new workflow between project stakeholders.

Although the lost formwork strategy allowed to get around experimental technical certification, further work should have to be conducted with certification authorities for the construction industry to define a legal and regulatory framework for 3D printed structures. The technological feats presented in this work are commercially available, but a legal framework and economic market are to be developed for the 3D printing technology to transfer into the mainstream construction industry.

Studying the conditions of design and fabrication of the Aix-en-Provence pillar also provides hints on possible improvements to the developed process, to push industrial assimilation forward.

Along with the paradigm shift of additive manufacturing comes the possibilities enabled by topology optimisation, which aims at attaining the most efficient structure geometrically possible for a given set of requirements. Optimality in terms of industrial design has become more and more critical due to scarcity of material resources and the need for lightweight structures.

A driving force for additive manufacturing is its ability to produce more complex 3D shapes in comparison to casting or subtractive processes. This complexity allows to design optimal structures based on topology optimisation techniques. One of the main current challenges is to modify optimisation algorithms to account for the additive manufacturing constraints, especially with regards to the processing parameters and structural stability while printing. A possible answer to these challenges would be to consider the multiphysics phenomenon aspect of 3D printing, which involves the elastic stability of the overall structure being built, the kinetics of hydration, the evolving viscoplasticity of fresh cement, the evolution of temperature within the printing environment, etc. As a matter of fact, all these physical problems, at multiple time and space scales, can be modelled on their own, but coupling them generates complexity and uncertainty regarding the process of 3D printing. Therefore, efforts should be concentrated on understanding and modelling the printing process in its multiple physical aspects, only then optimisation will be fully integrated with the processing, which would virtually change the way 3D printed structures are conceived today.

References

1. Bendsøe, M., Sigmund, O.: Topology Optimization. Springer, New York (2004)
2. Buswell, R.A., Soar, R., Gibb, A., Thorpe, A.: Freeform construction: mega-scale rapid manufacturing for construction. Autom. Constr. **16**, 224–231 (2007)
3. Buswell, R.A., de Silva, W.R.L., Jones, S.Z., Dirrenberger, J.: 3D printing using concrete extrusion: a roadmap for research. Cem. Concr. Res. **107** (2018)
4. Cesaretti, G., Dini, E., Kestelier, X.D., Colla, V., Pambaguian, L.: Building components for an outpost on the lunar soil by means of a novel 3D printing technology. Acta Astronaut. **93**, 430–450 (2014)
5. Crump, S.: Apparatus and method for creating three-dimensional objects US Patent US5340433A (1992)
6. Cui, C., Ohmori, H., Sasaki, M.: Computational morphogenesis of 3D structures by extended ESO method. J. Int. Assoc. Shell Spatial Struct. **44**(1), 51–61 (2003)
7. Duballet, R., Baverel, O., Dirrenberger, J.: Classification of building systems for concrete 3D printing. Autom. Constr. **83**, 247–258 (2017)
8. Duballet, R., Baverel, O., Dirrenberger, J.: Design of space truss based insulating walls for robotic fabrication in concrete. In: De Rycke, K., Gengnagel, C., Baverel, O., Burry, J., Mueller, C., Nguyen, M.M., Rahm, P., Thomsen, M.R. (eds.) Humanizing Digital Reality, Design Modelling Symposium 2017, pp. 453–461. Springer, Singapore (2017)
9. Duballet, R., Gosselin, C., Roux, P.: Additive manufacturing and multi-objective optimization of graded polystyrene aggregate concrete structures. In: Thomsen, M., Tamke, M., Gengnagel, C., Faircloth, B., Scheurer, F. (eds.) Modelling Behaviour, Design Modelling Symposium 2015, pp. 225–235. Springer, Cham (2015)
10. Gosselin, C., Duballet, R., Roux, P., Gaudillière, N., Dirrenberger, J., Morel, P.: Large-scale 3D printing of ultra-high performance concrete – a new processing route for architects and builders. Mater. Des. **100**, 102–109 (2016)
11. Khoshnevis, B.: Automated construction by contour crafting- related robotics and information technologies. Autom. Constr. **13**, 5–19 (2004)
12. Pegna, J.: Exploratory investigation of solid freeform construction. Autom. Constr. **5**, 427–437 (1997)
13. Roussel, N.: Rheological requirements for printable concrete. Cem. Concr. Res. **107** (2018)
14. Sachs, E., Haggerty, J., Cima, M., Williams, P.: Three-dimensional printing techniques, US Patent US08045632 (1993)
15. Zhou, S., Li, Q.: Design of graded two-phase microstructures for tailored elasticity gradients. J. Mater. Sci. **43**, 5157–5167 (2008)

Realization of Topology Optimized Concrete Structures Using Robotic Abrasive Wire-Cutting of Expanded Polystyrene Formwork

Asbjørn Søndergaard[1(✉)], Jelle Feringa[2], Florin Stan[3], and Dana Maier[4]

[1] Aarhus School of Architecture, Nørreport 20, 8000 Aarhus C, Denmark
asbjorn.sondergaard@aarch.dk
[2] Odico Formwork Robotics, Peder Skrams Vej 5, 5220 Odense SØ, Denmark
[3] UBE Design, Str. Izvorului no. 22, Alba Iulia, Romania
[4] Zünd Scandinavia, Knudsminde 4B, 8300 Odder, Denmark

Abstract. This paper presents a new method for cost-effective realization of topology optimized structures using robotic abrasive wire-cutting of expanded polystyrene formwork. Topology optimization has shown potential for generating material efficient designs and increasing performance in architectural structures. However, the method results in complex, structural morphologies which frustrate efficient construction of said structures. To overcome this, and make the realization of the potential of topology optimization feasible in general construction, new approaches are needed. We propose an integrated method of ruled surface rationalization and robotically controlled abrasive wire-cutting of formwork parts in Expanded Polystyrene. The method is demonstrated through robotic production of EPS formwork using a pilot abrasive wire-cutting end-effector on a containerized robotic work cell with an ABB IRB 6700 industrial manipulator, extended with external rotary axis. The usability of the formwork is demonstrated through the construction of a 21 m, prefabricated demonstrator structure using Ultra High Performance Concrete.

Keywords: Topology optimization · Concrete · Robotic abrasive wire-cutting Expanded polystyrene · Formwork systems

1 Background

1.1 Motivation

Current consensus within the scientific community of meteorological and atmospheric sciences outlines several imperatives to reduce anthropogenic emissions of carbon dioxide, mainly the mitigation of anticipated risk of climate change. Within this frame, the emissions associated with construction contribute significantly to global emission levels. Within the construction sector, construction of concrete structures – in particular the production and calcination of cement – represented in 2015 a combined 8% of global emissions [1], or more than four times the emissions generated by the global air

J. Willmann et al. (Eds.): ROBARCH 2018, *Robotic Fabrication in Architecture, Art and Design 2018*, pp. 473–488, 2019.
https://doi.org/10.1007/978-3-319-92294-2_36

traffic [2]. This figure is projected to more than triple until 2050 due to the increasing global consumption of concrete for construction purposes.

It follows from this outline that a technology capable of reducing the emissions associated with concrete construction could have significant impact on global emissions, assuming industry-wide adoption. To address this challenge, several solutions can be considered: switching to alternative fuels to heat the calcination kiln, such as natural gas, biomass or waste-derived fuels [3]; replacing cement with a negative emission substitute [4]; using engineered timber as substitute for concrete [5] or simply increasing the performance of concrete structures through optimization of the design geometry, hereby lowering general material consumption. The research effort presented in this paper is directed towards this last approach, while acknowledging the value of alternative strategies.

1.2 Hypothesis

Several studies indicate that various means of structural optimization can substantially increase the performance of concrete structures, in ranges that allow for reductions of material consumption [6–8]. Among the range of optimization strategies, topology optimization - commonly applied within the aeronautic and automotive industries – have recently seen an increasing interest as a form-finding strategy for architectural structures [9–14] (Fig. 1).

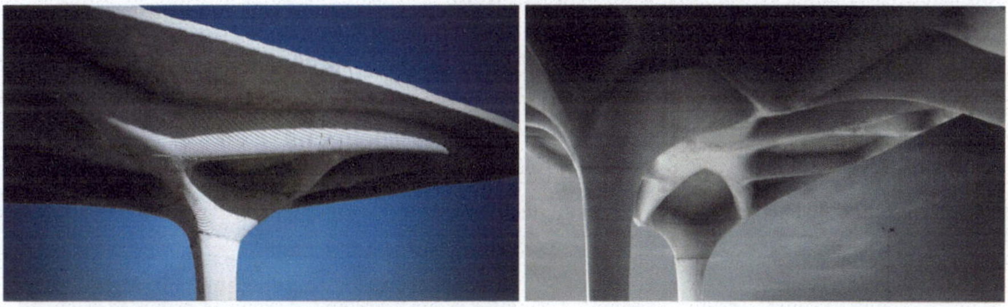

Fig. 1. Topology optimized concrete prototype from the Unikabetonproject, produced via robotic CNC-milling of EPS molds

While these studies indicate significant potential for increases in structural performance, this improvement is achieved through a simultaneous increase in geometric complexity, which challenge current means of construction. For concrete manufacturing, the primary expense for non-standard structures is formwork costs, which represent 50–70% [15] of the total costs of the concrete works. Since the material cost per weight unit is comparably low, there is little financial incentive to deploy reduction measures for concrete material consumption at the expense of increasing formwork costs [16]. It is our hypothesis that the main constraint prohibiting large scale adoption of topology optimization of concrete structures in the global construction sector is the non-availability of cost-efficient means of realizing advanced formwork.

2 Formwork Manufacturing

2.1 State-of-the-Art

Current industry standard for manufacturing of non-standard mold systems, falls generally in two categories: CNC-milling of timber/EPS molds (typically for designs of high complexity) and manually constructed formwork systems (typically flexible modules with bespoke auxiliary supports, or sheets of flexible material bend to match CNC-routed edges).

These methods are well proven, and capable of realizing complex structures, however with 2 key disadvantages. CNC-milling of advanced formwork is – while capable of producing designs of high complexity and high-end surface finishes – limited in production speed by its mechanical principle, which is defined by incremental subtraction of material from the work object, and thus speed is constrained as a function of the toolbit diameter relative to the target surface smoothness, defining the density of the toolpath and number of passes over a given surface. In practice, this typically translates to >240 min/m^2 of machining time to arrive at industrial grade surface smoothness. As such, the pricing per m^2 of milled formwork is predominantly defined by the required machining hours (as a function of the depreciation of the machining station), hence resulting in a premium pricing, available only to a low percentage of high-profile construction projects. Similarly, the construction of manual formwork systems is highly labor intensive, and constrained by the flexibility of the modular systems or formwork material applied.

To address the challenge of finding more cost-effective methods of realizing advanced concrete structures, a diversity of approaches has been suggested, including (1) actuator-based, flexible molds [17]; (2) actuator-based slip-casting [18]; (3) 3D concrete printing [19, 20]; (4) hybrid wax extrusion and CNC-machining [21]; (5) robotically manufactured stay-in-place formwork [22] and (6) robotic hot-wire cutting of EPS formwork [23]. A discussion of these proposals are given in [24]

Fig. 2. The realization of the Fjordenhjem Kirk Kapital HQ designed by Studio Olafur Eliasson, represents the first such commercial scale application of RHWC.

Robotic hot-wire cutting of expanded polystyrene formwork has been pioneered in commercial deployment by Odico, who have demonstrated as an international first the realization of large scale load-bearing structures using this manufacturing method.

For production of ruled surface geometries, hot-wire cutting holds a significant machining time advantage over traditional CNC-milling, as reaching the production surfaces can be achieved in a single pass operation, sweeping the entire surface. In practice this reduces production time of an equivalent surface to a few minutes (Fig. 2) compared to CNC-milling

Fig. 3a. Comparative samples produced by (from left: RCNC; RHWC; RAWC).

While the above developments demonstrates the viability of RHWC for construction purposes, the thermal cutting process of hot-wire cutting, in which polystyrene material is evaporated at 290–310 °C around an electrically heated wire, comes with a number of principal short-comings: firstly, cutting speed is a function of the material density of the polystyrene work object, and hence at high density material types, production speed is significantly lowered, whereas for CNC-machining it remains principally constant within the same classes of material. Second, the process is limited to thermally cut-able materials, which excludes a wide range of construction materials, including non-flammable foam types typically used in acoustic insulations, as well as common groups of solid construction materials, such as timber, natural and artificial stone and clay. Finally, the process is sensitive to maintaining a constant cutting speed across the entire length of the wire. This in practice excludes several classes of toolpath strategies, such as triangular sweeps and hyperbolic paraboloids, in which the Tool Center Point (TCP) remains relatively in-active, while the periphery of the wire moves at high velocity (Fig. 3a).

To address these challenges, Odico engaged in the development of a complementary process using a diamond-threaded abrasive wire rotating on electrically propelled flywheels, fixed within an 8-rod carbon fiber frame structure. Over the course of 4 prototype end-effectors, the process was refined to achieve an average cutting speed of 75 mm/s, hence reducing overall cutting time for the reference surface to <15 s, while leaving a finished surface of very high smoothness levels (Fig. 3b).

The identification of machining time is conducted through a comparative study of the 3 machining methods on 3 similar robotic workstations, using respectively IRB 6400R 2.8-200 (station 1), IRB 6400R 2.8-150 (station 2) and IRB 6700-300 (station 3) manipulators. Station 1 is equipped with a HSD ES-915a electrospindel with ø50 mm Schunk ball-end toolbit. Station 2 with a proprietary RHWC end-effector with ø 0.5 nickel-chromium wire suspended in a 2600 × 1600 × 150 mm carbon fiber frame.

Fig. 3b. Top row, from left: Workstation 1, Robotic CNC-milling. Workstation 2, Robotic Hot-Wire Cutting. Workstation 3, Robotic Abrasive Wire-Cutting. Bottom row: the corresponding machining result of sample 1 from each process.

Station 3 is equipped with a proprietary RAWC end-effector with electrically propelled ø1.5 mm abrasive wire (Fig. 2). The 3 stations produce a curved sample geometry of $1000 \times 1000 \times 150$ mm, with total ruled surface area of 1,187 m^3. We note that this effectively corresponds to a 9600% increase in production speed over robotic CNC milling, assuming product surfaces constructed from ruled surface geometries (Table 1).

Table 1. Comparative chart for machining times of sample 1.

Material	CNC	RHWC	RAWC
EPS S80	4 h 10 min 32 s	1 min 58 s	16 s
EPS S150	4 h 10 min 32 s	6 min 21 s	24 s
MX 300	4 h 10 min 32 s	19 min 32 s	36 s
Solid mat.	Yes	No	Yes

3 Prototype Design

The design freedom enabled by the above developments provides a manufacturing framework for achieving rapid production of advanced formwork for concrete construction. To assess the potential for using this capacity in cost-efficient realization of topology optimized concrete designs, a full scale prototype study was undertaken.

With the motivation to test the extremes of the solution space, an optimization design space was developed, constituting a blend between 3 primary architectural configurations: a wall situation; a canopy; and a corner. The design space was optimized in Altair Optistruct for dead-load and surface wind loads, using a volume constraint of 0.3. The resulting optimization provided the framework for the development and testing of a rationalization method and formwork system design capable of approximating the optimized mesh close enough to avoid compromising the reduction in material volume achieved through the optimization (Fig. 4).

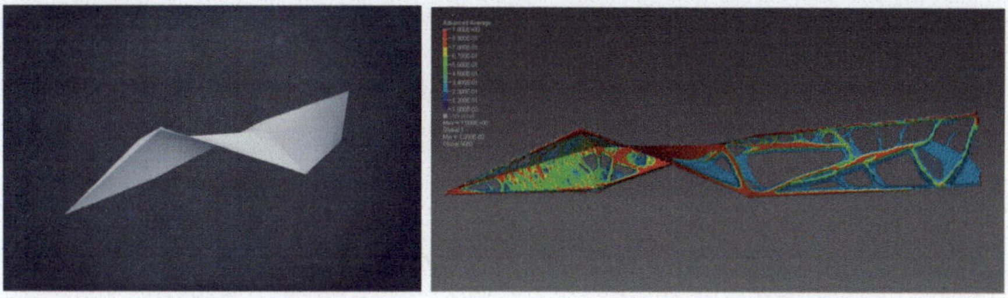

Fig. 4. Left: design space before optimization. Right: topological optimization result

3.1 Rationalization

As the mechanical principle of abrasive wire-cutting is limited to the production of ruled surface geometries, a translation of the optimization mesh to compositions of ruled geometry is necessary to enable manufacturing through this method.

Within this context, the main challenge is to approximate the optimized mesh close enough not off-set the performance gains achieved through the optimization, while remaining within the solution space of the constraints of the fabrication method.

To address this, a parametric rationalization scheme was developed, relying on (a) skeletonal tracing of the centerlines of the topology, and (b) a numeric approximation of the volume surrounding the centerlines through intersecting cells of ruled surface assemblies.

3.2 Formwork System

Through the rationalization procedure, the formwork system is now determined. The system is comprised of three overall constituents, which is repeated for all 6 concrete components: (1) the main formwork system in expanded S80 polystyrene, split in two main parts to accommodate the later insertion of reinforcement bars. (2) A surrounding steel casing from assembled standard Peri scaffolding modules. (3) A parametrically generated timber support structure consisting of connected cassettes of (1) OSB-plates ribbed with C24 coniferous beams to direct lateral pressure to the surrounding steel casing; (2) ribbed plywood panels to support directly the casting pressure exerted on the EPS formwork (Fig. 5)

The EPS formwork design is determined by inverting the intersecting surfaces to volumetric formwork cell units, segmented per surface level. The boundaries of the formwork cell volumes are determined by extending the intersection lines of the cell surfaces.

As the extension of the corner surfaces would collide with the volume of the linear extensions, the corners are separated as individual mold pieces, using the original intersection planes of the rationalization scheme. We have then an EPS formwork composition consisting of the following parts:

(A) corner segments, which hold the rounded hyperbolic paraboloid surface (highlighted red, Fig. 6). (B) main mold volumes, which hold the single ruled extension surfaces between the corners (parts in white, Fig. 6). (C) boundary molds, which

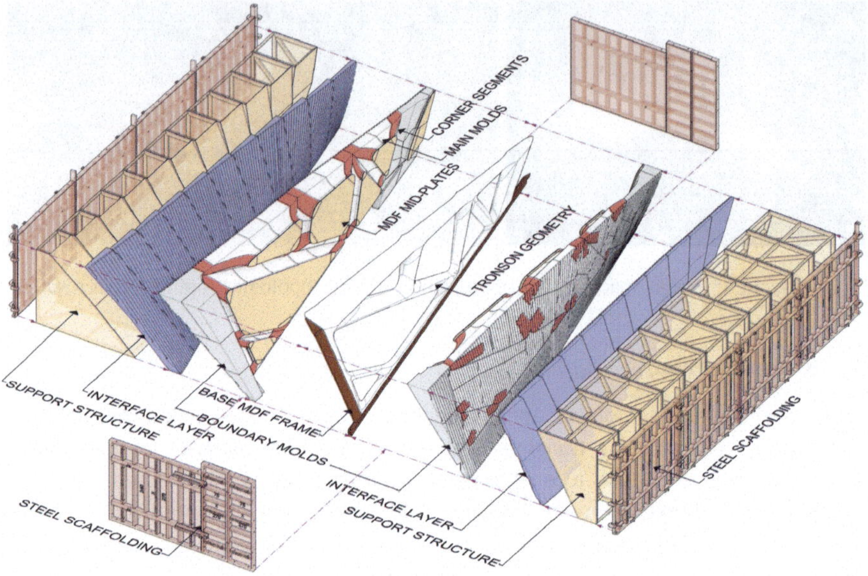

Fig. 5. Formwork system, exploded isometry.

close the outer boundaries of the formwork cell assembly to the bottom and sides, with the top remaining open for pouring of the liquid concrete (marked grey, Fig. 6). (D) an interface layer, which bridge the panelized segmentation of the timber support structure and part A-C of the EPS formwork, the depth of which is minimized to avoid self-intersection of the extension of the boundary surfaces. To guide the subsequent assembly process, the interface layer is designed with linear waffle profiles aligned with the isocurvature of the global interface surface

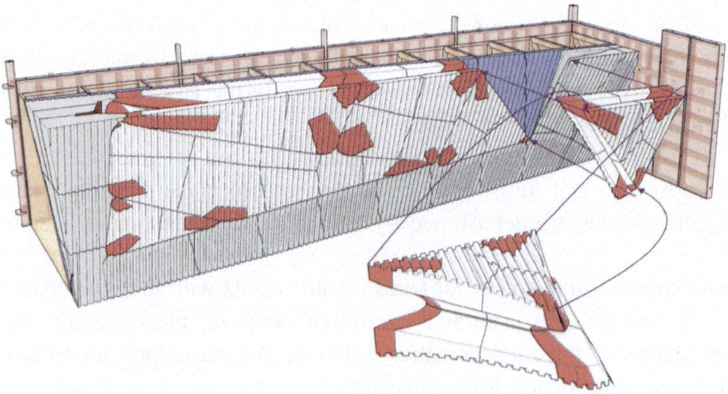

Fig. 6. Corner and main mold parts inserted into the global system

3.3 Robotic Production

With the rationalization and formwork design computed within a single workflow, we are now able to deduct the cutting motion for the robotic production cell. This system consists of (a) an IRB 6700-300 industrial manipulator, positioned inside a containerized transportable work cell frame and equipped with (b) a proprietary end-effector with an electrically propelled abrasive wire, fixed within an auxiliary carbon fiber frame, and (c) a 1300 mm high elevated work object fixture mounted on an external rotary axis (Fig. 7). Within this production cell, the surfaces are translated to the following cutting sequences:

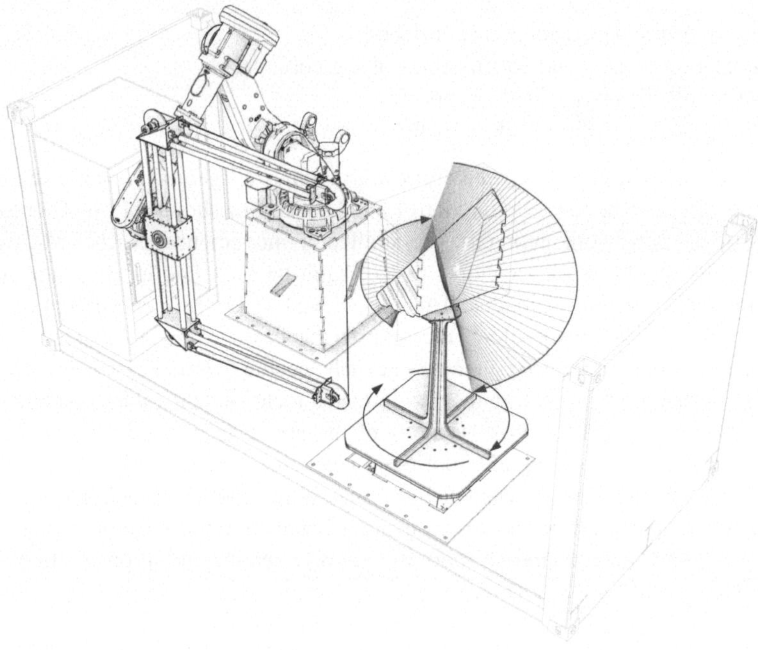

Fig. 7. Diagrams of the cutting motion for respectively corner hypar

Corner Mold Parts

The cutting sequence is initially discretized in (a) rotation of the work object on the external axis, and (b) surface cutting motion, with the Tool Center Point (TCP) of the abrasive wire end-effector following the line centers of the cutting surface isocurvature. While synchronized motion between external axis and end-effector was initially applied, the simultaneity of rotational and linear motion resulted in distortion of the linearity of the wire, hence deforming the product geometry. By segregating the motion of the external axis and the end-effector cutting in discrete time intervals, this problem is avoided. With this principal approach, cutting sequences are then computed in the following steps, describing all sides of the part in wire-motion:

(a) Top and bottom waffle profiling of the mold piece
(b) Side profiling
(c) Cutting of the hyperbolic paraboloid corner surface
(d) Back-side cutting and splitting of the piece in 2 halves, finally releasing the piece from the stock material.

Main mold parts
Similar to the corner parts, main mold parts are processes through the following sequence:

(a) Cutting of the 4 extended side surfaces
(b) Subtraction of the void for insertion of the corner part
(c) Cutting of the single ruled rib profiles
(d) Cutting of the waffle profiling of the formwork cell top and bottom surface

The cutting motion is deducted from the rationalized geometry using the Odicut proprietary framework for robotic control in McNeel Grasshopper [25]. Due to composition of the formwork geometries, end-effector movement describes the extreme regions of the robotic work envelope, hence requiring both fine-tuned control of tool orientation and collision avoidance. This is achieved by generating a global rotational volume based on the stock block or, in extreme cases using a local convex hull deducted from the current mold geometry at each specific external axis value. Collision avoidance is ensured by orienting the tool to match local tangent values computed from either horizontal or vertical cross sections of the above mentioned rotational/convex hull volume and thus avoiding any tool-mold geometry collisions. This further implies that during each transition movement from one cutting motion to another there is the above solution space, albeit specific to using a linear tool end-effector, which can be used to generate valid and smooth robot motion between the end of one cutting motion towards the start of next one (Fig. 8).

3.4 Formwork Assembly and Casting

The corner and main mold parts are assembled into larger formwork cells using polyurethane adhesive and acrylic based surface coating. The timber support structure, produced through digital routing and processing on a Hundegger machining station and files derived parametrically from the global geometry model, is manually assembled in two main parts. The interface layer is mounted to the timber support and subsequently, the formwork cells are mounted on the interface layer, using the waffle profiling as assembly guide (Figs. 9 and 10).

Global coating is added, and reinforcement is inserted. Following the assembly of reinforcement, the two formwork sides are closed, and concrete injected. After 24 h of

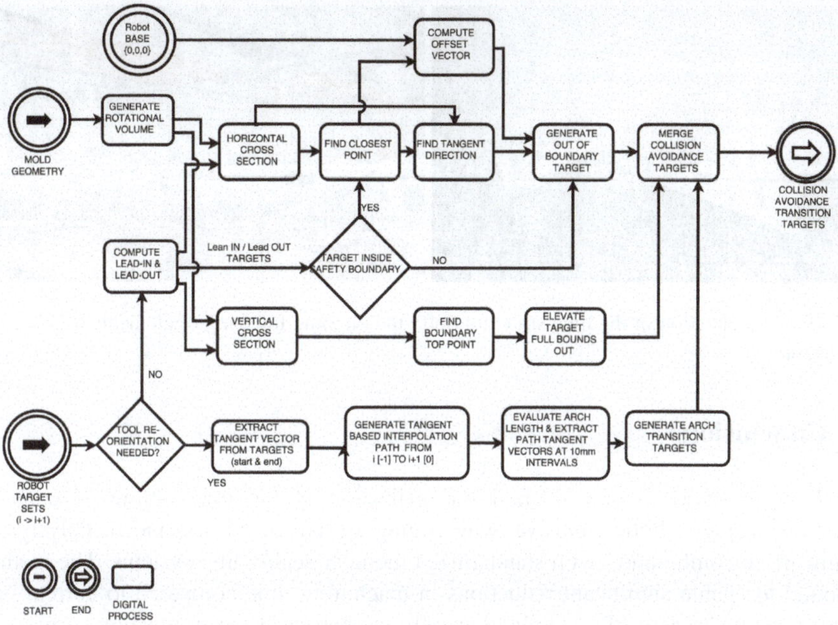

Fig. 8. Function diagram of the collision avoidance scheme

Fig. 9. Timber support of element C1 prior to EPS assembly (left). Assembled mold and finishing of element B1 (right).

curing, the formwork is demolded and demolished, completing the production of one structural component. The components are transported to site, and assembled using a combination of temporary auxiliary supports, and bolting[1].

[1] Note to reviewers: at the moment of writing 5 of 6 elements are cast, and on-site assembly scheduled for May 25th 2018.

Fig. 10. Visualization of the final structure (left) and ongoing finishing of elements C2, A2, and A1 (right)

4 Conclusions

We have presented a new method for realization of topology optimized concrete structures using robotic abrasive wire-cutting of advanced Expanded Polystyrene formwork in combination with standardized modular scaffolding systems. The method is found to enable significant reductions in machining time compared to current production methods, as well as facilitating full automation of the production of bespoke elements. The viability of the method is demonstrated through the production of a full scale, topology optimized, UHPC prototype structure, realized through prefabrication of 6 components to be joined on site.

While the overall feasibility of the method has been demonstrated, the complexity of the prototype geometry and the resulting construction process points to a number of workflow limitations to be addressed before achieving feasibility for large scale adoption. These limitations are outlined in the following.

5 Discussion and Further Work

The design choice of relying on doubly ruled, hyperbolic paraboloid corner sections enables a closer approximation to the optimized mesh than achievable when relying on faceted, single ruled geometries. However, it comes at the significant cost of needing to segment the formwork cells into corner and main sections. This adds substantial workload on the subsequent formwork assembly process, as well as increasing the room for production tolerance, in addition to requiring highly advanced motion planning and control of the part cutting sequences (Fig. 11).

The spatially advanced design of the formwork parts adds significant loss of formwork material, surmounting to an averaged 40–60% of the material stock block. As opposed to CNC-milling, which would granulate the cut-off material directly, the cut-off however here doubles in function as protective casing, which significantly eases transport logistics as parts can be shipped as regular, rectangular bodies. As a result of the segmentation scheme - determined by the geometry rationalization method – the overall volume of waste production is comparably higher with estimated 15–20% than achievable through CNC-milling. However, for the general discussion of waste

Fig. 11. Reinforcement of element B1 (top, left) and element B2 preparing for transport (top, right). Bottom: assembly of element A1 on A2, B1-B2

production it should be noted that (a) the production and consumption of expanded polystyrene formwork material requires a fraction of the energy associated with concrete manufacturing (Fig. 12).

Hence, even significant increases in formwork waste production is fully offset by reductions in concrete material consumption enabled by the use of polystyrene formwork; (b) compared to conventional formwork systems in solid and laminated timber or construction metals, the overall energy footprint of formwork production is significantly less for expanded polystyrene systems; (c) the material properties of polystyrene enables a direct and energy light recycling process, in which cut-off parts are granulated and re-used in forming of new stock block material.

The spatially warping configuration of the global design - and the resulting complexity of the component formwork systems – required as noted the design and manufacturing of bespoke timber support segments on a per element basis to reduce the general consumption volume. This step could be standardized in concrete component designs of less spatial and dimensional variability, and entirely avoided for component

Fig. 12. Casting result of element A1

designs with small formwork volumes, i.e. panelized configurations as. found in high-volume concrete prefabrication of general façade and structural components.

5.1 Further Work

With the prototype design intended to identify the practical boundaries and limitations of the proposed method, in addition to raising public awareness of the architectural possibilities of the merger of topology optimization and robotic manufacturing, current work is focused on the development of revised workflows that will enable a cost-range of production feasible within regular construction projects. The target is to achieve this through (a) the development of a slab component series within normative prefabrication dimensions; (b) a simplified production scheme that rely on nested, single-ruled formwork components to be used in combination with conventional prefabrication casting tables and (c) linear, parametrically produced reinforcement systems.

References

1. Trends in global CO2 emissions: 2015 Report. PBL Netherlands Environmental Assessment Agency, The Hague (2015)
2. International Civil Aviation Organization, Air Transport Bureau (ATB)."Aircraft Engine Emissions" Accessed 19 Mar 2008
3. Feiz, R., Ammenberg, J., Baas, L., Eklund, M., Helgstrand, A., Marshall, R.: Improving the CO2 Performance Of Cement, Part II: Framework for Assessing CO2 Improvement Measures in Cement Industry. Linköping University Post Print (2014)

4. Vlasopoulos, N.: Novacem carbon negative cement. In: Society of Chemistry and Industry Conference, Low Carbon Cements: Recent Research and Developments on Alternatives to Portland cements, London, UK (2010)
5. Börjesson, P., Gustavsson, L.: Greenhouse gas balances in building construction: wood versus concrete from life-cycle and forest land-use perspectives. Energy Policy **28**(9), 575–588 (2000)
6. Kulkarni, A.R., Bhusare, M.V.: Structural optimization of reinforced concrete structures. Int. J. Eng. Tech. Res. **V5**(07) (2016)
7. Søndergaard, A., Dombernowsky, P.: Design, analysis and realization of topology optimized concrete structures. Int. Assoc. Shell Spatial Struct. J. **53**(4), 209–216 (2012)
8. Adriaenssens, S., Block, P., Veenendaal, D., Williams, C. (eds.): Shell Structures for Architecture: Form Finding and Optimization. Routledge, Abington (2014)
9. Dombernowsky, P., Søndergaard, A.: Unikabeton prototype. In: Sheil, B., Glynn, R. (eds.) Fabricate: Making Digital Architecture, pp. 56–61. Riverside Architectural Press, Waterloo, CA (2011)
10. Søndergaard, A., Amir, O., Eversmann, P., Piškorec, L., Stan, F., Kohler, M., Gramazio, F.: Topology optimization and robotic fabrication of advanced timber space-frame structures. In: Reinhardt, D., Saunders, R., Burry, J. (eds.) Robotic Fabrication Architecture, Art and Design 2016, pp. 191–203. Springer, Switzerland (2016)
11. Stromberg, L.L., Beghini, A., Baker, W.F., Paulino, G.H.: Topology optimization for braced frames: combining continuum and beam/column elements. Eng. Struct. **37**, 106–124 (2012)
12. Rippmann, M., Liew, A., Block, P.: Structural 3D-printed floor. In: Yuan, P.F., Menges, A., Leach, N. (eds.) Digital Fabrication. Tongji University Press, Shanghai (2017)
13. Galjaard, S., Hofman, S., Ren, S.: New opportunities to optimize structural designs in metal by using additive manufacturing. In: Block, P., Knippers, J., Mitra, N.J., Wang, W. (eds.) Advances in Architectural Geometry 2014, pp. 79–93. Springer International Publishing (2015)
14. Aghaei-Meibodi, M., Bernhard, M., Jipa, A., Dillenburger, B.: The smart takes from the strong. In: Sheil, B., Menges, A., Glynn, R., Skavara, M. (eds.) Fabricate: Rethinking Design and Construction, pp. 210–217. UCL Press, London (2017)
15. Lab, R.H.: Think Formwork - Reduce Cost. Struct. Mag. (4), 14–16 (2007)
16. Sarma, K.C., Adeli, H.: Cost optimization of concrete structures. J. Struct. Eng. **124**(5), 570–578 (1998)
17. Jepsen, C., Kristensen, M., Kirkegaard, P.: Dynamic double curvature mould system. In: Gengnagel, C., Kilian, A., Palz, N., Scheurer, F. (eds.) Computational Design Modeling: Proceedings of the Design Modeling Symposium Berlin 2011, pp. 291–300. Springer, Berlin, Heidelberg (2011)
18. Lloret, E., Shahab, A.R., Linus, M., Flatt, R.J., Gramazio, F., Kohler, M., Langenberg, S.: Complex concrete structures: merging existing casting techniques with digital fabrication. Comput. Aided Des. **60**, 40–49 (2014)
19. Khosnehvis, B.: Automated construction by contour crafting – related robotic and information technologies. J. Autom. Constr. **13**(1), 5–19 (2004)
20. Lim, S., Buswell, R.A., Le, T.T., Austin, S.A., Gibb, A.G.F., Thorpe, A.: Development in construction-scale additive manufacturing processes. Autom. Constr. **21**(1), 262–268 (2012)
21. Accessed www.freefab.com
22. Hack, N., Lauer, W.V.: Mesh-Mould: robotically fabricated spatial meshes as reinforced concrete formwork. Architectural Des. **84**(3), 44–53 (2014)
23. Søndergaard, A., Feringa, J.: Scaling architectural robotics – realization of the Kirk Kapital Headquarters. In: Sheil, B., Menges, A., Glynn, R., Skavara, M. (eds.) Fabricate: Rethinking Design and Construction, pp. 264–271. UCL Press, London (2017)

24. Søndergaard, A., Feringa, J., Nørgaard, T., Steenstrup, K., Brander, D., Graversen, J., Markvorsen, S., Bærentzen, A., Petkov, K., Hattel, J., et al.: Robotic hot-blade cutting. In: Reinhardt, D., Saunders, R., Burry, J. (eds.) Robotic Fabrication Architecture, Art and Design 2016, pp. 150–164. Springer, Switzerland (2016)
25. Accessed http://www.grasshopper3d.com/

The Brick Labyrinth

Luka Piškorec[1(✉)], David Jenny[2], Stefana Parascho[2], Hannes Mayer[2], Fabio Gramazio[2], and Matthias Kohler[2]

[1] Aalto University, Helsinki, Finland
luka.piskorec@aalto.fi
[2] Gramazio Kohler Research, ETH Zurich, Zurich, Switzerland
{david.jenny,parascho,mayer,gramazio,kohler}@arch.ethz.ch

Abstract. This paper presents a project developed within the Master of Advanced Studies in Architecture and Digital Fabrication programme at ETH Zurich. The Brick Labyrinth is the first large-scale construction built in the Robotic Fabrication Laboratory, a unique multi-robotic setup for automated prefabrication at architectural scale. The project continues the tradition of robotic brick laying started at Gramazio Kohler Research but increases the significance of computational design and robotic control by solely working with a dry-stacked construction method. The paper introduces the design methodology, the computational framework and the robotic fabrication setup and addresses the structural challenges of the constructive system. It introduces strategies for an automated multi-robotic brick laying process at large-scale including custom-made robotic end-effectors to increase the speed of the process. The unique setup of the project and its material system demonstrate a fully reversible construction process at architectural scale, suggesting a new approach to physical prototyping, which could fundamentally change the way we design buildings. While this paper highlights the design explorations leading towards the final structure – featuring the development of a flexible dry-stacked brick bond and its potential to create unique spatial sequences – it also provides an outlook on how the integration of computational tools into automated fabrication processes can lead to new design typologies.

Keywords: Digital Fabrication · Computational design · Robotic fabrication
Robotic brick laying · Stability analysis · Dry-stacked masonry
Labyrinth

1 Introduction

The project presented here was devised and conducted over the course of three months as part of the Master in Advanced Studies in Architecture and Digital Fabrication curriculum at ETH Zurich, a programme that teaches fundamental methods and technologies of digital design and their 1:1 scale implementation in architecture and construction [1]. Through iterations of design studies and physical prototyping, the team of 17 students together with the support of tutors and researchers developed the project starting from the design brief and constructive system to the final design and its

J. Willmann et al. (Eds.): ROBARCH 2018, *Robotic Fabrication in Architecture, Art and Design 2018*, pp. 489–500, 2019.
https://doi.org/10.1007/978-3-319-92294-2_37

construction. Simultaneously, they optimised the robotic setup and mechanic end-effectors, as well as the structural simulation and fabrication procedure, to enable an efficient robotic construction process (Fig. 1).

Fig. 1. The Brick Labyrinth in the Robotic Fabrication Laboratory at ETH Zurich, 2017

With its dimensions of 8 m by 10 m in plan and a height of 2.8 m, the Brick Labyrinth is the first large-scale construction built in the recently opened Robotic Fabrication Laboratory (RFL) [2] at ETH Zurich with its four collaborating robotic arms suspended from two gantries and a work area of 45 m by 17 m. The project continues a long tradition of robotic brick laying started by Gramazio Kohler Research in 2006 [3, 4]. However, past projects such as the Programmed Wall or Structural Oscillations relied on adhesives to achieve structural stability. For the Brick Labyrinth, conceived as a temporary structure in the RFL, reversibility demanded for a dry-stacked construction method. With the focus on structures in equilibrium, the computational design process had to integrate structural simulation and pushed the potential of robotic control as placement of each single brick required a high level of precision.

The efficiency of the inherently slow robotic pick-and-place procedure was increased by the development of a new robotic end-effector, a brick magazine that can hold a stack of eight bricks at a time and dispense them one after the other. The simultaneous use of two synchronized robotic arms suspended from one of the two gantries in the RFL further increased the speed of the process[1]. The final structure

[1] Placing time per brick was reduced to an average of 15 s (4 min for 16 bricks). Two ABB robots each placing 8 bricks per cycle, with minimised gantry travel time. This speed was comparable to the process done in the group in 2006 setup (see [3]).

consisted of 10,135 bricks placed over the course of 19 days. Through a custom computational design and robotic control setup students developed all project stages and built multiple full-scale prototypes.

2 Design Research

2.1 Design Brief and Explorations

The assignment presented to the students asked for the design and construction of a labyrinth using the RFL robotic fabrication setup to dry stack bricks into a large-scale structure. The maximum amount of bricks was derived from the estimated construction speed of approximately 500 bricks per working day, totalling at a maximum of 12'000 bricks[2] for the planned fabrication time of four weeks. The construction area in the RFL and the maximal construction height were given as further constraints.

In a first phase, different conceptual approaches to the architectural typology of the labyrinth were explored. The conception of a contemporary labyrinth was inspired by topics such as the relation to the human body that can be found in some of Richard Serra's work [5], the labyrinth typologies based on the loss of orientation in never-ending self-similar spaces that can be traced back to the legend of the Labyrinth of the Minotaur and the Knossos palace complex [6], and the literary work of Jorge Luis Borges, which describes both physical as well as mental vastness of labyrinths [7].

This led to a series of different design proposals that were iteratively refined. Each included the design of a labyrinth typology as well as a constructive system, a specific brick-bond with its generative and structural logic. Simultaneously the structural performance limits of the different brick-bonds were tested in real-scale physical prototypes, while the overall labyrinth designs were 3D printed and their spatial qualities evaluated as scaled models.

The developed design strategies included both surface and volumetric design approaches. The former ones were articulated through surface arrangements, single- or multi-layered, which define differentiated spatial relationships. Strategies of realising these surfaces in dry-stacked brick structures included the stiffening of walls by introducing curvature in the horizontal plane, modulation of their thickness and cavities over the vertical development to distribute and control their mass, and adding buttress elements to improve horizontal stiffness. Volumetric design approach included design concepts treating the brick as a space-filling element, articulating space through a sculptural mass. Strategies featured false-arching techniques to create vaulted or enclosed spaces, experimentations with the orientation of the bricks so their perforated inner side is exposed to the view, or a field of bricks arranged in such way that it was only certain stable paths that allowed the visitors to walk on.

---

[2] Keller AG Ziegeleien donated 12000 standard bricks of the type kelesto, dimensions 24.5 cm × 11.5 cm × 7.0 cm. Keller AG Ziegeleien homepage, http://keller-systeme.ch/en, last accessed 2018/03/05.

2.2 Final Design

The final design chosen for the Brick Labyrinth – a synthesis of several investigations – was composed of two walls spiralling into each other. The spiralling concept generated a long winding path where the differentiated walls define a series of highly articulated spaces that have a strong impact on the physical experience of the visitor. The labyrinth has two entries at a slight offset after which the visitor is guided through a sequence of changing sections towards the central open area.

Driven by this overall idea, the constructive system of dry-stacking bricks became a main design driver and one of the biggest challenges of the project. Without any additional adhesive between the elements, the structural integrity of the walls is solely depending on gravity and friction, which makes the result all the more spectacular and surprising. To push the structural capacities and the spatial flexibility of the system to its limits, a special brick bond was developed that allows the transition between four different states of the system (see Fig. 2). A simple single-layer running bond (a) transforms into a loose running bond (b) with additional bricks placed in a 90° angle increasing the structural capacity to lean out locally (c), and finally to a doubled-up leaning wall counterbalanced by connecting ribs between the two layers (d).

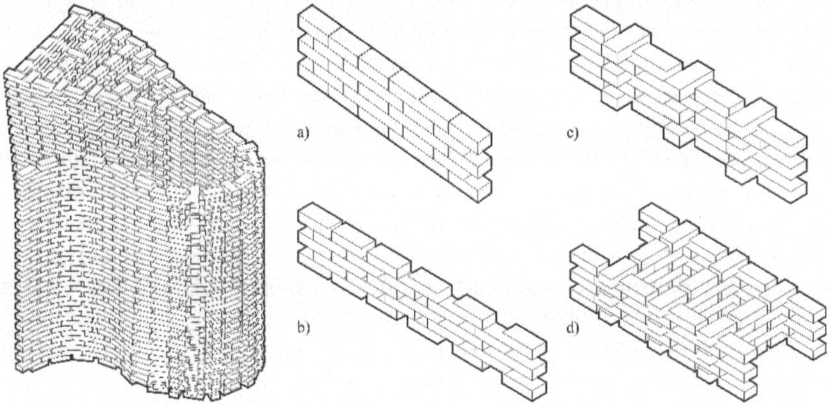

Fig. 2. Different states of the developed brick bond

Both walls start as a single layer running-bond, forming a nearly straight vertical extrusion at their ends. The lower part of the single-layered wall is undulating in order to increase the support polygon of the structure and allow for a curved development over the vertical axis. The doubling-up of both walls leaning towards each other creates a key moment in the spatial experience of the labyrinth (Fig. 3).

Fig. 3. **a.** Plan of the final labyrinth and **b.** Image of the leaning wall part with hidden transversal wall elements that provide counter-balancing.

3 Computational Setup

A custom computational setup specifically developed for the project integrates both design, structural and fabrication parameters in a single environment. The setup in Python is based on the COMPAS framework[3] developed by the Block Research Group at ETH Zurich, which recently became open-source [8]. This allows the integration of design, simulation and fabrication development in a single workflow. The geometric definition relies on the COMPAS data structures – mesh and network – enabling easy control of the connectivity between the single brick geometry and the overall assembly, and the development of a structural equilibrium analysis using the same underlying data structures. The COMPAS data structures are based on a graph system and allow to describe topological relations between elements through the definition of vertices and edges. In addition, the mesh data structure stores information about mesh faces and attributes. This enables a simple definition of both connectivity relations and geometric data of discrete element structures such as brick assemblies. COMPAS is CAD program independent, which allows for a generic definition of the design and analysis tools. However, the interface to Rhinoceros [9] and Grasshopper [10] environments allows for direct visual feedback and generation of the fabrication data that is sent to the robotic setup via an interface developed in Python and ABB Rapid.

3.1 Computational Design

By defining these interfaces, students were free to use any design tools they were more familiar with to facilitate their work, while still being able to rely on the advanced methods provided by our custom classes, such as stability analysis. The generation of the final geometry started from two drawn input curves that traced the central axis of

[3] COMPAS framework offers a pure Python base with flexible data structures and algorithms, independent geometry processing, interoperability with external libraries used in structural design, methods and solvers for numerical computation, high performance computing capabilities, visualization tools as well as interfaces to common CAD software and ecosystems. Full list of features and extensive documentation can be found on its GitHub webpage (see [8]).

the walls. The parametrisation of these curves allowed to precisely define the domain where the brick-bond would transition from one state to the other, while the bricks would be generated and distributed through an underlying algorithmic logic. For instance, the script could solve the smooth transition from the single-layered to the double-layered wall sections and generate the appropriate amount of buttresses that would stabilize them.

3.2 Structural Stability Analysis

On a computational level, the brick bond is building up on two of the main classes of COMPAS (see [8]). The brick modules as a sub-class of the COMPAS "mesh" data structure [11] and the brick assembly as a sub-class of the COMPAS "network" data structure [12]. The network is a connectivity graph meant for the representation of networks of vertices connected by edges. In our case, every brick mesh is represented by a vertice in a network connected to its vertical and horizontal neighbouring bricks through edges, allowing for a simple equilibrium analysis based on purely geometric parameters, in this case the mass centre point of each brick and their overlap areas. A brick assembly's stability is evaluated by iterating through its layers, starting with the top one, calculating the resultant force and its position on its supporting neighbours until the base of the wall. Multiple scenarios were analysed and abstracted in order to cover all options of the used bonds (see Fig. 4). This included a brick supported by one, two, three or four bricks. By checking if the resultant force lies in the supporting area of an underlying brick layer, it was clear at which element an unstable assembly would fail and could thus be easily adjusted. Abstraction strategies for the force transfer for situations with one, two, three or four support elements were developed. During the design process, the structural stability analysis was continuously verified with the behaviour of physical prototypes built in 1:1 scale. These prototypes helped to understand the influence of friction in geometrically unstable configurations.

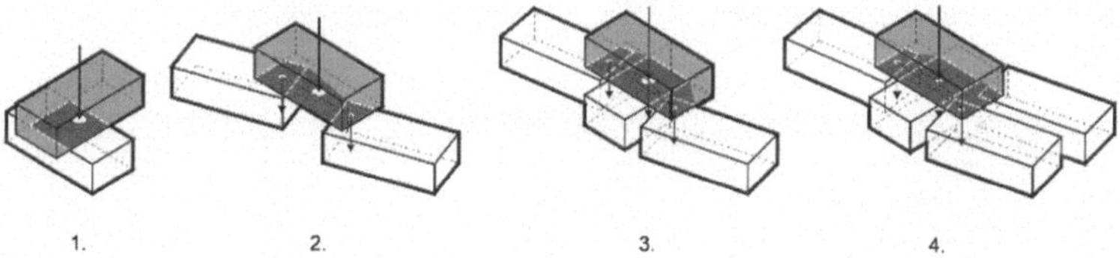

1. 2. 3. 4.

Fig. 4. Calculation of force distribution from one layer to the next for different support scenarios. **1**. One supporting brick, **2**. Two supporting bricks, **3**. Three supporting bricks **4**. Four supporting bricks

3.3 Robotic Fabrication Simulation and Online Robot Control

The fabrication sequence, robot and gantry movements, material pick-up and other production related processes were modelled and simulated in ABB RobotStudio [13], using a simplified version of the RFL setup (see [2]). The communication setup streamed the accurate robot and gantry simulation data from the company developed proprietary software to Grasshopper and visualized it in Rhinoceros. It simulated the handling of the bricks from the feeding station to their final placement, checked for possible collisions and minimised the gantry movement to reduce production time. A separate script controlled the two UR5 [14] feeding stations as their operation could be decoupled from the main production flow. The feeding stations would prepare the brick stacks and wait for the ABB robots to pick them up before resuming operation.

The computational environment described above was also used for the online control of the robot arms and the gantry system, whereas commands could be sent directly from the computer to the RFL system (see [2]). A server connection managed the bidirectional communication from digital to physical setup, translating robotic poses defined in the computational setup to Rapid code and directly sending them to the ABB controller. Synchronization of the two robots and the flow control of the process was achieved through feedback of the execution status of each robot task (Fig. 5).

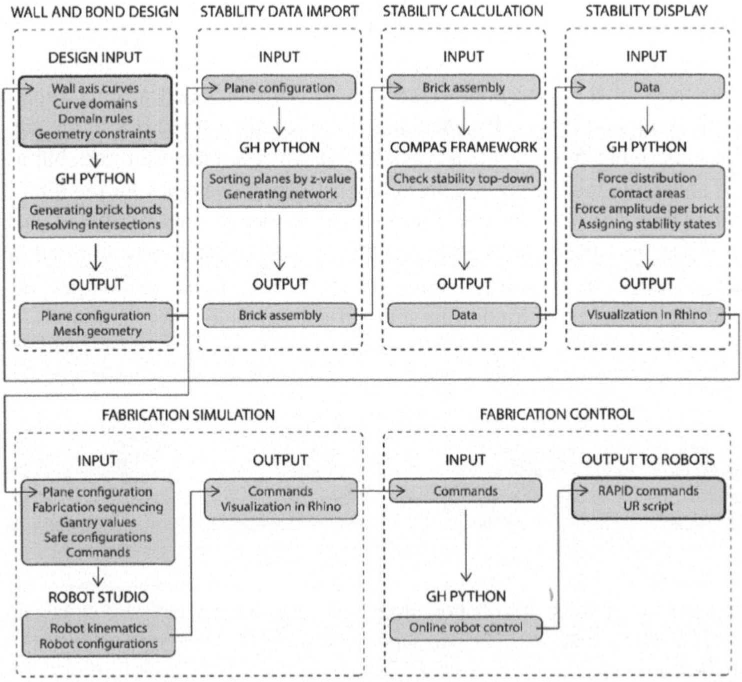

Fig. 5. Diagram of the computational workflow

4 Robotic Fabrication

4.1 Robotic Setup

Production took place in the RFL (see [2]) of the Institute of Technology in Architecture at ETH Zurich. It consists of two bridges with two telescopic z-axes from which a total of four ABB IRB 4600 2.55 robotic arms are mounted upside-down, enabling every robot base to have three additional semi-independent degrees of freedom. For the purpose of this project, we used only one bridge with two ABB robots in a work area approximating 100 square meters. To speed up the process, a custom developed brick magazine[4], able to pick up a stack of 8 bricks and deposit them one at a time, was mounted on each ABB robot. Two feeding stations placed on the periphery of the construction area completed the setup, where small-scale UR5 robot arms each assembled two stacks of eight bricks picking them from a feeding rail while the ABB robots were placing bricks in the structure. The decision of employing two different types of robots in the same process was taken in accordance with the robots' capabilities. UR5 robots are comparatively small and have a payload of 5 kg which only serves for manipulating a single brick. As a result, they lend themselves perfectly for pre-fabricating the brick stacks. In contrast, the RFL ABB robots reach the entire workspace and can carry up to eight bricks. Human operator is controlling the collaboration between the robots, ensuring that stacks are built during the time that the ABB robots are placing the eight bricks (Fig. 6).

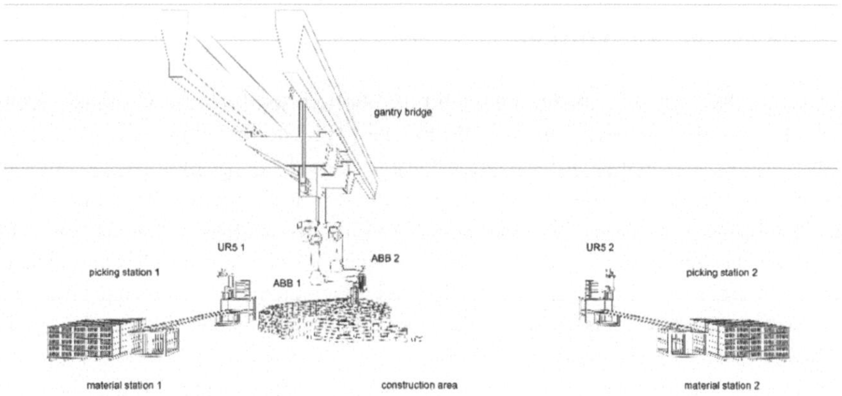

Fig. 6. Layout of the robotic construction setup in RFL with two UR5 picking stations and two ABB robots mounted on a gantry

[4] The 40 kg maximum payload of the robotic arm with approximately 15 kg weight of the tool itself limited the amount of bricks in the magazine to eight (8×3.2 kg + 15 kg = 40.6 kg).

4.2 Sequencing

The fabrication sequence of the walls was optimised for minimised gantry travel time along the X-direction, as this was the gantry's slowest movement direction by dividing each layer of bricks into segments of 16 bricks along similar X-positions. These 16 bricks were sorted according to their Y-positions and each robot arm would place eight bricks in its sector while synchronized with the other robot in order to avoid any collisions of the robotic arms and their gantry carts. Because no mortar or construction glue is used, the brick dry-stacking process is easily reversible. The robotic disassembly process was tested on the finished structure but not used in the end due to time constraints (Fig. 7).

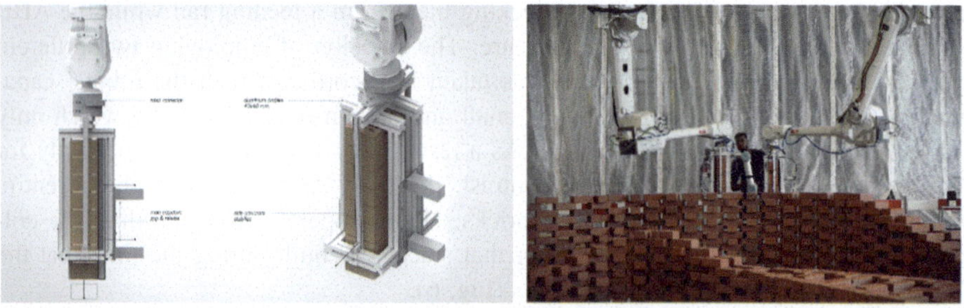

Fig. 7. **a**. Model of the brick magazine and **b.** ABB robots mounted on the overhead gantry picking up the preassembled brick stacks from the UR5 robot station

5 Conclusion and Outlook

Significant advances were made in the field of robotic brick laying at Gramazio Kohler Research since the first experiment in 2006 (see [3]). The technique was used for the construction of a non-standardised brick façade, as well as for multiple teaching and installation projects exploring its design space and developing new robotic brick bonds. This development is extensively documented in PhD dissertation of Tobias Bonwetsch [15]. ROB Technologies [16], a Gramazio Kohler Research spin-off, continued this research in close collaboration with Keller AG Ziegeleien [17]. In recent years there has been a strong development of in-situ robotic fabrication technologies, also applied to masonry structures [18–24].

This paper and the project described extends the existing research to robotic dry-stacking of bricks, thereby placing much emphasis on the structural challenges by including structural stability analysis into the design development. This strategy could be expanded in the future by extending the structural analysis to include friction between elements in the setup. In the project presented, friction was evaluated through physical tests at full-scale. Concerning the actual speed of construction (averaging 15 s per brick), it is evident that it is within the range of the projects done at Gramazio Kohler Research in 2006 (see [3]). However, the presented setup in RFL has a much larger building area[5]. Accordingly, one performance goal of the project was to retain a similar construction speed despite increased travel times to the feeding stations. Two strategies were applied to achieve this goal: The coordinated and synchronized control of a multi-robotic system and the use of a brick magazine that can deposit a stack of bricks. Overall, this represents a considerable improvement in efficiency compared to the 2006 setup, especially since the new setup requires only two human operators for the entire operation (one for the feeding station and another one for the construction process).

These concepts could be further improved by increasing the number of collaborating robots or the number of bricks in a magazine. Ideally, the brick magazine would be replaced by a continuous feeding system which would make the time consuming pick-and-place procedure obsolete. The project suggests a new type of rapid prototyping facility to test fully reversible structures in a completely automated setup, where design variations could be test-assembled, experienced at full architectural scale, and the material returned to its initial state "over-night", ready to be reused. This concept could inspire a new, more ecologically sound approach of prototyping in the age of 3D printing in different scales and material systems.

In addition to the technical and computational achievements, the brick labyrinth developed and built by the students of the MAS in Architecture and Digital Fabrication created a unique spatial experience for all visitors. It shows how the integration of computational tools in automated fabrication processes can lead to new typologies that extend our architectural vocabulary.

Acknowledgments. The Brick Labyrinth described in this paper was developed and built as a design and fabrication project of the Master of Advanced Studies in Architecture and Digital Fabrication at ETH Zurich in June 2017. The programme is run jointly by Gramazio Kohler Research and Digital Building Technologies and directed by Hannes Mayer.

The project was developed and executed by the students of the MAS year 2016/2017: Marco Caprani, Samuel Cros, Rodrigo Diaz, Ahmed Elshafei, Federico Giacomarra, Hakim Hasan, Marirena Kladeftira, Iakovina Kontiza, Thodoris Kyttas, Matthias Leschok, Coralie Ming, Wataru Nagatomo, Matteo Pacher, Maria Pachi, Sambit Samant, Theodora Spathi and Dai-Syuan Wu. The team of researchers and tutors consisted of Luka Piškorec, David Jenny and Stefana Parascho, with Hannes Mayer. Students were mentored by Prof. Fabio Gramazio, Prof. Matthias Kohler and Prof. Benjamin Dillenburger. Special thanks goes to the RFL technicians Michael Lyrenmann and Philippe Fleischmann and Andreas Reusser from IfB, as well as to our sponsor Keller AG Ziegeleien for supplying the bricks and for the long research partnership in robotic brick building.

References

1. MAS DFAB homepage. https://www.masdfab.com/program. Accessed 05 Mar 2018
2. Robotic Fabrication Laboratory (2010–2016). http://dfab.arch.ethz.ch/web/forschung/e/0/0/0/186.html. Accessed 05 Mar 2018

[5] The work area of RFL is 765 m^2 as compared to the 2006 setup (see [3]) of around 30 m^2.

3. Bonwetsch, T., Kobel, D., Gramazio, F., Kohler, M.: The informed wall: applying additive digital fabrication techniques on architecture. In: Luhan, G.A., Anzalone, P., Cabrinha, M., Clarke, C. (eds.) Acadia 2006: Synthetic Landscapes, Proceedings of the 25th Annual Conference of the Association for Computer Aided Design in Architecture, pp. 489–495. Louisville (2006)
4. Bärtschi, R., Knauss, M., Bonwetsch, T., Gramazio, F., Kohler, M.: The wiggled brick bond. In: Ceccato, C., Hesselgren, L., Pauly, M., Pottmann, H., Wallner, J. (eds.) Advances in Architectural Geometry 2010, pp. 137–147. Springer, Vienna (2010)
5. Photograph showing the work of Richard Serra: Inside Out. http://niezlasztuka.net/o-sztuce/richard-serra-rzezbiarz-blednika/. Accessed 05 Mar 2018
6. Plan of the Knossos palace complex from Baikie, James: The Sea-kings of Crete, Adam and Charles Black, London (1910)
7. Borges, J.L.: The Two Kings and the Two Labyrinths (1939), The Library of Babel (1941) and The House of Asterion (1947)
8. Van Mele, T., Liew, A., Méndez Echenagucia, T., Rippmann, M., et al.: COMPAS: a framework for computational research in architecture and structures, GitHub repository. https://compas-dev.github.io/. Accessed 08 May 2018
9. McNeel: Rhinoceros. https://www.rhino3d.com/. Accessed 05 Mar 2018
10. McNeel: Grasshopper. http://www.grasshopper3d.com/. Accessed 05 Mar 2018
11. The mesh is implemented as a half-edge data structure. It is meant for the representation of polygonal "surface" meshes. https://compas-dev.github.io. Accessed 05 Mar 2018
12. The network is a connectivity graph. It is meant for the representation of networks of vertices connected by edges. https://compas-dev.github.io. Accessed 05 Mar 2018
13. ABB: ABB RobotStudio. http://new.abb.com/products/robotics/de/robotstudio. Accessed 05 Mar 2018
14. Universal Robotics: UR5. https://www.universal-robots.com/products/ur5-robot/. Accessed 05 Mar 2018
15. Bonwetsch, T.: Robotically assembled brickwork, Manipulating assembly processes of discrete elements. Ph.D. thesis, ETH Zurich (2015)
16. ROB Technologies. https://rob-technologies.com/. Accessed 08 Mar 2018
17. Keller AG Ziegeleien. http://www.keller-ziegeleien.ch/de. Accessed 09 Mar 2018
18. Owano, N.: Bricklaying robot can make ergonomic, economic impact on construction sites. TechExplore, 21 April 2017. https://techxplore.com/pdf411970087.pdf. Accessed 07 May 2018
19. Dörfler, K., Sandy, T., Giftthaler, M., Gramazio, F., Kohler, M., Buchli, J.: Mobile robotic brickwork. In: Reinhardt, D., Saunders, R., Burry, J. (eds.) Robotic Fabrication in Architecture, Art and Design 2016, pp. 204–217. Springer, Switzerland (2016)
20. Giftthaler, M., Sandy, T., Dörfler, K., Brooks, I., Buckingham, M., Rey, G., Kohler, M., Gramazio, F., Buchli, J.: Mobile robotic fabrication at 1:1 scale: the In situ Fabricator: system, experiences and current developments. J. Constr. Rob. **1**, 3–14 (2017)
21. Keating, S., Spielberg, N., Klein, J., Oxman, N.: A compound arm approach to digital construction. In: McGee, W., Ponce de Leon, M. (eds.) Robotic Fabrication in Architecture, Art and Design 2014, pp. 99–110. Springer, Switzerland (2014)
22. Keating, S., Leland, J., Cai, L., Oxman, N.: Towards site-specific and self-sufficient robotic fabrication on architectural scales. Sci. Rob. **2**(5) (2017)
23. Lim, S., Buswell, R.A., Le, T.T., Austin, S.A., Gibb, A.G.F., Thorpe, T.: Developments in construction-scale additive manufacturing processes. Autom. Constr. **21**(1), 262–268 (2012)
24. Minibuilders: IaaC Barcelona (2014). http://robots.iaac.net/. Accessed 09 Mar 2018

Author Index

J. Willmann et al. (Eds.): ROBARCH 2018, *Robotic Fabrication in Architecture, Art and Design 2018*, pp. 501–503, 2019.
https://doi.org/10.1007/978-3-319-92294-2

The manufacturer's authorised representative in the EU is Springer
Nature Customer Service Centre GmbH, Europaplatz 3, 69115 Heidelberg,
Germany. If you have any concerns regarding our products, please
contact ProductSafety@springernature.com

Printed and bound by CPI Group (UK) Ltd, Croydon, CR0 4YY
05/10/2026
02228141-0016